THE PHYSIOLOGY OF STANDING

(Das Stehen)

Postural Reactions and Equilibrium with Special Reference to the Behavior of Decerebellate Animals

by
G.G.J. Rademaker

Edited, with a Foreword, by D. Denny-Brown

Translated for the Public Health Service, National Library of Medicine, U.S. Department of Health, Education, and Welfare, and the National Science Foundation, Washington, D.C. by Franklin Book Programs, Inc., Cairo, 1975.

The publication of this study was supported through the Special Foreign Currency Program of the National Library of Medicine, National Institutes of Health, Public Health Service, U.S. Department of Health and Human Services, Bethesda, Maryland, pursuant to an agreement with the National Science Foundation, Washington, D.C. by the Al Ahram Publishing House, Cairo, A.R.E.

University of Minnesota Press, Minneapolis

Published by the University of Minnesota Press,
2037 University Avenue Southeast, Minneapolis, MN 55414
Printed in Cairo, A.R.E., by Al Ahram.

Library of Congress Cataloging in Publication Data
Rademaker, Gijsbertus Godefriedus Johannes, 1887-1957
The Physiology of Standing = Das Stehen.

(Translated for the Public Health Service, National Library
of Medicine, U.S. Department of Health, Education,
and Welfare, and the National Science
Foundation, Washington, D.C., by Franklin
Book Programs, Inc., Cairo, 1975.)

Bibliography: p.
Includes indexes.
    1. Equilibrium (Physiology).   2. Standing Position.
    3. Cerebellum.   4. Posture.   I. Denny-Brown, Derek, 1901.
II. Title.   III. Title: Stehen
[DNLM: 1. Equilibrium.   2. Cerebellum.   3. Reflex.
    4. Decerebrate State.   5. Labyrinth.   6. Posture.
WL 320 R 127 a]
QP 471. R 321 3 599.01'8   80-23983
ISBN 0-8166 - 0857-1

Originally published as **Monograph in General Neurology**
and Psychiatry, No. 59, by Julius Springer, Berlin, 1931

PROF. Dr G. G. J. RADEMAKER

G. G. J. Rademaker was born in 1887 at the Hague in Holland. On graduation in medicine in 1912 he returned to the Hague to practice until 1916, when he went to Surabaya in the Dutch East Indies, where he practiced surgery for the next six years. Returning to Holland in 1922 he became fascinated by the advances being made in physiology of the nervous system. After a preliminary study of the detailed anatomy of the brain stem with Winkler, he joined the group working with Rudolf Magnus in the department of pharmacology in Utrecht, who had already elucidated the essential features of the labyrinthine and righting reflexes. Rademaker began work on the formidable problem of the detailed anatomy of the centers responsible for the righting reflexes. His crucial experiment of section of the rubrospinal tract at the decussation of Forel appeared to settle the question, and in his monograph *Die Bedeutung der roten Kerne* in 1926 he concluded that the essential structure was the red nucleus. Subsequent investigators, reviewed by Fulton (1943), showed that the question is more complex; the red nucleus is but one of a series of levels of elaboration of the righting reflexes. Rademaker's monograph nevertheless greatly clarified the issues.

In 1924, using his considerable surgical skill, he set out to study the long term survival of animals following total removal of the cerebellum, and later, of animals from whom he had removed both cerebral hemispheres. Finally he was able to secure survivals in two animals after removal of the cerebral hemispheres and cerebellum. With the help of his devoted wife he was able to keep decerebellate dogs and cats in good condition for as long as six years and decorticate dogs for eighteen months, an extraordinary feat before antibiotics and modern neurosurgery.

There resulted the remarkable monograph *Das Stehen*, published in 1931, primarily an analysis of cerebellar defect, but with much wider significance in relation to the neurological mechanisms of righting, standing and equilibrium. Rademaker devised ingenious clinical tests to investigate the exact nature of the complex abnormalities of posture and movement that had hitherto been loosely categorized as «incoordination,» «asthenia,» «dyssynergia» and «atonia.» Some of the basic reactions that Rademaker established, such as placing and hopping, have become well known to physiologists and clinicians since the work of Bard (1933). Nevertheless, few in the English-speaking world have been able to read in detail the 476 pages of heavy German in the original monograph, which is a mine of carefully documented information defining precisely the part played by the many cutaneous, proprioceptive and labyrinthine reactions that are the fundamental features of regulation of posture and equilibrium. That the cerebellum is not in fact an essential part of the mechanism of any of the labyrinthine reactions was an extraordinary finding. The exaggeration of various reactions following removal of cerebellum revealed its essentially modulating function. Subsequent work on primates (Carrea and Mettler, 1947; Denny-Brown and Gilman, 1965) has shown the universal applicability of Rademaker's findings made in decerebellate dogs.

In the last fifteen years the investigations of Stella, Moruzzi, Brodal, Sprague and others on the functions and connections of the roof nuclei of the cerebellum, and of Granit and his associates on the relation of cerebellum to the gamma system of regulation of proprioceptive input have added greatly to knowledge of cerebellar function, admirably reviewed by Dow and Moruzzi (1958). More recently Eccles and his associates have brilliantly elucidated the synaptology of the cerebellar cortex, providing a new view of cerebellar function at the neuronal level. For the full understanding of these contributions the classical work of Rademaker on the mass effects that follow cerebellar ablation is still essential. The complete objectivity that Rademaker achieved gives them a timeless quality that also distinguished the investigations of Gordon Holmes on the human cerebellar syndrome ten years earlier. Curiously, Rademaker discusses the findings of Holmes hardly at all, basing his criticisms chiefly on the earlier contributions of Luciani, Babinski and André-Thomas.

In the year 1927, as the experiments for *Das Stehen* were nearing conclusion, Rudolf Magnus, director of the Institute at Utrecht, died suddenly and unexpectedly. This great loss to physiology was a particularly devastating blow to Rademaker. It was thought by many that Rademaker was the logical successor to become director of the new Institute then nearing completion. But it was not to be; he was appointed to the distinguished chair of physiology in Leyden, succes-

VI

sor to Einthoven. There he set to work to complete *Das Stehen*, and in 1935 to write a companion study of labyrinthine function, equally remarkable for its clarity of analysis, published in France as *Rèactions Labyrinthiques et Equilibre*. The investigation of human patients that formed part of this study had been made at the Salpêtrière with Raymond Garcin, a collaboration that became the basis of an enduring friendship. After a difficult period in the German occupation of Holland in the second world war, during which he played a part in the Dutch Resistance, he continued to contribute papers on subjects such as lengthening reactions (1947). visual placing and optokinetic reactions (1948) and clinical discussions of ataxia, nystagmus and related subjects. Gradually he spent more time in the clinic. In 1953 his associates presented him with a Festschrift which fills one number of *Folia psychiatrica, neurologica et neurochirurgica neerlandica* with a handsome acknowledgement of his clinical skill as a neurologist as well as of his scientific attainments. He died April 9, 1957.

The initiative and foresight of the National Library of Medicine in generously providing this translation of *Das Stehen* is a tremendous service to the many investigators in neurology, neurophysiology and in the special senses whose interests lie in motor coordination, cerebellar and labyrinthine function. We wish to acknowledge the kind assistance of Professor Henk Verbiest of Utrecht in providing a portrait of Rademaker and also in clarifying some biographical details.

D. Denny-Brown, Editor
Prefessor of Neurology, Emeritus
Harvard University
Cambridge, Massachusetts

VII

References:

Anonymous (1957) G.G.J. Rademaker. Rev. Neurol. 96:134-135

Bard, P. (1933) Studies on the cerebral cortex. 1. Localized control of plac
ing and hopping reactions in the cat. Arch. Neurol. Psychiat., Chicago,
30:40-74

Carrea, R.M.E., and Mettler, F.A. (1947) Physiologic consequences
following extensive removal of the cerebellar cortex and deep cerebellar
nuclei and the effect of secondary cerebral ablations in the primate. J.
Comp. Neurol. 87:169-288.

Denny-Brown, D., and Gilman, S. (1965) Depression of gamma innerva
tion by cerebellectomy. Trans. Amer. Neurol. Association 90:96-101

Dow, R.S., and Moruzzi, G. (1958) The Physiology and Pathology of the
Cerebellum. Minneapolis, Univ. Minnesota Press.

Eccles J.C. Ito, M., and Szentágothai, J. (1967) The Cerebellum as a
Neuronal Machine. Berlin, **Julius** Springer.

Fulton, J. F. (1943) Physiology of the Nervous System, 2nd Edition. New
York, Oxford Univ. Press.

Rademaker, G.G.J. (1926) Die Bedeutung der roten Kerne und des
Übrigen Mittelhirns für Muskeltonus, Korperstellung und Labyrin-
thinereflexe. Berlin, Julius Springer.

Rademaker, G.G.J. (1931) Das Stehen:Statische Reaktionen,
Gleichgewichtreaktionen und Muskeltonus unter besonderer
Berücksichtigung ihres Verhaltens bei kleinhirnlosen Tieren. Berlin,
Julius Springer.

Rademaker, G.G.J. (1935) Réactions Labyrinthiques et Équilibre:
L'Ataxie Labyrinthique. Paris, Masson.

Rademaker, G.G.J., and Ter Braak, J.W.G. (1948) On the Central
mechanism of some optic reactions. Brain 71:48-76.

Ter Braak, J.W,G., Storm van Leeuwen, W., and Verbiest H. (1953)
Miscellanea medica in honorem viri clarissimi, Gysberti Godefridi
Ioannis Rademaker. (With list of publications.) Folia Psychiat.
Neurolog. Neurochirurg. Néerland. 56 (4):393-565.

Verbiest, H. (1957) In memoriam Prof. Dr. G.G.J. Rademaker, Ned.
tschr. Geneesk. 101:849-851.

XIII

INTRODUCTION

If the human skeleton is set upright on the floor, it will collapse; the same will happen if one tries to set erect the corpse of a man before the onset of rigor mortis. Bones, joints, ligaments and muscles are not capable of maintaining the upright position of the body. A certain state of muscle tension, regulated from the central nervous system, is necessary; the connection of the muscles with the spinal cord alone is not sufficient. When the spinal cord is severed from the brain, mammals lose the ability to stand. After lesions of the mesencephalon and the cerebellum, too, the ability of standing is more or less impaired.

The investigations reported in this book concern the following questions :

1. How does standing take place, and which reactions participate in normal standing ?

2. Which parts of the central nervous system have to be intact for these reactions to occur ?

3. What will happen, if one or several of these reactions are disturbed?

The following animals were investigated, apart from a number of decerebrated dogs[1] and cats :

8 decerebellate dogs

Erik, extirpation of cerebellum : 9 Oct., 1924, still alive, i.e., lived for 72 months.

Piccolino, extirpation of cerebellum : 24 Dec., 1924, died 7 Jan., 1929, lived for 49 months.

---

(1) Hitherto, in the literature, the extirpation of the cerebrum above the thalami optici and Sherrington's cross-section through the brainstem between the anterior and the posterior corpora quadrigemina of the mesencephalon were termed decerebration. This often led to misunderstandings. In this volume, I mean by decerebrate animal an animal whose midbrain is transversely severed and only a remnant of midbrain is preserved attached to the pons. An animal from which the cerebrum was removed, leaving the thalami optici, I shall call a decorticate, or thalamus animal.

2

Caesar, extirpation of cerebellum : 3 Feb., 1925, died 15 March, 1927,
lived for 25 months.

Moor, extirpation of cerebellum : 5 Jan., 1926, still alive, i.e.,
lived for 57 months.

Pim, extirpation of cerebellum : 27 Feb., 1926, died 26 Oct., 1926,
lived for 8 months.

Wolf, extirpation of cerebellum : 26 Apr., 1926, still alive, i. e., lived
for 54 months.

Tommy, extirpation of cerebellum : 1 May, 1926, died 8 Sept., 1926,
lived for 4 months.

Susanne, extirpation of cerebellum : 7 May, 1926, died 23 Jan., 1927,
lived for 8 months.

6 decerebellate cats

Fridel, extirp. of cerebellum : 27 Oct., 1924, died 27 Apr., 1925, i.e.,
lived for 6 months.

Carolus, extirp. of cerebellum : 19 Nov., 1924, still alive, 1925, i.e.
lived for 70 months.

Pierette, extirp. of cerebellum : 27 Feb., 1925, died 1 June, 1928,
i.e., lived for 39 months.

Nikker, extirp. of cerebellum : 8 Dec., 1926, died 24 Aug., 1927, i.e.,
lived for 8 months.

Esperance, extirp. of cerebellum : see below

Peggy, extirp. of cerebellum : see below

1 decerebellate monkey

Corrie, extirp. of cerebellum : 12 Jan., 1926, died 20 Apr., 1927, lived
15 months.

4 unilaterally decerebellate dogs

Mops, unilat. extirp. : 11 Feb., 1926, died 8 Sept., 1926, i.e., lived
for 7 months.

Peter, unilat. extirp. : 4 Feb., 1926, died 12 Apr., 1927, i.e., lived for
14 months.

Fox, unilat. extirp. : 16 Feb., 1927, still alive, i.e., lived for 43 months.

Tip, unilat. extirp. : 9 Nov., 1927, still alive, i.e., lived for 35 months.

1 unilaterally decerebellate cat

Josephine unilat. extirp. of cereb. : 3 June, 1924, died 12 May, 1930, lived for 76 months.

2 dogs and one cat, whose right half of the cerebrum was removed in addition to the cerebellum.

dog Vici, extirpation of cerebellum and one half of cerebrum : 7 Apr., 1925, died 19 Oct., 1926, i.e., lived for 18 months.

dog Daumling, see below.

Cat Esperance, extirpation of cerebellum and one half of cerebrum : 17 March, 1925, died 19 Oct., 1925, i.e. lived, $6\frac{1}{2}$ months.

2 dogs whose cerebrum was removed in addition to the cerebellum :

dog Daumling :	dog Robbie :
extirp. of cerebellum : 30 Dec., 1925,	extirp. of left half of cerebrum : 9 Dec., 1926,
extirp. of right half of cerebrum; 20 Feb., 1926;	extirp. of right half of cerebrum: 29 Dec., 1926
extirp. of left half of cerebrum ; 13 May, 1926;	extirp. of cerebellum : 7 Feb., 1927;
died 10 June, 1926	died 17 March, 1927

5 decorticate (thalamic) dogs

dog Daumling see above	dog Robbie see above
dog Miesel :	dog Bob :
extirp. of l. half of cerebrum 18 Dec., 1926.	extirp. of l. half of cerebrum 9 Feb., 1927.
extirp. of r. half of cerebrum 7 Jan, 1927; died 19 March, 1927.	extirp. of r. half of cerebrum 4 March, 1927; died 28 March, 1927.

dog Vos :

extirp. of l. half of cerebrum : 28 Jan., 1927.

4

extirp. of r. half of cerebrum :

21 Feb., 1927; died 8 Nov., 1927; (thus was decorticate for 8 months)

1 cat both whose labyrinths were extirpated in addition to cerebellum.

cat Peggy, extirp. of both labyrinths : 17 May, 1926, (by Dr. De Kleyn) extirp. of cerebellum : 8 June, 1926; died 13 Aug., 1926.

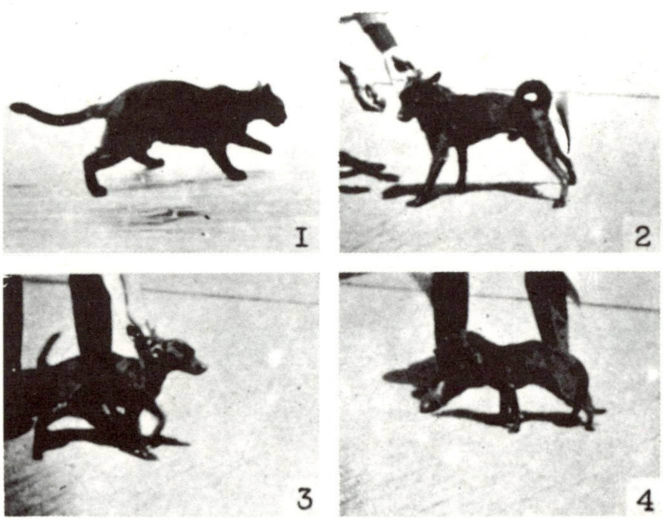

Fig. 1 1. Cat Pierrette, photographed 13 May, 1926.   2. Dog Erik, photographed 10 June, 1925.   3. and 4. Dog Piccolino, photographed 10 June, 1925.

Fig. 2 Decerebellate monkey Corrie. Extirpation of cerebellum 13 Jan., 1926. Photo taken 22 May, 1926.

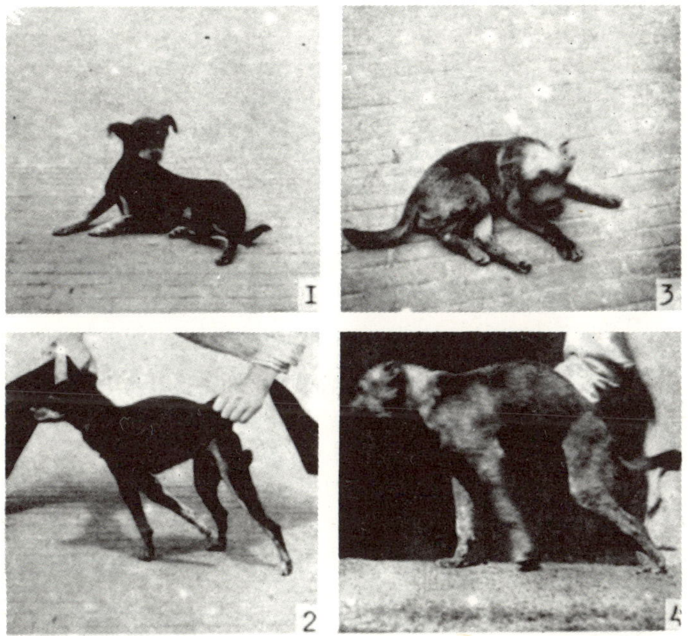

Fig. 3   Dogs Vici and Daumling (decerebellate,   right half of cerebrum removed). 1 and 3.   The two dogs in their usual posture in which they lay or crawled around.   Sometimes they   stood up on all fours and took a few steps,   while   the rear part of their bodies fell sometimes to the right side and sometimes to the left.   Dog Vici always lay with the rear part of his body on the right side   (1), dog Daumling   always on the left (3)   lateral position. In Figures 2 and 4 both dogs are set up passively on their paws.

1 *dog both whose labyrinths were   extirpated in addition to   decortication of left cerebrum.*

dog Jenny : extirp. of both labyrinths : 17 Feb., 1927,

            extirp of left half of cerebrum : 2 March, 1927;

            died 8 Feb., 1930.

To this material were added a number of animals, dogs and cats, in which either one half only of the cerebrum was extirpated, or the spinal cord was severed, or the extirpation of both labyrinths was carried out (De Kleyn), further animals in which the spinal roots be-

6

longing to one or both hindlegs were severed, or whose four paws were rendered anesthetic by severance of the cutaneous nerves.

Finally, some animals were investigated after a partial extirpation, or lesion of the cerebellum; median section of the cerebellum, extirpation of the vermiform process, severance of the peduncle of the cerebellum, etc.

I. GENERAL INFORMATION ON STANDING AND POSTURAL REACTIONS

In the older physiological textbooks the question as to which mechanism enables man to stand in an upright position without the lower extremities giving way to the weight of the trunk was already discussed in detail. The course of the lines leading from the various parts of the body to the center of gravity was accurately determined, as was the position of the joints while standing. Bone-structure, joints, and ligaments were held responsible for the supporting function of the legs, while the muscles were given importance only in movement, but not in standing, for the reason that they show only little fatigue in the standing position, which seemed incompatible with an effective muscular activity.

Duchenne (58, 59) was the first to perceive, in 1867, that the ligaments alone could not possibly suffice for the fixation of the bones in the various positions of standing, but that rather a continuous muscular contraction was necessary for standing. He drew attention to the fact that muscles in the standing position feel hard and strained and extensors as well as flexors of the legs participate in the process of fixation. He further pointed out the fact that standing may be considerably impaired by muscular atrophy. In 1896 Sherrington (273) found that muscular reflexes, particularly reflexes stimulated in the muscles themselves (proprioceptive impulses), exert an influence on standing. He observed that an extensor stiffness (decerebrate rigidity) occurred in animals after a transverse severance of the mesencephalon between the anterior and posterior corpora quadrigemina. The animals were able to «stand» passively on their limbs, i.e., the stiffly extended limbs were able to support the body. After section of the posterior spinal roots pertaining to the limbs, the rigidity disappeared instantly. Sherrington spoke of a «reflex standing» of decerebrate animals; the stiff limbs possess a static tone influenced by the «standing reflexes».

Magendie (187), Longet (176), Schiff (268), Vulpian (315), Christiani (40), Munk (214) and especially Goltz (107; 1892) had shown previously that decorticate animals (thalamus animals) are able to stand. In contrast to the standing of decerebrate animals, it here seems to be a question of reflex standing with *apparently normal distribution of muscle tone.** Several investigators ascribe to the cerebellum an essential role in standing. Edinger considered the cerebellum as the organ of «statotonus».

Luciani (181) observed astasia in decerebellate animals, i.e., these animals are not able to stand quietly as a result of the incomplete summation of the individual impulses leading to the muscles. Von Bechterew (18) considered the cerebellum as the central organ of static coordination and defines the astasia occurring after extirpation of the cerebellum as «a disturbed state of harmony between the separate movements and muscle tensions and the posture of the body».

«By 'static coordination' says v. Bechterew, we understand that complex reflex mechanism which guarantees a normal coordination of muscular activity with the position of the body in the sense of a perfect stability of the latter under all possible conditions. Its most essential prerequisite consists in that, at each given moment, the effect of the body's center of gravity is counteracted by a force which prevents the natural development of the effect of the body's weight, i.e., the tendency to fall forward. Let us imagine a quadruped in standing position. The vertical line of gravity of its body will fall somewhere in the range of the rectangle formed by its extremities. The means for the maintenance of body balance are in this case the bones and the joints with their muscles which retain the extremities in a certain position. The forces developed by the tension of the muscles of one side must be kept in balance by the tensions developed by the muscles of the other side; likewise the muscle tension of the forelegs has to correspond to that of the hindlegs. Otherwise the

---

* Throughout this volume Rademaker uses the word "tone" in the sense of any sustained contractile process in the muscles, estimated by their resistance to stretch and by their firmness (Ed.).

equilibrium of the body would be disturbed and the animal would fall in the direction of the weaker side.»

"If an animal stands with all its four legs on a movable surface, each inclination of the surface to the right, with the concomitant displacement of the point of gravity of the body to the same side, will cause a strong tension in the muscles of the right extremities, particularly the abductors, while the abductors of the left extremities will slacken and only the abductors will contract simultaneously with the abductors of the right extremities. If the supporting plane is inclined forwards, the forward displacement of the centre of gravity of the body will cause a strong tension in the muscular system of the forelegs, particularly of the extensors, while the extensors of the hindlegs slacken and the flexors of these extremities come into action so that finally the animal lowers the rear part of its body into a sitting position. Exactly the opposite movements occur if the supporting plane is inclined backwards. The extensors of the hindlegs are subject to tension, while those of the forelegs slacken . . . . Let us place an animal in standing position and lift one of its forelegs, for instance the right one, from the ground. As a result of this movement the center of gravity is shifted forwards and to the right and now, to maintain the balance of the body, the muscles of the other three legs have not only to take over the burden of the whole trunk, which was hitherto distributed to the four extremities, but they have also to rearrange the trunk so that the center of gravity falls within the limits of the triangle formed by the three legs resting on the ground. To this purpose the thorax is inclined backwards and towards the side opposite to the lifted extremity".

According to v. Bechterew, these reactions are regulated by the cerebellum. After removal of the cerebellum, they are altered; the harmony of the particular movements with the position of the body is disturbed, the animal becomes "astatic". Also according to v. Bechterew, this reflexo-static coordination presupposes a particular static perception which permits a precise determination of the position of body and head in relation to the vertical plane. The cerebellum is the central organ of this static perception; the organs of deep sensation, the semicircular canals and structures related to the third cerebral ventricle are the peripheral organs of perception. According to v. Bechterew, the importance of these organs of perception is proved by the following observations :

1. "When the skin was removed from the feet of frogs and pigeons, these animals lost to a considerable extent the ability to preserve the balance of the body and to maintain static coordination. When the soles of the feet of a perfectly healthy man are artificially anesthetized, he is no longer able to keep himself solidly on his legs and distinctly staggers to and fro (Vierordt)." Also the balance of the body and the static coordination are instantly disturbed after severing the posterior spinal roots.

2. "After elimination of the cerebellum, all forces otherwise affecting the semicircular canals prove ineffective, i.e., they do not produce the normal results, while lesions of the cerebellum remain effective even after the destruction of the semicircular canals." Likewise, the lesions of certain parts of the cerebellum cause almost the same phenomena as the severing of a semicircular canal.

3. As a result of lesions in the third cerebral ventricle, conditions occur *"which correspond to the highest degree with conditions occurring after affections of the semicircular canals"*. Therefore, according to v. Bechterew, we may assume that the structures related to the deep part of the third ventricle represents a special organ of postural coordination, as do the semicircular canals.

It is supposed that from these three organs of perception impulses are sent to the cerebellum and to the cortex of the encephalon. Against v. Bechterew's very interesting theory one may object that after the extirpation of the cerebellum (1) the labyrinthine reactions are not lacking (see p. 495) and (2) that the postural reactions, too, are not abolished, as we shall see later.

Numerous investigations have been carried out by Sherrington and his collaborators, and by other investigators, on "standing " and "standing reflexes" of decerebrate animals. But, to this day, we lack an exact investigation on the reflexes which enable a normal animal to stand, and on the behavior of static reactions in animals without cerebrum or cerebellum. The results of the following investigations shall fill this gap.

Reflex reactions which bring about an active body posture and are released by a certain position of the body or individual parts of the body are termed static or postural reactions.

In the study of posture we appropriately distinguish between the behavior of the body at rest and in movement. Reflexes released by a position or posture are called static, those released by active or passive movements, statokinetic reflexes. These two groups cannot sharply be separated. Frequently, for instance, a reaction which starts as a statokinetic one, changes later into a static reaction. For practical purposes one distinguishes between passive and active posture. When a dog lies lazily, i.e., with completely relaxed muscles, on its side in front of the warm fireplace, it shows a passive posture. In standing it presents an active posture. Reflexes which bring about an active posture are called positive static, or, in short, static reactions, while those processes which, after cessation of the active posture, cause a passive one, are termed negative static reactions.

Therefore, reflexes not classed as (positively) static reactions are : (1) reflexes which cause passive postures or positions; (2) reflexes not leading to any posture; and (3) reflexes released by a movement.

Of the countless static reactions, a number of reflexes with specific functions are combined into a sub-group.

Postural reflexes are those static reflexes which, by the release of a certain distribution of tension in the muscular system, control the position of the different parts of the body in relation to each other and the fixation of the joints necessary to this purpose. To this group belong, among others : the tonic cervical and labyrinthine reflexes of Magnus and De Kleyn, the supporting, magnet and hopping reflexes, and other reactions which all participate in the occurrence of normal standing that will be discussed in detail.

Magnus termed a number of reflexes which adjust the posture of the body in relation to its environment as *righting* reflexes.

By preparation for standing (*Stehbereitschaft*) we understand a number of reflexes which bring the distal ends of the limbs into the position necessary for standing and maintain them in that posture.

It is difficult to decide to which kind of reflexes the *decerebrate* rigidity belongs. In decerebrate animals the large proximal articulations of the limbs are fixed into extended position by the tension of the muscles. Sherrington therefore calls this decerebrate rigidity a "postural reflex". It is certain that the decerebrate rigidity represents a complex of postural reflexes; but are those reflexes static, i.e., are they reflexes of position ? What releases the stimuli causing this rigidity is not exactly known.

As is well known, decerebrate rigidity is caused by the absence of stimuli which are normally relayed through the midbrain and is therefore, according to Sherrington (276), a "release phenomenon". Decerebrate rigidity, however, does not only represent a deficiency symptom, but is also a positive phenomenon. The occurrence and continuation of decerebrate rigidity, indeed, proves that certain stimuli, certain centers, certain reflex mechanisms which influence the muscle tone are active. Decerebrate rigidity is not a statokinetic reaction, it is present also when the animal lies still. But is it a static reaction? Is it caused by posture? As is well known, rigidity appears in every position, in ventral, lateral and dorsal position, in the air as well as on a support. Therefore, it is not subject to a certain posture; however, its intensity can be altered by a change of posture.

In the dorsal position it is usually most strongly marked, due to the influence of the tonic labyrinthine reflexes of Magnus and De Kleyn, but it is not absent in the ventral position, the minimum posture of tonic labyrinthine reflexes. It is likewise distinctly present when labyrinthine reflexes are totally absent due to bilateral extirpation of the labyrinth (Magnus and De Kleyn, 146) or after severance of the nervi octavi (Sherrington).

The position of the head in relation to the thorax also influences rigidity (tonic neck reflexes of Magnus and De Kleyn). With the head ventrally flexed, the rigidity of the forelimbs in dogs and cats is less than when it is lifted dorsally, but does not disappear. Rigidity of the hindlimbs is least when the head is lifted dorsally, but sometimes does not disappear completely. In most cases it even persists when the head is kept simultaneously in the minimum posture for tonic labyrinthine and tonic cervical reflexes (Fig. 4). Likewise, it occurs in animals which have been decerebrated after the extirpation of both labyrinths a few weeks previously and whose posterior roots C1, C2, C3, were severed to counteract the tonic neck reflexes (Magnus and Storm van Leeuwen, 190).

Sherrington (278) noted that the rigidity of a limb disappears after severance of the spinal root belonging to this limb. Thus it was proved that proprioceptive stimuli from the muscles of the extremity in question are of great importance to the maintenance of rigidity.

According to Liddell and Sherrington (173), the proprioceptive stimuli are released by the muscles being stretched through passive tension which causes a reflex contraction (myotatic reflexes). These stimuli, however, cannot cause rigidity in a lateral position as no stretch or muscle traction then occurs. Magnus and Liljestrand (175) further observed that a distinct increase in the tonic resistance to stretch of the hindlimbs occurs after decerebration, even if the posterior spinal roots that supply these paws have been cut several weeks previously.[1] The rigidity of the hindlimbs, therefore, is not caused exclusively by stimuli emanating from the limbs themselves, although it disappears when the posterior spinal roots are cut after decerebration. Leiri (158) reported that decerebrate cats whose labyrinths had been extirpated and the nervi trigemini severed showed a distinct rigidity even under water.

The elimination of the various stimuli emanating from the neck and the extremities, which normally release the postural reflexes, is thus not able to completely abolish decerebrate rigidity, but only to reduce it. Consequently, it is probable that the rigidity is caused by a conjunction of various postural reflexes and must, therefore, be ranked among the general static postural reflexes. It can be demonstrated, however, by suffocation of a decerebrate animal, that internal stimuli may also stimulate rigidity.

The balancing reactions which maintain the line of the center of gravity of the body within the basis of support or draw it back to it, occupy a special position. Static, as well as statokinetic reactions, for instance those released by falling movements, may function as balancing reactions. Among the static reactions, the righting reflexes and some postural reflexes may in certain cases represent balancing reactions, as we shall see later.

---

(1) In 1931 when this book was published the type of extensor rigidity released by the high level of decerebration of Pollock and Davis (Brain 50 : 277-312) which damaged anterior lobe of cerebellum and released tonic labyrinthine reflexes caused great confusion. Such labyrinthine rigidity is not affected by dorsal root section (Stella, Atti Soc. Med. Chir., Padova 23 : 5-16, 17-21, 1944, Granit, Holmgren and Merton. J. Physiol. 130 : 213-224, 1955). Ed.

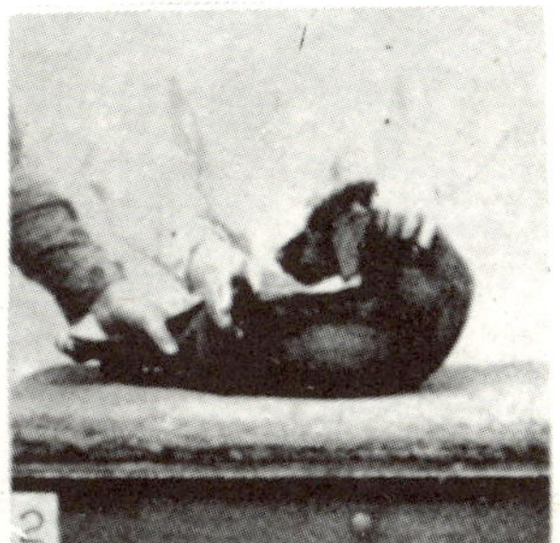

Fig. 4 Two decerebrate dogs in dorsal position on a support. Head ventrally flexed so that the buccal cleft is below the horizontal and forms with it an angle of 35-45°. Thus the head is in the minimum position of the tonic labyrinthine and neck reflexes on the forelimbs. There was definite rigidity of the forelimbs which are extended upwards in the air (1) and offer a distinct resistance to passive flexion by pressure on the soles (2).

When a dog is roused from sleeping in a lateral position, one sees :

I.   that the animal, influenced by various righting reflexes, assumes a sternal or ventral position;

II. that the animal then puts its paws on the ground in the position required for standing, extending its limbs and raises itself into standing posture. If now the supporting surface is slanted or the animal's back is laden with a sack of sand, one sees :

III. that the limbs adjust to the new circumstances, to the increase of the load.

In the following chapters, we shall devote particular attention to those static reactions which produce the standing position (II) and which adjust this position to the static situation or to an increase in the load (III).

II. THE DISTURBANCES OF THE ABILITY
TO STAND AFTER TRANSVERSE SECTION
OF THE SPINAL CORD

After transverse section of the spinal cord in the midthoracic region a number of reflexes reappear in dogs, after the fading away of shock symptoms, as for instance, the ipsilateral flexor reflex and the crossed extensor reflex (Goltz and Freusberg 106, Sherrington 276). Tapping of the patellar tendon then again elicits the patellar reflex and often even *patellar clonus* and crossed reactions. Likewise, a *distinct muscle tone is again present.* The muscles no longer feel flaccid; they react to tapping and other mechanical stimuli. According to circumstances, sometimes the flexor, sometimes the extensor muscles show tonus.[1] When, for instance, the animals are suspended with the head upwards, the hindlimbs are usually extended and a distinct resistance, a clear extensor tonus, is felt when attempting to passively flex them.

In a dorsal (supine) position, on the other hand, the animals flex their hindlimbs and passive extension then almost always encounters a marked resistance. Brondgeest (32) has observed that the muscles of the hindlimbs, after cutting of the spinal cord, are in a certain reflex conditioned medium tension which disappears after the severance of the posterior spinal roots belonging to these limbs. Based on corresponding findings, Sherrington (278) holds responsible, in the first place, the proprioceptive stimuli emanating from the muscles in question. However, a number of other sensory stimuli, too, affect the centers of the spinal cord promoting the muscle tone. A bitch, whose thoracic cord I had severed, when held up in the air in a normal position while urinating, kept her flexed hindlimbs flexed in during the discharge; after emptying the bladder the paws were suddenly extended with force and held for some time in an extended posture. The animal thus showed a change in muscle tone due to stimuli which probably emanated from the bladder. The hindlimbs reacted also to brushing of the skin, pinching and various other stimuli sometimes with extension and heightened stretch reflex, sometimes with flexion

---

(1) Definition of muscle tone : p. 452

and fixation in flexed posture. After the severance of the spinal cord, the muscles not only showed brief phasic tensions *but the hindlimbs were capable of maintaining postures for a certain time because of tonic after-discharge.* Upon pinching the toes, the limb is pulled in by the ipsilateral flexor reflex, the contralateral limb is extended by the crossed extensor reflex and the limbs then remain in this position as long as the pinching continues. Sherrington (279) observed the development of an ulcer on the foot of a dog with a severed spinal cord; the animal kept the diseased leg permanently flexed, the healthy one extended for weeks. After daubing the ulcer with cocaine, the compulsory attitude ceased instantly; it was thus caused by a tonic reflex muscle contraction.

The hindlimbs of spinal animals are not only capable of performing complicated movements, such as scratching movements when the abdominal skin is irritated, *but these movements are sometimes performed with considerable strength.* In a lying position the animals sometimes have fits of rhythmic running movements. When the foot was held fast during such a paroxysm, the hindlimb of the dog whose spinal cord was severed some weeks previously, made such powerful flexion and extension movements that the trunk was pushed to and fro on the supporting surface.

With this animal we even succeeded, by means of exteroceptive stimuli, to release an extensor tonus in the hindlimbs which made possible the lifting and carrying of the posterior part of the body. On pinching the toes, the crossed hindlimb was extended so forcibly that the hindquarters were lifted up from the prone posture. When the pinching was stopped, the limb was able to carry the hindquarters for about 20 seconds (Sherrington observed a continuance of the crossed extensor reflex for 10-15 seconds in spinal animals). Although the muscles again have some muscle tone, and the hindlimbs perform vigorous movements and are able to remain in an extended position, even to carry the body by means of certain exteroceptive stimuli, yet the rear part is not able to stand.

Philippson (224) published cinematographic pictures of dogs whose thoracic spinal cord he had severed and which were able to stand, as well as run and even gallop. Magnus, who observed similar phenomena, stated "that this was an illusion insofar as dogs learn, after some time, by an appropriate posture of the head and by a vigorous contraction of the shoulder muscles to maintain the posterior part of the body above the ground in a horizontally suspended position, so that the feet sometimes just touch the floor and, only hanging, so to speak, from the haunches, simply perform reflex running movements, so that it seems as if the animals were able to stand on their hindlegs. But numerous observations on dogs with severed spinal cord, a number of which were kept alive for years, have shown that the hindlegs do not receive a true static tonus and are not capable of carrying the weight of the posterior part of the body by themselves"[1]

When the toes of the flexed hindlimbs are set with the balls of the foot upon a support, no extension of the hindlimbs occurs in spinal dogs and the hindquarters are not lifted up (absence of extension upon touching the soles). When the hindlimbs are passively extended and brought into a posture suitable for standing, they give way after a short time under the weight of the posterior part of the body (absence of fixation of the extended posture upon static stress). The hindlimbs are not set upon the ground in a position suitable for standing, when the animals are held suspended in the air by the skin of the nape of the neck and of the back and slowly lowered to a surface (absence of preparation for standing). The various reactions which normally prevent the rear part of the body from falling from the standing posture, too, are totally absent.

Bastian (14) in 1890, reported his precise observations made on patients with transverse lesions of the spinal cord. Based on his own and on similar findings in the literature, Bastian came to the conclusion that in man the severance of the spinal cord in the upper segments results in an absolute, continuous, flaccid paraplegia with hypotonia, anesthesia and absence of all reflexes. He maintained that the reoccurrence of muscle tone, spinal reflexes and spontaneous movements as well as the return of sensibility are an absolute proof for the incompleteness of transection.

Bastian's report was followed by a series of publications with conforming observations which, supported by Bruns (36) in Germany, Dejerine (49) in France and Van Gehuchten (92, 93) in Belgium, contributed to the almost general recognition of "Bastian's Law" by neurologists. It was inferred that the centers of the tendon reflexes and other reflexes, spinal in inferior mammals, had moved and become

---

(1) See however, Denny-Brown and Liddell (57), on spinal stretch reflexes.

cephalic in phylogenetically more highly developed animals; Van Gehuchten assumed them to be in the midbrain in man.

These observations in man, for a long time, led to the hypothesis that the findings of experimental neurophysiology could not be simply applied to human beings. For this reason, many clinicians believed that they could justifiably neglect the results of animal experiments.

Little by little, however, cases were published (von Schultze, Senator, Jolly, Brouwer (34) and others) which disproved Bastian's Law. Although in several of these cases the connecting tissue was evidently not conducting, yet the transverse section was not complete. One of Kausch's (1901) cases, however, in which the spinal cord was completely severed, showed distinct patellar reflexes. Finally, it was established by Head (116), Riddoch (116 and 244) and Lhermitte (171) on war-wounded patients that even in man in cases of total transverse lesion there may occur a distinct muscle tone, spinal reflexes, such as plantar [1] and other dermal reflexes, patellar and Achilles tendon reflexes, periostial reflexes, ipsilateral flexor- and crossed extensor reflexes and likewise also automatic extension and flexion movements, just as in spinal dogs. Even hypertonia of the leg muscle with increased patellar reflexes and patellar clonus was observed in several of the wounded (Riddoch 244, Claude and Lhermitte 41). Although the force of the automatic movements was sometimes considerable, the legs were not capable of carrying the trunk. The patients were not able to stand.

Thus, as was already emphasized by Riddoch and Lhermitte, there is an absolute conformity between observations in man and the results of animal experiments.

In man, the spinal reflexes usually return much later than they do in the dog. This may be connected with the fact that shock symptoms last longer, the higher the species of animal (Sherrington). To this should be added that in animals the transection is carried out aseptically, while in war-casualties the wounds usually become infected and, as observations on animals have shown, the spinal reflexes

---

(4) By stimulating the sole of the foot, the big toe moved now dorsally, now the reverse; sometimes Babinski's phenomenon was positive for a prolonged period, but later there was constant plantar flexion.

are absent much longer when the healing of the wound does not proceed smoothly. Furthermore, after severance of the spinal cord, there easily occur other intercurrent diseases such as decubitus (bedsores) or cystitis, and this might have been the case particularly in war-wounded men, where under the circumstances, initial treatment could only be very insufficient. In animals, it was sometimes striking to observe how, due to a small non-infected decubitus injury, the reappearance of spinal reflexes was retarded, reflexes already restored disappeared again and the muscle tone, too, decreased. This was the case to a much more obvious degree when the complications caused fever.

Often the retarding influences of intercurrent diseases are not sufficiently taken into consideration and the cause for the decrease in muscle tone and for the weakening or total lack of reflexes is sometimes too exclusively sought in the central nervous system. Riddoch, however, was able to observe released patellar reflexes in some war-wounded as early as three weeks after the severance of the spinal cord. Thus the spinal reflexes may, in certain cases, reappear rather rapidly in man also.

In the older cases published to support Bastian's Law, the transverse lesion of the spinal cord was caused either by proliferating or infectious processes, or was sometimes accompanied by feverish complications. Such cases do not permit any conclusion on the physiology of the central nervous system.

III. THE "STANDING" OF DECEREBRATE ANIMALS

Sherrington showed, in 1896, that after a transverse section of the mesencephalon between the anterior and the posterior quadrigeminal bodies an extensor rigidity (decerebrate rigidity) appears in animals and that these animals are able to stand when they are set up on their limbs. The rigidly extended limbs possess a static tonus and are able to carry the body by means of their "stretch reflexes".

The standing of decerebrate animals, however, differs from normal standing in essential points, it represents, so to speak, a "caricature of normal standing" (Magnus). It differs from normal standing in so many points that one may ask oneself if one should really still call it "standing".

Decerebrate animals show an entirely different distribution of tonus, a hypertonia of extensor muscles and the joints of the extremities are in an extremely extended position.

Richter (243) objected to my previous work (234), in which I demonstrated the strength of the extensor tonus in an exaggerated way. In his opinion, there is an extensive similarity between the extensor tonus and the position of joints in decerebrate cats in the standing position with those of normally standing cats. Here, I should like to point to the photographic pictures in Sherrington's original work which best show how remarkably strong this extensor rigidity can be in the decerebrate animal. It may sometimes also be less strong as shown by the illustrations of Magnus and myself. (Fig. 5 and Fig. 6).

I also mentioned in the publication in question that in a number of decerebrate animals, the tonic labyrinthine reflexes exerted so strong an influence that in the dorsal position the animals showed a marked decerebrate rigidity with exaggerated extensor tonus and that in the ventral position, however, this extensor tonus was sometimes so much reduced that the paws were no longer able to carry the body. In other decerebrate cats with excessive dorsiflexion of head and neck, the influence of the tonic neck reflexes was so great that the hindlegs showed an extreme flexor tonus so that they were pulled in to a position of flexion in all of the animals' postures. These animals were only able to stand with their forelimbs.

In different preparations, extensor rigidity shows considerable differences, for which the plane of the cross section plays an important role.

Intact animals in standing posture usually show a strong contraction in triceps as well as in biceps of the forelegs. Occasionally, when the forepaw is turned under, only the biceps feel rigid and contracting whereas the triceps feel flaccid. According to Sherrington and others, decerebrate animals in standing posture show a strong tonus only in the extensor muscles while the flexors have only a very slight tonus or none whatever. More recent investigations with string galvanometers, however, showed that in decerebrate animals the biceps, too, have a distinct tonus which increases simultaneously with that of the extensor muscles. The stronger the extensor rigidity of the standing decerebrate animal, the stronger is also the flexor tone.

Another difference is the fact that in decerebrate animals, the extensor tonus; which makes standing possible, is not caused by contact or by pressure on the supporting surface. When an intact animal is held suspended in the air in a ventral position, the limbs usually hang down in a half extended posture and are easily moved; but as soon as they touch the surface, they are extended, fixed in an extended position by tension of extensor and flexor muscles and thus rendered into solid pillars.

If, on the contrary, a decerebrate animal is suspended in the air in a ventral position and then brought down on a surface, the distribution of tonus remains unchanged. It is true that the extensor tonus sometimes increases a little if one sets the paws carefully with the soles on the surface in the posture required for standing. This is particularly the case if one pushes the shoulders at the same time.

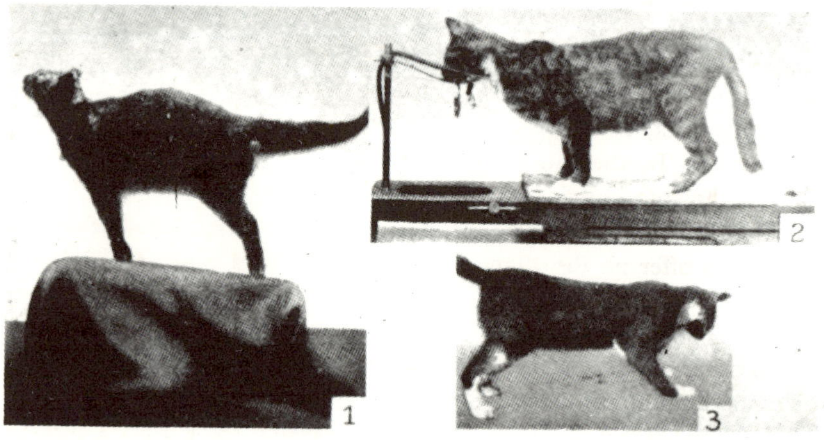

Fig. 5 Decerebrate cats standing. 1. From Sherrington (C. S. Philosophical Transactions, Series B. Vol. 190, Fig. 1, p. 187). 2. From Magnus (R. Magnus, Korperstellung, Fig. 1) 3. From Rademaker (G. G. J. Rademaker, Die Bedeutung der roten Kerne. Fig. 188).

Fig. 6. Two decerebrate dogs in standing posture. Dog A is decerebrate caudal to the red nuclei, dog B through the red nuclei and just in front of the point of exit of the Nn. oculomotorii. Note the much stronger rigidity of dog A.

In decerebrate animals, the extensor posture appears not only on standing but also in the lateral and dorsal positions. In the decerebrate animal, the extensor tonus is increased to a maximum in the dorsal position (Fig. 7). That is, if a decerebrate animal is laid on its back with its muzzle pointed upwards at an angle of $\pm$ 45° to the horizontal, then the paws are extended to a maximum under the influence of the labyrinthine reflexes and the resistance to passive flexion is stronger than on standing (Fig. 7, No. 3 and Fig. 8).

In this respect, the decerebrate animal behaves differently from the intact one. In the intact animal, resistance to passive flexion is strongest in the standing posture while in the dorsal position the paws are flexed. In the intact animal, in the dorsal position, the passively stretched paws offer no resistance to flexion, and even may be actively flexed so that some resistance is felt in passive extension.

Still another difference is the inability to influence the distribution of tonus by static changes. In the standing of a normal animal, the distribution of tonus is immediately adjusted to static change, while in the decerebrate animal, it is in no way influenced by such change, for example by slanting of the supporting surface (Fig. 9). Also, if an animal is carefully set up on its limbs and its body is moved to and fro, forwards, backwards or sideways, no reaction or corrective movements of the limbs will occur (Fig. 10).

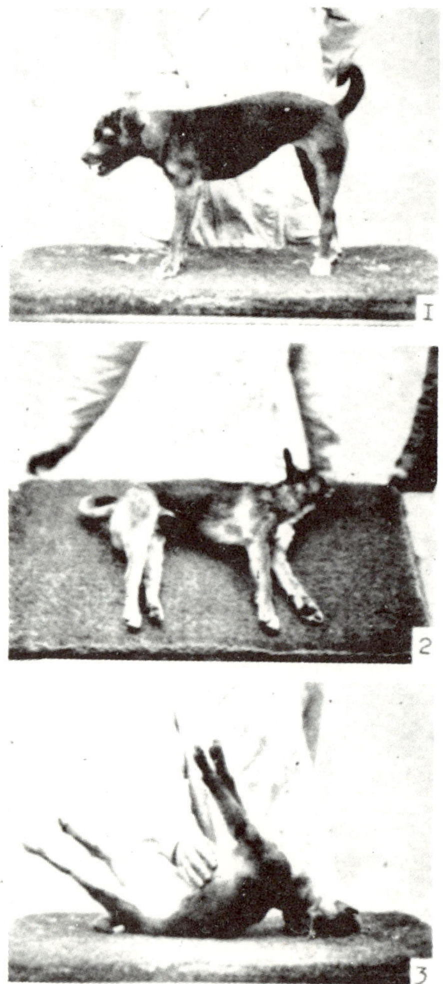

Fig. 7. Decerebrate dog A. 1. Passively set upon its paws. 2. In lateral position, too, the animal shows a distinct rigidity. 3. In dorsal position, maximum extension and rigidity of legs. The muscles of neck and spinal column are also in a state of maximum tension. By the contraction of the neck muscles the anterior part of the trunk is lifted from the support.

The so-called static tonus of decerebrate animals, therefore, is neither caused by the standing posture, nor adjusts to static changes. It even is usually strongest in the dorsal position.

In addition, the standing of decerebrate animals shows other differences. When a decerebrate animal is held suspended in the air in a ventral position and then lowered till the paws touch a surface, the animal cannot stand because its limbs occupy a totally wrong position. The animal puts down its paws with the toes turned under, i.e., in a position totally unsuited for standing (Fig. 11, No. 1 and 2). If the animal is released, its paws slide backwards and the animal drops on its chest and belly and does not make a single movement to prevent the fall (Fig. 11, No. 3).

Thus, the decerebrate animal is able to stand only when the limbs are carefully put into the correct position. Even then, it falls over at the slightest push, at the least displacement of the surface or by the smallest movement it makes and it does neither try to prevent the fall nor to rise from a lateral position. The displacement reactions and the righting reflexes are totally absent (only the neck righting reflexes can still be produced in the decerebrate animal). *Thus, the standing of decerebrate animals occurs only through an abnormal distribution of tonus, which does not adjust to static conditions. The limbs are not set down in a posture necessary for standing, and there is no correction of any abnormal posture. Reactions for the maintenance and readjustment of balance as well as the labyrinthine and body righting reflexes are totally abolished.*

Fig. 8. Decerebrate dog B in dorsal position. 1. The limbs are left untouched. The forelegs are extended upwards in the air, the hindlegs stretched backwards. 2. The limbs are subjected to static stress. The pressure on the soles meets a strong resistance. (Compare this Fig. with Fig. 6 No. 2) .

Fig. 9. Decerebrate dog A. Passively set up on its feet. 1. The support of the animal is raised at the head. The legs do not move back at the joints of shoulders and hips; thus the animal would fall backwards if not held at the neck. 2. The support of the animal is raised posteriorly. The forelegs are not moved forwards as in the intact animal, the hindlegs are not flexed so that the animal would fall head foremost if not held back by the tail.

The decerebrate animal stands, so to speak, like a toy animal with only the difference that in the decerebrate animal there are several factors, for instance, change of position of head to trunk which may influence the distribution of tone.

In addition to the tonic neck and labyrinthine reflexes of Magnus and De Kleyn, produced by postural changes of the neck, decerebrate animals also show spinal reflexes such as the patellar reflex, the ipsilateral flexor reflex and the crossed extensor reflex. The labyrinthine progressive reactions, such as preparation for jump (Sprungbereitschaft) and lifting reactions, can still be produced. Besides, the extensor tonus gradually decreases after some time (usually after 5 to 10 minutes) in the decerebrate animal set up on its legs, and the paws give way more and more. The standing of decerebrate animals thus differs from the standing of intact and that of decorticate animals which are able to stand for hours without their limbs giving way.

Recently, Sherrington and Liddell (173) demonstrated the importance of the muscle reflexes which are caused by stretch (myotatic

Fig. 10. Decerebrate dog B. 1. The animal set upon its legs. 2. The trunk
is passively moved forwards. The limbs are not shifted forwards as in
the intact animal; they show neither righting reactions nor correcting
movements, but go passively more   and more backwards from the
shoulder and hip joints so that the animal finally lies on the support
with its chest and belly, with its legs extended backwards.

Fig. 11. Decerebrate dog B. 1. Held in the air by head and tail in ventral position. 2. The four paws are brought in touch with the surface.
The paws are not put down in a way as to enable the animal to stand
(lack of preparation for standing). 3. With further lowering the paws
glide backwards and the animal lies on the surface in a thoracic-
ventral position with the limbs in an abnormal posture.

reflexes), in the standing of decerebrate animals. They observed that
in the decerebrate animal, the stretching of a knee extensor produces
a reflex increase of contraction in this muscle. The knee flexors,

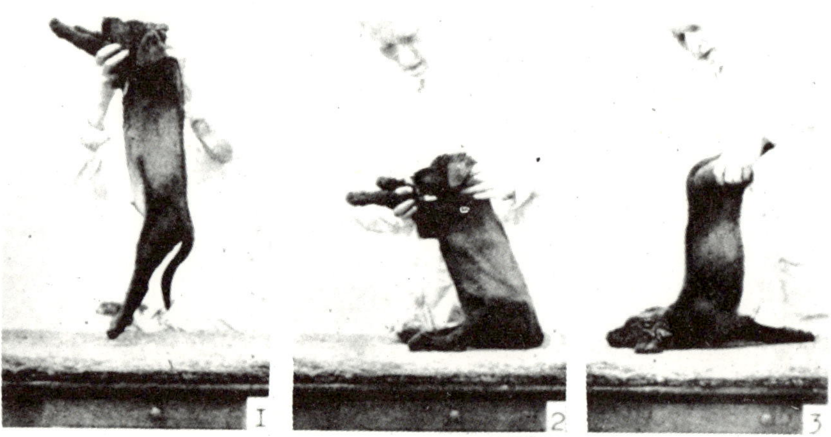

Fig. 12. Decerebrate dog B. 1-2. In lowering the animal from a suspended position, head upwards, the hindlegs in touching the support are not set in a position suitable for standing. 3. The forelegs fail in the same way in lowering from suspension head downwards.

among them the m. semitendinosus, did not show this tension caused by stretch.[1]

The increase in tension was proportional to the degree of stretch. In certain cases, a minimal lengthening of the muscle (by 1%) was sufficient to cause a distinct increase in tension which disappeared instantly as soon as the stretch ceased. On the contrary, if the muscle was retained in the stretched position, the increase in tension continued for some time, to gradually decrease, however, after about 6 minutes and finally disappear.

Liddell observed that the increase in tension of the knee extensor produced by a stretch of 3 mm continued steadily after 30 minutes but, on the contrary, the increase in tension caused by a stretch of 7 mm had already disappeared after 5-10 minutes. Only the stretched muscle or part of muscle shows the increase in tension; the tension of the other muscles remains unchanged.

---

(1) Experiments carried out together with Dr. Hoogerwerf showed that in the forelimb of the decerebrate animal, the elbow flexors (m. biceps and m. brachialis) as well as the elbow extensor (m. triceps) react to stretch with a reflex increase of tension.

Stretching of the knee extensor on one side sometimes excites a simultaneous contraction of the knee extensor on the other side. In the spinal animal, a reflex tension can also be produced by stretch of the extensor muscles.[1]

After severance of the posterior roots of the spinal cord belonging to the hindlegs, the myotatic reflexes of the knee extensors disappear.

The stretch reflexes are also influenced by tonic labyrinthine reflexes. With an equally strong extension, the increase in tension was less pronounced in normal postures of the head (minimum position of tonic labyrinthine reflexes) than in dorsal postures of the head (maximum position for these reflexes, Liddell and Sherrington 174).

Sherrington and Liddell are of the opinion that the stretch muscles of the decerebrate animal set up on its paws are extended by the weight of the trunk. In this way, myotatic reflexes appear which prevent the joints from giving way and thus bring about "standing".

Against this supposition by Sherrington and Liddell, it can be objected that, in setting up decerebrate animals on their legs, no prolonged extensor tonus can usually be observed. Neither could the stretch reflexes explain extensor rigidity in the lateral position, i.e., when the influence of the force of gravity is absent. These reflexes, however, can indeed play a role which is shown when downward pressure is exerted on the shoulder girdle of the decerebrate animal set up on its forelegs. In this case, the extensor rigidity of the forelegs and the resistance to passive flexion usually increase distinctly.[2]

Sherrington (272, 276) indicated earlier that various stimuli, which in spinal animals produce brief reflexes, can cause in decerebrate animals long lasting extensor reflexes (among others up to the duration of 20 minutes in decerebrate monkeys) which can continue even after

---

(1) In a publication in 1927, Denny-Brown and Liddell (52) state that, in dogs whose spinal cord had been severed in the 1-3 lumbar segment 2-3 months previously, the m. gastrocnemius indeed showed a strong tonic stretch reflex, the knee extensor developed a weaker response and m. tibialis anticus showed none.

(2) By this pressure also the fingers in the metacarpo-phalangeal joints are passively moved in a dorsal direction. As shown in experiments jointly carried out with Dr. Hoogerwerf, the passive dorsal movement of the fingers in which the finger flexors are extended, causes an increase of rigidity in the forelimbs. See also Denny-Brown, Proc. Roy. Soc. 104 : 252, 1929.

the cessation of the stimulus (afterdischarge). The importance of this continuation of the reflex for the standing of decerebrate animals is obvious. It is strange, however, that this afterdischarge contraction is absent in myotatic reflexes after cessation of stretching; contraction stops instantly within a tenth of a second.

The other stimulations which influence the decerebrate rigidity and thus "standing" have already been discussed in detail.

Many decerebrate animals are particularly irritable and show a strong immediate increase in rigidity on pinching the tail, shaking to and fro, even on beating with the hand on the supporting surface. Of course, similar stimulations are almost unavoidable in setting up the animal on its legs, thus they sometimes participate in the "standing" of the decerebrate animal.

As decerebrate rigidity still occurs following a transversal lesion, situated at the level of the medulla oblongata rostral to the nervi octavi, the centers of rigidity should lie caudal to this plane. They are probably situated in the medulla oblongata because the rigidity disappears after transverse section through the lower half of the pons (Sherrington, Magnus and others). The afferent stimulus does not reach the centers of rigidity by way of the cerebellum, because a decerebrate rigidity lasting for 8 hours has been observed after decerebration and total extirpation of the cerebellum (Magnus and Beritoff, 20). From this observation, it can be deduced that the cerebellum cannot play the role in "statotonus" which was attributed to it by Edinger (see chapter XIX). The pyramidal tracts, rubrospinal tracts, median longitudinal fasciculus (234) and the dorsal spinocerebellar tracts (Thiele 299) likewise do not participate in the mechanism of decerebrate rigidity.

On the other hand, Sherrington observed that, after severance of the ventrolateral columns, the rigidity always disappeared. Probably the reflex arc of the reflex "standing" of decerebrated animals (whether with its descending or with its ascending part) proceeds in these columns.

In cases of human decerebrate rigidity (Kinnier Wilson 332, Violet Turner 311, Gordon Holmes 119, Walshe 317). the mesencephalon was totally or almost totally severed by tumors at the level of Sherrington's decerebration. The clinical symptoms, too, were similar to those of animals : extensor rigidity of the entire body, stiffness of limbs, dorsiflexed neck and curved, stiff back.

After transverse lesion of the mid-brain, the clinical observations also coincided totally with the results of animal experiments. In addition to the result of a transverse section of the mid-brain, decerebrate rigidity in animals also occurs when only Forel's decussation is split (234)*. Raymond and Raymond Cestan (241, 242) observed a man in whom, as shown in a post mortem examination, a small tumor had destroyed the decussation of Forel. The man was not able to sit up in bed and lay motionless. He further showed a hypertonia of the musculature and absent righting reflexes in accord with experimental results in animals.

The various cases of decerebrate rigidity that have been reported always showed an extended posture of the legs. Conversely, the stiff arms sometimes were flexed, in other cases, however, were extended. Walshe's case, among others, kept the arms flexed. Walshe pointed to the fact that, in man, the anterior extremities have another function, and that the flexor muscles of the arms, in opposition to the behavior in the animal, usually counteract the force of gravity. Thus, the distribution of rigidity in this case conformed to Sherrington's statement that the muscles counteracting the force of gravity show an increase of tonus in decerebrate rigidity.

Walshe, therefore, considers flexion of the upper limbs a characteristic symptom of decerebrate rigidity in man, other authors, on the contrary, the extended position with pronation (Kinnier Wilson and others). Thus, the position of arms plays an important role concerning differential diagnosis for decerebrate rigidity, but in my opinion wrongly so.[1] As will be shown in the following observation, probably neither the flexion nor the extended position of the arms excludes decerebrate rigidity.

Simons observed in hemiplegics that a dorsal extension of the head and neck causes some flexion of the paretic arm. The reflexes produced by a retracted posture of the head may thus individually produce

---

* This conclusion was later discredited. See Ingram Ranson and Barris Arch. Neurol. Psychiat., Chicago 31, 768-786, 1934 and Fulton, J. F. Physiology of the Nervous System, 2nd. ed. New York 1943, Oxford Univ. Press. (Ed.).

(1) see footnote on page 12. Ed.

modification of rigidity. This perhaps explains the fact that patients with decerebrate rigidity, which usually caused the head and neck to be strongly extended backwards, showed a partly flexed and a partly extended posture of the arms.

Various animal species exhibit similar differences. In decerebrate dogs and cats, the tonic neck reflexes produced by a raising of the head cause a decrease in the extended position of the hindlegs, and in rabbits, on the contrary, an increase.[1] Besides, in animals with decerebrate rigidity, the posture of the extremities, particularly of the hindlegs, often varies. There are decerebrate dogs and cats with all four limbs extended, while others only extend the forelegs and keep the hindlegs in a position of flexion. The flexed posture of the hindlegs appears only when the neck is very stiff and kept dorsiflexed very strongly. In this case, the posture of the neck sometimes produces such powerful tonic neck reflexes that the hindlimbs have a flexed tonus instead of an extensor tonus.

The posture of the back also plays a role (see Chapter X). In animals with flexed hindlimbs the back is hollowed, whereas the extended position of the hindlegs is associated with a dorsal convexity of the back.

Furthermore, it is possibly of importance that, in the cases of human decerebrate rigidity, just as in decerebrate animals, the rigidity of the neck muscles differ in intensity and sometimes is entirely absent. In animals with a lack of stiffness in the neck, the forelimbs sometimes show a stiffly extended position only when the head is passively bent backwards. With a ventrally bent head, the forelegs often are easily flexed and, with passive extension of the limb, a flexor tonus can be felt. Contrary to other cases, Walshe's case showed no rigidity of the neck muscles.

Sometimes, it can be observed in decerebrate animals that, due to respiratory difficulties or other stimuli, rigidity increases in spasms and with each spasm the position of the extremities changes. With the occurrence of a strong rigidity of the neck muscles, the hindlegs are often drawn into a position of flexion between the spasms; when the animal lies quietly in a lateral position, the limbs are found more or less extended. Similar attacks have also been observed in man; rigidity suddenly increased, the head was moved vigorously backwards

---

(1) See footnote on page 12. Ed.

and the flexed arms changed over to an extended position and pronation. Just as in the decerebrate animal, an alternating flexed and extended posture was observed in one and the same case.

From these observations, it can be derived, in my opinion, that the posture of flexion of the arms is diagnostically neither for nor against a decerebrate rigidity.

IV. PREPARATION FOR STANDING
("*STEHBEREITSCHAFT*") : THE PLACING REACTIONS

Normal standing is made possible by :

1. Preparation for standing; by this we understand the ability to place the extremities in a position required for standing;

2. a sufficient supporting tonus.

The preparation for standing in the limbs is caused by : 1. stimuli released from various parts of the body surface 2. optic stimuli.

A. PREPARATION FOR STANDING ON THE FORELEGS UPON STIMULATIONS OF THE BODY SURFACE (TACTILE PLACING)

An intact dog, blindfolded and with the head hanging downward keeps its forelegs, flexed or extended, directed backwards. When the ventral surface of the muzzle is brought into contact with a surface, for instance the edge of a table, the forelimbs are instantly moved forwards and the paws placed on the table beside the muzzle (preparation for standing, tactile placing.) The animal then extends its forelimbs, thus lifting the head and the anterior part of trunk from the surface (Fig. 13).

Preparation for standing also occurs, even if not as promptly, by touching the edge of the table with the dorsal side of the muzzle; in this case however, the paws are not always correctly placed on the surface. They often only make unsuitable movements and push their paws in the air along the muzzle without touching the edge, but they also sometimes put them correctly on the table.

The preparation for standing on the forelegs is absent in decorticate animals (thalamus animals, Fig. 21, No. 2 and Fig. 258), in de-

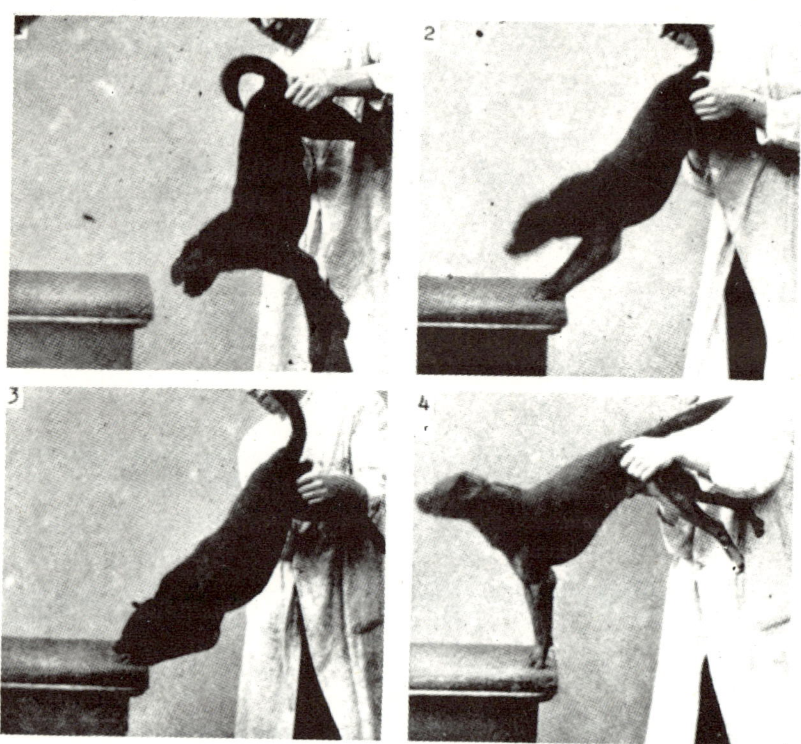

Fig. 13. Preparation for standing in the decerebellate dog Moor.
1. The animal with blindfold in hanging position, head downwards.
2. On contact of the muzzle with the support the forepaws are moved
forwards and the soles are set down on the support beside the muzzle
(preparation for standing). 3. and 4. By extension of the forelegs,
the anterior part of the body and the head are raised from the surface
(supporting reaction).

cerebrate (Fig. 12, No. 3) and in spinal animals. After extirpation
of cerebellum, too, the preparation for standing is suspended (Fig. 14,
No. 1) *but only temporarily*. This condition persists in various dura-
tions from a few days (sometimes only 2-4) to several weeks. Generally,
the delay was greatest in those animals which showed symptoms of
rigidity after extirpation of the cerebellum. A considerable time
after extirpation, placing appears promptly in the forepaws in all decerebellate dogs and cats. (On the behavior of readiness to stand
after unilateral extirpation of the cerebellum, see Chapter XVIII).
Placing is absent in new-born dogs in the first 3-6 weeks after birth.

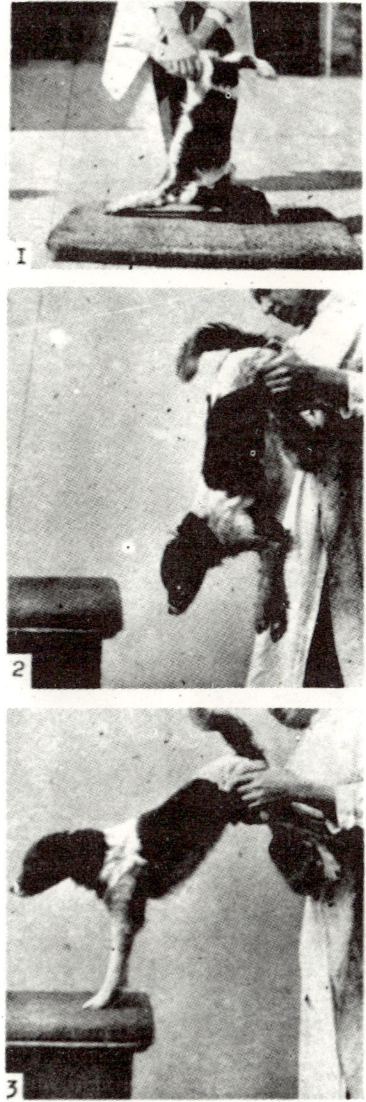

Fig. 14. Decerebellate dog Wolf. 1. Twelve days after the extirpation of cerebellum. The animal suspended head downwards, is lowered till the muzzle touches the ground. The forepaws are not brought into standing position even when the chest and the anterior side of the forelimbs touch ground (the eyes are not blindfolded). 2-3. A considerable time after the extirpation of cerebellum. On contact of the muzzle with the edge of the table the forelimbs are now immediately set down on the table beside the muzzle and then extended.

34

The cerebrum extensively controls preparation for standing on the forelegs since, first of all, it is absent after a total extirpation of the cerebrum and, in the second place, in animals with unilateral destruction of the cerebrum it is only present in the foreleg on the side of extirpation. By contact of the muzzle with the supporting base, only this unaffected paw is placed beside the muzzle while the contralateral paw remains suspended in the air (Fig. 15 and Fig. 22A).

Fig. 15. Preparation for standing in the forepaws after unilateral extirpation of cerebrum. 1. Dog Sep, 1 month after extirpation of the left half of cerebrum suspended head downwards, the eyes blindfolded. 2. On contact of the muzzle with the support only the left forelimb is set down while the right one remains suspended in the air. 3. Dog Daumling after extirpation of the right half of cerebrum and of the cerebellum, with closed eyes suspended, head downwards. 4. On contact of the muzzle with the edge of the surface, only the right forelimb is set down. The left one does not show any placing. Extirpation of cerebellum 30 Dec., 1925. Extirpation of the right half of cerebrum 20 Feb., 1926. Photos taken on 6 May, 1926.

For testing placing after unilateral extirpation of the cerebrum, it is advisable to wait till the animal suspended in the air is totally calm and only then to bring the muzzle in contact with the base. When the animal is agitated and makes running movements with the forelegs, the contralateral forepaw may be placed on the table quite accidentally. When the animal is calm, the placing of the ipsilateral forelimb may sometimes also lead to alternating movements of the opposite limb which usually are performed in the air (Fig. 16B), and bring about a setting down of the paw on the surface thus stimulating preparation for standing. In order to incontestably check this, the right and the left forelegs must be restrained alternately while the muzzle touches the surface. One will then see that the foreleg ipsilateral to the extirpation is always promptly placed on the surface, the contralateral, however, always remains hanging in the air without any reaction.

As we shall see later, contact with the back of the foot or the ventral side of the chest with a surface will also cause preparation for standing in the forelegs.

B. PREPARATION FOR STANDING ON THE HINDLEGS UPON STIMULATION OF THE BODY SURFACE

When an animal with blindfolded eyes is slowly moved downwards while suspended head upwards, the hindlegs are moved backwards and placed with the soles of the feet on the supporting surface as soon as the tip of the tail or the back of the thigh touches it (preparation for standing, Fig. 17, A and B.) Then follows the extension of tarsal- and knee joints with fixation in the extended posture (supporting reaction) so that the hindlegs support the hinderpart of the body.

On touching the tip of the tail with the hand, too, a dog suspended head upwards often extends his hindlegs and moves them backwards as if towards a surface. If, however, with this movement the paws do not touch a surface, they are drawn in again instantly even if the tail remains in contact with the hand (Fig. 17C). If in the course of this extension movement only one paw comes into contact with the surface it remains in an extended position while the other paw returns

Fig. 16. Impairments in preparation for standing on contact of the muzzle with a support in decerebellate, (on the right side decorticate) dog Vici. A. The animal, blindfolded, is held suspended head downwards and then the muzzle is brought into contact with the edge of the table (A2). In this contact the right forelimb is at first moved forwards (A 3-7) and set down on the table (A 8), then it is extended (A 10-14) so that the anterior part of the body is raised. The left forelimb remains without reaction hanging backwards. B. In repetition of the experiment the right forepaw shows the same reactions (B 1-6), now, however, the extension of the right leg is accompanied by a movement of the left one, which is at first flexed (B 7-11) and then again extended, after which it remains hanging in the air directed backwards.

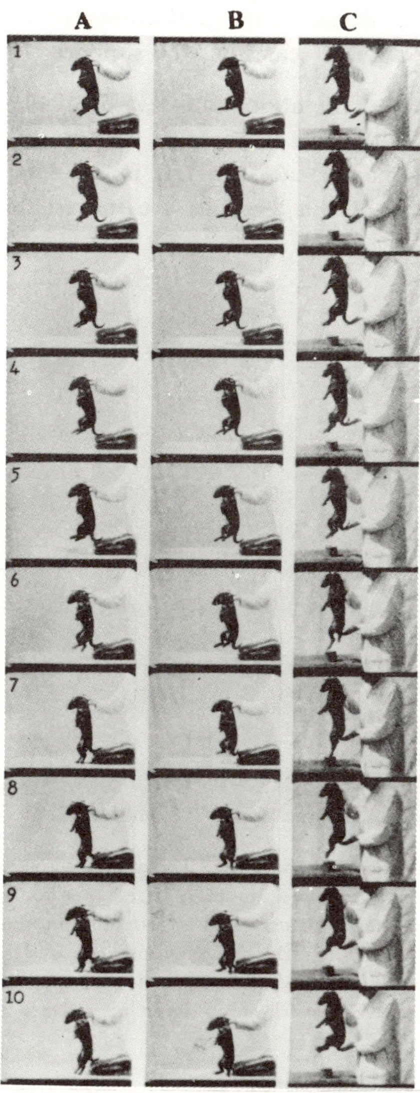

Fig. 17. Preparation for standing of the hindlimb on contact of the tail with a support. Decerebellate dog Piccolino, blindfolded, suspended head upwards, above a table on which a doubly folded mat is placed. A and B. The animal is lowered till the tip of the tail touches the edge of the mat. After contact (A 5 and B 6) the h ndlimbs are moved backwards, extended and set down on the table (A 6 and 7, B 8). The limbs are fixed in extended position and support the trunk (A 8-10, B 9 and 10). C. The tip of the tail of the suspended animal is touched with the hand (C 5) followed by an extension of the hindlimbs (C 7) which, however, are instantly flexed again (C 10) as they do not make contact with a surface.

to the position of flexion after performing several flexion-extension movements as if it were searching for support (Fig. 18).

Also, on contact of the abdominal skin or of the dorsum of the hindfoot with a surface, the foot is set down on it. In this case, however, the paws are not moved backwards as in touching the tip of the tail or the caudal side of the legs but are moved forwards.

Fig. 18. Decerebellate dog Piccolino.   1. The animal suspended head upwards, keeps its hindlegs drawn up in a flexed position.   2. By touching the tip of the tail with one hand an extension movement of both limbs occurs, in which the right hindpaw only contacts the support.   This paw is fixed in extended position and supports the trunk.   The left hindpaw, however, is drawn up again in a flexed position.

These reactions of the hindlegs are "conditioned reflexes" according to Pavlov (217-220). If the animal is kept hanging in the air and the tip of its tail touched repeatedly (7-8 times) while the hindpaws never find a support with their extension movements, then the subsequent touchings of the tail do not produce any more extensions of the hindlegs. The extension reflexes caused by touching of the tail can be induced again by lowering the animal several times in such a way that first the tip of the tail and then the soles of the hindfeet touch a surface, or by leaving the animal without stimulation for some time, for instance for a day. The extension reflex caused by touching of the tail thus possesses the typical characteristics of a "conditioned" reflex.

The same is true for the reactions of the forelimbs. A dog is held suspended head downwards. If we touch the ventral side of its muzzle with a board so narrow that there is no place beside the muzzle, the animal will move its paws forwards sev eral times in vain and finally will leave them hanging again in the air. When the contact is repeated, the animal again tries to set its paws on the board beside the muzzle. If, however, after 6-10 repetitions, the paws never reach a support, the reactions cease to appear even when the muzzle touches the edge of a table where it could set down its paws beside the muzzle. These reactions, too, will return if one waits for a while or if the muzzle is brought into contact with the edge of a table several times consecutively and the anterior part of the body and the forepaws are pushed forwards on the table, i.e., when the conditioned stimulus is followed by an unconditioned one.

It is proved by these characteristics as well as by the absence in decorticate, decerebrate and spinal* animals (the reactions of the hindlegs are also absent, Fig. 11, No. 3; Fig. 21, No. 1) that preparation for standing is a cerebral reflex.

After extirpation of the cerebellum, placing of the hindlimbs is temporarily abolished; it promptly reappears, however, together with the reactions of the forelimbs, a certain time after extirpation. While placing of the forelegs on contact of the muzzle with a surface is nearly always present in new-born dogs, 3-4 weeks after birth when the animals begin to run, placing of the hindlegs on contact with the tail usually manifests itself only much later. If, however, the animals are lowered every day regularly several times from the day of birth from suspension head upwards in such a way that first the tail and then the hindlegs touch the ground, preparation of these limbs to stand, will also be found 3 - 4 weeks after birth. Probably the animal must

---

* Under special circumstances it has recently been shown that simple tactile placing of this type can recover after complete spinal transection, see Forssberg, Grillner and Sjöström, Acta physiol. Scand 92: 114, 1974, even in the human infant, according to Paine and Oppé, *Neurological Examination of Children*, Heinemann Medical Books, New York, 1966. Ed.

"learn" this reflex by landing on the hind quarters or, to speak in Pavlov's terminology, be conditioned by the repeated succession of such contacts.

Testing of placing of the hindlegs in the dog Vici, whose right half of cerebrum had been removed in addition to the cerebellum, resulted in some observations worth mentioning. The animal suspended head upwards, held the left hindleg strongly extended while the right one was drawn in a position of flexion (Fig. 19, No. 1). When the animal was slowly moved downwards, with or without a blindfold, it extended the right hindleg as soon as it contacted the surface with the hairs of the tail, placed the sole down correctly on the ground and supported the trunk with this foot fixed in extended position. Strange

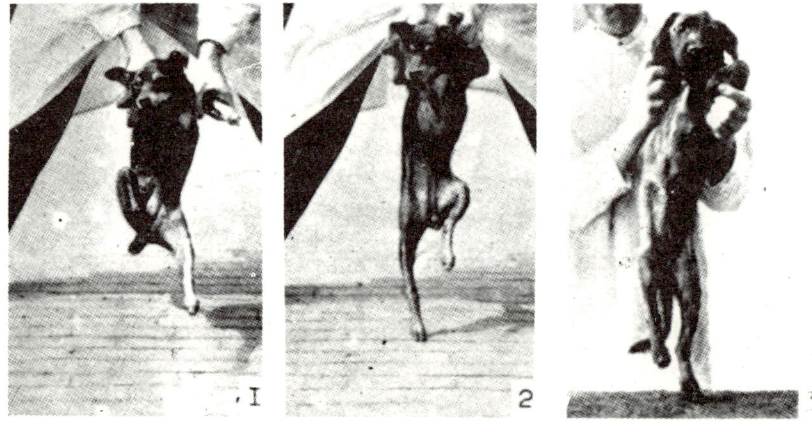

Fig. 19. Preparation for standing of the hindlimbs after a unilateral decortication. 1. Dog Vici after the extirpation of the right half of cerebrum and of the cerebellum, suspended, head upwards. The animal keeps the left hindleg maximally extended, the right one strongly flexed. 2. In moving downwards, on contact of the tip of the tail with the ground, the right hindlimb, with an extending movement, is set down on the ground and fixed in extended posture, simultaneously the left one is strongly flexed (see also chapter XX). 3. Dog Sep after left decortication, set down on a surface from suspension head upwards. The somewhat flexed left hindlimb on contact with the surface, was instantly extended into standing posture while the formerly fully extended right hindpaw was simultaneously somewhat flexed.

to say, the left hindleg at the same time was drawn into a position of flexion and remained in this posture (Fig. 19, No. 2) as long as the animal did not change its posture. When the animal was lifted up from the ground, the left hindleg immediately returned to its extended position and the right hindleg to its flexed position.

The animal showed similar reaction when the skin of the right hindpaw was brought into contact with the ground; but following contact with the left, however, both paws remained without any reaction.

When the tip of the tail of the dog Vici, suspended head upwards, was touched with the hand, the same extension of the right hindleg with the simultaneous flexion of the left hindleg occurred; when the right paw did not touch a support with this stretching movement, both paws returned to their initial posture in spite of repeated touching of the tail. After several repetitions, movement of the paws did not occur at all. The reaction returned when the animal was lowered several times in such a way that first the tip of the tail *and then the sole of the right hindfoot* came in contact with the surface.

Similar manifestations are shown by dogs from which only one half of the cerebrum was removed : in these cases also the hindleg ipsilateral to the extirpation only shows preparation for standing, the simultaneous flexion of the opposite limb, however, is much less pronounced (Fig. 19, No. 3).

C. THE CORRECTION OF ABNORMAL POSTURE OF THE LIMBS ON EXTEROCEPTIVE STIMULATIONS, AND MUNK'S TOUCH REFLEX.

Intact dogs show placing of the fore- and hindlimbs also on contact of the skin of the paw with a surface.

When an intact dog held in the air in a ventral position, blindfolded, is brought into contact with the edge of a table, for example with the back of the foot of one of its forelimbs, it will first set down this paw and then usually the other paws on the surface. When, on the contrary, the skin of the back of the fore- or hindpaw of a decorticate dog is brought into contact with the edge of a table, placing is absent. After extirpation of one, for instance the right, half of the cerebrum, it is also absent on contact of the skin of a left limb with the edge of a table : however, it is present on touching with the skin of a right limb. When a decorticate animal is set down on a surface on one limb and the dorsum of the paw instead of the sole is brought into contact with the surface, the decorticate animal will then sometimes correct this posture, but less quickly than the intact one. After unilateral decortication, animals in similar circumstances will indeed correct such a passive inversion of foot posture on the side of the extirpation but less promptly with the contralateral paw particularly in certain experimental conditions (Fig. 20).

Goltz, Munk, Dusser de Barenne and others have already referred to the imperfect correction of faulty posture in totally or unilaterally decorticate animals, but they interpreted them as impairment of the deep sensibility in analogy with the astereognosis with preserved tactile sense observed in humans with lesions of the cerebrum. Decorticate animals still react to diverse skin stimulation (among others, to stroking, blowing upon, pinching of the skin, sprinkling with water, flea bites). According to our conception, however, this impairment of correction of posture at least partly, is to be regarded as an impairment of the preparation for standing on stimulations of the body surface. (On the behavior of this correcting capacity to proprioceptive stimulation in decorticate animals see Chapter XIII) .

On touching the back of the foot of an intact animal held in the air, a peculiar flexing-extending movement of the paw touched, and often of the opposite paw, occurs. This reflex is known as Munk's touch reflex. After a total extirpation of the cerebrum, it is entirely absent (Munk), after unilateral extirpation of the cerebrum, however, it lacks only in the contralateral side. Munk's reflex is probably the expression of preparation for standing, because when the dorsal side of a paw of an animal held in the air with blindfolded eyes is stroked with two fingers and those fingers then remain in touch for awhile, the paw, by a reflex movement, is eventually put on the fingers. The reflex discontinues if one strokes the back of the foot several times in succession and this time removes the finger quickly from the back of the foot; it appears again, however, if one waits for awhile or when the back of the foot is passed several times on the edge of the table upwards from below till at the end of the movement the sole of the foot reaches the edge of the table. Thus Munk's reflex in its appearance and disappearance, behaves like a placing response.

Fig. 20. Dog Sep (left decorticate) with blindfold, set down on a surface with the back of the right forepaw (1) or the right hindpaw. (2) The abnormal posture is not corrected. (The animal is supported passively and exerts no pressure on the paws).

When the fore- or hindlegs of intact animals are left to dangle over the edge of a table they will be promptly withdrawn and set down with the soles of the feet on the table, but the forelimbs do not move forwards, as on touching the muzzle with a surface, but move backwards. Similarly the hindlegs are not extended backwards as on touching the tail but are flexed and move forwards towards the table. Decorticate animals lying on a table in a ventral position, contrary to intact and decerebellate dogs, do not draw up the paws hanging down over the edge and do not place them with the sole on the table (Fig. 21). The absence of this "correcting movement", too, was wrongly interpreted by Munk and other authors as an impairment of the deep sensibility.

The absence of preparation for standing in decorticate animals in a ventral position upon stimulation of the surface of the trunk must also be emphasized for the reason that when animals in lateral position are stimulated on the surface of the trunk, they will instantly shift the trunk into a ventral posture even if the head is passively fixed in a lateral position (Magnus body righting reflex on the body). In a previous publication (234), I discussed the relation of these righting reflexes to the function of the red nuclei of the midbrain. The contact stimuli as well as other stimulations emanating from the body surface are capable of producing a number of reflexes in decorticate animals which occur by way of various parts of the central nervous

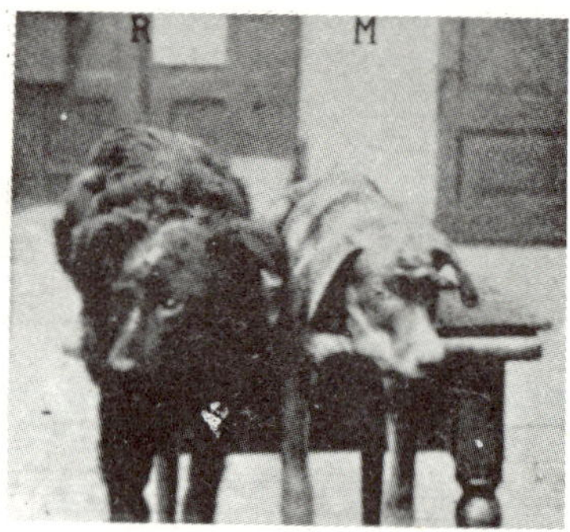

Fig. 21. Decorticate dogs Robbie (R) and Miesel (M). 4-6 weeks after extirpation of cerebrums in ventral position on a table. The hindlegs (1) and forelegs (2) are hanging down over the edge. Although the animals are capable of apparently normal running (Fig. 257) they do not correct this posture.

system. (1) From the presence of these reflexes, however, it should not be concluded that absent reactions, such as Munk's touch reaction, should have no relation to contact stimuli.

The impairment of preparation for standing becomes particularly obvious when the animals are placed on a lattice. The decorticate animal instantly treads through the lattice with all four paws and lets them hang between the bars while animals with a unilateral extirpation of cerebrum tread through with only the contralateral limbs (Fig. 22A). Without blindfold, unilaterally decorticate animals are sometimes very uneasy on a lattice, make attempts to liberate themselves and make running movements in the course of which the contralateral paws sometimes step on the bars by a mere chance. These movements usually are inhibited by a blindfold.

----

(1) By exteroceptive stimulation are produced :

1. By way of the spinal cord : the spinal flexor and extensor reflexes on touching, stroking and pinching of the skin; Sherrington's extensor thrust, etc.

2. By way of the medulla oblongata and pons : the pinna reflex of the external ear on touching the hairs of the ear, the reflex of the hair of the ear and whiskers, the corneal reflex, reflexes on stimulation of the surface of mouth and nasal cavity as the sneezing reflex, etc.

3. By way of the midbrain : Magnus' body righting reflex on the body and head.

Totally or unilaterally decerebellate dogs, cats and also monkeys are capable of running on a lattice just as well as intact animals. When a limb is pulled passively through the bars, it is drawn up again instantly and set down on the bars in a natural position (Fig. 22B and C).

We have seen that after a unilateral extirpation of the cerebrum, only the preparation for standing of the contralateral limbs was absent. This, however, is not the only impairment shown after unilateral decortication. When the body surface, contralateral to the extirpation, comes into contact with a surface, no preparation for standing whatever is caused in either the ipsilateral or the contralateral limb. For instance, when after extirpation of the right half of the cerebrum, the back of the foot of the right forelimb is brought into contact with a surface, the animal immediately sets down the right forelimb on the surface while the left remains hanging in the air. The same occurs when the ventral side of the muzzle is brought into contact with the surface. Contact with the back of the foot of the left forelimb with the surface, on the contrary, produces no placing and both forelimbs remain hanging in the air. Thus the animal shows the following after extirpation of the right half of the cerebrum :

1. On contact with the body surface situated on the median line (muzzle, chest, belly, tail) : Readiness to stand of the right limbs +, of the left limbs —.

2. On contact with the body surface of the right side : Readiness to stand of the right limbs +, of the left limbs —.

3. On contact with the body surface of the left side : Readiness to stand of the right limbs —, of the left limbs —.

Great importance for the physiology of the cerebellum is attributed by various authors to reactions due to the impairment of preparation for standing and of the correcting movements which appear at first after a total or a unilateral extirpation of the cerebellum. Therefore, we must enter into particulars of these impairments.

If one considers how easily the mechanism of conditioned reflexes is retarded by various influences (see Pavlov 217-220), it is not surprising that after such an important operation as the extirpation of

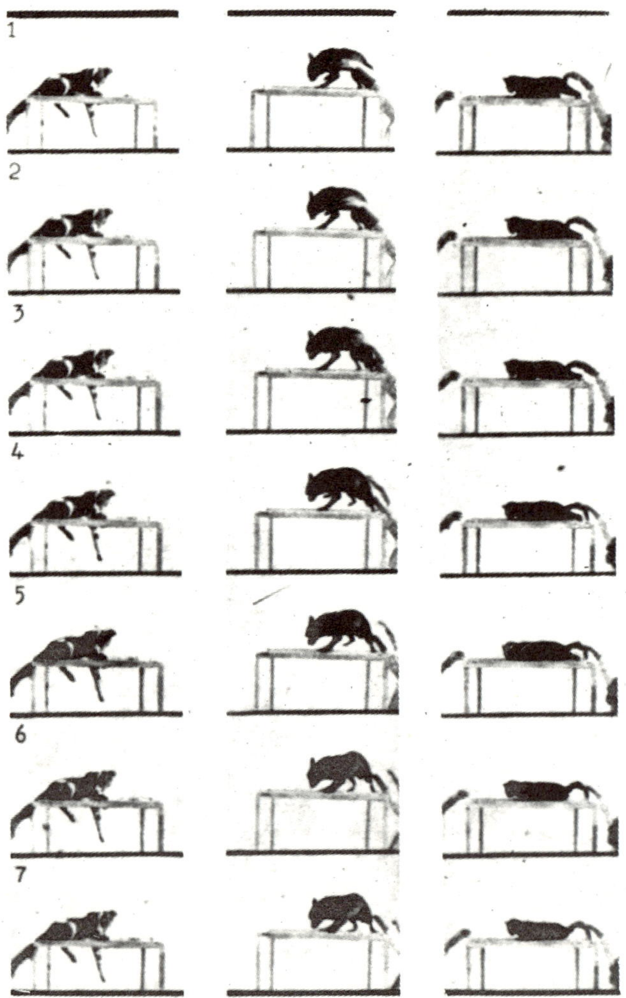

Fig. 22. A. Dog Nero after extirpation of the right half of cerebrum, on a lattice. The left paws have stepped through, the right ones are positioned correctly on the bars. The posture of the left limbs is not corrected. B. Decerebellate dog Erick running on the lattice (B4). The animal sets down all four paws correctly on the bars. Sometimes they temporarily step through, but the error is corrected instantly. C. Decerebellate cat Pierrette. The paws are set down correctly on the bars and do not step through, even when the animal is pulled to and fro by the tail.

cerebellum, preparation for standing and reactions to correct abnormal postures of the limbs are absent for a certain time (Fig. 23, see also Fig. 14). Later, after varying intervals of time (4 days to several weeks), the preparation to stand is constantly recovered.

After the unilateral extirpation of the cerebellum, too, the reactions are impaired immediately after the operation in the limbs of both sides. After some time, however, they are again evident, first in the limbs of the contralateral, then on those of the ipsilateral side; though with the difference that on contact of the muzzle with the surface both paws do not simultaneously set down as in intact dogs, but almost always the contralateral paw is the first placed on the surface. Later both paws place simultaneously, and sometimes either one or the other paw is put down first.

Placing of the ipsilateral as well as the contralateral forelimb also appears when the opposite paw is held fast. Thus, the reaction of one limb does not depend on the movement of the other.

Fig. 23. A. Dog Tommy, 4 days after extirpation of cerebellum. Absence of placing on contact of tail and the sitting legs with the surface. B. Dog Suzanne, 4 days after the extirpation of cerebellum. The animal does not correct the dorsal position of the forepaws.

The preparation for standing of the ipsi- and contralateral paws, as well as Munk's touch reflexes no longer show distinct differences in this stage, Lewandowsky (167), who also observed the impairments of correcting movements and of Munk's reflexes after a total or unilateral extirpation of the cerebrum and cerebellum, drew the conclu-

48

sion from the results that the cerebellum as well as the cerebrum are the central organs of muscle sense [1]. By muscle sense he means "in general the ability to perceive position and movement of the parts of the body, be it by the sensibility of the muscles themselves or by way of the skin or the joints". He does not mean by that the capacity of conscious perception because, as he quite rightly says, that we cannot decide with certainty whether, in an animal, the conscious sense perception is absent or not. Based on the disappearance of Munk's touch reflexes, he assumes that stimulations of the skin, too, take part in the occurrence of muscle sense. In his opinion, the unilateral extirpation of cerebellum causes impairment of muscle sense and the sense of position only in the ipsilateral part of the body.

Dusser de Barenne (61) and before him Munk (213, 214) have objected to the conclusion that the sensibility of the skin is not impaired after the extirpation of cerebellum and that a temporary absence of Munk's touch-reflexes is not to be interpreted as an impairment of the sense of touch. In my opinion, this objection is correct : the temporary absence of a reaction caused by contact or tactile stimulation is not proof for an impairment in the sense of touch. Strange to say, however, Dusser de Barenne was led to believe by the temporary impairment of correcting movements that the cerebellum was indeed a central organ of the perception of deep sensation. Dusser de Barenne takes these impairments of correction for "an expression of the partial interruption of proprioceptive stimulations which, coming from the periphery, are conducted to the central nervous system from bones, joints, muscles and ligaments; namely these proprioceptive stimuli which arrive by way of the cerebellum".

In agreement with Sherrington [2], he considers the cerebellum as a central organ, for an important reflex apparatus relating to proprioceptive stimuli. According to Dusser de Barenne, a unilateral ex-

---

(1) Lussana (183) had earlier (in 1850) expressed his opinion that the cerebellum was the central organ of muscle sense. By muscle sense, however, Lussana does not understand it as Lewandowsky, but as the ability to estimate the resistance to be surmounted in voluntary movement and to regulate the strength of muscle contraction necessary to overcome this resistance. ("Le sens musculaire sert aux animaux pour connaître la résistance de la matiére et pour régler, dans leur mouvements volontaires, la force de la contraction des muscles pour la vaincre").

(2) According to Sherrington the cerebellum is "the headganglion of the proprioceptive system" (276).

tirpation of cerebellum causes an impairment of proprioceptive sensibility of the homolateral half of the body.

The following speaks against Dusser de Barenne's conceptions :

1. the impairment of the ability to correct postures does not depend only on a loss of reactions to deeper proprioceptive stimuli but also partly on the absence of reactions to stimulations of the skin;

2. the loss of corrective movements is only of a temporary nature; it sometimes lasts only for a few days and

3. in the first days, after a unilateral or total extirpation of cerebellum, the ability to correct the defect by visual perception, too, is impaired although the animals can see very well (Fig. 14, No. 1).

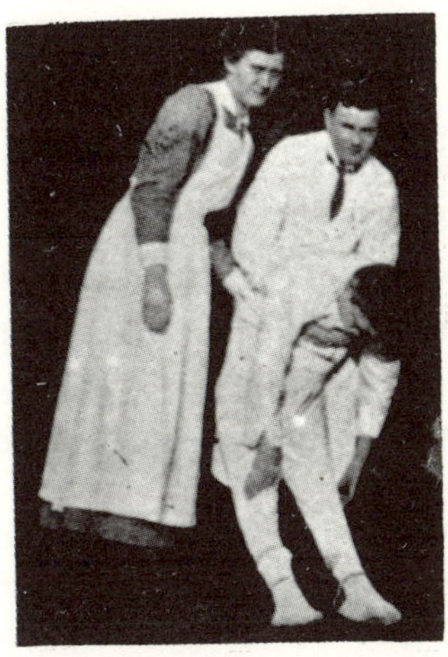

Fig. 24. Tumor of the midbrain. Following contact of the heel with the floor, the feet are not set down with the soles on the ground. After E. J. Abrahams, Hypokinesis by een Tumor in den Aquaeductus Sylvii. Utrecht, 1925. Observations from the Psych. and Neurol. Clinic of the State University at Utrecht.

It is thus highly probable that the impairment of the ability to correct abnormal postures and of the ability to stand after the extirpation of cerebellum are not due to a loss of sensory input but probably to a general inhibition or effect of shock, since in almost all serious, and particularly in feverish, diseases, such impairment of these reactions can also be observed.

Moreover, it is to be borne in mind that in the testing of these reactions, the eyes are blindfolded and that even in intact animals the blindfold sometimes can have an inhibiting effect on the occurrence of reflexes and may even cause their not taking place at all. In decerebellate animals which give an impression of uneasiness because they constantly knock against something, this inhibiting influence is surely not less.

Finally, it seems that the abnormal distribution of tonus also plays a certain role; this view is supported by the fact that corrective movements and the preparation for standing are absent longer in dogs which were rigid after the extirpation.

Dusser de Barenne's assertion that after a unilateral extirpation of cerebellum, the correcting ability shows an impairment only on the side of the extirpation is also not quite correct. Following a unilateral extirpation, the correcting movements and placing are at first absent on both sides, as was previously observed by Ducceschi and Sergi in Luciani's laboratory; on the contralateral side, however, they are more quickly regained.

Thus, the temporary absence of the ability to correct posture and of Munk's touch reflex is in no way proof that by extirpation of the cerebellum the reflex arcs of proprioceptive stimulations are permanently interrupted. Naturally, this does not mean that the reactions released by proprioceptive stimulation normally are not influenced by the cerebellum, just as it cannot be said that they appear in a normal way in decerebellate animals. It is, however, certain that after a total extirpation of the cerebellum, preparation for standing, ability to correct postures and Munk's touch reflex are again promptly manifested on both sides after a certain time.

To what extent stimulation from the body surface participates in the correct setting down of the feet *in man* has not yet been systematically investigated. It is only known that in conformity with observations in new-born dogs, in children in the first half year of their life, placing is absent*. If an infant is held by the armpits and lowered

---

* See footnote to page 48. Rademaker did not recognize the presence of certain contrary, interfering reactions which we now call tactile and visual avoiding. Ed.

till the feet touch the ground, the contact does not cause a setting down of the feet with the soles on the floor but the legs are drawn up and lifted from the ground.

Under pathological conditions, preparation for standing can also be absent in adults, as is well known to every neurologist (see preceding page, Fig. 24).

D. THE OPTICAL PREPARATION FOR STANDING (VISUAL PLACING)

The setting down of limbs in the correct posture with the soles of the feet on the ground can also occur as a result of optical stimuli. When an intact dog, suspended head downwards, is moved in such a way that the head is brought near a surface, for instance the edge of a table, the forelegs are set down at once on the table as soon as the muzzle approaches the edge, *thus before proper contact has taken place.*

The optical preparation for standing, too, is a "conditioned" reflex because when one brings the head to the edge of a table several times in succession and withdraws it as soon as the animal tries to set down its paws on the table, the reactions of the forepaws cease to occur.

In totally or unilaterally decerebellate animals, the optical preparation for standing (visual placing) is present, though not immediately after extirpation (see among others Fig. 14, No. 1 : the decerebellate dog Wolf 12 days after extirpation of cerebellum, lowered to the ground with open eyes, did not set down the paws in a position to stand). After a unilateral extirpation of the cerebellum, the optical preparation for standing usually returns first in the contralateral, then in the ipsilateral. Immediately after a total or unilateral extirpation of the cerebellum, the optical reaction shows impairments quite similar to those of the readiness to stand on contact stimulations, although the animals can see[1].

---

(1) One cannot reliably decide whether an animal feels, sees, hears, thinks or is conscious. It can only be established whether it shows reactions to tactile and pain stimulations, to optical and acoustical impulses. "Seeing" in the physiological sense

After an extirpation, for instance of the *right* half of the cerebrum, when the *left* half of the head is brought close to a table, no preparation for standing occurs. This failure is caused by the leftsided hemianopsia. Bringing the right side of the head to the table, on the other hand, excites preparation for standing, but in the sense that only the right forepaw is placed on the table while the left remains hanging in the air. The same is true when the front of the muzzle is brought to the edge of the table : in this case also placing occurs only in the right, and not in the left, forepaw (Fig. 25).

Since no blindfold was used in this investigation, all defensive movements and the alternating coordination of one paw with the movements of the opposite one were not always inhibited (see p. 42). Consequently, the left forepaw is often put on the table quite by chance. For the satisfactory testing of the preparation for standing, it is particularly advisable to restrain alternately the left and the right forelimb. On bringing the muzzle close to a table, the right-sided decorticate animal will now always show placing of only the right foreleg and the intact animal of both.

Thus after a unilateral extirpation of the cerebrum, the optical preparation for standing behaves exactly the same as the reactions to contact stimulation.

Preparation for standing, after extirpation of the *right* half of the cerebrum :

	Right paws	Left paws
Response to stimulation of surface		
of the right side of the body	+	—
of the left side of the body	—	—
Response to optical stimulation		
of the right field of vision	+	—
of the left field of vision	—	—

is responding to optical impulses with conditioned reflexes. Responding with unconditioned, subcortical reflexes, as for instance contraction of the pupils or conjunctival reflexes on incident light, such as those shown also by decorticate animals, are not seeing in the physiological sense. Decerebellate animals followed with their eyes animals running near the cage and reacted sometimes with a wagging of the tail if someone approached the cage, thus they could see.

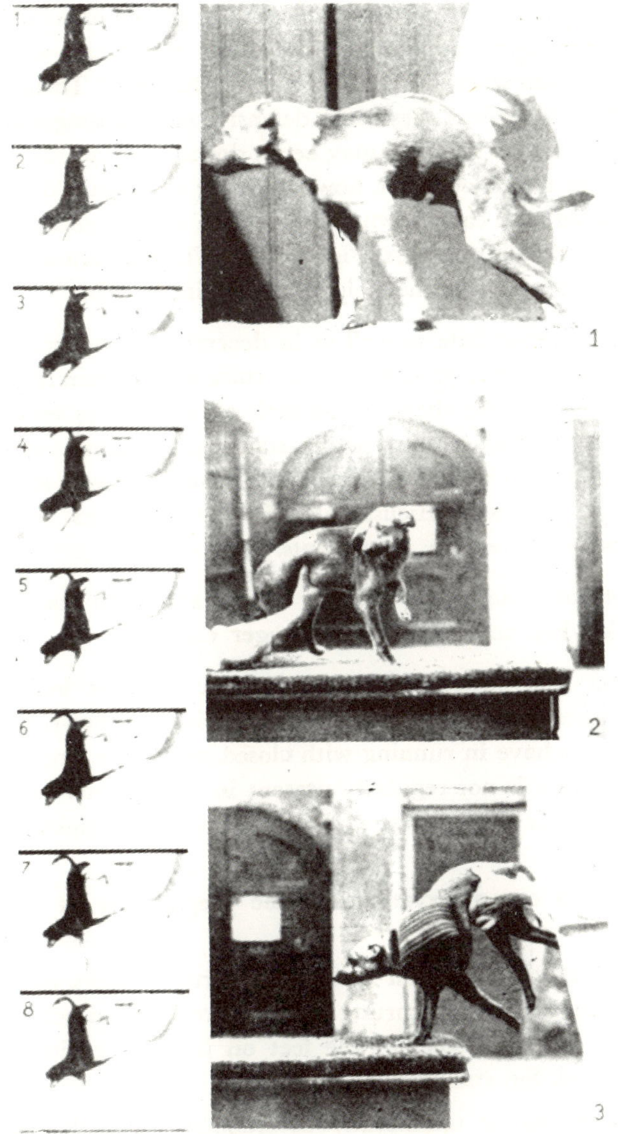

Fig. 25. A, 1-8 : The decerebellate, right decorticate dog Vici without blindfold, slowly lowered from a hanging position suspended head downwards. When nearing the surface, the right foreleg moves forwards and is set down while the left remains extended backwards in the air. B, 1, 2 and 3: decerebellate, right decorticate dog Daumling with open eyes, held passively by the hindquarters in standing posture. The animal does not set down the left forelimb in a position correct for standing.

When unilaterally decorticate animals are placed on a lattice, roof of a Hans Meyer-cage, the paws opposite to the extirpation will step through the bars even when the eyes are not blindfolded. Then the animals are usually very restless, they make defensive and running movements in the course of which the paws may be set down quite accidentally on the bars; with the next movement, however, they will again tread through. Finally, the animals calm down and leave the contralateral limbs hanging through the lattice bars (Fig. 22A). They are not able to correct this position even with the help of visual impressions.

In totally decorticate as well as in decerebrate and spinal animals, of course, the optical preparation for standing is entirely absent. It is noteworthy that the thalamus animals almost always set down their limbs in the correct posture while standing, running, stumbling or falling, although placing is absent on visual as well as on tactile stimulation. We will discuss later how this occurs by means of proprioceptive correction movements.

In new-born dogs, also placing is still absent in the second and third week when the eyes are no longer closed.

The fact that optical stimuli are of the greatest importance for the correct posture of the feet in human beings is shown by the difficulty tabetics have in running with closed eyes. The optical preparation for standing is not regularly present in human infants in the first half year of their life and in adults in certain pathological conditions (Fig. 23).

To sum up our observations on the preparation for standing we conclude :

1. that *intact* animals on contact of certain places of their body surface with a supporting surface are able to set down their limbs with the soles of the feet on the surface in the posture required for standing;

2. that totally or unilaterally decerebellate animals react in the same way, though they show a temporary impairment immediately after the extirpation of the cerebellum;

3. that preparation for standing on stimulation from the body surface is constantly absent in *thalamus* (decorticate) *animals* as well as in decerebrate and spinal animals. It follows that the centers of elaboration of the necessary stimuli are located in the cerebrum, the more so as the reactions show thecha racter-

istics of Pavlov's conditioned reflexes (according to Pavlov and Krasnogorsky the "cutaneous analyser" i.e., the centers for conditioned reflexes on stimulation of the skin are formed by the gyrus coronarius and the gyrus ectosylvius);

4. that after *unilateral removal of the cerebral cortex* the preparation for standing is constantly impaired : thus it is totally absent on stimulation from the body surface of the side of the intact half of the cerebrum, on the other hand, on stimulation from the muzzle, tail and the half of the body on the side of extirpation, it is present only in the ipsilateral paws, while the contralateral paws remain suspended in the air;

5. that a unilaterally decorticate and totally decerebellate dog shows the same impairment in the reactions that are shown by an animal after a unilateral decortication (however the setting down of the hindpaw ipsilateral to the decortication was accompanied by a strong flexion of the other hindpaw);

6. that Munk's touch reflex represents an expression of preparation for standing;

7. that the impairment in correcting abnormal postures of the limbs, observed in totally or unilaterally decorticate animals, is also an expression of impairment of preparation for standing (for corrective movements on proprioceptive stimulation in totally or unilaterally decorticate animals see Chapter XIII);

8. that intact, decerebellate and unilaterally decerebellate animals are also capable of optical placing of the paws in a posture necessary for standing;

9. that after a total or unilateral extirpation of the cerebellum optical preparation for standing is *temporarily* impaired although the animals can see;

10. that after unilateral extirpation of the cerebellum, optical placing is temporarily absent at first on both sides, then unilaterally only on the side of extirpation, and later is restored on both sides;

11. that after a unilateral decortication, the readiness to stand on optical stimulation from the contralateral field of vision is to-

tally absent and from the ipsilateral half of the field of vision only affects the contralateral limb;

12. that the optical preparation for standing is absent (impaired, Ed.) in decorticate and in new-born dogs as well as in infants during the first year of life and in adult human beings under certain pathological conditions;

13. that the optical preparation for standing (visual placing), too, represents a conditioned reflex of the cerebrum.

V. THE SUPPORTING REACTIONS

For normal standing in addition to the correct posture, a sufficient "supporting tonus" is necessary which is produced by the so-called "supporting reactions"[1]. The supporting tonus, i.e., the distribution of tonus required for normal standing, is produced, by the counter-pressure of the supporting surface on the soles of the correctly placed hands[2] and feet. In the following, we shall speak of a "static stretch"

Fig. 26. Normal dog in standing posture

(1) Sherrington named the reflexes which take part in the decerebrate rigidity and make possible the "standing" of decerebrate animals "posture reflexes", attitude reflexes and "standing reflexes". For the distinction from these reflexes and from the "static tonus" of decerebrate animals, the distribution of tonus in *normal* standing is termed "supporting tonus" and the reactions producing it are called "supporting reactions" (see footnote p. 7).

(2) In analogy with the nomenclature in human beings we will distinguish the following, upper arm, forearm, hand and digits of the forelimb.

when a pressure is exerted on the sole of the foot after the distal phalanges of the extremities have taken up the position necessary for normal standing (in the forepaw the wrist joint extended, the digits directed forwards and nearly vertical to the metacarpus; in the hindpaw the toes directed forwards and vertical to the metatarsus, (see Fig. 26 and Fig. 42). It is immaterial here whether the pressure is exerted by a supporting surface, by pressure of a hand or a board, in a ventral, lateral or dorsal position of the animal.

An intact or decerebellate dog in a dorsal position keeps its paws flexed (Fig. 28, No. 1). If one grasps a limb and performs some passive flexing and extending movements, one may usually feel some resistance to passive extension but none or almost none to passive flexion; a distinct extensor tonus is absent. Thus, the position of the joints as well as the tension of the muscles are totally unable to support a load.

In a ventral position in the air, the animals keep their limbs drawn in towards the body or let them hang down in a half extended position (Fig. 27, No. 1). By passively moving the limbs a slight resistance to passive flexion may sometimes be felt which, however, is in no way sufficient to carry the trunk. *When, on the contrary, the animals are set down on a base, the legs instantly change into solid columns and show a strong supporting tonus* (Fig. 27, No. 3 and 4).

Investigations of this phenomenon showed that the following five factors participate in the production and maintenance of the supporting tonus :

1. touching of the sole of the foot (magnet - or exteroceptive supporting reaction);

2. the posture of the distal phalanges (metatarsus, metacarpus, and digits);

3. pressure on the sole of the foot;

4. the relationship of the proximal joints of the extremity to each posture (in the forelimb the position of shoulder, elbow and wrist joints, in the hindlimb the hip- , knee- , and tarsal joints);

5. the passive flexion of proximal joints of extremities by static stretch.

The first three factors together form the *positive supporting reaction* (Magnus 197).

Fig. 27. Decerebellate dog Wolf (body weight 9½ kg).   1. The animal held in the air by head and tail in ventral position.   The limbs are somewhat flexed and easy to move passively.   2. The back is loaded with a sandbag of 5.3 kg which makes it appear strongly depressed.   3 and 4. The animal is set down on a surface.   The limbs are extended and fixed in an extended posture. The muscles of the extremities are distinctly bulging and feel hard and strained. The extremities are changed into solid columns which carry not only the trunk (3) but also the sandbag (4) (note the change in the curvature of the back).

IA. THE MAGNET REACTION (EXTEROCEPTIVE SUPPORTING REACTION) OF THE HINDLIMB

The exteroceptive stimulations induced by touching of the sole of the foot may cause an extension of the hindlegs with fixation in extended position. This influence of exteroceptive stimulation can best be observed in decerebellate dogs in dorsal (supine) posture. Such an animal in a dorsal position with the muzzle pointing vertically upwards keeps its four limbs flexed (Fig. 28, No. 1).   *With the slightest touch on the ball of the toe, an extension of the limb occurs by which the touching finger is pushed away. If one slowly withdraws the finger the hindlimb will follow, as if attracted by a magnetic force, until it is maximally extended.*

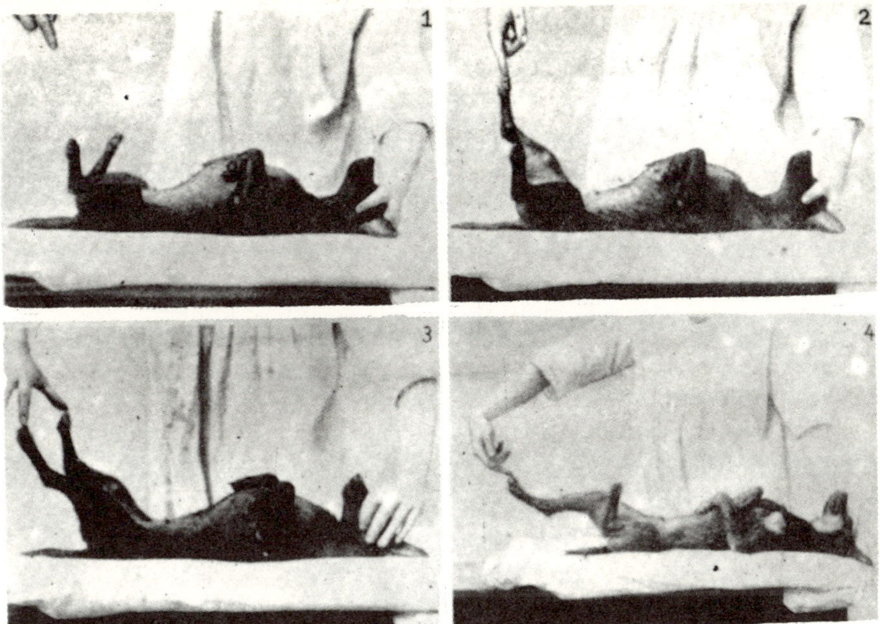

Fig. 28. Magnet reaction of the hindlimb. 1. Decerebellate dog Piccolino in dorsal position with the muzzle pointing vertically upwards. The hindlimbs are totally flexed. 2. On the slightest touch of the sole maximum extension of the right hindlimb. 3. On touching the soles of both hindlimbs extension of both. 4. Decerebellate dog Moor : magnet reaction of the right hindleg.

Touching of the soles of both hindpaws leads to an extended posture of both hindlimbs (Fig. 28, No. 3).

The extension persists as long as the contact continues, even if the contact would last for 45 minutes and longer. When the contact is discontinued, the paw instantly returns to the flexed position. The magnet reaction appears on touching the sole of the foot as well as of the tip of the toe, sometimes on touching the claw or the hair growing between the toes, but never on touching the back of the foot (Fig. 29).

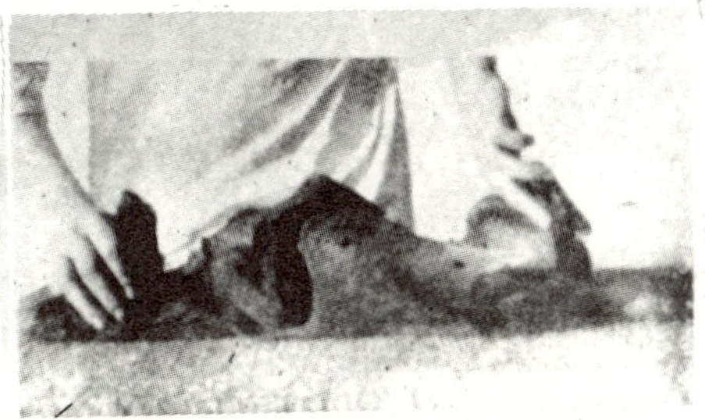

Fig. 29. Decerebellate dog Piccolino in dorsal position with muzzle pointing vertically upwards (in the same posture as Fig. 28, No. 1). On touching the back of the foot no extension of the hindlimb.

The most sensitive areas are the balls of the soles of the distal phalanges; touching of these balls with a piece of cotton is sometimes enough to elicit the magnet reaction. The magnet reaction can be produced when the eyes of the animal are open as well as when they are closed. It not only occurs in a dorsal position but also when suspended head upwards, or in ventral and other positions in the air (Fig. 30).

In the limb extended by a magnet reaction, the muscles feel hard and strained, the bellies of the extensor, abductor, adductor as well as the flexor muscles curve distinctly outwards. By passively moving the thigh at the hip a resistance is felt by forward and backward movement, in abducting and similarly, though to a lesser degree, in adducting. With the slightest movement of the thigh, the pelvis moves too, showing that the hip joint is fixed. The flexors and extensors of knee- and tarsal joints are subject to a strong contraction. If one now clasps the metatarsus and tries to flex the knee- and tarsal joint, one encounters a strong resistance. Thus, all joints are now fixed and the limb turned into a rigid column.

The influence of the magnet reaction extends, in addition, not only to the muscles of the touched limb but also to the muscles of the lumbar and the thoracic parts of the spine and to the neck. Intact and decerebellate dogs held horizontally in the air with the head and tail in natural positions have a hollow, flaccid, sagging back. When the soles of the hindlimbs are touched, not only the touched paws are

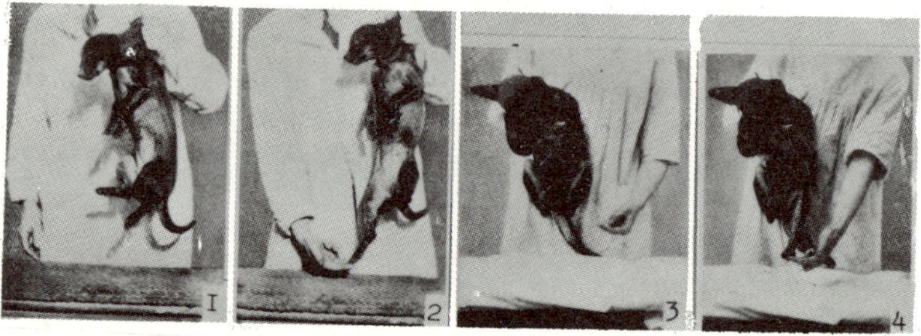

Fig. 30. Magnet reaction of the hindlimbs in suspended posture in the air. 1. Decerebellate dog Piccolino suspended, head upwards. The animal keeps both hindlimbs flexed. 2. On touching the sole, instant maximum extension of the hindlimb. 3. Decerebellate dog Moor with blindfold in ventral position in the air, one hand supports the thorax. The hindlimbs are flexed. 4. On touching the sole, maximum extension of the hindlimb, and the touching finger is pushed away.

extended but also the lumbar and thoracic muscles contract and straighten the back. Straightening even occurs when the back is burdened with a sack full of sand (Fig. 31, see also Fig. 27 and Fig. 33). In decerebellate dogs, following contact with the soles, the unburdened hollow back is sometimes not only extended but even somewhat curved convexly upwards (Fig. 31, No. 3).

Fig. 31.   Influence of the magnet reaction on the tonus of vertebral and cervical muscles. 1 and 2. Decerebellate dog Erik with blindfold, with and without sandbag on its back held in the air by tail and head in ventral posture. The animal is suspended with sagging back. (cervical and back muscles relaxed.) 3 and 4. Touching the soles of the hindlimbs with two fingers. This causes not only extension of the hindlimbs but also of neck and back muscles. The back lifts the sack upwards. (4). The muscles on both sides of the spine feel hard and contracted. The cervical muscles, too, are contracted causing the head to press actively on the ventrally held hand. 5 and 6. The same phenomenon in the decerebellate dog Piccolino.

If one supports the abdomen with one hand and then touches the soles of the feet, in decerebellate dogs there also occurs simultaneously with the extension of the hindlimbs a powerful contraction of the vertebral muscles so that the anterior part of the body is raised (Fig. 32), and sometimes even falls backwards. The trunk, which at first was resting limply on the hand, is suddenly fixed by the contraction. On removing the fingers, it again falls limply forwards. Thus ,contact with the hindlimbs, according to conditions, can counteract a dorsal concavity (Fig. 31) as well as a dorsal convexity of the back (Fig. 32).

Fig. 32.   Influence of magnet reaction on the extensors of the back.   1. Decerebellate dog Piccolino with blindfold, the ventral surface supported with one hand on which the anterior part of the body rests. Spine relaxed and in its upper part convexly curved. 2. On touching the soles, extension of the hindlimbs and of the back.   The anterior part of the body is raised and the back straightened and fixed, while head and neck are moved ventrally by labyrinthine righting reflex.   On taking away the hand, the anterior part of the body falls forwards (as in 1) and the spinal muscles relax.

The reaction of the vertebral muscles can also be observed after touching only *one* hindpaw. The back is then also extended straight and does not show any real lateral curvature (Fig. 33). Unilateral contact thus causes a symmetrical tension of the extensors of the back on both sides. It seems, however, that contact with one side under certain circumstances, can also cause an asymmetrical tension of the vertebral muscles (see p. 128).

Fig. 33. Decerebellate dog Piccolino with loaded back, held in ventral position in the air by head and tail.   2. On touching only one sole, the back is stretched completely straight.

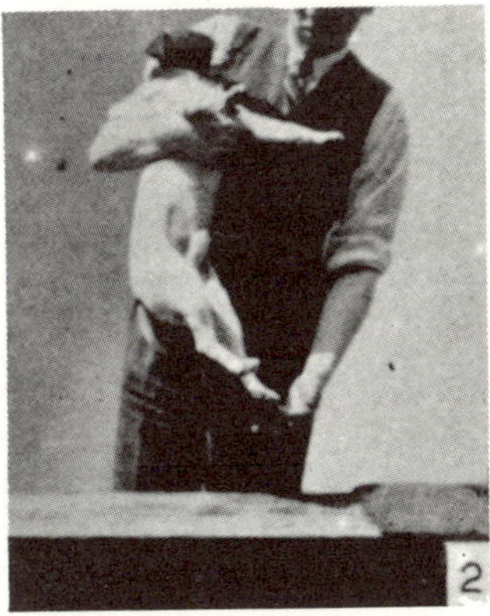

Fig. 34. 1. Dog Fox (decerebellate on the right side) suspended, head upwards with extended hindlimbs.   2. On touching the sole of the left hindpaw the right hindlimb is somewhat flexed.

In certain cases, contact with the soles of the hindlimbs also causes a contraction of the *cervical muscles*; depending on the posture of the neck before contact and sometimes contraction of the cervical flexors or sometimes the extensors of the neck is produced. If the animal, supported under the lower jaw, is held suspended in the air with a raised head, touching of the soles of the feet will cause a tension of the cervical flexors (Fig. 31 and 32). If the neck is held flexed the contrary occurs, a contraction of the extensors of the neck.

In the postural change of the back and neck on touching the foot, sometimes labyrinthine righting reflexes also take part (see Fig. 32), but only as a secondary result. Often a direct influence on the tonus of the cervical muscles can also be observed, as seen in labyrinthectomized dogs.

If, on touching the sole of one hindfoot, one tests the resistance against passive flexion of the other statically extended hindlimb, a distinct reduction of extension or support tonus in this latter paw can sometimes be felt (Fig. 34). Contact with *one* paw thus results in an increase of supporting tonus of the limb touched, and a decrease in extension- and supporting tonus of the crossed hindlimb. The decrease in tonus of the contralateral limb also occurs when the touched paw was already fixed in an extended position before contact; in this case it is not caused by the stretch reaction of the touched paw, but directly by contact stimulation. This decrease of tonus is particularly distinct after unilateral extirpation of cerebellum or cerebrum. After unilateral extirpation of cerebellum it occurs in the hindpaw of the side of extirpation on touching of the contralateral paw (Fig. 34). After unilateral decortication of cerebrum (see Fig. 37) it is found in the contralateral paw on contact with the hindpaw of the side of decortication.

As we have seen, after the unilateral decortication, the hindlimb contralateral to the extirpation sometimes also shows a decrease in extensor tonus after contact with the tail.

Finally, the touching of the soles of the hindfeet in certain cases also influences the muscle tonus of the forelimbs. If a dog is lifted from the ground by its forelegs one will notice that it flexes the elbow- and shoulder joints as soon as the hindlimbs have left the floor, and draws itself up with its anterior extremities (Fig. 35, No. 1 and 2). On touching the soles of the hindpaws, however, the contraction of the elbow- and shoulder flexors instantly relaxes and the animal drops down (Fig. 35, No. 3), to lift itself up again on cessation of the contact.

Under these circumstances, contact with the soles of the hindpaws exerts an inhibiting influence on the tension of the *flexors* of the shoulders and elbows.

When the animal is moved upwards by its forelegs in such a way that the shoulder- and elbow joints are passively flexed more and more (Fig. 36) the animal also raises itself as soon as the hindpaws leave the ground and sinks back again when the soles are touched with two fingers. Strange to say, raising occurs here by *an extension* of the elbow- and shoulder joints, while contact with the sole causes a slackening of the *extensor* tension and a flexion of these joints (Fig. 36).

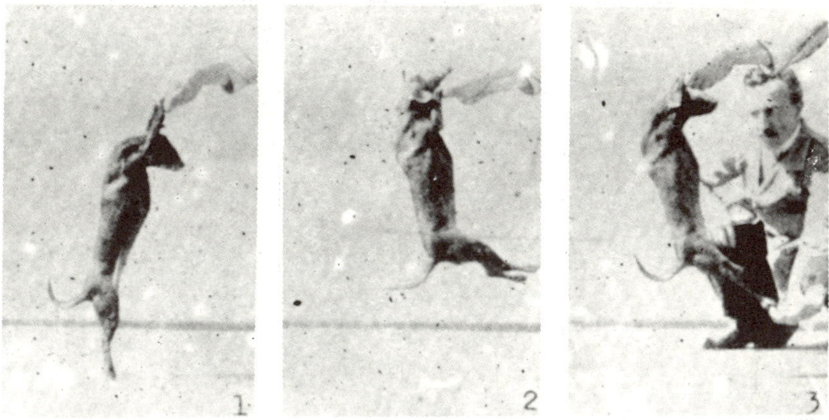

Fig. 35. Influence of contact of the soles of hindfeet on the tonus of the fore-limb muscles in the decerebellate dog Piccolino. 1. The upper part of the body of the animal, lifted from the ground by the forelimbs, stands only on the hindlegs. Forelegs are stretched, flexors show no contraction. 2. On further lifting, flexion of the forelegs in shoulder- and elbow joints. Flexor muscles of the upper arm bulging and strained.   3. On touching the soles of the hindpaws, relaxing of tension in the flexors, the animal drops downwards and hangs on totally extended forelegs.

Fig. 36. Influence of contact with the soles of the feet on tonus of the forelimb muscle system in the decerebellate dog Piccolino.   1. The upper part of the animal's body is lifted from the ground by the forearms, the upper arm is held backwards so that shoulder and elbow joints are brought into flexed position. 2. When the animal is lifted up, elbow and shoulder joints are extended as soon as the hindpaws leave the ground.   3. On touching the soles of the hindfeet with two fingers the tension of the extensor muscles is relaxed and the animal sinks down again.

In summing up, we can say that the exteroceptive stimulations produced by touching the soles of the hind paws will cause the following :

1. a tension of all muscles of the touched hindlimb which are extended and then fixed in an extended posture;

2. a contraction of the lumbar-, thoracic- and also in certain cases the cervical vertebral muscles;

3. a slackening of the extensor and supporting tonus of the contralateral hindlimb;

4. alterations of tension in the muscles of the forelimbs.

A simple touching of the sole, in certain cases, can cause alterations in the tension of almost all muscles of the body.

The occurrence of a magnet reaction of the hindlimbs is dependent on various factors which sometimes act as inhibiting and sometimes promote extension, thus among others, the position of the head in space (see Chapter IX), the position of the head at the atlanto-occipital joint to the neck (see Chapter VII), the position of the neck in relation to the trunk (see Chapter VII), the curvature of the vertebral column (see Chapter X), the posture of the pelvis relative to the vertebral column (see Chapter X), the posture of the forelimbs (see Chapter XII), the posture of the contralateral hindlimb (see Chapter VI), the posture of the hip joint on the side of the touched paw (see Chapter XIII). In addition, stimulations from the skin of various parts of the body surface can inhibit the occurrence of magnet reactions, further affective and psychic influences (the "internal conditioned inhibitions" of Pavlov), sensory impressions, etc. (for details see Chapter XIV).

After operations on the central nervous system, magnet reactions show widely varying *behavior*. Immediately after *extirpation of cerebellum* they are absent, but sometimes appear again the next day or a few (3-4) days later. Some time after the operation they are always evident and particularly vivid. They were initially observed by us in decerebellate dogs. In a ventral position of the animal in the air, in a hanging position head upwards *and also in a dorsal position with the muzzle directed upwards* the slightest touch (in Piccolino even contact with a piece of cotton) always causes an instant forceful maximal extension which continues as long as contact persists. The influence of the position of the head, pelvis, etc. on the magnet reaction is regularly and distinctly evident in decerebellate dogs, and the alteration of tension of the vertebral muscles is particularly pronounced in decerebellate animals. *The magnet reaction in decerebellate dogs is characterized by their extremely high sensitivity, by the constant, automatic occurrence, by the strong extension and fixation of the hindlegs with active participation of the vertebral muscles, and by the almost machinelike influence of the position of the head, pelvis, the contralateral paw, etc. on the reaction.*

In intact dogs, the magnet reaction appears in a widely varying intensity; even in one and the same dog this intensity may change. The extension response is sometimes lively and strong, almost like in the decerebellate dog and sometimes it is only a token effect or totally absent.

In general, the magnet reaction of the hindlimbs is suppressed much more by other changes in posture in intact dogs than in decerebellate ones. Particularly conspicuous in intact animals is the inhibiting influence exerted by the dorsal posture.

After *a complete decortication*, the magnet reaction is totally absent. Strange to say, a magnet reaction, not only of the ipsilateral but also of the contralateral hindpaw, was observed after *a unilateral decortication*. This was shown particularly distinct in the decerebellate, right-sided decorticate dog Vici. When the animal was held in the air by the shoulders suspended head upwards, it kept the left hindlimb extended and the right one flexed (Fig. 37, No.1). On contact with the sole of the left foot, this posture of the limb was not altered (Fig. 37, No. 2). Conversely, on touching the sole of the right foot, a maximal extension of this limb and a simultaneous flexion of the left limb instantly occurred (Fig. 37, No. 3). As long as contact with the sole lasted, the right leg remained extended and the left flexed. On contact with the left foot, the flexed left leg also promptly showed the magnet reaction and both hindlimbs remained extended and fixed in an extended posture as long as the finger was kept in contact with the sole (Fig. 37, No. 4). In a dorsal position of the animal, the magnet reaction of the left hindlimb could be induced in the same way.

Thus, in the dog Vici, the crossed left hindlimb showed a typical magnet reaction although Munk's reflex and preparation for standing

in this limb were absent, as is always the case after a contralatera
decortication.

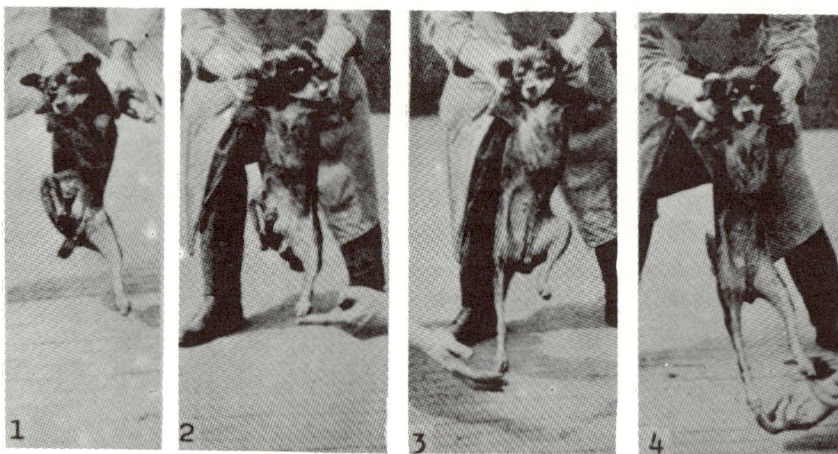

Fig. 37.   Magnet reaction of hindpaws in dog Vici (for more than a year decerebellate and decorticate on the right side).   1. The animal suspended
head upwards, keeps the left hindlimb fully extended, the right
flexed.   2. The posture of the limbs does not change on touching the
sole of the left hindfoot.   3. Conversely, a contact with the sole of the
right hindfoot promptly causes a magnet reaction in this limb while the
left is flexed at the same time.   4. The flexed left limb now also shows
a magnet reaction on contact with the sole.   As long as contact with
the roles continues, bothhindlegs remain extended and fixed in stretched
posture.

On the strength of these observations, one can say that the presence of *one* half of the cerebrum is sufficient for the production of the
magnet reaction in both hindlimbs.   Whether the magnet reaction
after a unilateral decortication is brought about by means of the preserved half of the cortex and thus represents a reaction of the cerebrum,
this is another question.

As we have seen, the magnet reaction is always absent after total
decortication (observations on 5 thalamus dogs which lived from 4
weeks to 8 months) which, however, is not an absolute proof that it
is a reaction of the cerebrum, especially as observations after unilateral
extirpation of the cerebrum are by no means unequivocal.   While
the dog Vici showed a facile magnet reaction on both sides, it was
absent in the contralateral paw of dogs which had only one-sided decortication (except in the female dog Jenny, where it was weakly pres-

ent in the contralateral paw). Findings after the unilateral extirpation of the cerebrum thus vary and caution in interpretation is advisable.

It remains to be discussed whether the magnet reaction, similar to preparation for standing, is perhaps a "conditioned reflex". It could possibly be even conceived as an expression of preparation for standing, as it occurs on contact with a part of the body surface. We are inclined to reject this idea because the magnet reaction and placing show various differences in their occurrence. First of all, after the unilateral decortication, preparation for standing is always absent, as we have seen, on contact stimulation of the contralateral body surface, while touching the contralateral paw may sometimes cause a magnet reaction. In dog Vici, for instance, preparation for standing was absent in the left hindlimb even when the sole of this paw was brought in contact with a surface and on the contrary, a magnet reaction still occurred. Secondly, placing the animal in a dorsal position has a different influence on magnet reaction and on preparation for standing. While animals in a dorsal position do not show any preparation for standing on contact of the muzzle with the edge of a table (Fig. 38, No. 1 and 2; the forelimbs sometimes make defensivemovements and may in this case push against a table), a contact of the soles of the hindlimb with a hand or with a board will also instantly produce magnet reactions in this position, particularly in decerebellate dogs (Fig. 38, No. 3 and 4).

Thus the magnet reaction certainly does not represent an expression of preparation for standing.

This, of course, does not mean that normally the touching of a sole of the foot does not cause preparation for standing in addition to the magnet reaction, and thus contributes to placing on the surface in the position required for standing. In any case, touching of the sole of one foot produces preparation for standing in the other extremities; as soon as the sole of one foot touches the surface, the others, too, are set down on it. The magnet reaction not only is not an expression of preparation for standing, but it also does not behave like a "conditioned reflex". On repeated contact of the sole of the foot with a finger, the magnet reaction is not extinguished, every time the extension of the paw occurs anew. This makes it all the more doubtful that the magnet reaction comes about by way of the cortex. To determine whether the central mechanism utilizes the cerebral cortex or is entirely subcortical, further experiments are needed and these, if possible, for longer observation times and with animals which had the cortex as well as the cerebellum removed.

In *decerebrate dogs* the magnet reaction is absent. However, *it is present after the bilateral extirpation of the labyrinth* and is influenced in a characteristic way by a change of posture of the head in relation to the trunk. *After transverse section of the spinal cord*, Sherrington observed an extension of the hindlimb and a protraction of the toes, sometimes even only on the slightest touching of the sole, for instance with a piece of paper ("extensor thrust" of Sherrington). The extensor thrust is always followed instantly by a flexion, even when the touch continues. Sherrington never succeeded in getting an extension to continue for half a second.

Could the extensor thrust be regarded as the phasic component of the magnet reaction ? After section of the spinal cord, the extensor thrust appears only after several weeks, and at first only by pinching of the skin folds between the balls of the toes. This treatment does not produce any magnet reaction in normal or decerebellate animals. Extensor thrust and magnet reactions apparently have different adequate stimuli. Later on, the extensor thrust occurs also on spreading the balls of the toes and finally even on touching of the sole. At this stage, the animals are usually extremely susceptible to irritation so that a contact at other points, for instance the back of the paw, may cause an extensor thrust. The most sensitive spot for the release of the extensor thrust, however, is the skin fold between the balls of the toes, and, for the magnet reaction, the plantar surface of the tip of the toes. The magnet reaction occurs only on contact with the sole, while the extensor thrust has been observed also after touching other points. Therefore, it seems very doubtful whether the extensor thrust represents a phasic component of the magnet reaction.

After severing the proper posterior roots or the cutaneous sensory nerves, there is naturally no reaction to touching of the sole. *In new-born dogs* the magnet reaction is absent in the first days after birth.

In different animal species the magnet reaction is not uniformly distinct. Cats, whether intact, decorticate (only investigated in acute experiments) unilaterally decorticate or decerebellate, never showed a distinctly pronounced magnet reaction, and monkeys and rabbits even less. Investigations are very difficult in cats and monkeys. When grasped, cats usually start back instantly, the support tonus diminishes and the limbs are drawn in a strongly flexed posture. In this condition a touching of the sole does not cause extension. If one lifts a cat from the ground by the skin of the neck or back, one remarks that the legs are often extended so as to remain as long as possible in contact with the floor (Fig. 39). It is, however, not quite certain whether this occurs through a magnet reaction.

Fig 38. Deecrebellate dog Piccolino with blindfold in dorsal position in the air. 1 and 2. In this position contact of the muzzle with a board does not produce preparation for standing of the forelimbs. 3 and 4. Touching of the soles, however, brings about a magnet reaction.

In the decerebellate female monkey Corrie, too, the experimental conditions caused a strong inhibition and the animal, held in the air by head and tail, held its limbs also forcefully flexed. Although no magnet reaction occurred under these conditions, touching of the sole of the foot sometimes caused a grasping movement of the digits. In cats, too, such a movement could be observed on touching the sole of the foot.

Fig. 39. Decerebellate cat Pierrette. 1 and 2. The animal is lifted from the surface by the skin of the neck and the tail. The legs are extended and touch the ground as long as possible. 3. On leaving the ground the paws are more or less flexed.

This can perhaps be explained by the fact that the hindlimbs of cats and monkeys not only serve for standing and running, but also for climbing. In standing and running cats usually keep their legs only slightly extended, the knee- and ankle joints are always much more flexed than in dogs. Cats are also less capable of carrying loads; the supporting functions are much less pronounced.

The hindlimbs of rabbits, too, show no magnet reactions, since the posterior part of the body of these animals does never truly stand, but sits or is moved forwards by jumping. Thus probably the occurrence of magnet reactions in the various species of animals is dependent on the function of the extremities. We shall return to this subject in the discussion of magnet reactions of the forepaws.

No systematic investigations are yet available on the presence of a magnet reaction in man. Normal adults in a dorsal position show no extension of the legs on touching the sole of the foot. When infants are held by the shoulders and lowered till their feet touch the ground, the legs are not fixed in extended position but drawn up in a flexed posture. Thus in infants (as well as in new-born dogs) the magnet reaction is absent.

Unfortunately, I have not yet had the opportunity for a systematic investigation of humans with brain diseases or on infants who have just started to stand and to run.

IB. THE MAGNET REACTION OF THE FORELIMB

When a decerebellate or an intact dog is held in the air in a ventral position, the forelimbs are usually more or less strongly flexed (Fig. 40, No. 1 and 3). On touching the sole, however, they are instantly extended and changed into rigid columns, similar to the hind-limbs (Fig. 40, No. 2 and 4). In this extension, the digits are usually moved dorsally at first, then the wrist- and elbow joints are extended while simultaneously, the upper arm is moved forwards in the shoulder to the intermediate position[1]. If one grasps the extended limb with two fingers laterally at the wrist or by the forearm and tries to flex it, one has just as little success as by pressure on the sole. Wrist- and elbow joints are fixed in maximally extended positions and on passive movements one encounters an almost insurmountable resistance. The flexor and extensor muscles of the upper- and forearm feel rigid and stressed.

Also, in passive movements of the upper arm at the shoulder forwards, backwards, medially or laterally a distinct, if less pronounced, resistance can be felt.

These movements are accompanied by an associated movement of the scapula. The scapula is fixed not only to the upper arm but also to the thorax. One succeeds with difficulty in sliding the scapula to and fro over the ribs. In addition, a contact with the sole of the

---

(1) Often, it cannot be distinctly determined whether the change of position of the upper arm is brought about only by a movement in the shoulder joint or by a movement of the scapula, or by both together. Usually the latter is the case. All movements which cannot be localised with certainty to the shoulder joint are termed "movements in the shoulder".

forepaw brings about a tension of the muscles of the vertebral column particularly in its upper, thoracic part (Fig. 41) and, in certain cases also, of the muscles of the neck.

In contrast to the hindlimbs, the forelimbs do not show any magnet reaction in the dorsal position of the animals (Fig. 42). The different behavior of the magnet reaction in the forelimb in a ventral and dorsal posture is caused by the position of the wrist joint, i.e., owing to the maximal flexion of this joint in the dorsal position (Fig. 42), while in a ventral position in the air it forms an open angle of 90° or more directed backwards (Fig. 27, 33, 40). When in the dorsal position the wrist joint is passively extended for 90-110°, in this position, too, a magnet reaction of the forelimb promptly occurs on contact with the sole.

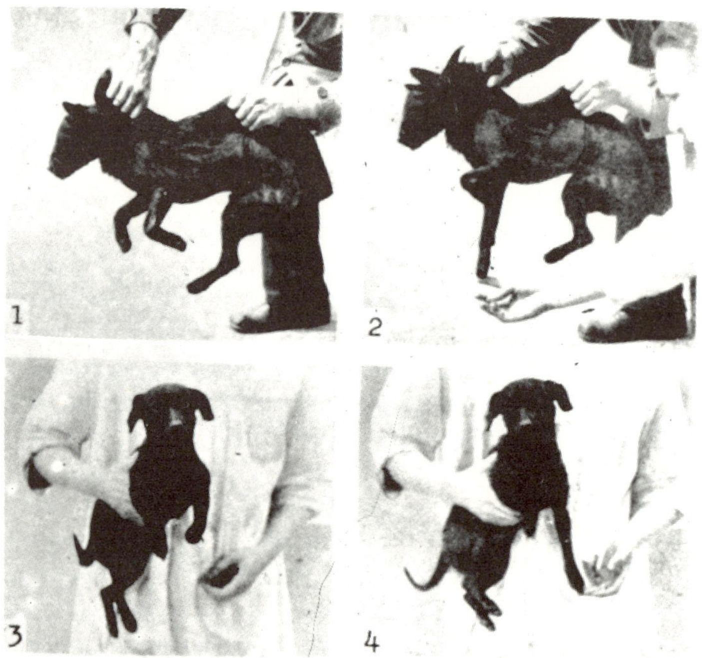

Fig. 40. Magnet reaction of the forelimb. 1. Decerebellate dog Erik, held in the air by the skin of back and neck. The animal keeps its limbs in a half extended posture. 2. On touching the sole, the right forelimb is instantly maximally extended. 3. Decerebellate dog Piccolino held in a ventral position in the air supported by the thorax. The animal keeps its forelimbs also half extended. 4. In this animal, too, contact with the sole produces a strong magnet reaction of the forelimb.

The occurrence of a magnet reaction of the forelimb depends also on the posture of the head, the neck, the contralateral limb and on the position of the limb in relation to the trunk. Lifting and lowering of the head exerts a contrary influence on the occurrence of magnet reaction in hind- and forelegs.

Concerning the occurrence of the magnet reaction after extirpation of the cerebellum, after the total or unilateral removal of the cerebral cortex and after decerebration, the reaction of the forelimbs will be the same as that of the hindlimbs.

On touching of the sole of the forelimb of cats and monkeys, a grasping movement, a flexing of the digits and a flexion of the arm occurs. In addition to standing and running, the forelimbs of cats serve particularly for grasping and climbing. In the life of monkeys, too, standing and running only play an inferior role as these animals predominantly hang in trees, sit on branches or jump from branch to branch. It is remarkable that decorticate cats and monkeys (Karplus and Kreidl 136, Magnus 195, 196) already show, a short time after decortication, very facile grasping movements on touching the sole of the foot, while in decorticate dogs the magnet reaction cannot be produced for weeks, even months, after decortication.

Magnus observed the grasping reflex in decorticate monkeys as well as in thalamus-cats already within the first 24 hours. After decerebration, the grasping reflex as well as the magnet reflex is absent. Based on our observations in decerebellate *dogs*, Schoen and Pritchard

Fig. 41 Influence of contact with the sole of the forepaw on the tension of the vertebral back muscles 1. and 3. Decerebellate dogs Piccolino (1) and Erik (3) with blindfold, held in the air by the tail and skin of the neck. The back sags and is relaxed. 2 and 4. On touching the forepaws with two fingers, the limbs are extended, the vertebral muscles are strained, first in the anterior, then in the posterior part, which makes the back rigid and straight, sometimes even curved convexly upwards.

Fig 42. Decerebellate dog Erik with blindfold in dorsal position. Touching the sole of the forepaw, kept flexed at wrist-, elbow and shoulder joints, does not produce a magnet reaction.

investigated the influence of rubbing of the sole of the foot on the muscles of the extremities by the graphic method. The investigated muscle was severed from its distal insertion and connected with a recording lever. As experimental animals, however, they chose *cats* (thalamus-cats and intact cats in a light ether narcosis). Schoen (269) observed on rubbing the sole of the forepaw a distinct contraction of the shoulder muscles, the m. supraspinatus, teres major, deltoideus, pectoralis major, rhomboideus and m, trapezius, and considered these contractions as components of the supporting reaction, and neglecting completely the grasping reflex. Whether they represent components of the grasping reflex of the exteroceptive supporting reaction (magnet reaction) is not directly evident. Pritchard (231) observed on rubbing the sole of the hindpaw contraction of the flexors with simultaneous relaxation of the extensors (ipsilateral flexor reflex) but sometimes also a simultaneous tension of extensor- and flexor muscles. Whether, as Pritchard asserts, the simultaneous contraction can be considered as an extensor reaction (magnet reaction) seems very questionable, since the joints were fixed with needles during the investigations and the registered muscles severed from their insertions. It can, therefore, not be decided whether the simultaneous contraction of flexors and extensors would have caused, in normal conditions, an extension or a flexion or perhaps only a stronger fixation. It is not impossible that, as in the magnet reaction in dogs, in the grasping reflex, too, the flexor and extensor muscles are simultaneously contracted. As is well known, man, by strongly clasping an object, simultaneously contracts flexor and extensor muscles of the forearm.

It is certain at least, that on touching the sole of the foot with a finger a grasping reflex almost always occurs in cats and monkeys. Whether the magnet reaction can also occur under certain circumstances, be it on touching certain parts of the sole of the foot, be it on account of a certain posture of the animal, be it in certain positions of the joints of the distal extremities, has not yet been ascertained. Strange to say, a contact of the sole with a surface does not produce a grasping reflex in monkeys and cats. In this, perhaps, the places of contact play a certain role. While the grasping reflex appears most vivid in cats on touching the *proximal part of the sole* of the hand and leg, in monkeys on touching the center of the palms of the hand or the sole of the foot, the magnet reaction in dogs makes itself felt most strongly on touching the balls on the plantar surface of the distal phalanges.

In man, whose arms have no longer any standing functions, a touch of the palm of the hand also releases a grasping reflex. Healthy infants always show this reflex till the end of the first year of life (Peiper-Isbert 221) : in older children it is difficult to distinguish it from the grasping response to visual stimuli. Gamper (90, 91) observed grasping movements on touching the palm of the hand of his anencephalic "midbrain" child. Under pathological conditions (frontal lobe tumors) the grasp reflex again automatically occurs in adults, in this case the object is sometimes forcibly clasped with the hand. Kleist (138) recently published interesting films of pathological grasping movements. Unfortunately, he does not state whether the patients reacted with a magnet reaction on touching the sole of the foot. Such patients as well as people with cerebellar diseases would be particularly suitable to investigate the question of occurrence of magnet reactions in the legs of man.

Exteroceptive stimulation produced by contact of the sole of the foot with a supporting surface does not alone cause support tonus. Other stimuli also take part since, firstly, decorticate dogs are capable of standing in spite of the absence of the magnet reaction, and, secondly, since cats whose feet have been made insensitive by severing the nerves, can still stand and run (Sherrington 279). Also, human beings with artificially anesthesized feet are able to stand, though staggering a little (Vierordt).

II. THE INFLUENCE OF THE POSTURE OF THE DISTAL PHALANGES ON SUPPORTING TONUS : THE PROPRIOCEPTIVE SUPPORTING REACTIONS

The position of the distal phalanges of the extremities, the metacarpus and digits of the forelegs and toes on the hindlegs, as well as the contact of the soles of the feet with a surface are of critical importance for the production of the supporting tonus.

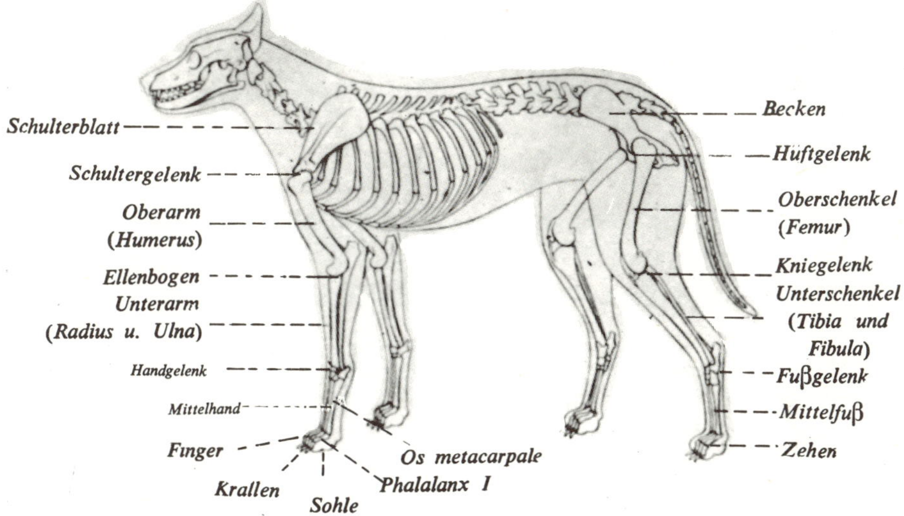

Fig. 43. Skeleton of a dog in a normal standing posture. In a normally standing dog, the wrist joints are always maximally extended, they even sometimes form an open angle forwards (Figs. 26 and 43), the tarsal joints of the hindlegs also show an extended posture which, however, is not maximal and of an alternating degree. Only the phalanges, provided with balls on their volar or plantar sies, rest on the surface, but not on the *metacarpal-* nor the *metatarsal bones*. The proximal basal phalanges are directed forwards and form an almost right angle with the metacarpal and metatarsal bones, while the interphalangeal joints are almost extended. On standing decerebellate (Fig. 44) and decorticate animals (Fig. 257) the phalanges show a similar posture.

We have already seen how in intact and decerebellate dogs the preparation for standing leads to appropriate posture for standing. We now come to the question in which way this posture of the phalanges is able to influence the distribution of tonus in the leg muscles.

When an intact or a decerebellate dog is brought into a dorsal position and the wrist joint passively extended without touching the sole, the elbow is then also extended and the upper arm is moved in the shoulder joint towards the intermediate position[1] . At the same time the muscles of the upper- and forearm become hard and contracted (Fig. 49, No. 1 and 2). When the wrist joint is fixed in an extended position the elbow joint offers a strong resistance against flexion which cannot be overcome in further attempts to flex it (Fig. 49, No. 3). In the same way a passive movement to flex the upper arm encounters a distinct resistance and is instantly accompanied by an intermediary movement of the scapula. It is almost impossible to move the scapula passively to and fro, as it is strongly fixed to the thorax. The muscles of the back and neck, too, sometimes show a slight change in tension. The stretched position of the elbow and the fixation of the different joints distinctly increase when, without touching the sole, the digits are passively extended dorsally.

Conversely, the elbow joint goes into a flexed position on passive flexion of wrist joints and digits, while all resistance against a passive flexion of the paw disappears (*negative supporting reaction*). On a strong passive flexion of the interphalangeal and wrist joints the limb sometimes is actively drawn into a flexed position, particularly in decerebellate dogs. In the maximal position of flexion of the wrist joint, the elbow joint can easily be extended to $90°$ then, however, an extraordinarily strong resistance against extension occurs so that it is not possible to extend the elbow joint completely.

With the wrist joint fixed in an extended position, the forelimb forms an extended, rigid column; with a passively flexed wrist joint, however, it has a flexor tonus and thus cannot be maximally extended. With a loose wrist joint the limb is easily movable and can be maximally flexed and extended without any resistance worth mentioning.

The conditions are similar in the hindlimbs. When the hindlimbs are passively extended and the toes are passively directed backwards,

---

(1) In the shoulder joint we distinguish a flexed, intermediary and extended position. The movement of the forearm from the intermediary position forwards (protraction) is termed an extension and backwards (retraction) a flexion. The hip joint, however, is extended when the thigh is directed backwards, flexed when it is directed orwards (Fig. 45).

Fig. 44. Decerebellate dogs in a standing posture.   1. Decerebellate dog Erik.
2. Decerebellate dog Piccolino.   3. Decerebellate dog Moor.   4. Decerebellate dog Wolf.

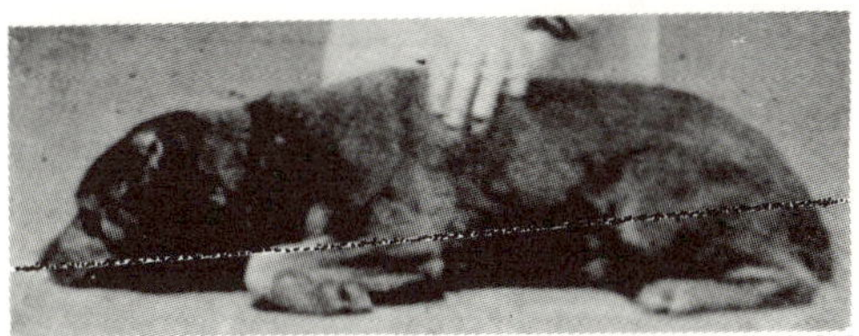

Fig. 45. Normal dog in a prone position with flexed limbs.   The upper arm is
directed backwards in the shoulder joint, the thigh directed forwards
in the hip joint.

without touching the sole, the limbs also offer a distinct   resistance
against passive flexion.   After release of the toes this resistance ceases
(Fig. 46, No. 5), while in passive flexion of the toes the hindlegs are
sometimes drawn in to a flexed position (negative supporting reaction); in this case they offer a more or less resistance to a passive
extension.

Fig. 46. Positive and negative supporting reaction of the hindlimb (1 and 3) and the forelimb (4-6). 1. Decerebellate dog Wolf in a dorsal position, the muzzle pointing vertically upwards. On static stress, the hindlimb is stretched, fixed in extended position and offers strong resistance against pressure on the sole (*positive supporting reaction*). 2. The hindlimb shows the same phenomenon on static stress on the sole by the hand while the animal is standing on the other three paws. 3. On the passive flexion of the toes, the resistance stops and sometimes the paw is even actively drawn into a flexed position (*negative supporting reaction*). 4. Decerebellate dog Moor in a standing position. Extension of the left forelimb by static stress: strong supporting tonus. A passive flexion due to pressure on the sole hardly succeeds (*positive supporting reaction*). 5. After removal of the hand, reduction of the supporting tonus. 6. On passive flexion of the wrist joint, the elbow and shoulder joints also go into flexion (*negative supporting reaction*).

When we ask how the dependence of the position of joints with one another comes about, we have to consider that by passive extension of the wrist joint and dorsal movement of fingers the flexors of these phalanges are stretched (Fig. 47). It could be now possible that this tension produces stimuli in the stretched muscles which cause reflex contractions not only of the stretched muscle but also of the remaining leg muscles, and thus bring about the change of position and fixation of elbow- and shoulder joints. On the other hand, it could also be possible that the stretched muscles (Fig. 47, A and B) which originate on the humerus only behave like elastic ligaments, and that the increased passive tension produced by the stretch causes the extension and fixation of the elbow joint.

That this last possibility should not be neglected is shown by observations on the forelimbs of cats in rigor mortis.

However, the elasticity of wrist joint and finger flexors cannot explain the fixation of the scapula on the thorax and the tension of the vertebral muscles on static stretch on the sole of the forelimb; these can be produced only by reflex reactions. Also, the observations on animals in rigor mortis cannot be directly applied to living ones. As shown in animals under deep narcosis, the elasticity of the muscles plays only a subordinate role in the reciprocal dependence of position in the joints of the extremities; in a deeply narcotized dog it is possible, for example, to bring about total flexion of the elbow with passively extended wrist joint. The hindlimbs, too, after a transverse section of the spinal cord and similarly after severing the posterior roots belonging to these limbs do not show any extension and fixation in extended positions on the dorsal movement of the toes. These observations strongly support the participation of reflex reactions. They further furnish a proof that the position of the phalanges of the forelimb has a distinct influence only on the position of the proximal joints when the muscles of the forearm are in rigor mortis or have a distinct tone.

Based on these findings Schoen (269) and Pritchard (231) investigated in cats the influence of the position of phalanges on the muscles which were detached from their places of insertion and attached to a recording lever. In these experiments the humerus, radius and ulna, as well as the femur, fibula and the tibia were fixed immovably with drills so that a passive movement of other joints was impossible. Schoen

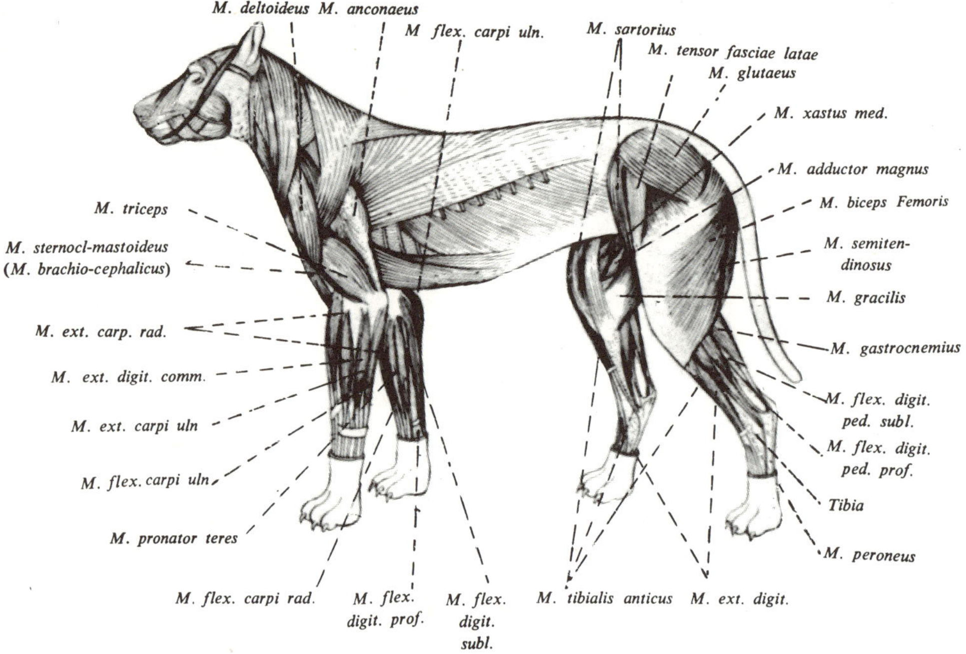

Fig. 47. Muscular system of the extremities of the dog. The following muscles of the digital, wrist, toe and tarsal joints are stretched by static stress on the legs. A. *Wrist joint palmar flexors* : m. flexor carpi radialis: from epicondylus int. *humeri* → os metacarpi II and III; M. flexor carpi ulnaris: from epicondylus int. humeri and proximal terminal piece of unla → os pisiform. B. *Wrist joint and finger flexor*: m. flexor digit. sublimis: from epicondylus int. humeri → phalanx II of the 4 fingers; m. flexor digit. prof.: from epicondylus int. humeri → phalanx III of the 4 fingers (m. radialis volaris : from med. edge of radius → the joint tendon of m. flex. digit. prof.); (m. ulnaris volaris : from the dorsal side of the unla → the joint tendon of m. flex. digit. prof.) C. *Tarsal joint extensors* : m. gastrocnemius: from planum poplitaeum ossis femoris → tuberositas calcanei m. biceps femoris, m. semitendinosus : which from jointly a tendon which terminates on the calcaneus. D. *Tarsal joint extensors and toe plantar flexors*: m. flexor digit. ped. sublimis: from epicondylus lat. femoris → phalanx II of the 4 toes; m. flexor digit. ped. profundis: from the dorsal side of fibula and tibia → phalanx III of toes (m. soleus is absent in the dog).

and Pritchard found that under these conditions the static stretch on the soles was able, in intact and slightly narcotised cats as well as in thalamus cats, to produce reflex contractions of almost all muscles of the extremities, the flexors, extensors, abductors and adductors. This was still the case even when, after severing of all cutaneous nerves, the magnet reaction was eliminate. Schoen was able to show that the proprioceptive stimulations which produce reflex contractions in the anesthetized forepaws, emanate from the stretched palmar- flexors of digital and wrist joints (Fig. 47, group B). After severing these flexor muscles, the static stress on the sole of the foot no longer produced reflex contractions of the other muscles of the extremities, while these still showed reflex contractions on static stress after the elimination of the extensor muscles.

For the production of these reflex tensions, not all the flexor muscles of hand and fingers had to be intact; it was sufficient when the deep flexor, m. flexoris digit. prof. was entirely or at least partly preserved, and even after severance of this muscle and preservation of the other flexor muscles still reflex contraction occurred, but always greatly weakened, with the extension of the intact m. flexoris digit. profund. by static strain on the sole the following muscles showed distinct reflex contractions :

I. the remaining flexors of

    phalangeal and wrist joints :   m. flex. digit. sublimis
                                                        m. palmaris longus
                                                        m. flex. carpi ulnaris
                                                       m. flex. carpi radialis

II.   the extensors of   m. ext. digit. communis
      phalangeal and wrist   m. ext. digit. lat.
      joints :   m. ext. carpi uln.
                                                          m. ext. carpi rad.

III. the forearm flexor :   m. biceps, m. brachialis
IV. the forearm extensor :   m. triceps, cap. long., cap. lat.
                                                           and cap. med.

V. the shoulder flexors :   m. teres major
                                                          m. deltoideus
                                                        m. latissimus dorsi

VI. the shoulder extensor : m. supraspinatus
VII. the upper arm adductor : m. pectoralis major
VIII. the fixators of the scapula : m. serratus ant.
                                                           m. levator scapulae
                                                        m. trapezius
                                                     m. rhomboideus

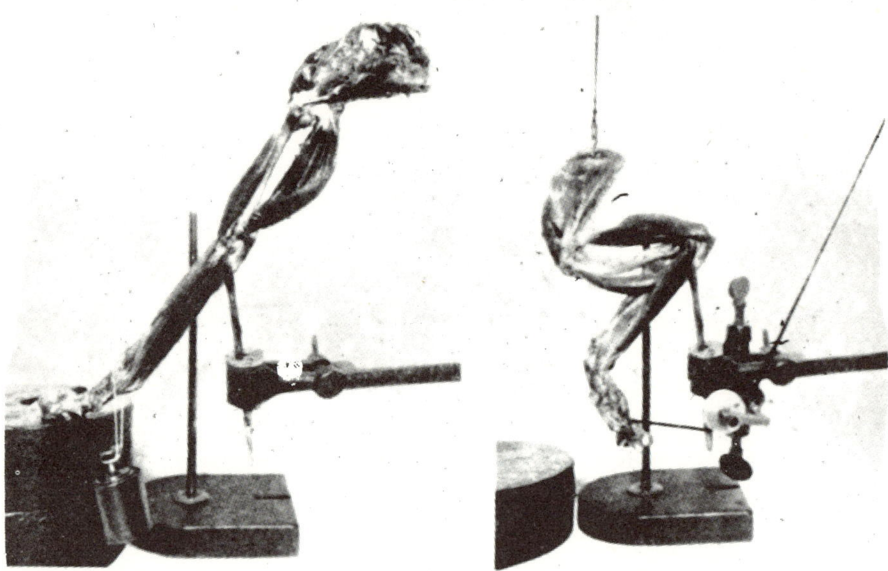

Fig. 48. Forepaws of a cat in rigor mortis, from medial side.   The biarticulate muscles are preserved, i.e. the long extensors and flexors of the toes, the biceps and the elongated head of triceps, in addition to a part of the subscapularis on the scapula. The preparation is fixed by a clamp on the ulna.   1. The wrist joint is extended by a load (static stress). The result is an extension of elbow - and shoulder jo nts. The passive stretching of wrist joint and finger flexors causes a stretching of the elbow joint. The stretching of the elbow joint caused a stretch of m. biceps and thus an extension of the shoulder joint. 2. The wrist joint is flexed by pulling on a string and at the same time the shoulder is prevented from passive movement by a counter pull. The result of flexion of the wrist joint is a flexion of elbow- and shoulder joint.   According to R. Schoen, Die Stützreaction. 1. Mitteilung. Pflügers ulmaut Archiv. Vol. 214, p. 21, 1926.

These contractions also occurred when the sole of the foot was previously made insensitive by severance of the cutaneous nerves.

Thus the change of position of the phalanges produced a reflex contraction in almost all muscles of the forepaws, the flexors, extensors, adductors and abductors, and by this contraction the digits[1], the wrist-, elbow- and shoulder joints are fixed and, in addition, the scapula is fixed to the thorax.

---

(1) Probably due to this fixation the phalanges do not lie flat on the supporting surface in normal standing (see Fig. 26).

Similar reflex reactions owing to proprioceptive stimulation have been observed by Pritchard on static stress on the anesthetized hindlimb. As shown by Pritchard on cats, these stimuli emanate from the stretched flexors of the toes and to a much stronger degree, from the gastrocnemius muscles.

Bringing the phalanges into the position required for standing thus produces reflex contractions which probably participate in the formation of the supporting reactions.

Reflex reactions also participate in the negative supporting reactions. Schoen observed in decorticate and intact, slightly narcotized cats on a passive flexion of the wrist joint a reflex relaxing of triceps, and serratus anticus, a reflex contraction of teres major and latissimus dorsi, i.e., a relaxing of the elbow extensor and of the fixator of the scapula as well as a contraction of the shoulder flexor. A contraction of the biceps could not be detected on exclusive flexion of wrist joint and fingers with fixed upper- and forearm; it occurred only on passive flexion of the elbow. The reflex reaction on flexion of the phalanges observed by Schoen can explain the relaxation of the paw and the retraction into a flexed position in the shoulder joint. The proprioceptive stimuli here emanate from the stretched digital and wrist joint extensors.

III. PRESSURE ON THE SOLE OF THE FOOT

When one places an intact or decerebellate dog in a dorsal position and without touching the sole of the foot, brings the fingers and wrist joint into the position of standing, the foreleg is extended and fixed in an extended position (Fig. 49, No. 1-3). On *touching* the sole of the foot, both the extended position as well as fixation distinctly increase (Fig. 49, No. 4). Fixation is further increased when a *pressure* is exerted on the sole (Fig. 49, No. 5). The hindlimb reacts in exactly the same way.

Pressure on the sole of the foot acts in the first place as an exteroceptive stimulation similar to touching, but the intensity of stimulation is much stronger. In the second place pressure on the sole is

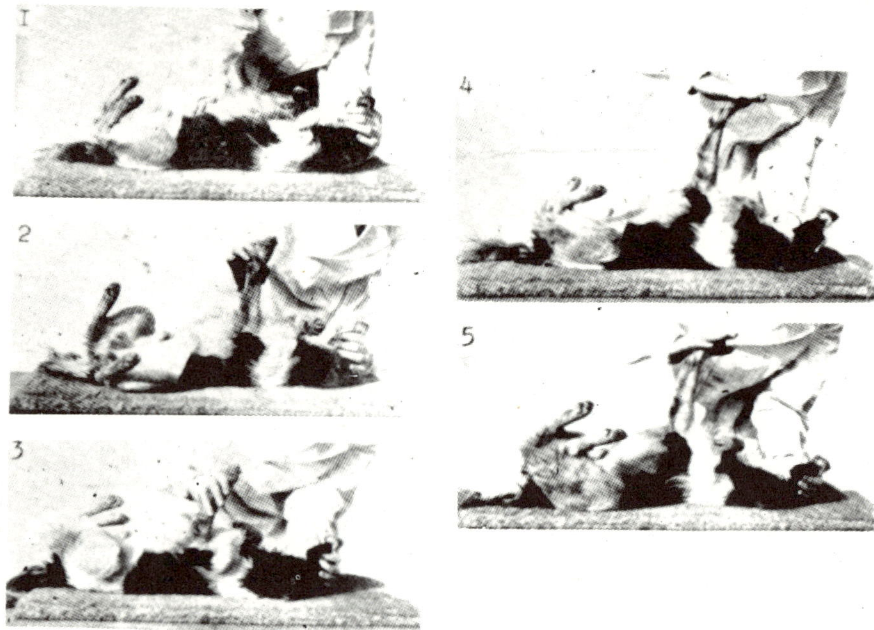

Fig. 49. Positive supporting reactions of the forelimbs of the decerebellate dog Wolf. 1. The animal in dorsal position, the muzzle pointed upwards in an angle of 45-60° to the horizontal. The forelegs are flexed in all joints. 2. On passive extension of the wrist joint (without touching the sole) the elbow extends more than 90° and the upper arm is moved forwards in the shoulder joint. 3. With the wrist joint held passively extended, the elbow is passively flexed only 10° (so that the elbow forms an angle of ± 80°. 4. On touching the sole, extension increases (compare position of elbow and of upper arm in shoulder joint with that of No. 2). In grasping the limb and passively flexing the elbow, one encounters an even stronger resistence. 5. Pressure on the sole (see the over extension of the wrist joint) causes the resistance to be at a maximum and it is almost impossible to flex the elbow.

capable of stretching the wrist joint even more, even as to over-stretch it and move the fingers and toes even more backwards (see in Fig. 50, arrows V_1, V_2, and H_1). The already strongly stretched flexor muscles of these phalanges are even more extended, which probably causes an increase in reflex tension, for Liddell and Sherrington showed that the strength of the reflex contraction of a muscle is dependent on the amount of stretch as well as on the initial tension. With a certain initial tension m. rectus femoris showed on an extension of only 1 mm an increase of tension of more than 2 kg.

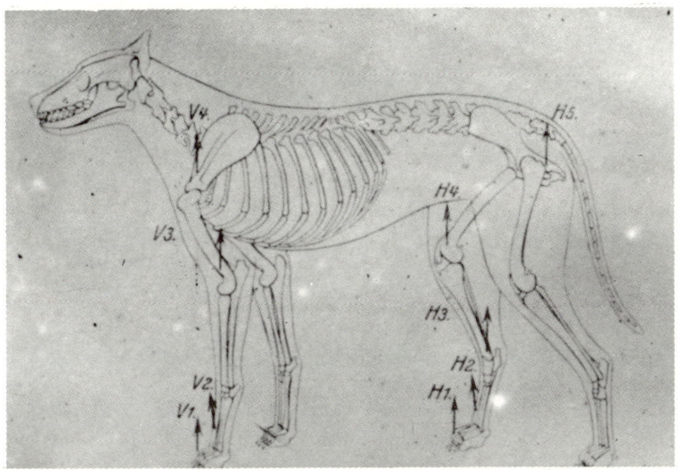

Fig. 50. The changes of position of joints of the extremities which are caused by the counter pressure of the supporting surface.

Pressure on the sole of the foot also causes in the forelimbs a flexion of elbow- and shoulder joints (Fig. 50, V_3, and V_4) and in the hindlimbs of the tarsal-, knee- and hip joints (Fig. 50, H_2, H_3 and H_4). By this, the extensors of these joints are stretched and this stretching, too, produces reflex contractions. As Schoen demonstrated, an extension of the triceps not only causes a reflex tension of the muscle itself but also of other muscles, among others the biceps, supraspinatus, pectoralis major. These tensions, for instance the simultaneous tension of biceps and triceps intensify the fixation of the joints and contribute to enabling the limbs to carry the trunk. If the back of a standing animal is stressed by an added load, the counter pressure of the surface on the sole of the foot increases, this brings about an even stronger extension of flexors of the phalanges and of extensors of the proximal joints which produces again a stronger reflex tension and therewith an adjustment to the load.

Pressure on the sole of the foot leads to an extension of the flexed limb, which can be observed in thalamus dogs in which the magnet reaction is absent. The three factors: contact with the sole of the foot, position of the phalanges, and pressure on the sole, together form the *positive supporting reaction*. They are the reason that the limbs of intact and decerebellate dogs, when set down on their feet, are changed into rigid columns which are not only able to carry the trunk but also twice the weight of their bodies (Fig. 51).

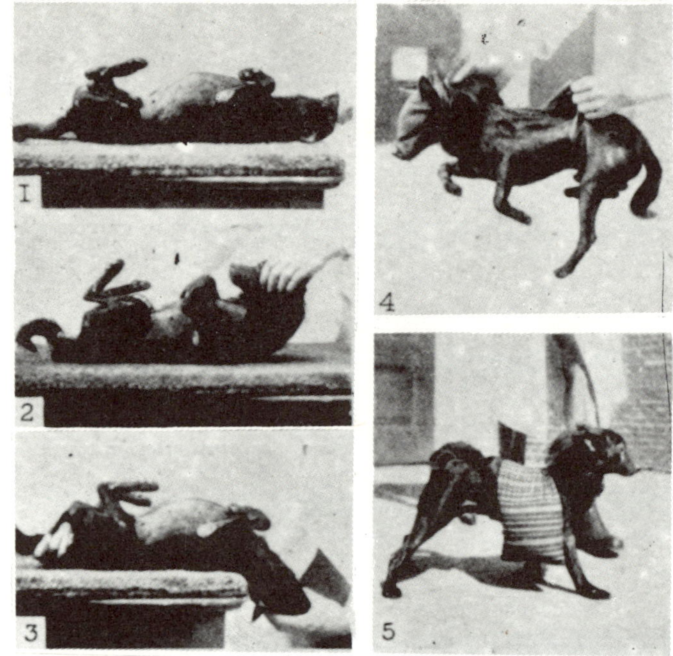

Fig. 51. Decerebellate dog Erik. 1. In a dorsal position, muzzle pointing upwards at an angle of 45° to the horizontal (maximum position of tonic labyrinthine reflexes) the animal keeps the limbs flexed. 2 and 3. Change of head posture does not cause any distinct change in tonus or position of the legs. 4. Also in ventral position in the air, position and distribution of muscle tonus in the extremities is totally unsuited for standing and carrying of the body. 5. Set down on the legs, however, the limbs are not only able to carry the body but also twice the body weight. Body weight of dog 6 kgs. Weight of sand bag 5.5 kgs.

IV. THE RECIPROCAL POSTURAL DEPENDENCE OF THE PROXIMAL JOINTS

The fourth factor participating in the derivation of support tonus is the reciprocal postural relationship of the proximal joints. We have seen that in the preparation for standing on contact of the muzzle with a surface, the forelegs move forwards at the shoulder, and on contact of the tip of the tail the hindlegs move backwards at the hip joints. Similarly a magnet reaction and static stress cause an extension of the legs in the shoulder- and hip joints. Just as with the phalanges, the extension of the proximal joints brings about an extension of the other joints. When, for instance, in an intact or decerebellate dog, the upper arm is passively moved forwards at the *shoulder joint*, then the elbow- and wrist joints are also extended (Fig. 52, No. 1) and cannot be passively flexed as long as the upper arm is held forwards. Simultaneously the digits move dorsally (Fig. 52, No. 1) but they can still be passively flexed without much effort as well as moved further dorsally. In this movement an elastic resistance can be felt and on release the fingers spring back into their initial position.

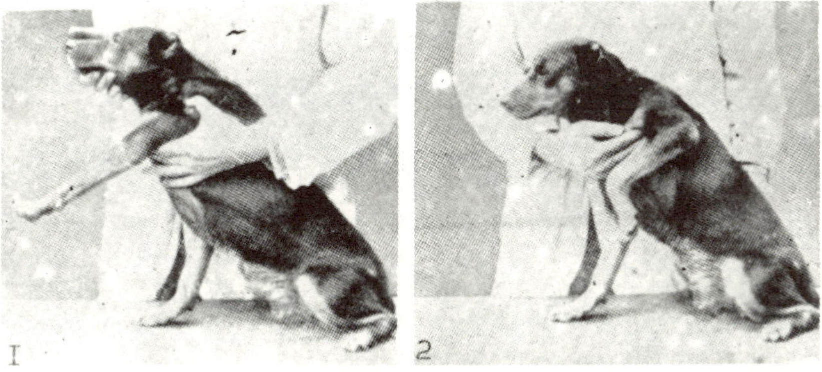

Fig. 52. Reciprocal postural dependence in the foreleg of a normal dog. 1. In passive extension of the upper arm in the shoulder joint forwards, the elbow- and wrist joints are also extended and the digits dorsiflexed. 2. On passive flexion of the upper arm, there appears a simultaneous flexion of the elbow- and wrist joints and a plantar flexion of the digits.

When the upper arm is passively moved backwards (Fig. 52, No. 2) fixation disappears, elbow- and wrist joints change into flexion, while the digits plantarflex. When the shoulder joint is kept in flexion, elbow- and wrist joints are easily movable to a certain degree but they cannot be maximally extended. Extension and flexion of the *elbow joint*, too, cause a change in posture and fixation in the remaining joints. When the elbow of a flexed forelimb is passively extended, the upper arm moves forwards in the shoulder joint, the wrist joint extends, the digits dorsiflex. With a fixation of the elbow joint in maximally extended position a passive flexion of the shoulder joint is only possible with resistance while the wrist joint is absolutely fixed and cannot be passively flexed. In addition, the digits show a strong elastic fixation. With passive flexion of the elbow joint the shoulder joint as well as the wrist joint and the digits are also flexed. In this case the wrist joint is freely movable but only within certain limits : with the elbow kept passively in total flexion an extension of the wrist joint of more than 90° encounters an insurmountable resistance. The shoulder joint, too, is in this case only freely movable in an extensive passive backward movement of the upper arm and a resistance can be felt. In the digits, a strong extension encounters considerable resistance; on flexion or moderate extension movements no or only a slight elastic fixation can be felt. Corresponding conditions appear in the hindlegs. When the thigh is stretched backwards from the *hip joint*, knee- and tarsal joints are also extended (Fig. 53, No. 1 and 3) and fixed in an extended position. With a passive extension of the *knee joint*, the thigh goes backwards in the hip joint and the tarsal joint is stretched and fixed in an extended position. When the knee joint is kept passively extended no passive flexion of the tarsal joint is possible.

Conversely, the flexion of a proximal joint in the hindleg causes a simultaneous flexion of the other joints and a relaxation of the fixation. Thus, for instance, the passive forward movement of the thigh from the hip joint brings about a more or less simultaneous strong flexion of the tarsal- and knee joints (Fig. 53, No. 2 and 4) and these joints are easy to move passively so that only a maximal extension of these joints meets a distinct resistance when the thigh is held forwards. When the knee joint is also passively flexed, the tarsal joint goes into flexion and can be moved easily within certain limits, i.e., the tarsal joint can be totally flexed without any resistance and extended up to an angle of 90°. On further extension, however, a distinct resistance occurs and a maximal extension encounters a complete and insurmount-

Fig. 53. Reciprocal postural dependence of the proximal joints of the hindlimb in an intact (1 and 2) and in a decorticate dog (3 and 4). 1 and 3. The thigh is passively moved backwards in the hip joint. Extension and fixation in an extended position of the knee- and tarsal joints and plantar flexion of toes which are fixed elastically (the phalanges form an almost straight line with the metacarpus). 2 and 4. The thigh is passively moved forwards at the hip joint. More or less strong flexion and cessation of fixation in knee- and tarsal joints. The foot hangs down limply at the tarsal joint. Plantar flexion of toes in the interphalangeal joints and reduced fixation.

able resistance. In these passive movements, the changes in posture of the toes are essentially different from those of the digits on the passive movement of the proximal joints of the forelimb. With a maximally extended hip- and knee joint, the phalanges and metatarsal bones form an almost straight line instead of an angle open towards the front (Fig. 53, No. 1 and 3) and are strongly and elastically fixed in this extended posture. With a flexion of the legs in the hip- and knee joints, fixation diminishes and the toes usually move dorsally at first and then plantarly. In a maximally flexed paw, the digits are plantarflexed in the interphalangeal and also in the metatarsophalangeal joints and fixed slightly elastically in this posture (Fig. 54).

Thus a passive extension of a proximal joint of the fore- or hindleg causes an extension and a more or less strong fixation in the extended posture of the re-

maining joints, while a passive flexion of a proximal joint is accompanied by flexion and a slackening of the fixation of the remaining joints.

This applies to intact, decorticate and decerebellate dogs.

Fig. 54. Position of toes in maximum flexion of hindlimb. The limb is passively flexed in the tarsal-, knee and hip joints. In so doing, the toes go into plantarflexion in the inter- as well as in the metatarsophalangeal joints.

We have described how, with an extension of the distal phalanges (wrist-, interphalangeal-, tarsal- and toe joints) the other joints of the extremities are also extended and fixed by a reflex reaction. It is natural that we now ask ourselves whether reflex processes also participate in the reciprocal postural dependence of the proximal joints.

It is certain that this dependence is due at least partly to the anatomical arrangement of the bi- and multiarticulate muscles, as was shown in observations on dead dogs with beginning rigor mortis, in which a similar reciprocal dependence on posture of the joints as well as their fixation could be demonstrated.

With dogs in deep narcosis, the extension of a proximal joint is also accompanied by an extension even if a less strong fixation of all the remaining joints occurs. When the animals are almost or completely narcotized to death, associated movements may even occur which, however, can be prevented with a very slight effort. These animals also lack a distinct fixation; the wrist joint, for instance, can be easily flexed with a maximally extended elbow joint. Associated movements and fixation could also be observed *after transverse section of the spinal cord* but usually less strongly than in intact dogs. In the decerebrate animal, on the contrary, the changes of position and fixation of the other joints after a passive movement of the elbow- and knee joint are particularly strong [1].

The observations on dead or deeply narcotized dogs show that the reciprocal postural and fixation dependence can be explained by the anatomical arrangements of the bi- and multiarticulate 'muscles and that the differences in fixation are not necessarily the result of a deficiency of special reflex reactions but can be traced back to the different force of muscle tonus prior to the passive movement.

Nevertheless, the possibility that some special reflex processes play a role has not yet been sufficiently investigated. Schoen investigated the influence of the passive extension of the elbow joint in cats on the muscles of the upper arm, which were detached from their insertions. When part of the triceps was detached from its insertion and attached to a recording lever, the other part, however, is stretched in situ by the passive movement of the elbow joint there occurs on passive extension of the elbow a tonic contraction of the detached part of the triceps and also of the other[2] *shoulder flexors* : teres major, deltoideus and latissimus dorsi. Schoen thus observed, on the passive extension of the elbow, definite reflex contractions of the shoulder flexors which, however, cannot explain the extension of the shoulder joint which normally occurs on passive extension of the elbow. On passive flexion of the elbow, he observed a reflex contraction of the triceps and supraspinatus (extensor of the shoulder joint).

The *reciprocal postural dependence* of the proximal joints explains the increase in supporting tonus of the hindlegs which occurs when in the standing animal the anterior part of the body is lifted from the ground (Fig. 55). Similarly, the decrease in supporting tonus of the hindlimbs on a downward movement of the anterior part of the body can be brought about as well as the increase or decrease of supporting tonus of the forelegs on lifting and lowering movements of the posterior part of the body. By these changes in supporting tonus, an adjustment of limb posture of fore- and hindlegs to the respective position of the supporting surface can be brought about (Fig. 55, No. 3 and 4).

---

(1) On the contrary, a passive flexion of a shoulder - or hip joint scarcely brings about any change in the posture of the other joints of decerebrate animals.

(2) The m. triceps is not only an elbow extensor but also a shoulder flexor.

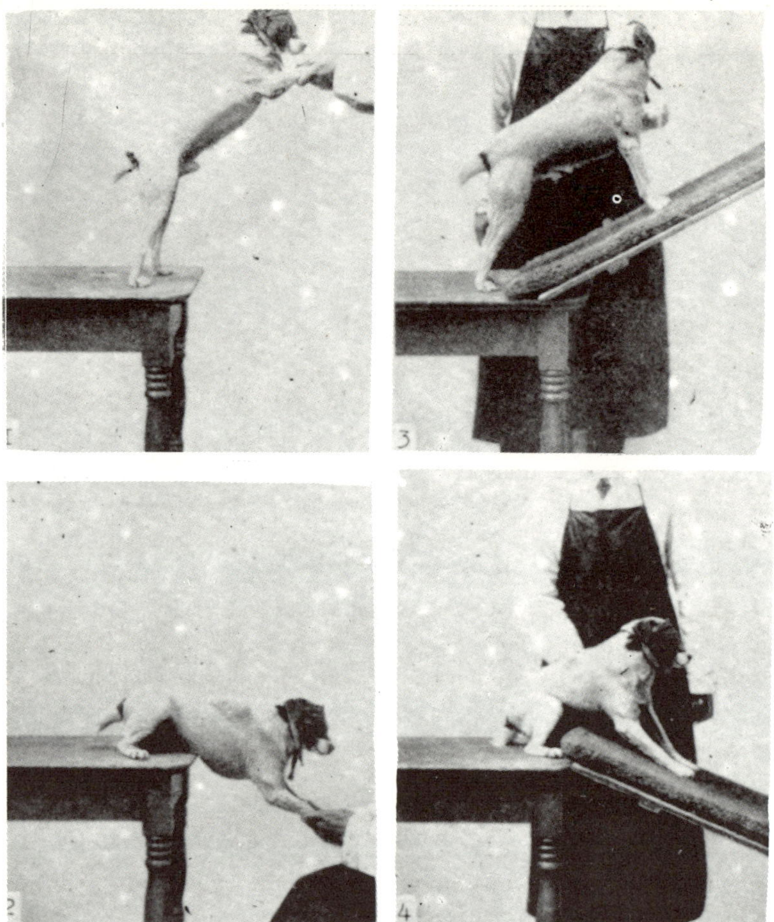

Fig. 55. 1 and 2. A dog standing on the hindlimbs, the upper part of the body is alternately moved upwards and downwards so that the hip joint is passively extended
and flexed.   When the thorax is lifted, the hindlimbs are extended are fixed
in an extended position; when the thorax is lowered, they are flexed and fold
up. 3 and 4. By this change in position and fixation of the hindlimb joints, an
adjustment to static conditions is brought about when the animal stands with
the forepaws on a support which is moved up and down, while the support
of the hindlimb remains fixed.

The reciprocal postural dependence of the joints of the extremities plays
a role in the correction of abnormal limb postures. If a decorticate
dog is set down on a surface, on the dorsum of a foot, it will be seen
that this posture is corrected as soon as he starts to run. On the forward movement of the limb in the shoulder joint, the wrist joint is
extended and the fingers move dorsally so that the paw is set down
with the sole on the surface. The same occurs in the hindlimb when
it is lifted from a posture with foot turned down, moved forwards and

then set down again (Fig. 56). Due to the reciprocal postural dependence, decorticate animals, a short time after ablation and in spite of the absence of preparation for standing, always set down the paws correctly when running.

In man, a similar reciprocal postural and fixation dependence exists between the joints of the lower extremities and between the hip and knee joints. When the leg of a man lying supine is lifted by the heel, the knee joint is flexed when the hip joint is flexed more than 90°. If one tries to passively counteract the associated movement of the knee joint, one encounters on further flexion of the hip joint an insurmountable resistance which, however, instantly ceases when the knee joint is released. In conformity with observations made after operations on the central nervous system in animals, the reciprocal postural and fixation dependence is variably affected by pathological conditions. In meningitis, for instance, it is increased; in tabes, mongoloid idiocy, polyneuritis, etc. it is lowered;

In summing up, we see that the following factors participate in the production and maintenance of supporting tonus :

    I.   exteroceptive stimulations, emanating from the sole, which are produced by contact or pressure;

    II. proprioceptive stimulations derived from the flexor muscles of interphalangeal-, wrist- and toe joints which are brought about by stretching of these muscles;

    III. proprioceptive stimulations which are caused by stretching of the extensor muscles of elbow- and shoulder joints and tarsal-, knee- and hip joints;

    IV. the reciprocal postural dependence of the proximal joints.

The principal role in the maintenance of support tonus is played by the proprioceptive stimulations brought about by muscle stretch; after cessation of exteroceptive stimulations, the animals are still able to stand. Recently, Liddell and Sherrington (173) have stressed the importance to the standing function of myotatic reflexes caused by proprioceptive stimulation due to muscle stretch. The effect of these reflexes, however, should be sharply distinguished from that of the

Fig. 56. Correction of a misplaced foot by reciprocal postural dependence of the
joints of extremities in the decorticate dog Fuchs. 1. The left forelimb of the
animal is passively set down on the ground on the dorsum of the foot. 2. On
moving the limb forwards at the shoulder joint, the wrist joint is extended,
the digits move dorsally so that the paw (3) is set down in a correct posture
with the sole on the ground. 4. The left hindlimb is set down passively with
the back of the foot on the ground. 5. In passive flexion of the limb from the
hip joint, the toes move dorsally.   6. Thus the limb, when released, is set
down correctly on the ground.

supporting reactions. Liddell and Sherrington observed in decerebrate cats that a stretching of an extensor muscle, particularly by the
knee extensor (m. quadriceps) causes a reflex contraction of the
stretched muscle. The contraction still occurs after severance of the
sensory cutaneous nerves and also when all the other muscles of the
extremities are detached from their insertions; it is, however, always
absent after section of the posterior roots of the spinal cord belonging
to the stretched muscle. Stretching causes a phasic reaction. By

maintenance of the stretched condition, a tonic reaction appears ("postural reaction" or "static stretch") which sometimes lasts for as long as 6 minutes. After a total transverse lesion of the spinal cord, only the phasic reaction usually recovered[1]. With a definite initial tension, the reflex contraction is already considerable when the stretch amounts to only 1% of the length of the muscle.

The myotatic reflexes of decerebrate animals described by Liddell and Sherrington thus agree in some points with the proprioceptive supporting reactions both are produced by muscle stretch, both show with persistent stretching a tonic continuance, remain in existence after elimination of the sensibility of the skin, disappear after severance of the posterior roots of spinal cord, belong to the extremities and are impaired by total transverse lesion of the spinal cord. But they differ from each other in many points, for instance myotatic reflexes occur only after stretch of extensor muscles while the reflex contraction of the proprioceptive supporting reactions are brought about by the stretching of the *flexors* of phalanges. In myotatic reflexes, only the stretched *muscle contracts*, on stretching even only a part of a muscle[2]. In the case of proprioceptive supporting reactions, not only the stretched muscle but also the remaining muscles of the extremities show a reflex contraction.

The stretching of a flexor muscle, for instance m. semitendinosus, causes no myotatic reaction of this muscle but even inhibits the myotatic extensor reflex. According to Liddell and Sherrington, a passive extension of the knee joint brings about a cessation of the myotatic contraction of the quadriceps, due to the stretching of the knee flexors. On the other hand, the supporting reaction of the hindlimb causes anextension of the knee which is not accompanied by a decrease in tension but, on the contrary, by an increase in tension of the thigh muscles.

---

(1) In a publication which appeared after the preparation of this manuscript, Denny-Brown and Liddell (52) report that in further investigation on dogs with severed spinal cord, it appeared that the tonic continuance on stretching of the quadriceps was poorly sustained owing to low level of spinal transection. On stretching the gastrocnemius, however, a prolonged stretch reflex was distinctly present.

(2) Liddell and Sherrington observed that, on stretching the m. quadriceps of one side, an associated contraction of the same muscle of the crossed side sometimes occurred.

In the opinion of Sherrington, only the extensor muscles[1] are stretched in the "standing" of decerebrate animals and only these muscles are capable to counteract gravity, since only they show myotatic reflexes. In the standing of intact as well as decerebellate or decorticate animals, not only the extensor-, but also the flexor muscles are stretched and the tension of the flexors is surely of greatest importance for the maintenance of the various standing postures in man and animal.

When, for instance, a dog is standing with its carpus or tarsus directed backwards, the m. triceps feel flaccid, the m. biceps, however, feel hard and contracted. Nevertheless, a dog is very well able to stand with its foot directed backwards and can even withstand a rather strong pressure on its back without its knees giving way. In this limb posture, the strained ms. *flex.* carp. uln.; *flex.* carpi rad.; *flex.* digit. prof. and *flex.* digit. sublim. (flexors) originating on the epicondylus humeri int. (Fig. 47) probably participate in the fixation of the elbow in an extended condition. A man standing with the upper part of his body bent forwards has no tension in the ms. quadriceps; it is easy to passively move the patellae to and fro. After a bilateral total paralysis of the quadriceps, a human is able, even with bent knees, to stand, to run and to climb stairs (Fig. 57).

Just as by the magnet reaction and by the supporting reaction, the flexed muscles are also *stretched* and fixed in a stretched position; the stretched muscles are strained and *shortened*. However, the reaction caused by a static stretch is not quite identical in its effect with that produced by contact. When a decerebellate dog in a dorsal position with the muzzle directed vertically upwards is touched on the sole of the hindpaw, the limb is maximally extended and the extensors also show, in addition to an increase in tension, a maximal contraction. If the paw is under static stretch, i.e., a pressure is exerted on the sole, the extensors of the hip-, tarsal and knee joints are stretched, i.e., the muscles are lengthened and the extended posture decreases. The increase in contraction produced by stretching will indeed prevent much stretching, perhaps even lessen stretching, but will not stop it and the flexion displacement will decrease more and more with in-

---

(1) As shown in yet unpublished investigations made in collaboration with Dr. S. Hoogerwerf, static stretch of the forelegs in decerebrate cats not only produce the action currents of the triceps, but also of the biceps.

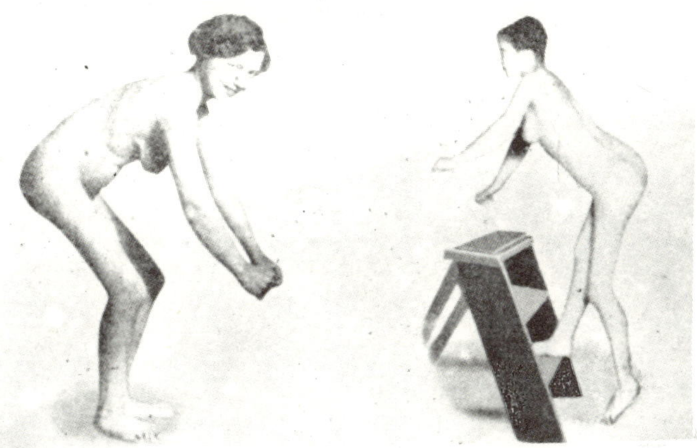

Fig. 57. Girl with total paralysis of the knee extensors (bilateral paralysis of quadriceps). a. In an attempt to bend the knees. b. Climbing the stairs (from F. Lange, Lehrbuch der Orthopädie, Fig. 88. P. 399. Verlag von Gustav Fischer, Jena 1914.)

crease in pressure. When a certain pressure is exerted on the sole of a maximally extended limb the extended posture will decrease and the length of the extensors increase till the fixation contraction is equal to the applied pressure. Thus, the fixation contraction is not strongest in the maximal stretch posture but in a certain intermediate position[1].

When a totally *flexed* hindleg comes under static stress, it is extended and, as with touch stimulus, the extensor shortens but not to a maximum contraction, particularly under heavy pressure. As a result, not a maximum but an intermediate posture occurs. *Thus, the strong static stress on a flexed leg causes a stronger contraction of the extensor muscles, but a lesser extension than on simple contact.*

---

(1) This agrees with Sherrington and Liddell's observation that the reflex tension of extensors increases with the increase of stretching, i.e., with the increase of muscle length. The conditions in static stretch are complicated by the fact that by the pressure on the sole not only uniarticulated but also multiarticulated muscles are stretched, which are at the same time extensors of one and flexors of another joint. In flexion of the tarsal joint by pressure on the sole, for instance, the extended m. gastrocnemius, the extensor of the tarsal joint as well as flexor of the knee will pull the knee joint in a flexed position due to its topographic- anatomical arrangement. The flexion of the knee joint will increase due to the reflex contraction of the extended gastrocnemius, however, it will be inhibited or lessened by the reflex tension of the quadriceps. By this process, the quadriceps are stretched by the pressure on the sole as well as by the passive extension. By means of these different factors, the knee joint is fixed in a certain intermediate position.

102

In agreement with this fact, the joints of a dog's extremities do not show a maximum extension in standing and the extensor posture decreases with a loading of the back although the extensor muscles have a stronger tension.

Static stress does not always cause a slighter extensionthan contact, for under certain conditions a contact produces only an incomplete, slight extension. On static stress when the pressure is not too great, the extension distinctly increases (Fig. 58). The increase is caused by the greater intensity of the response to touch and by the reflex contractions produced by the change of posture of the phalanges. In comparison with this influence, the passive extension of extensors of knee- and hip joints is insignificant.

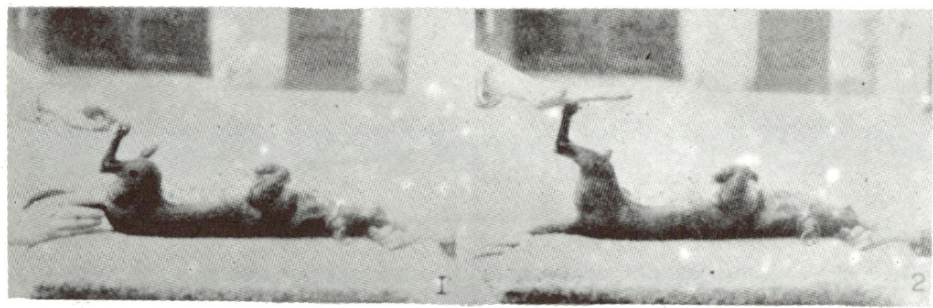

Fig. 58. Decerebellate dog Piccolino in a dorsal position with muzzle extended horizontally.   1. In this posture of the head, the hindlimb on touching the sole only shows a slight extension.   2. On static stress with not too strong a pressure, a distinctly stronger extension occurs, the muscles of the extremities are more stretched and the extensor muscles more shortened than on a simple touching of the sole of the foot.

Static stress on the forelegs causes extension of wrist and elbow joints while the upper arm is moved forwards at the shoulder joint almost to the intermediate position. In addition, contraction of the muscles of the scapula and the anterior part of the vertebral column is brought about (Fig. 64) and also in certain cases, of the neck. The fixation of joints of the extremities due to static stress is so strong that the forelegs of intact or decerebellate dogs in standing position do not buckle even under a powerful pressure (Fig. 59). The animals are able even to stand on one foreleg .with the back loaded with a sack of sand amounting to almost the weight of their bodies without the leg giving way (Fig. 59, No. 5).

The fixation of the upper arm at the shoulder joint, too, is distinctly stronger than in response to touching the sole only. Passive movement of the upper arm backwards, forwards, linwards and outwards encounters the greatest resistance and every movement, even if ever so slight, is accompanied by a corresponding movement of the scapula.

The scapulas, too, are particularly strongly fixed and it is almost impossible to shift them on the thorax. The vertebral muscles have also a stronger tension on static stress than on a simple contact with the sole of the foot. This is most clearly demonstrated in a dog held suspended in the air in a ventral position; the hollow and sagging back becomes particularly rigid and straight on static stress, in decerebellate dogs it is usually curved convexly (Fig. 60).

In the dorsal position, the stronger influence of static stress on the forepaw can be sometimes observed in the contracted condition of the cervical and dorsal muscles (Fig. 61).

In static stress on the hindlegs, tarsal- and knee joints are stretched and the thigh is moved backwards at the hip joint almost to the intermediate position. In decerebellate dogs, it will usually exceed this (Fig. 46, No. 1, Figs. 28, 31). See also the position of hindlegs in standing decerebellate dogs (Fig. 44).

At the same time, the postures are so strongly fixed that one can only with difficulty bend the hindlegs by pressure on the pelvis (Fig. 59). Passive movement of the thigh forwards, backwards, outwards, or inwards encounters a distinct resistance and is accompanied instantly by associated movements of the pelvis.

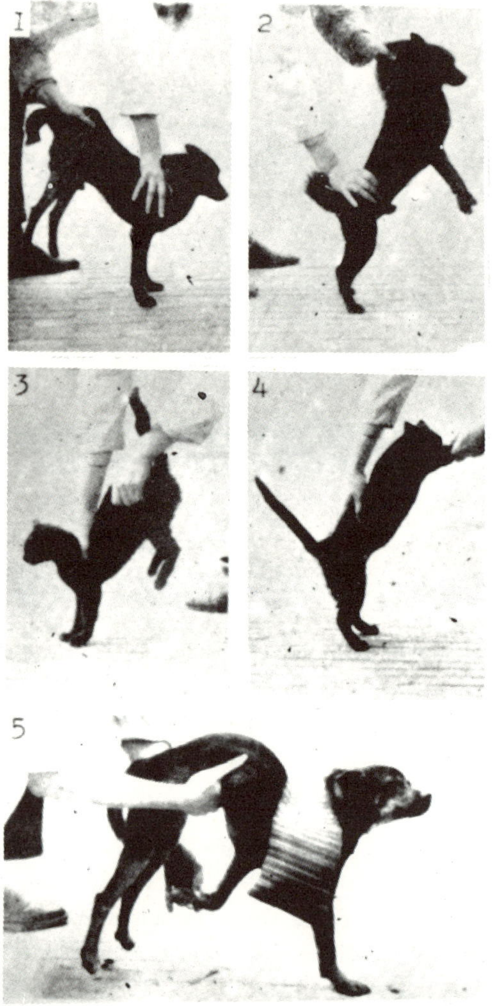

Fig. 59. 1 and 2. Decerebellate dog Erik. 1. The animal is set down on its forefeet. Under heavy shoulder pressure, it is practically impossible to make the legs give way. 2. The animal stands on its hindlegs. Heavy pressure on the pelvis encounters an almost insurmountable resistance. 3 and 4. Decerebellate cat Pierrette. In this animal also and under similar conditions, a strong resistance against pressure on the shoulder and pelvis is shown. 5. Decerebellate dog Moor, standing on one forelimb the animal is able to carry on its shoulder a $5\frac{1}{2}$ kg sandbag without the limb giving way. The body weight of the dog is $9\frac{1}{2}$ kgs.

Fig. 60. Decerebellate dog Erik. 1. The animal is held in the air in a ventral position. On loading the back with a sandbag, the limp back has sagged. 2. On touching the soles, the back becomes straight and stiff. 3. With static stress (upward pressure) on the paws, the contraction of the vertebral muscles distinctly increases, the back curved convexly upwards. See also Fig. 27.

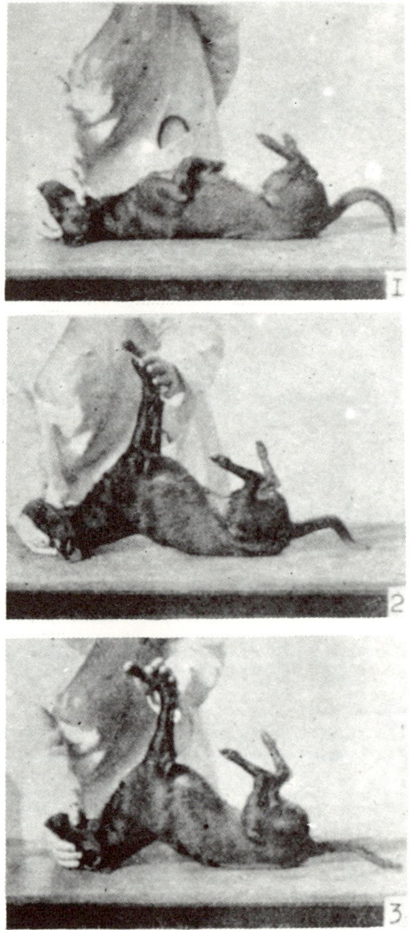

Fig. 61. Influence of static stress on the forepaws on the tension of cervical and vertebral muscles in dorsal position of the animal.   1. Decerebellate dog Piccolino in a dorsal position with he muzzle pointed upwards at an angle of 45° to the horizontal. Neck and back lie flat on the support.   2. On passive extension of the foreleg joints and touching of the soles, the long back muscles in the thorax and the neck dorsiflexors contract.   The lower cervical and the upper thoracic spinal column are lifted from the support and the head pulled caudally. Cervical and dorsal part of the spine form an angle of 135°.   3. With pressure on the soles of the feet, the tension and change of posture increase even more.   Cervical and thoracic part of spine now almost form a right angle.

In addition, static stress on the hindlegs gives rise to a tension in the muscles of the pelvis, vertebral column and sometimes of the neck, as in the case on touching of the soles of the feet but to a much stronger degree. The contraction of the lumbar muscles is particularly well developed. By this tension, the pelvis is moved dorsally in relation to the caudal intervertebral joints of the lumbar part of the spine, but only when the hindlegs are not passively fixed and can move backwards. When this is prevented, a dorsally directed movement of the caudal half of the vertebral column occurs due to the tension of the lumbar muscles, the back is straightened or convexly curved upwards as is the case of decerebellate dogs.

Finally, static stress on one hindleg, in the certain cases, brings about a decrease in stretch and support tonus of the crossed hindleg. For instance, when a dog is held in the air suspended head upwards and static stress is applied successively to one and the other hindleg, there sometimes appears in intact as well as in decerebellate dogs a decrease of support tonus in the leg first stressed when a strain is put on the second. This phenomenon is particularly distinct in unilaterally decerebellate or unilaterally decorticate animals (Fig. 62; see

Fig. 62. 1. Dog Fuchs (decerebellate on the right side) suspended in the air head upwards. With static stress, the right hindpaw shows a strong supporting tonus and offers vigorous resistance against the hand pressing the sole. 2. With static stress on the sole of the left hindpaw, the resistance suddenly ceases.

also Fig. 34 and Fig. 37, No. 1 and 3). After the unilateral extirpation of cerebral cortex, this decrease in support tonus is particularly strong in the paw contralateral to the extirpation, while in the other hindpaw it will at most be only slight.

The decrease in supporting tonus on static stress of the opposite hindleg appears particularly strong when the latter has been in a flexed position prior to the stress. In this case, static stress brings about an extension of this limb and this stretching movement probably furthers the decrease in supproting tonus in the opposite hindleg (crossed flexor reflex). A static stress on an extended hindleg also causes a decrease in supporting tonus in the crossed limb. In this case the decrease is not, as perhaps in the first case, a secondary but a direct reflex result of the static stress.

We have now seen that static stress on the limbs causes a contraction of the dorsiflexors of the neck (Fig. 31, No. 4 and 6). These dorsal muscles, too, according to circumstances, react differently to static stress in the limbs. Sometimes the convex curve of the back is abolished (Fig. 32), sometimes a concave one (Fig. 60, No. 1 and 2) and often a concave curve is changed into a convex one (Fig. 60, No. 1 and 3). The same stimuli thus cause, according to circumstances, different changes of tension in the cervical and vertebral muscles (examples of the "coordination" of Magnus). Strangely, asymmetrical tension of these muscles has never been observed, not even on static stress on one paw only or both paws on the same side.

The question that now suggests itself is whether static strain cannot, under any circumstances, cause an asymmetrical tension of these muscles. This question is of great importance to understanding the maintenance and the recovery of balance. When an animal is set down in an inclined posture, for instance on a support that is tilted to the right, it will turn its head towards the left and a concavity of the spinal column to the left appears (Fig. 112). Are the asymmetrical muscle tensions here brought about by the fact that an unequally strong counterpressure is exerted on the four limbs by the inclined supporting base ? An old observation by Magnus and De Kleyn on unilaterally labyrinthectomized rabbits is worth mentioning in this connection. They observed that the rotation of the head made by animals in standing and sitting increases sharply when the animals are lifted from a surface and held in the air in a ventral position, and instantly decreases when they are set down on their legs. Static stress in these animals thus causes a rotation of the head towards the normal position. However, this is no proof that static stress also in intact animals produces an asymmetrical tension of the cervical muscles. It cannot be excluded that the muscle centers of the left- and right rotators of the neck have become differently responsive to the stimuli emanating from the limbs due to the unilateral labyrinthine extirpation.

Magnus observed further that an asymmetrical tension of the neck muscles could be produced by the unilateral stimulation of sensory nerves of the body surface. When a labyrinthectomized, blindfolded animal is held in the air by the pelvis in a lateral position (Fig. 230) the animal will let its head droop in a lateral position. However, as soon as the animal in the same lateral position of the trunk is laid down on a supporting surface, the head rotates instantly to the left into the normal position (Magnus' body righting reflex on the head, Fig. 230). Magnus was able to demonstrate that this rotation was brought about by the one-sided counter pressure of the body on the surface. Asymmetrical contraction of the longissimus dorsi muscles can also be released by one-sided counterpressure. When an animal is laid in a lateral position on a surface and its head also fixed in a lateral position, the counterpressure will instantly cause a rotation of the trunk into a ventral position (Magnus' body righting reflex on the body, Fig. 229). However, an asymmetrical contraction due to unilateral static stress or touching of the sole of the foot was never observed, in ventral, or in lateral position in the air, and only slightly in the dorsal position on a supporting surface (Figs. 28, 30, 33 and 40). Neither does it occur on turning the head or the pelvis, even when only the outside or the inside of the sole is touched or placed under static stress. Asymmetrical tensions of the vertebral muscles due to static stress can indeed be observed, for example, when a dog held in the air in a lateral position by both forepaws and the tail is lowered towards the ground. The dangling hindleg is extended as soon as it touches the ground, the pelvis then turns into the normal position, while at the same time the extended hindleg is moved laterally (Fig. 63). In this case the rotation of the pelvis is not brought about by the asymmetric tension on the muscles of the lumbar spine and pelvis. By extending the hindleg, the rear part of the body is raised and moved laterally and at the same time is pulled into an adduction posture by its weight (Fig. 63, No. 2). By this adduction posture a lateral

Fig. 63. Decerebellate dog Piccolino with blindfold held by forelimbs and tail in the air in a lateral position. 2. The animal is lowered till the dangling left hindpaw touches the ground. Instantly the left hindlimb is extended and holds the body upwards to the left. The body weight causes an adduction of the left hindlimb. 3. Due to the adduction posture (see Chapter XIII) the left hindlimb makes a step laterally and at the same time the pelvis, trunk and head turn into a normal posture.

correcting movement of the limb is produced (leg-slackening reaction, see Chapters XI and XIII) accompanied by a passive associated movement of the pelvis. Other turning and side leteral movements of the head, neck or of the vertebral column by static stress on the limbs also occur which can be shown to be always dependent on changes in position of the legs in the shoulder- or hip joints. Asymmetrical tension of the cervical or vertebral muscles as a direct result of static stress on the limb could not be established.

It is remarkable that in simultaneous static stress, both hindlegs are stressed simultaneously and both not only show a similar supporting tonus but also that the sum of the maximum force of supporting tonus for both hindlegs amounts to twice the maximum support tonus of one hindleg. Thus, in simultaneous stress, the supporting tonus is not reciprocally reduced.

To sum up, the investigations showed the following changes caused by static stress :

A. In stress on the forelimb :

    1. extension and fixation in extended position of the stressed limb;

    2. fixation of the scapula;

    3. change of posture and fixation of the vertebral column and

    4. in certain cases, a change of contractile tension of the cervical muscles.

B. In stress on the hindlimb ;

    1. extension and fixation in the extended position of the stressed limb;

    2. change of posture and fixation of the pelvis;

    3. change of posture and fixation of the vertebral column;

    4. in certain cases a change in contractile tension of the cervical muscles;

    5. in certain cases an increase in stretch- and supporting tonus of the opposite limb;

    6. in definite cases change of tonus in the forelimb (Fig. 35, 36).

Thus, static stress causes the same reflex postural changes as does a simple contact the changes, however, are much more prominent. Only the extension of the limbs is less vigorous with a strong pressure on the sole than with a simple touching.

From the observations we have presented it should be clear that precise investigations concerning the strength of muscle tonus are exceedingly difficult. Through simple processes like the response to the slightest touching of the sole of the foot or extension of the phalanges, the tension of almost all the muscles of the body can change and a great many centers in the central nervous system become active. In general, the different components of the positive[1] and negative supporting reactions in intact dogs can be easily observed when they are not too strongly inhibited.

Cats are less suited for these investigations. As soon as they are grasped by the investigator they draw back, draw in their limbs and show no trace of supporting tonus. In the dorsal position they also either keep their limbs flexed or make defensive movements. In cats the supporting function is less developed than in dogs; they are usually less able to carry loads on their back and they run and stand almost always with strongly flexed limbs. Nevertheless in cats, a strong supporting reaction can occur, for instance, when they are set down on their fore- or hindlegs only and a pressure is exerted on the shoulders or pelvis (Fig. 59). The rabbit has a distinct supporting reaction only in the forelegs. In *new-born animals*, dogs, cats and rabbits, supporting reactions are absent. In the first 2 to 3 weeks .after birth, new-born dogs do not react with an extension of the limbs on touching of the sole, or on extension of the phalanges and only as little on pressure on the sole. The passively extended leg, too, instantly gives way in cases of static stress. In the first weeks, the animals are not able to stand or run, but lie and crawl with their bellies on the ground.

In moderately strong narcosis the supporting reactions are present and in deep narcosis they disappear.

After the *total and unilateral extirpation* of cerebral cortex, the proprioceptive positive and negative supporting reactions are found in

---

(1) The behavior of exteroceptive components, the magnet reaction, in different species of animals, also after operations in the central nervous system, have already been mentioned above (p. 73-99).

all four limbs. In dogs they are usually present again in all four legs within the first two hours after extirpation of one half, as well as after extirpation of the other half of the cerebrum carried out a few weeks later. The supporting reactions of decorticate dogs, however, are not quite normal. First of all, no distinct exteroceptive component (magnet reaction) can be observed, secondly as we shall see later (Chapter V) the strength of the supporting tonus of the hindlegs is diminished.

The investigation of positive supporting reactions in *decerebrate animals* is very difficult because of the extensor rigidity of the legs. As we have seen, the magnet reaction is absent. On the other hand, a pressure on the sole, particularly of the forepaws, usually causes a distinct increase of fixation in the extended position[1]. Passive flexion of the phalanges sometimes causes a decrease, but not a disappearance of the extensor tonus (Fig. 64) and not a definite flexed posture.

Fig. 64. Decerebrate dog with moderate rigidity (dog B of Fig. 6) with the dorsum of toes (1) and (2) on a support.   1. Although the toes are plantar flexed, the limbs are able to carry a counter pressure of almost 7 kgs without giving way.   2. With the wrist joint flexed in a right angle and passively plantar flexed toes (compare posture of toes on 1 and 2), the joints of the elbow and shoulders are able to carry a pressure of more than 6 kgs without giving way. Body weight of dog 6.5 kgs.

---

(1) As was shown by experiments carried out together with Dr. S. Hoogerwerf, pressure on the sole of the forepaws as well as dorsally directed movement of the digits cause a distinct increase of action currents in both the biceps and triceps.

Thus, the negative supporting tonus appears distinctly impaired.

Maximum passive flexion of the wrist joint in decerebrate animals is accompanied by a forced associated flexion of the elbow joint (similar to animals in rigor mortis) (Fig. 48). Further flexion of elbow- and shoulder joints may then follow but still encounters a distinct resistance even when the digits are passively flexed. The extensor tonus can indeed disappear and the limb is drawn into a flexed posture when toes are pressed strongly into a flexed position. Similarly, on pinching the toes, Sherrington's ipsilateral flexor reflex appears.

Investgations on the behavior of supporting reactions after the *total transverse section of the spinal cord* can be made only with greatest difficulty owing to the continuous alterations in tonus in these animals. Through the passive postural changes of the pelvis, through alteration of the position of the animal, by stimulations of the surface of the caudal half of the body, stimuli arising in the internal organs as for instance from the bladder, etc., there sometimes appears in the hindlimbs of spinal animals a distinct extensor tonus, usually of short duration, and sometimes a flexor tonus. Thus the hindlegs show, for instance during urination, a flexed posture with flexor tonus and after urination an extension of short duration; in the dorsal position the paws are drawn in into a flexed posture; but when suspended head upwards, however, they are usually extended and offer a distinct resistance to passive flexion.

Our investigaton of spinal animals resulted in the following conclusions :

1. In static stress on the flexed hindlegs when the animal lay in a dorsal position there never occurs an extension.

2. Static stress on the extended hindleg never causes a tonic fixation in the extended position which persists as long as the stress is maintained.

A dog with a severed thoracic cord was able to carry the posterior part of its body for half a minute on its hindlegs which were at first passively extended and then set down on a surface with the soles in the position required for standing (even with lifted head, when the weight resting on the hindlegs was not reduced by a downward flexion of the head and tension of the muscles of the neck and the anterior part of the back). After a short time, however, (the longest duration was 30 seconds), they gave way.

3. Touching of the sole never causes a typical tonic magnet reaction, but sometimes a short extensor contraction (extensor thrust). It seems, however, that the position of the toes exerts a certain influence on the muscle tension in the spinal dog. The passively extended hindleg with dorsally directed posture on the toes shows sometimes a distinct, though short, resistance to passive flexion which instantly disappears on flexion of the toes. Passive flexion of the toes (without pinching!) even causes sometimes a flexion movement of the hindlegs. Whether these changes in tension represent a reflex component of the supporting reaction could not be established up till now due to the lack of recording the contraction of the muscles detached from their insertions. As we have previously discussed in detail, the extensor thrust is probably not a phasic component of the magnet reaction.

After severance of the posterior roots belonging to the limb any reaction on static strain is absent.

The behavior of the supporting reaction after *extirpation of the cerebellum* is remarkable. Immediately after extirpation, the limbs are rigid in some of the animals and relaxed in others. In the first case, the reactions on static strain behave as in decerebrate animals and, in the latter, the supporting reactions are totally absent. When such animals are set down on their legs, these instantly give way (Fig. 65). Neither in the air nor on a surface can extension of the legs be observed on contact or on causing static stress on the soles of the feet.

The supporting reactions return almost always within the first week, with their exteroceptive as well as with their proprioceptive components, and even become strikingly lively and uninhibited. This is also the case in rigid animals after cessation of rigidity. In these cases, static stress causes a particularly intense extension as well as a strong fixation in the extended posture so that standing decerebellate dogs are able to carry on their backs a bag full of sand weighing the same as their own bodyweight (Fig. 51, No. 5 and Fig. 59, No. 5). Conversely, the supporting tonus instantly diminishes with the cessation of the static stress and the paws can now be passively moved with ease.

Fig. 65. Behavior of supporting reaction in the first days after extirpation of cerebellum. 1-3. decerebellate dog Pim on the second day after extirpation of cerebellum. When the animal is set down on its forelimbs (1) or hindlimbs (2) they immediately give way. No magnet reaction of the paws on touching the sole. (3). 4-6. The same dog on the fourth day after extirpation. The forelimbs are able to carry the trunk (4), the hindlimbs however, not yet, (5), although on contact with the sole, they already show a distinct magnet reaction(6).

In decerebellate dogs, not only exteroceptive but also proprioceptive stimulations produce strong changes in posture and tension. Passive stretching of the wrist joint and dorsal movement of the toes (without touching the sole!) is always followed by a vigorous extension of the limb together with a considerable resistance to passive movement (Fig. 49, No. 2 and 3). Conversely, the extensor tonus usually disappears with passive flexion of the wrist joint and toes, in this case, the

Fig. 66. Decerebellate dog Piccolino in a dorsal position.　1. The rear part
of the body is lifted by the feet, the　hindlegs are totally extended.
2. On passive　flexion of the toes, the hindlimbs are somewhat flexed
in the knee- and hip joints, although　counteracted by　the weight of
the body.

limb sometimes is even actively flexed, i.e., against the effect of gravity
and in the ventral position of the animal.

As shown by these observations, in decerebellate dogs not only the
positive but also the negative supporting reactions appear particularly facile and one can say that they are increased. Decerebellate
dogs always stand with exaggeratedly extended legs (Fig. 44), which,
in running, they lift abnormally high and extend with abnormal force;
manifestations which can perhaps be traced back to the vigorous
positive and negative supporting reactions. The constant appearance
of supporting reactions in decerebellate dogs in the most varied situa-
tions (Figs. 28 and 30) is also remarkable. In intact dogs, the reactions appear less regularly, they may indeed be lively, but may also be
totally inhibited by the experimental situation, particularly on investigating animals in the dorsal position. In decerebellate animals,
however, even in the dorsal position, muzzle directed upwards,
the supporting reactions are always constantly present and lively;
touching as well as static stress then produce strong and vigorous extension and fixation of the joints of the limb. If a board loaded with
several kilograms is laid on the soles of the feet of these animals, they
usually extend the legs, lift the board and hold it high (Fig. 67). The
decerebellate badger-dog Piccolino, whose weight was only 4½ kgs,
carried in a dorsal position with one hindpaw a board loaded with
2 kgs and with both paws one with 3 kgs (Fig. 92).

In decerebellate dogs, the changes in contraction of the vertebral muscles and the change in posture of the back caused by them are also remarkably clear. While in intact dogs held in the air by the tail and the skin of the neck, the concavity of the hollow back is compensated in static stress on the legs, in decerebellate dogs there even usually appears a convex upward curve of the back (Fig. 27, No. 3 and 4, Fig. 60), even when the back is burdened with a heavy sandbag. On standing, the backs of decerebellate dogs are almost curved upwards (Fig. 44) and also strongly fixed. Passive movements of the neck of decerebellate dogs also encounter a strong resistance. Due to the vigorous fixation of the cervical and thoracic spine, decerebellate dogs appear extremely rigid on standing and running.

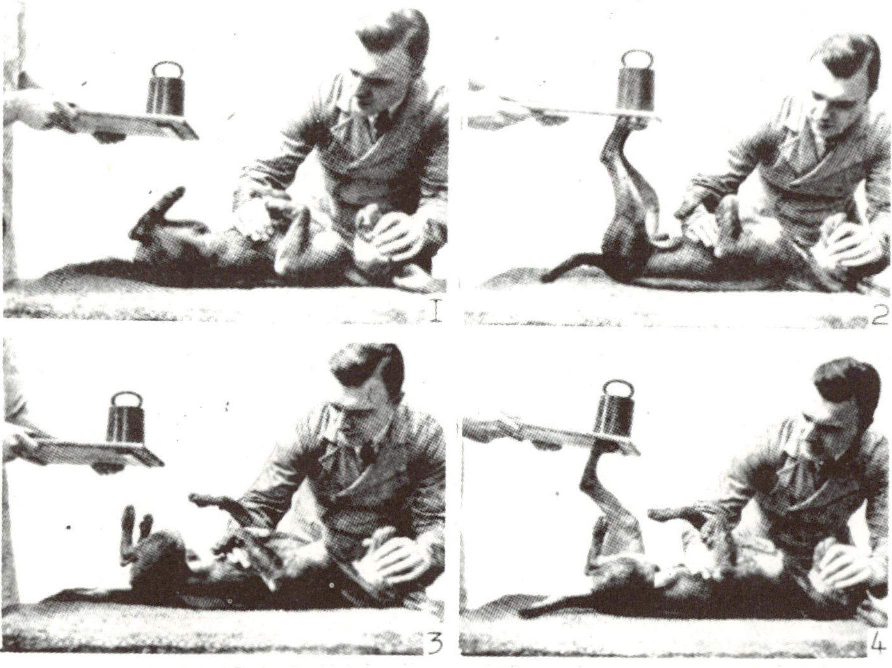

Fig. 67. Supporting reactions in the decerebellate dog in a dorsal position. The dog Moor in a dorsal position on a surface, muzzle pointing vertically upwards, kept its hindlimbs totally flexed. (1) on loading the soles with a board weighing 2 kgs, the hindlegs are extended, lift the board and keep it lifted in the air. (2) When the board is placed on one paw only it is also able to lift and carry it (3 and 4).

"Ils marchent comme soudé et l'animal n'a plus sa souplesse" (They walk as if their joints were soldered and no longer have their agility) says André-Thomas. The animals also show a strange combination of rigidity and abrupt and exaggerated movements, and a motor restlessness which does not permit them to stand quietly but induces them constantly to make dance-like movements.

The strong fixation of the vertebral column in decerebellate animals is also evident when the head is passively flexed downwards with a rapid movement. The rear part of the body is then sometimes lifted from the ground as with a lever (Fig. 93). When one tries to lift the head with one hand by the lower jaw it will not be moved upwards like the head of an intact dog so that the hand slides off, but head and trunk are lifted together as if they were one compact mass (Fig. 68); neck and back are here quite stiff and usually curved convexly upwards. When the rear part of the body is now lifted, for instance, by the tail, the fixation diminishes as soon as the hindlegs leave the ground and the back becomes hollow and relaxed.

Fig. 68. The decerebellate dogs Wolf and Piccolino. When one tries to lift the head of the standing animal by hand from the lower jaw, usually the whole trunk is lifted as if head and thorax formed a solid mass.

Thus static stress in decerebellate dogs causes particularly widespread changes in posture and strong reflex contraction of the muscles of the extremities and spine. The muscles feel hard on palpation and cause a vigorous fixation and a strong supporting tonus. Not a trace of atony can be observed in standing decerebellate dogs.

In the following chapters we will see how the occurrence of supporting reactions is automatically influenced to a strong degree by the position of the head, the back, etc. As in dogs, the supporting reactions are more distinctly present in decerebellate cats than in intact animals, although investigations are usually rendered difficult by defensive movements. In pigeons also, findings after partial extirpation of the cerebellum conform in general to the observations made on decerebellate dogs. Bremer and Ley (31) extirpated the cerebrum in pigeons and, an hour later, destroyed on both sides the lobus ant. cerebelli. They observed that the legs hung down in a half extended position (Fig. 69, No. 1) when the birds were held in the air in a ventral position. However, when the pigeons were set down on a base,

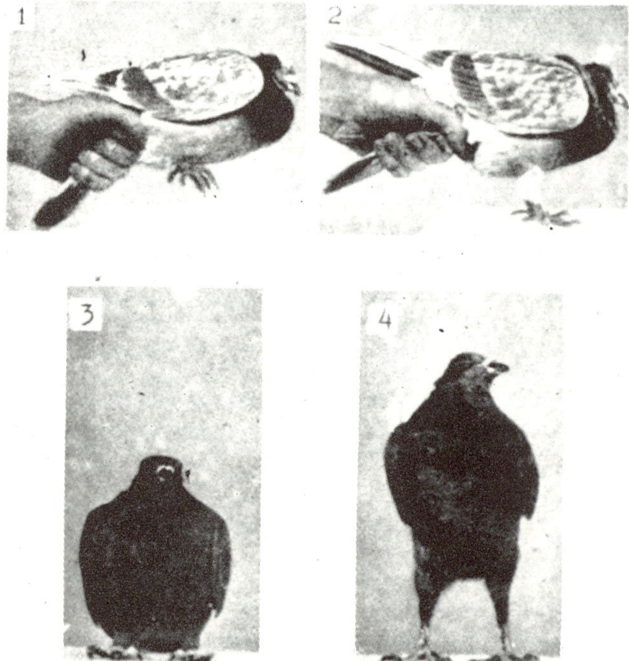

Fig. 69. 1 and 2. *Thalamus* pigeon, lobus ant. cerebelli bilaterally destroyed. 1. The animal kept in the air suspended by the tail in ventral position. 2. The animal set down on a surface with its paws. 3. Thalamus pigeon, 1 hour after removal of the cerebrum in a standing posture. 4. The same animal after destruction of lobus ant. cerebelli set down on its paws ($1\frac{1}{4}$ hours after extirpation of the cerebrum, $\frac{1}{4}$ hour after destruction of lob. ant. cerebelli). According to : F. *Bremer* and R. *Ley*: Recherches sur la physiologie du cervelet chez le pigeon. Bulletin de l'Academie Royale de Medecine de Belgique. 1927. p. 60-101.

the legs instantly changed to an exaggeratedly extended posture and felt stiff[1] (Fig. 69, No. 2). In these birds, the extended posture as well as fixation in the joints of the extremities were more pronounced on static stress than in thalamus pigeons in which the lobus ant. cerebelli was intact (Fig. 63, No. 3 and 4). Groebbels (110) observed increased supporting reactions in pigeons after lesion of nuclei of cerebellum.

Immediately after *unilateral extirpation of the cerebellum* in dogs the supporting reactions are usually impaired on both sides, but they reappear promptly in the limb contralateral to the extirpation in one or two days; in the ipsilateral limb they are absent a few days longer. When these animals are set down on a supporting surface on the contralateral limbs, these can very well carry the body while the ipsilateral limbs give way under the body weight. When the contralateral limbs are passively flexed and then set down with the soles on the surface, they are extended immediately and lift the trunk while the ipsilateral limbs, when passively flexed and set down on the base, show no extension. This phenomenon is all the more remarkable as the ipsilateral limbs, in the various positions in the air and in the dorsal position on a base, are kept in an extended posture (Fig. 70, No. 1) and show exaggerated extensor tonus and resistance to passive (non-static) flexion. In spite of the increased extensor tonus, however, they are not able to carry the trunk. The explanation for this contradictory behavior is found in the varied effects of static stress. While the somewhat flexed contralateral limb reacts to static stress with a brisk extension and vigorous fixation in the extended posture (Fig. 70, No. 2), no distinct changes in tension and no evident increase of resistance against passive movements can be observed in the ipsilateral limb on static stress. (Add to this that the extensor tonus of the ipsilateral limb in the normal position of the animal is weaker than in the dorsal position, due to the influence of tonic labyrinthine reflexes).

When the animals in this stage are also set down on both hind- (Fig. 187) or forelimbs or on all four limbs, the ipsilateral limbs will give way (Fig. 185) and the animals fall to the side of extirpation.

---

(1) This phenomenon was only temporary, a month later the animals no longer showed any distinct anomaly.

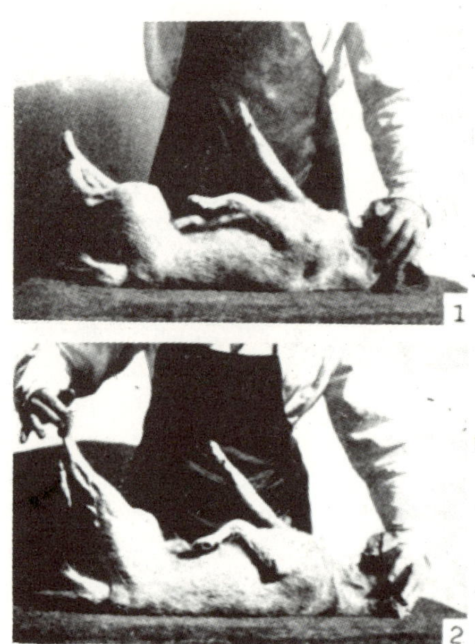

Fig. 70.   1. Dog Fox (2 weeks after extirpation of the right side of the cerebellum) in a dorsal position with the head held symmetrically to the body, muzzle pointing upwards.   The animal keeps the right paws extended, the left flexed.   2. On static stress of the left hindlimb there was an extension with fixation in extended position of this limb, but a reduction of extension of the right hindlimb.

This is not only caused by the lack of supporting reactions in the ipsilateral limbs but also, for example, by the considerable decrease in supporting tonus of these limbs on static stress on the contralateral paws (Fig. 70 and 71).   When the animals are set down from the air on a supporting surface in a ventral position, the increased extensor tonus of the ipsilateral paw instantly disappears as soon as the contralateral paw is set down on the base.

Thus, unilaterally decerebellate animals show at this stage :

1. impaired supporting reactions in the ipsilateral limbs,

2. increased extensor tonus in the ipsilateral limbs,

3. distinct supporting reactions in the contralateral limbs,

4. disappearance of increased extensor tonus in the ipsilateral limb on static stress of the contralateral limb.

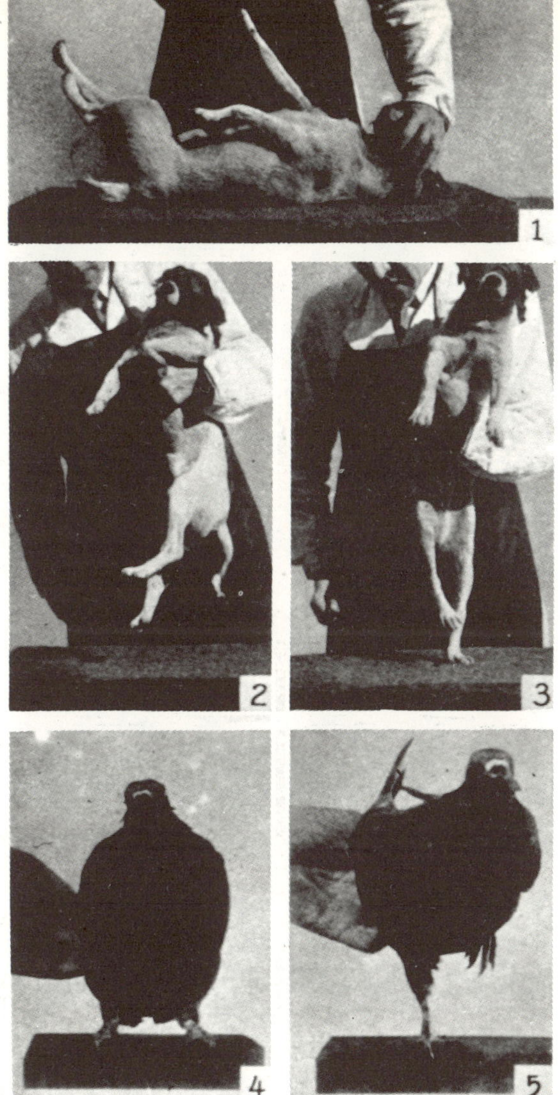

Fig. 71. 1 and 2. Dog Fox (2 weeks after extirpation of the right half of the cerebellum). In dorsal position on a surface (with head symmetrical to trunk) as well as suspended head upwards the animal keeps the right limbs extended, the left more or less flexed. 3. When the animal is set down on its hindlegs, the left hindlimb is extended and fixed in extended posture, the right, however, is drawn into flexed position. The animal supports itself only with the left hindlimb. 4. Thalamus pigeon, $\frac{1}{2}$ hour after extirpation of cerebrum held by the tail in standing posture. 5. The same pigeon $\frac{1}{4}$ hour after destruction of the left half of the lob. ant. cerebelli supported in the same way as in No. 4. The bird keeps the right leg fixed in an exaggeratedly extended position. The left in half flexion. No. 4 and 5 from F. Bremer and R. Ley: Recherches sur la physiologie du cervelet chez le pigeon. Bulletin de l'Academie Royale de Médecine de Belgique. 1927, p. 60-101.

Whether similar impairments of supporting reactions after the unilateral extirpation of the cerebellum occur in pigeons has not yet been investigated.

It is, however, remarkable that Bremer and Ley after unilateral destruction of lobus ant. cerebelli in pigeons observed that when they were held passively in a standing position, they showed the same limb posture which appears in dogs in the first period after extirpation of cerebellum when the animals are set down on the hindlegs after suspension head upwards (Fig. 71).

Gradually the magnet- and supporting reactions also reappear in the ipsilateral limbs and, usually already within the second or third week after extirpation, cause so strong a fixation that the limbs are again capable of carrying the trunk. When the dogs are brought into a dorsal position with the muzzle pointing upwards and the hindlegs are alternatively put under static stress, the ipsilateral limb sometimes shows a stronger supporting tonus and offers the strongest resistance to pressure on the sole of the foot. However, at this stage the supporting tonus of the ipsilateral limb is also greatly reduced by static stress on the contralateral limb so that, although ipsi- and contralateral limbs, each separately, are able to carry the trunk, the dogs are not yet able to stand on all fours[1].

About 4-5 weeks after extirpation, the supporting tonus is again of a normal strength, the dogs are able to stand freely and even to carry on their back loads equal to their weight (Fig. 105, No. 1). When the animals are set down on a surface alternately with one or other paw, no distinct difference appears in the supporting tonus of the left or the right limb, and by setting down on all fours the static strain on the contralateral limbs no longer causes a distinct decrease in supporting tonus in the limbs on the side of extirpation.

The proprioceptive supporting reactions as well as the magnet reactions are lively on both sides and according to the posture of the head and the position of the animals, appear stronger sometimes on one side, sometimes on the other.

After *unilateral extirpation of the cerebral cortex* in dogs, the proprioceptive supporting reactions are almost always present on both sides

---

[1] Other circumstances, too, participate in this phenomenon, as we shall see later (see Chapter XIII.).

within the first 2-3 hours; these animals are already able to stand and run again, in opposition to those that are unilaterally decerebellate. In spite of the preserved ability to stand and run, the contralateral legs show distinct impairment in distribution of stretch tone and in supporting reactions in the first days after extirpation.

When the animals are held in a ventral position (Fig. 72 No. 1)

Fig. 72. Dog Jenny (left side decorticate, bilaterally labyrinthectomized) 1 - 3. In ventral position (1) as well as suspended head upwards in the air (2) and in dorsal position on a surface (3) the right limbs show a more extended position and a more vigorous extensor tonus than the left. 4. When the animal is set down on a surface on only the sole of the left hindfoot, this limb is extended and fixed in an extended posture so that it does not give way under the weight of the posterior part of the body (the right limb is only slightly supported by one finger). 5. Conversely, if the animal is set down on the extended right hindlimb it collapses and is not able to carry the rear part of the body. In order to show that these phenomena are not produced by tonic labyrinthine reflexes, the pictures of not only a unilaterally decorticate but also of a bilaterally labyrinthectomized dog were chosen.

or suspended head upwards in the air (Fig. 72, No. 2) or laid in a dorsal position on a supporting surface (Fig. 72, No. 3), the contralateral limbs are more extended and offer a stronger resistance against passive and not static flexion, and the extensor tonus is then somewhat higher. On the other hand, the supporting tonus is reduced when the animals are set down on a surface alternately with the left and with the right leg. While the hindleg of the side of extirpation is capable of carrying the body (Fig. 72, No. 4), the contralateral limb gives way under the weight of the trunk (Fig. 72, No. 5).

Following the unilateral extirpation of the cerebrum, the contralateral limbs thus show an increased extensor tonus[1] and a reduced supporting tonus.

With a passive extension and flexion of the phalanges, the positive and negative supporting reactions in the contralateral limbs are not entirely absent, but are only weak.

In the ventral position in the air, static stress applied with one hand indeed causes a distinct extension and a vigorous support tonus in the paws of the side of the operation (Fig. 73, No. 1 and 2); however, it does not bring about a distinct increase in fixation in the contralateral, extended legs so that in this position, too, the legs on the sides of the operation offer a stronger resistance to passive pressure on the sole of the foot. In the dorsal position, on the other hand, pressure on the soles encounters the strongest resistance in the contralateral hindleg. The cause is that in this position the legs on the side of the operation (like the legs of intact dogs) show a weak supporting reaction or none at all (Fig. 73, No. 3-4), so that the influence of the abnormal extensor tonus of the contralateral leg predominates. Thus, the resistance to pressure on the sole is stronger at one time in the ipsilateral and at another time in the contralateral limb, depending on the position of the animal.

Further, it is worth mentioning that the abnormal extensor tonus in the contralateral hindleg decreases on static stress on the hindleg on the side of the operation (Fig. 74).

The impairments in the contralateral limb grow less gradually, the increased extensor tonus diminishes or disappears; the supporting

---

(1) The increase in supporting tonus varies in strength in different animals and sometimes is almost completely absent.

tonus increases so that after a few days there is scarcely any difference between the supporting tonus of the two forelegs.

The supporting tonus in the contralateral hindleg, however, is distinctly reduced for a considerable time (the longest observation extended for over 7 months). In addition, the magnet reaction is absent or at most, only a token response.

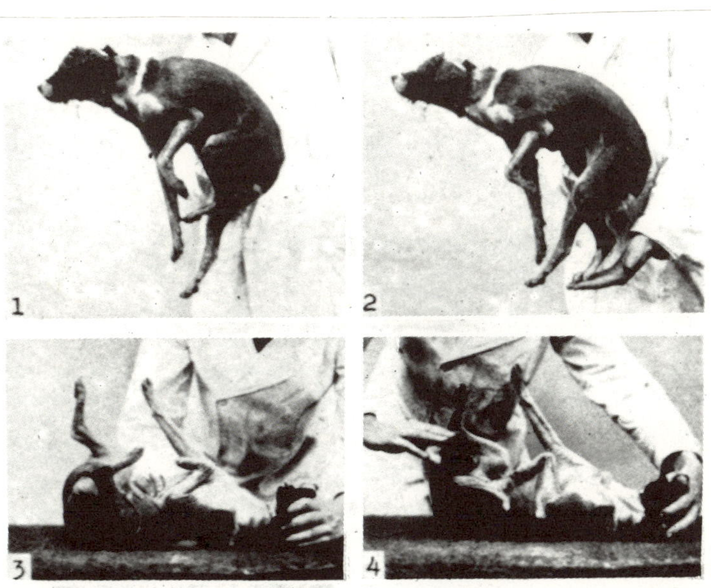

Fig. 73. Dog Jenny, left side decorticate, bilaterally labyrinthectomized in a ventral position in the air. The animal keeps the left limbs flexed, the right extended. 2. The left hindleg is extended and fixed in an extended position on static stress (through which it shows a stronger resistance against passive flexion than the right one). 3. In a dorsal position with the head symmetrical to the trunk, the animal keeps the left limbs flexed, the right extended. 4. In this posture the left hindlimb does not go into extension on static stress (only the tarsal joint is somewhat extended).

The animals show the following lasting impairments :

1. impairment or total absence of the magnet reaction in the contralateral limbs;

2. a reduced supporting tonus in the contralateral hindleg;

3. as seen in Chapter IV, impairment of the preparation for standing.

Unilaterally decorticate and bilaterally labyrinthectomized dogs showed the same changes in supporting tonus that followed unilateral decortication alone.

A totally decerebellate, right-sided decorticate dog showed the same increased extensor tonus (Fig. 74, No. 3 and 5) and reduced supporting tonus in the left legs that followed extirpation of the right half of cerebrum alone. In this animal also, the abnormal extensor tonus of the right hindleg disappeared with static stress on the left hindleg (Fig. 74, No. 4 and 6). It showed, however, probably due to the extirpation of the cerebellum, distinct dissimilarities in the following points:

1. The increased extensor tonus of the contralateral left legs remained permanent.

2. The contralateral left legs showed lively magnet reactions (Fig. 37).

3. The supporting tonus was permanently and strongly reduced not only in the left hindleg but also in the left foreleg ! (The supporting tonus of the left limbs was much weaker than after an extirpation of the right half of the cerebrum alone; the supporting tonus of the right legs seemed normally strong).

4. Touching and static stress on the feet caused a lively extension of the legs and a strong fixation in extensor posture, *also in the dorsal position.*

The lively appearance of magnet- and supporting reactions in the dorsal position corresponds with observations made after the extirpation of the cerebellum alone. Particularly striking in this animal was the influence of static stress on the right hindleg on the distribution of tonus in the left. Simultaneously with the appearance of supporting tonus in the right hindleg, the abnormal extensor tonus of the left not only disappeared but it was even vigorously drawn into a flexed position (Fig. 74, No. 4; several factors participate in this flexion, as we shall see later).

In a dog operated in three phases : *first one half, then the other half of the cerebrum and finally the whole cerebellum was extirpated,* the legs behaved in the first week after the last extirpation exactly like those of a decerebrate dog. (Fig. 79, No. 3 and 4).

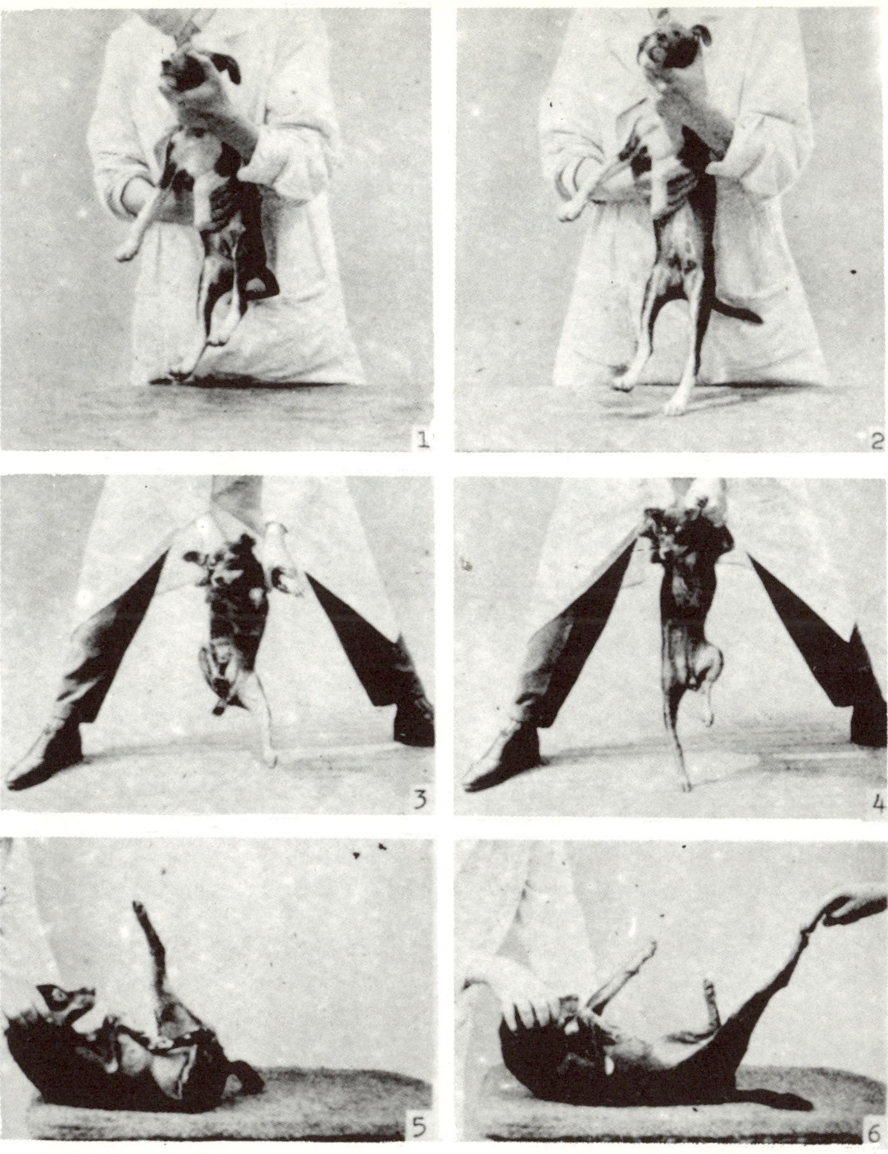

Fig. 74. Dog Black-and-white (left half of the cerebrum removed) suspended head upwards in the air. The animal keeps the right hindlimb extended, the left drawn into a half flexed position. 2. On static stress of the left hindlimb, it is extended and fixed in extended position, the right hindlimb shows a decrease of extensor tonus and is slightly flexed. 3. Dog Vici (extirpation of the right half of the cerebrum and total extirpation of the cerebellum) suspended head upwards in the air. The left hindlimb is maximally extended, the right flexed. 4. On static stress in the left hindlimb in this animal, too, the stressed limb goes into an extended posture while the right does not only show a strong decrease in extensor tonus but also is drawn into flexion (compare No. 3 and 4 with Fig. 37, No. 1 and 3). 5 and 6. In dorsal position of the animal, the left hindlimb also shows extended posture with increased extensor tonus (5) which disappears on touching and static stress on the sole of the right hindlimb (6).

Rigidity gradually decreased, in the third and fourth week, however, the extensor tonus distinctly increased. The supporting tonus, on the contrary, was reduced, particularly in the hindlegs. When the animal was set down on its hindlegs, they gave way under the weight of the body. Magnet reactions were absent, proprioceptive positive supporting reactions, however, were distinctly present.

(The animal died 5½ weeks after the last extirpation from distemper, i.e., still in the stage of shock manifestations, seen after extirpation of the cerebellum alone).

In man no systematic investigations have yet been carried out on supporting reactions of the legs. However, there are various observations which point to the fact that the reactions resemble those of dogs. In normal standing human beings flexors as well as extensors of the thigh and tibia are stressed (Duchenne de Boulogne), i.e., a distribution tonus similar to that of an intact standing animal. As in dogs, it is scarcely ever possible to force the legs of a standing man to give way under pressure on the shoulders, while this is easily possible when they are extended in a man lying on his back. It is well known that the legs of dockworkers are able to carry 200-300 kgs in addition to their body-weight without giving way and adjustment to the load takes place more or less mechanically without much thought.

Standing and running motions of sleep-walkers also strongly point to reflex influences.

As in dogs, it is evident that in man, too, the ability to stand is abolished by deep narcosis and by the total transverse lesion of the spinal cord, while in lesions of the ventral part of the midbrain the same impairments to the ability to stand are shown (decerebrate rigidity). In hemiplegics, a pressure on the balls of the toes often causes the occurrence of a foot clonus. In man, too, reflex changes in the tension of the leg muscles can be produced by pressure on the balls of the toes. Conversely, the extensor tonus can be made to disappear by plantar flexion of the toes (negative supporting reaction). This influence of passive plantar flexion of the toes was already known by the nurses of Charcot's clinic who made use of plantar flexion for the elimination of extensor spasms.

Pierre Marie and Foix observed in hemi- andparaplegics, a flexion of the leg in the hip-, knee- and ankle joints on a forced plantar flexion of the big toe. These phenomena have probably to be interpreted as expressions of Sherrington's ipsilateral proprioceptive flexor reflex. The difference between the ipsilateral flexor reflex which appears not only on forced plantar flexion but also on pinching and on forced dorsal movements of the toes, and the negative supporting reaction which is produced by non-forced flexion of phalanges of the extremities, has not yet been thoroughly analysed.

Whether exteroceptive stimulations, too, participate in the occurrence of positive supporting reactions in man is not known. In normal persons, the touching of the sole of the foot does not produce any magnet reaction.

As in animals, exteroceptive stimulations in man are not absolutely necessary for the occurrence of supporting tonus sufficient for carrying the body. When the soles are anesthetized as in cases of Leprosy, Tabes, Beriberi, etc., human beings are still able to stand and to run. After the artificial anesthetization of the sole of the foot, standing, even if unsteady, is possible. When one leans only on one leg while standing, the muscles of the other leg instantly become soft and flaccid even when the sole remains in contact with the floor. Thus the proprioceptive stimulations seem to be the most important.

Contrary to observations in cats (Pritchard), it seems that in man the stretch of the Achilles tendon does not play an exclusive principal role for the supporting tonus, because a supporting tonus sufficient to carry the body also appears when the Achilles tendon is slackened by static maneuver. When the heels are set on the ground with flexed ankle joints one feels how the thigh muscles are strained as soon as one places oneself on the balls of the heels. Thus, these tensions appear although the bodyweight tends to extend the ankle joint and to abolish the stretched state of the Achilles tendon. With a totally severed Achilles tendon man is still able to stand well. When one places oneself on the balls of the heels with a dorsiflexed ankle joint, the muscles of the soles of the feet and the plantarflexors of the ankle joints are stretched. From which of these two groups of muscles emanate the stimulations producing the supporting tonus in this case has not yet been investigated. The muscles of the soles of the feet are also stretched in normal standing position and also on standing on the balls of the toes. It is therefore probable that proprioceptive stimuli from these muscles play an important role, although stretch is not absolutely necessary for the occurrence of a sufficient supporting tonus because persons with flat feet (Negroes) are also able to stand and to carry loads.

When one places oneself on the balls of the toes so that the Achilles tendon is stretched and the flexors of the soles of the feet are stretched by the counterpressure of the surface, a supporting tonus also appears; it seems, according to circumstances, that the proprioceptive stimulations producing the supporting tonus emanate from different muscles.

(In the development of foot clonus the principal role is probably played by stretch of the Achilles tendon. In addition to dorsal movement of the foot it can also be produced by tapping on the tendon. On the other hand, a passive plantar movement of foot and toes sometimes also causes foot clonus. In this case, the stimulations assumedly emanate from the stretched dorsiflexors of the ankle joint and toes).

Under pathological conditions in man, too, abnormally strong contraction of the leg muscles on static stress has been observed. Thus a patient suffering from Wilson's disease instantly got a cramp in the muscles of the calf as soon as he was placed on his feet, so that he always stood on his toes (Howard and Royce). Based on these observations, it is probable that in man also the stretch of the muscles of the lower leg can cause a reflex contraction of the stretched muscle (myotatic reflex) as well as of other muscles.

Myotatic reflexes of the muscles of the lower leg can be particularly distinctly observed when a man standing on both legs moves alternately forwards and backwards on the ankle joints. In the forwards movement of the body, the calf muscles are stretched and the heels are lifted from the ground due to the reflex contraction of these muscles. When the body moves backwards the dorsiflexors of ankle and toes (m. tibialis ant., m. ext. digit. log., m. ext. hallucis long. etc.) are stretched and due to the reflex contraction of these muscles, the muscle tendons stand out, the balls of the toes are lifted from the ground and the toes move dorsally. Foix and Thevenard observed in two patients with cerebellar atrophy how the body rocked forwards and backwards in broad based standing and that in this case the tension of the extended muscles was particularly strong. The tension became stronger and stronger, the balls of the toes and heels were lifted alternatingly higher and higher so that the patient had finally to shift his feet in order not to fall. Normal persons, too, never stand totally still but rock slightly forwards and backwards.

The fact that the stretch of the Achilles tendon probably also causes reflex contraction of the antagonists in extrapyramidal disease in man is shown by the contraction of m. tibialis ant. on the strong passive dorsal movement of the foot (paradoxical contraction of Westphal). As in animals, static stress normally causes an influence on the tension of the pelvic and vertebral muscles in man. This can be clearly felt when one palpates the psoas muscle during alternate standing on the left and on the right leg. While standing on the right leg the muscles on the left side of the lumbar part of the spine (m. sacro-spinalis) are hard and strained, the same occurs on the right side while standing on the left leg. Recently, Thévenard (298) published in a very interesting work several cases [1] in which the influence of static stress on the muscles of the lumbar part of the spine was impaired. Thus in a woman (Mme. le So . .) there appeared after an encephalitis, a lordosis on standing similar to paralysis of mms. sacrolumbales. Investigations showed, however, that the psoas muscles were neither paralysed nor that they only remained relaxed. He reported further on a child (Francoise G.) with encephalopathy that would lie in bed totally straight and was hypotonic on the right side. When she was placed on her feet she showed a scoliosis with the concavity to the right and, in addition, a hypertonia of the muscles of the right arm and the right side of the face. The left m. sacrolumbalis felt strained and hard, the right, however, relaxed. Based on these findings by palpation Thevenard supposes that the scoliosis is due to the absence of contrac-tion of the *right* psoas muscle[2] .

---

(1) Cases of torsion neurosis or torsion spasms and dystonia lordotica progressiva (Oppenheim's disease) usually with encephalitic etiology.

(2) To support his conception Thévenard points out that patients with paralysed psoas muscles have a hollow back on standing. A concavity to the right, therefore, can be caused by paralysis or a lack of tension of the right psoas muscle. In my opin-ion, however, his findings after palpation do not exclude that the scoliosis is due to an abnormally strong shortening contraction of the right psoas muscle. It is not im-possible that in placing on the legs, the right muscles exerted a stronger traction and pulled the upper part of the boly to the right and that they relaxed when the upper part of the body had yielded to a certain extent. By the force of gravity on the one hand, and the tension of the extended left psoas muscle on the other, scoliosis could then be maintained.

134

In Other cases, Thévenard observed an abnormally strong contraction in the vertebral extensors on standing. In addition to his own observations, he mentioned, among others, a case of postencephalitis described by Mourgue (210, 211). On lying down and crawling on hands and feet, the man showed hypotonie of the psoas muscles and a lordotically flaccid back while in bed he would lie straight and flat. On standing, however, the upper part of the body and the occiput were extended spastically backwards to the right, the right leg was also stiff and spastic and the foot in an equinovarus posture.

Thévenard further observed cases of spastic postencephalitic torticollis in which standing caused changes in the tension of the cervical muscles; torticollis appeared only in standing posture.

Supporting tonus is absent or insufficient in persons in deep narcosis, in diseases in which the spinal reflex arcs are interrupted, after transverse lesion of the spinal cord, in transverse myelitis, in severe tabes (degeneration of the posterior roots and spinal ganglia), in severe cases of anterior poliomyelitis (degeneration of the anterior horns), in polyneuritis and in diseases with severe degeneration of muscles with serious muscular dystrophies and also in chorea and double athetosis.

As shown by the reported observations in man, the conditions resemble those in dogs in several points.

Intact persons lying supine have no distinct supporting reactions (as mentioned above, the hindlegs of intact dogs show no supporting reactions or only weak ones). Investigation on supporting reactions in man when suspended head upwards encounters great difficulties.

Under pathological conditions, however, similar reactions in the legs can sometimes be detected in man lying supine. For example, in a woman observed by Stenvers and myself, suffering from myotonia, a passive flexion of the knee joint was found almost impossible with dorsally directed stretching of the foot and toes; a passive flexion, however, was very easy when the foot and toes were plantarflexed and sometimes even an active flexion could be brought about. Similar observations were recently reported by Foerster and by Schwab (270). A 12-year-old boy who showed the clinical picture of a severe cerebellar lesion showed in a test of muscular resistance an abnormally reduced resistance against passive flexion (hypotonia) in all joints of the arms and legs. When the leg was first passively flexed at the hip- and knee joints, however, and the foot and toes moved dorsally, the whole extremity was vigorously reflexly extended after 2 seconds and remained fixed in an extended position as long as the foot and toes were held dorsally. The extremity could not be flexed now even with the greatest effort by the investigator (Fig. 76, No. 1). The leg, however, could easily be moved in all the joints when the foot was held plantar-flexed; there was now even less resistance to passive flexion than can usually be felt in normal persons. In this patient also, a magnet reaction could be observed.

Another patient with a large, inoperable tumor on the ventral side of the *right* half of the cerebellum (pontine angle tumor) showed when lying supine a distinct positive and negative supporting reaction in the *right* leg. On passive plantar flexion of the toes, the fixation in the extended posture disappeared, there even appeared a sharp flexion of the leg in the knee and hip joints with a tonic contraction of the flexors. A dorsal passive movement of the foot caused an extension with a strong fixation in the extended posture.

Observations show that in man, too, supporting reactions occur and that in cases of cerebellar lesion, as in animals, they appear more lively and unrestrained and can even be detected in the supine posture.

Although, in man, arms and hands are not used for standing, yet reactions in them are evident which resemble the supporting reactions of the forelimbs in dogs and cats. Leri (162) has already observed that the biceps muscle in intact persons is strongly contracted and the elbow joint is brought into a flexed position (Fig. 75) when the forearm is passively supported and, with the other hand, fingers and wrist joints are passively flexed ("signe de l'avant bras").[1]

In the case of myotonia already mentioned, a passive flexion of the wrist joint and fingers not only caused a vigorous flexor movement but also such a strong fixation of the elbow joint in the flexed position such that, to a certain degree, the upper part of the body could be pulled up by the arm without the elbow giving way. Conversely, the

---

(1) Leri states that the "signe de l'avant bras" is negative or reduced in organic, spastic or flaccid hemi- or diplegia, in tabes, in Friedreich's disease, in neuritis of the brachial nerves (lead palsy) and also frequently in multiple sclerosis, syringomyelia and others; it is, however, positive in normal persons, in cases of hysterical hemiplegia, in true paraplegia and also in cerebellar diseases.

elbow joint remained totally extended and it was almost impossible to bend passively when hand and fingers were held dorsally. Stenvers and I observed in a boy with a cerebellar abscess a particularly distinct change in elbow fixation and in the tension of the muscles of the upper arm on passive dorsal- and volar flexion of the wrist joint and fingers.

Foerster and Schwab observed in the two above mentioned cases of cerebellar disease, that with dorsiflexion of the wrist joint and fingers (in the first case on both sides, in the second with right sided cerebellar pontine angle tumor only on the right side) the elbow joint became extended and remained fixed in an extended position as long as the wrist joint and fingers were in a dorsal posture; passive flexion of the elbow joint was almost impossible (Fig. 76, No. 2). When the hand was released, resistance diminished after a short time and ceased instantly on a passive flexion of wrist joint and fingers. In the second case, there even occurred an extraordinary sharp flexion of the arm at the elbow joint with a strong tonic contraction of the m. biceps.

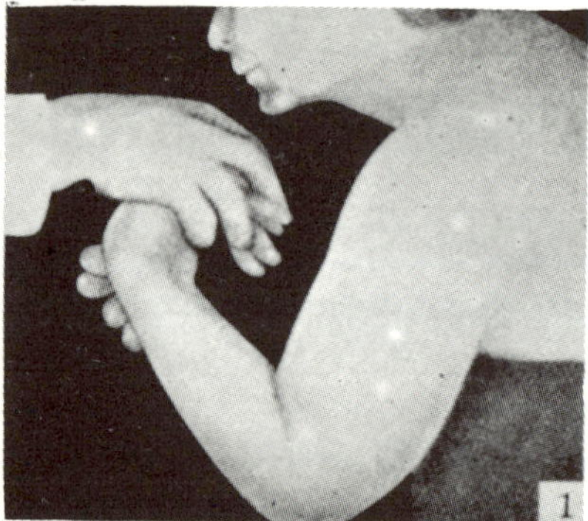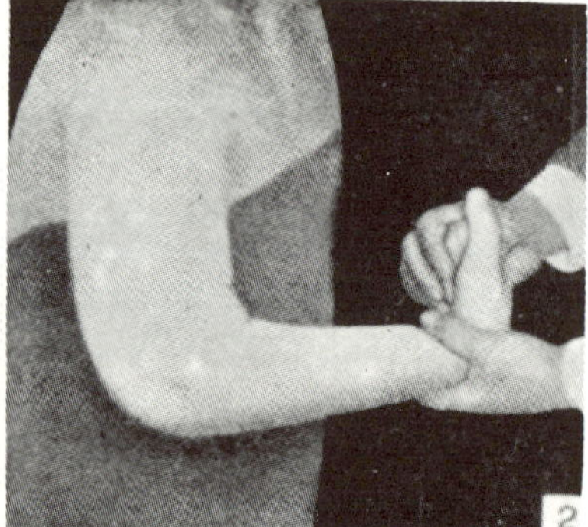

Fig. 75. "Signe de l'avant - bras" by Leri. Woman with organic hemiplegia. 1. Positive reaction on the healthy side.   2. Negative reaction on the diseased side.

Thus these observations teach that in man, reactions are evident in the upper as well as in the lower limbs which resemble the proprioceptive positive and negative supporting reactions. In patients with cerebellar lesion these reactions can appear in an increased manner, in conformity with observations made on decerebellate animals.

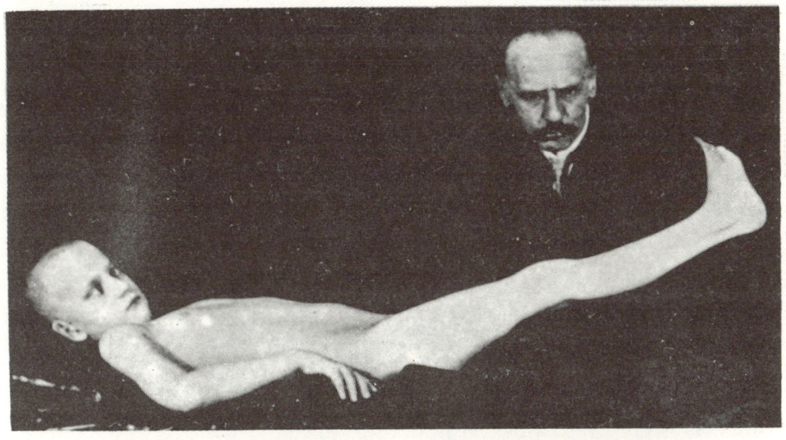

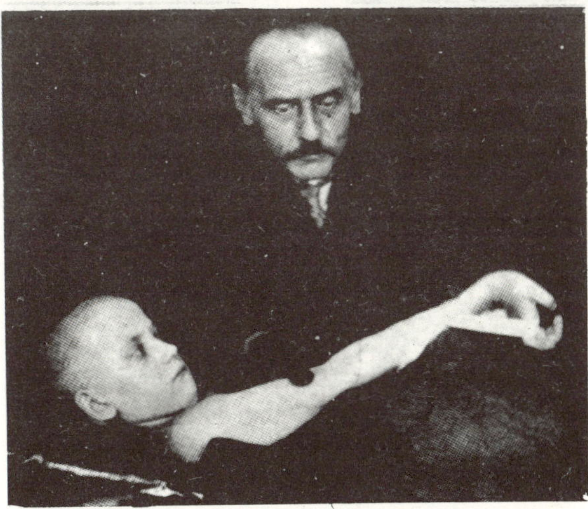

Fig. 76. Positive supporting reaction in the right leg and arm of a boy with right sided pontine angle tumor (after O. Schwab : Z. Neur. 108, p. 585, 1927).

Leri's observation that this forearm phenomenon is negative in tabes and in Friedreich's disease is in agreement with the absence of supporting reactions in animals after severance of the posterior roots to the extremities.

In decerebrate animals a flexion of the wrist joint and digits does not cause a disappearance of the extensor rigidity and flexion of the forelimb. This, too, conforms to a certain degree with Leri's observations, who in 30 cases of hemiplegia in adults and in 6 cases of cerebral unilateral infantile paralysis found the forearm phenomenon on the diseased side to be negative; in 5 out of 6 cases of spastic diplegia it was absent on both sides, and in the sixth case feebly positive only on one side. It is remarkable that in hemi- and diplegia Leri's forearm phenomenon is only feeble or absent while the "phénomène des raccourcisseurs" (Marie-Foix) is usually clearly present in paraplegics. The question has already been discussed whether the flexion caused by Marie-Foix's vigorous plantar flexion of the toes could perhaps be compared to the ipsilateral flexor reflex on pinching the toes. Leri's forearm phenomenon, however, is allied to the negative supporting reactions.

In children in the first half year of their life, any distinct supporting reactions as well as distinct supporting tonus *is absent.* The legs give way immediately when they are placed on them; this cannot only be due to a weakness of the leg muscles, as infants are able to perform kicking and defense movements with great vigor.

Schaltenbrand, in an interesting investigation (266), describes how the process of sitting up and standing up from a supine lying (dorasl) position is brought about in children of different ages. He observed that the human being in the second and third half year of life undergoes a "quadruped stage" and, in order to stand up, has first to turn into a ventral postion. Later, however, this mode of standing up is altered and a turning into a ventral position no longer takes place. Schaltenbrand concludes from this that the body righting reflexes on the body, observed by Magnus in animals, are present in primitive form also in infants and are only fully developed later on. He is of the opinion that new anatomical systems, perhaps the corpus striatum, participate in the transformation of righting reflexes and that they are not brought about by way of the red nuclei like the "primitive" body righting re-flexes. According to Schaltenbrand, the primitive righting reflexes are no longer used by the adult man[1] . This conclusion, however, is unjustified because he could not prove in any way that the change in standing up is caused by a transformation of body righting reflexes.

As we have seen in infants in the first half year of life as well as in new-born dogs, preparation for standing as well as the supporting

---

(1) Schaltenbrand does not give any explanation for the fact that old persons again show the "quadrupedal stage". Surely he is of the opinion that, later, an involution of the righting reflexes takes place and these again are brought about by way of the red nuclei.

reactions are absent. Hopping and tilting reactions are also absent as we shall see later. The center of gravity of the body is then shifted caudally in the first year of life as the trunk and the extremities grow more rapidly than the head which probably also influence the mode of standing up. At any rate, it is not admissible to conclude on a transformation of the righting reflexes before the different reflex contractions such as the static and balance reactions, which develop only after the first year of life, have been thoroughly investigated and their influence on the mode of getting up is known.

It would be desirable that by means of the study of these reflexes in newborns, and adults and animals, one could get away from the method of inferring from the comparison of the capacities of anencephalics to those of adult decorticate animals regarding the different functions in the mesencephalon, striatum and cerebrum relative to the standing and walking functions, as was done by Fischer and Edinger (63) and recently by Jakob (125) and Gamper (91).

Gamper observed an arhinencephalic who remained alive for three months and in whom cerebrum, corpora striata and thalami optici were totally absent. On the observed capacities of the child he writes: "We asked ourselves while the child was alive which parts of the central organ he had at his disposal, we expected at least the presence of the thalamopallidae system. *In his general behavior the child approached so strongly that of a normal infant* that we were totally justified to assume in him the central organs which, according to the prevailing ideas, form the cortical basis of the capacities of an infant. All the more surprising were the anatomical findings which revealed the child as a mesencephalic".

Nevertheless, Gamper concludes, since the child could neither stand nor run, the same as a newborn infant or a newborn dog : "The child was not capable of any complicated achievements such as are possible in the mesencephalic rabbit, cat or dog. In this regard, mesencephalic man is undoubtedly at a disadvantage compared to the mesencephalic animal". Gamper was first surprised that the child achieved so much, almost as much as a normal infant, and later on was astonished that he did not achieve more, since "a mesencephalic system, developing at the outset according to its own laws, functionally reaches a greater and thus in itself more complete efficiency than under the conditions of dependence of normal development". According to his conception, the mesencephalon of the arhinencephalic, deformed infant of 3 months, should thus reach a greater efficiency than in an adult, and as great as the entire brain of a normal infant, as regards standing, adjusting and walking functions!!

Finally, there remains the question how one should imagine the central mechanism of the proprioceptive supporting reactions. As we have seen, these reactions are brought about by stimulations which are produced by the *stretch of muscles*[1] . In the spinal animal, these stimuli cause an increase in fixation of a passively extended limb of only short duration; thus they only cause a phasic tension of the muscles[2]. In decerebrate animals, however, tonic tensions are produced, but here only the extensor muscles react to stretching and a reflex contraction only of the extended muscles is brought about, and not an associated reaction of the other muscles (Liddell and Sherrington). In thalamus and mesencephalic animals, however, the stretching of extensor, as well as flexor muscles causes a tonic reflex contraction of the stretched muscles and a tonic contraction of the other muscles of the extremities.

One could imagine the process as follows : that proprioceptive stimuli emanating from stretched muscles, when they can engage exclusively spinal centers (as in the spinal animal), are only able to produce phasic contractions, that the tonic component of muscular tension appears only when the stimulations proceed also by way of centers in the medulla oblongata; while only through the use of midbrain centers the reflex tensions of the stretched flexor muscles and the associated reaction of the other muscles is brought about. This conception agrees with Sherrington's assumption that the tonic stretch reflexes produced by proprioceptive stimulation are produced in the decerebrate animal by way of a central mechanism situated between the midbrain and the medulla oblongata. However, the above mentioned findings also admit another interpretation,t i.e.,[that the different results of stretch in the decerebrate and in the   halamus anima

---

(1) It seems that stimulations from the surface of the joints and periostial stimulations do not participate in these reactions, since after severance of all flexor muscles of the wrist joint and the digits, the reactions to static stress were always totally abolished in the forepaws (Schoen).

(2) As mentioned before, tonic contractions on stretching of the m. gastrocnemius have recently been observed in the spinal animal (Denny - Brown and Liddell 52).

are based on a different initial tonus of the muscles. In the spinal animal only the stimulations of spinal reflexes reach the muscles and influence the tonus while in the decerebrate animal labyrinthine stimulations are added (tonic labyrinthine reflexes) which change the tonus and cause a different initial tonus. At any rate, in decerebrate animals the tonus of the extensor muscles predominates. This perhaps explains the tonic reaction of extensor muscles to stretching.

Besides, in thalamus and mesencephalic animals, a number of other stimulations reach the muscles, among others, stimulations from the labyrinths (stimulations of the labyrinthine righting reflexes) and from the body surface which in addition to righting reflexes also produce a different distribution of tonus[1]. This different distribution of tonus is perhaps the reason that the flexor muscles, too, react on stretch and an associated reaction of the other muscles is produced.

The conception that the different muscle reactions in spinal, decerebrate and decorticate animals depend on a different initial tonus as well as Sherrington's conception that proprioceptive stimulations engage higher centers, is naturally only pure hypothesis. Some obervations, however, rather support the first conception. Girndt (95), among others, could show that the different behavior of the so-called spinal reflexes, for instance the ipsilateral flexor reflex and the crossed extensor reflex in spinal, decerebrate and decorticate animals can be sufficiently explained by the different levels of tonic discharge of the individual centers of the spinal cord. On faradic stimulations of a sensory nerve, there appears in the thalamus animal a reflex contraction of the extensor muscles in the crossed leg which is soon followed by a relaxation (reverse relaxation) of the muscle in question, in which the residual tonus is much weaker than the initial tonus; in the decerebrate animal, however, the crossed stimulation produces a reflex contraction of the extensor muscles which persists after cessation of the stimulation (tonic after-discharge) and finally returns gradually to the level of the initial tonus.

Girndt, however, observed that in animals in which decerebration did not cause a distinct rigidity, there also sometimes appeared phasic extensor contractions swinging rapidly to relaxation, i.e., the same reaction as in the thalamus animals.

---

(1) See footnote on page 12. (Ed).

In the thalamus animal, crossed stimulation causes a contraction of the extensor muscles at the same time with the relaxation of the flexor muscles followed by a brief reverse contraction, while in the decerebrate animals the flexor muscles show no reaction at all. When, however, a flexor tonus is produced by means of a painful stimulus (for instance pinching of the toes) the flexor muscles also show a relaxation with reverse after-contraction as in the thalamus animal.

We thus see that in changes of initial tonus, the decerebrate animal can show reflexes of extensor and flexor muscles similar to those of the thalamus animal, and that for the occurrence of these reflexes the use of centers situated rostral to the level of decerebration is not necessary.

Beritoff (21) showed that the toneless flexors of decerebrate animals in these cases sometimes reacted to crossed stimulation when the tendency to flexion of the spinal centers was produced by tonic cervical and labyrinthine reflexes. Jonkhoff's observation that through injection of picrotoxin, the extensor rigidity of the decerebrate animals is transformed into a flexor rigidity and that the tonic cervical and labyrinthine reflexes in this case do not, as normally, release an increase or a decrease of extensor tonus, but the increase or decrease of flexor tonus, confirms the conception that according to the initial tonus the same stimulations can cause quite different reflexes. While the decerebrate cat in maximum position of tonic labyrinthine reflexes shows a vigorous extensor tonus and in the minimum position shows a reduced one, there appeared in the decerebrated picrotoxin-cat in the minimum position a vigorous, in the maximum position a reduced flexor tonus.

Similar observations on the occurrence of Babinski's reflex support this conception. On brushing the sole of the foot in a normal adult man, a plantar flexion of the big toe occurs, in hemiplegics and in newborns, a dorsal movement. Hitherto, this phenomenon was explained by the fact that by the interruption of the pyramidal tract in hemiplegics and by the incomplete maturation of the pyramidal fibres in newborn babies, the primitive spinal reflex occurs. This is masked in adults by another reflex which causes the plantar flexion by way of the cerebrum and along the pyramidal tract. It became evident, however, that the positive Babinski reflex cannot be observed constantly in hemiplegics and that, under certain conditions, sometimes even in one and the same patient, a plantar flexion and sometimes a dorsal movement of the big toe can be produced. Walshe reports a case in which the positive Babinski reflex could be transformed into a negative one when, by means of the tonic cervical reflexes by passive or active rotation of the head, the distribution of tonus was changed in the paralysed extremity. Guillain and Barré, Boveri, Bychowski and others observed the transition of the positive Babinski reflex into a plantar flexion when the patients were turned from a dorsal into a ventral position, which was similarly observed by Mankowski and Beder when they passively flexed the knee- and hip joints. Even after the *transverse lesion of the spinal cord,* either a plantar flexion or a dorsal movement of the big toe could be observed at different times in the same patient (Riddoch, Lhermitte). It is thus natural to assume that the negative and positive Babinski reflex can be brought about spinally *without participation of higher centers.* These observations further prove that the same stimulations can also sometimes cause a flexor reflex, and sometimes an extensor reflex in the spinal man. Following Uexkull's and Jordan's findings in invertebrate animals, in which the centers of stretched muscles can be "engaged" (i. e., a stimulation in network of nerves always flows towards the stretched muscles), Magnus observed that also in spinal mammals, a certain stimulus as, for instance, the tapping of the patellar tendon, sometimes causes an extensor reaction and sometimes a flexor reaction according to whether the extensor or the flexor muscle was previously stretched ("reflex inversion", "reversal").

From different aforementioned observations it can be inferred that the centers of the flexor and extensor muscles are not only "sensitized" by the state of extension but probably also by the state of tension caused by other stimulations.

In the supporting reactions, the simultaneous tensions of agonists and antagonists occur. When an animal is injected with strychnine or tetanus toxin after the transverse section of the spinal cord, the muscle tonus of the hindlegs is changed to such a degree that even tactile stimulations produce tonic contractions in all muscles, flexors, extensors, abductors and adductors (Sherrington 277, and others)[1]. Under pathological conditions it is, therefore, possible that simultaneous tension

---

(1) Claude and Lhermitte report similar observations in war casualty who was shot through the spinal cord and infected with tetanus.

of agonists and antagonists occur also in the spinal animal like those in supporting reactions.

From the above-mentioned observations it follows, first : that a certain stimulus can cause different reactions depending on the initial tonus and secondly, that it is possible to produce a simultaneous muscle tension in all extremities of the spinal animal, however, they cannot prove that the simultaneous contractions of the muscles of the extremities on static stress are to be considered purely static reflexes which do not appear normally in the spinal animal due to an unsuitable initial tonus. They do not exclude the participation of the centers rostral to the level of decerebration in the production of supporting reactions. The establishment of the central mechanism for supporting reactions still requires more precise investigations. Considering the general underestimation of the functioning of the spinal centers and the trend to look for a special center and a special reflex arc for each variant reflex, we have found it necessary to discuss the above observations exhaustively.

Summing up the results of the investigations on the *proprioceptive supporting reactions* we found the following :

1. positive supporting reactions are caused by the stretch of digital, wrist, and toe plantar-flexors;

2. the stretch causes a reflex extension movement and a fixation in the extended position of the other joints of the extremity; all the muscles of the extremity participate in the reflex fixation;

3. due to the stretch, not only the muscles of the extremities go into reflex contraction but also those of scapula, spine and pelvis and, in certain cases, also other muscles;

4. in static stress on the extended leg, stimuli emanating from the stretched extensor muscles of elbow- and knee joints or of tarsal-, knee- and hip joints participate in the fixation; due to these stimuli here, too, not only the stretched but also other muscles of the extremity contract;

5. the proprioceptive positive supporting reactions are always present after extirpation of the cortex as well as that of the cerebellum;

6. in the ventral position, the supporting reactions in decorticate dogs produce extensor movements as well as fixation, the flexed hindlegs, in a dorsal position, however, do not show a distinct extensor movement on static stress; in this position, the fixation in extended position of the passively stretched fore- or hindlimb is usually far less than that on static stress in a ventral position;

7. intact dogs, also, show in the dorsal position and on static stress a much less facile extension movement; sometimes it is even totally lacking in spite of the distinct presence of magnet reaction. Similarly, the fixation in extended position of the passively extended limbs by static stress is less strong in the dorsal position of the animal than in the ventral position;

8. after extirpation of the cerebellum (in dogs), the proprioceptive supporting reactions may temporarily be absent, but appear later strikingly exaggerated; in the dorsal and ventral position of the animal the extension movements and fixation in an extended position are particularly strongly marked. A very vigorous contraction of the vertebral muscles and an accentuated change of posture of the back is produced;

9. after unilateral extirpation of the cerebellum, the proprioceptive supporting reactions in the ipsilateral leg are temporarily absent but appear later in a very active form; static stress then produces a more vigorous supporting tonus according to circumstances, sometimes in the ipsilateral and sometimes in the contralateral limb (Chapter VI);

10. after the unilateral extirpation of the cerebral cortex, the positive proprioceptive supporting reactions are again soon evident in the contralateral legs where they, for at least some time, cause a more feeble fixation than in the ipsilateral legs. This was also the case in a unilaterally decorticate, totally decerebellate dog, although the supporting reactions and the magnet reactions were very active in the contralateral legs;

11. in decerebrate animals, the behavior of the positive proprioceptive supporting reactions is very difficult to investigate due to decerebrate rigidity;

12. in spinal animals, the flexed hindleg shows on static stress

as well as other stimulations sometimes an extensor flexion movement (the extensor thrust), while static stress on the passively extended hindleg sometimes causes a fixation of short duration[1] ;

13. after severance of the posterior roots of the spinal cord dis'ributed to the limb, a reaction on static stress was found lacking;

14. in certain cases, static stress on one leg caused a decrease in the extensor supporting tonus of the crossed leg. This decrease appears asymmetrically particularly clear immediately following unilateral extirpations (after a unilateral extirpation of the cerebellum, the decrease of the raised extensor tonus is evident in the ipsilateral leg and after a unilateral extirpation of the cerebrum, in the contralateral leg);

15. the negative supporting reactions are produced by stretching of the extensors of the phalanges, leading to reflex changes which sometimes cause a reduction or disappearance of supporting tonus, especially in decerebrate animals in which the limbs are drawn into a flexed position;

16. negative supporting reactions are distinctly present in intact, decorticate and decerebellate dogs;

17. conversely, the flexing of phalanges in decerebrate animals causes only a slight reduction of extensor rigidity (in this reduction probably the anatomical arrangement plays the principal role; the extensors of digital and wrist joints originate on the humerus and those of the toes, on the femur);

18. in spinal animals, a flexion of the toes causes an instant disappearance of the extensor tonus produced by any stimulation whatever;

19. in man, too, reactions similar to proprioceptive supporting reactions have been observed in arms and legs and in patients with cerebellar diseases lying supine;

20. the absence of supporting tonus in new-born children as well as impairments of supporting tonus in cases of tabes, polyneuritis and related conditions agree with observations made on animals.

---

(1) The stretch - flexion movement as well as the fixation were absent in my ani mals when they were in a dorsal position.

VI. THE STRENGTH OF SUPPORTING TONUS AND THE ADAPTATION OF SUPPORTING TONUS TO THE LOAD

It is possible to load the back of a dog standing on all four legs with a rather heavy burden without the legs giving way. The question is now raised whether the legs are able to carry this load by means of an excessive supporting tonus produced by standing, or whether an adaptation to the load occurs. It is difficult to exclude the existence of an excessive supporting tonus, but certainly an adaptation also participates. This becomes evident when a sandbag is dropped suddenly on the back of a standing dog from a moderate height. On impact, the legs are slightly flexed but are instantly extended again. Thus, the supporting tonus at first was not sufficient for the increased load. When, however, the sandbag is slowly lowered on the back, the legs do not give way due to the adaptation to the load.

How does this adaptation occur ? Due to the increase in load, the counterpressure of the supporting surface also increases, thus augmenting the intensity of exteroceptive stimulations on the soles of the feet. In addition, the higher counterpressure causes an increase in stretch : I. of the plantar flexors of the distal phalanges, II. of the extensors of elbow-, knee- and tarsal joints and III. of either extensors of flexors of shoulder- or hip joints, according to the position of the limbs (Fig. 50).

When the forelegs stand directed forwards, the flexors of the shoulder joint are stretched, but when they are directed backwards, the extensors of the shoulder joint are stretched. In the hindlegs, on the contrary, the flexors of the hip joint are stretched when the legs are directed backwards and with limbs directed forwards the extensors of the hip joint are extended (see Fig. 47).

As mentioned already, Liddell and Sherrington proved, in the decerebrate animal, that the reflex tension of a stretched extensor muscle is the stronger the greater the extension. In intact, decerebellate and decorticate animals the effects of an increase in extension have not yet been investigated. Similarly, it is still unknown whether all the above-mentioned muscles react at all on stretching and thus exert an influence on the state of tension of the other muscles.

The stretching effects of flexors of the distal phalanges, m. triceps brachii and the muscles of the Achilles tendon have already been discussed. However, it is very probable that an important role is played by stimuli produced by the stretch of muscles and emanating from the muscles themselves, since the adaptation is impaired after severance of all the posterior roots distributed to the limb, while it remains preserved when only the paws are made insensible by severance of cutaneous nerves. Whether the adaptation to the load in the vertebral muscles is produced by extension of these muscles is unknown. Surely an important role in adaptation is played by the strong counterpressure of the supporting surface, because in animals suspended in the air in a ventral position, the fixation of the back and the tension of the vertebral muscles is much more vigorous with a powerful pressure on the soles than with a slight one.

Even if it is quite certain that the ability to carry loads is made possible by the adaptation of muscle tensions, this does not at all exclude that a surplus of supporting tonus was present prior to the loading. When a decerebrate dog with a pronounced rigidity is set up on its paws, the limbs will not give way on a sudden loading stress because these animals have at their disposal a higher extensor tonus than necessary to carry the trunk. Similar conditions are found in dogs in the first days after extirpation of the cerebellum, whole limbs show a maximum rigidity. The force of extensor tonus in all four legs on standing, as well as in the dorsal or lateral position and in ventral position in the air, is greater than the weight of the trunk, which can be proved, among other means, by the sudden pressure of a spring balance against the soles of the feet.

In other dogs, after the extirpation of the cerebellum, the limbs are totally limp and by no means capable of carrying the body, and also the limbs of rigid decerebellate animals after cessation of rigidity. Later, however, the supporting tonus again increases in all decerebellate dogs and after 4-6 months is even exceptionally strong. A decerebellate dog standing freely and without any load usually has its legs excessively extended, the bellies of the muscles of the extremities are prominent, the joints strikingly fixed (Fig. 44) and the spine, too, is rigidly curved upwards. Standing decerebellate dogs produce an impression of excessive supporting tonus; on loading the back, however, they show distinct reactions of adaptation.

The strength of supporting tonus, i.e., maximum adaptation to the heaviest load which the paws are able to carry without giving way, behaves quite differently in the various intact, decerebellate and decorticate animals. As we shall see in the following, it first of all depends on the point of application of the load and on the posture of the animals.

A. STRENGTH OF SUPPORTING TONUS OF INTACT AND DECEREBELLATE DOGS

Most of the intact dogs are able to carry on their backs a load nearly equal to the weight of their body (see Table I). In this, however, the various breeds show considerable differences; young intact dogs usually can carry less than adult ones of the same bodyweight.

In decerebellate dogs (some time after extirpation) the strength of supporting tonus is not reduced; they are even able to carry on their backs a load amounting to more than their bodyweight (Table I).

With increased loading of the back, the supporting tonus of the hindleg is the first to give way, that is, in intact as well as in decerebellate dogs. After the yielding of the hindlegs, the forelegs are usually flexed and the animal lies down flat on the ground.

When the dog is set on two balances of equal height in such a way that the two forelegs stand on one, the two hindlegs on the other, one notices that fore- and hindlegs carry exactly half the bodyweight and that on burdening the center of the back, the load is equally distributed on both fore- and hindlegs. Since the hindlegs still give way first, the strength of supporting tonus here must be less than in the forelegs.

How does this falling off of supporting tonus come about ? The extensors of the proximal joints are passively stretched by the load, i.e., the extensors are passively lengthened thus flexing the joints. With an increasing load, the stretch, however, does not grow so great as to force the joints passively into a maximally flexed posture[1] .

---

(1) This is indeed the case with reduced supporting tonus (Figs. 65 and 78).

TABLE 1

	Body weight		Possible load on back of animals	Limbs give way with a load
	in kgs		in kgs	of kgs
Intact dogs				
Grey dog	13		12	15
Fox	10		12	15
Black dog	7.6		5.3	8.8
Intact young dogs				
Small black dog	7.6		5.3	7
Yellow dog	7.4		1.7	3.3[*]
				5.3[*]
Decerebellate dogs				
Erick	7	very well	8.7	
		but still	9.5	10.6
Moor	9.1	very well	12	
		but still	14.2	15.3

(*) With a load of 3.3 kg on the back the hindlegs give way, with a load of 5.3 kg all four legs give way instantly.

Starting from a certain load, the animal grows unsteady and with increasing load suddenly *lifts* one of the hindlegs and sets it down on the ground in a flexed position simultaneous with the lifting of the other hindleg.

Instead of an increase in fixation, the heavier load thus produces a lifting and flexing movement of the leg. In the following chapters we will repeatedly occupy ourselves with the phenomenon that a moderate stretch of certain muscles produces extension with fixation, a very strong stretch of the same muscles, however, produces a flexion movement. When an animal is set down on its forelegs only and the back above the shoulders is loaded, the forelegs do not give way when loaded with a sandbag of the same weight as the body. Sometimes, the animal is even capable of carrying a load of the same weight on the shoulders while standing on one foreleg only. For instance, the dog Erick (body-weight 7 kg) standing on one foreleg, was able to carry a load of 10 kg on its shoulder without diminishing the supporting tonus of this leg. With a load of 10.7 kgs, the limb began to waver, with an even higher load it was moved forwards and flexed. It is peculiar and remarkable that dogs standing on one leg only, are able to carry just as much as they carry while standing on all fours, because it should be expected that the four legs together carry four times as much as, at least, the weakest leg, i.e., the hindleg. The investigation revealed, however, that, for instance, the decerebellate dog Moor (bodyweight 9.1 kg) when standing on all four legs, was not able to carry on its back more than 14-15 kg, while standing on one hindleg, however, it could carry on its pelvis far more than half this amount, i.e., 10.6 kg.

The reduced ability to carry loads when standing on all four legs can be explained by the fact that the lordosis produced by loading the spine causes a reduction in the supporting tonus in the hindlegs (Chapter X). On burdening the other parts of the body, the determination of total force of supporting tonus in the four legs shows quite different amounts. When, for instance, the dog Moor was put on a balance and at the same time a pressure was exerted on the pelvis and shoulders, the four legs were able to carry four times as much as one hindleg (i.e. 40-45 kg).

Fig. 77. Determination of support tonus by means of a balance in the decerebellate dog Erik. As shown in the picture the supporting tonus does not yield to a counterpressure of the balance amounting to 15 kgs.

The determination of strength of supporting tonus in hind- and forelegs, as shown with the dog Moor in Fig. 77, shows that it is able to withstand a counterpressure of 10 ½ kg while standing on one hindleg and on both hindlegs a counterpressure in the balance of 20 kg, while standing on one of the forelegs they do not give way to a counterpressure of 13-15 kg and on both together can carry 30 kg.

The strength of supporting tonus of various decerebellate animals determined by this method is listed in Table 2.

The counterpressure shown in Table 2 can be very well carried by the limbs; they can even carry a somewhat higher load for a short time; however, they give way instantly to a counterpressure given in Table 3.

It appears from these tables that supporting tonus in decerebellate animals is not reduced when examined a considerable time after the extirpation.

Furthermore, we deduce from them that the strength of supporting tonus of both hindlegs amounts to twice that of one leg so that the supporting tonus is not reciprocally reduced by a simultaneous pressure on both legs.

TABLE 2

	Strength of supporting tonus in standing					
	Body Weight	All Four	2 fore-legs	1 fore-leg	2 hind-legs	1 hind-leg
	kgs	kgs	kgs	kgs	kgs	kgs
Normal dog-	8	30-35	24	10-12	15-18	9
Decerebellate dog Wolf	9.5		35	17	27	12
Decerebellate dog Moor	9	40-45	30	13-15	20	10-13
Decerebellate dog Erik	7	30	23	13	15	8-9
Decerebellate dog Piccolino	4.65	20	18	9	9	$4\frac{1}{2}$
Decerebellate cat Carolus	4.35		11		10	
Decerebellate cat Pierrette (*)	4		10		7.5	
Decerebellate cat Nikker (2 mos. after extirpation)	2.1		4.5		3.8	

(*) All animals except the cat Nikker had the cerebellum extirpated for more than a year.

TABLE 3

Determination of counterpressure which causes a lessening
of supporting tonus in standing on

	All four	2 forelegs	1 foreleg	2 hind-legs	1 hind-leg
	kgs	kgs	kgs	kgs	kgs
Normal dog	40	28	15	22	11
Decerebellate dog Wolf		40	18	30	15
Decerebellate dog Moor	50	36	17	22	15
Decerebellate dog Erik	32	25	15	17	11
Decerebellate dog Piccolino	22	22	10	10	5
Decerebellate cat Carolus					
Decerebellate cat Pierrette				8	
Decerebellate cat Nikker				4	

Just as fixation in extended position on static stress of the hindlegs is reduced when the hip joint is passively moved forwards, and increased with a backwards movement (Fig. 55), the strength of supporting tonus also is distinctly influenced by changes of position. For instance, the strength of supporting tonus of the hindlegs, with the trunk kept almost horizontal, amounts to 10-12 kg in the dog Moor and in the dog Erik to 8 kg and with an obliquely raised trunk in the dog Moor and in the dog Erik, this increases to 15 and 10 kgs respectively. It is, however, considerably reduced when the trunk is lowered below the horizontal till it finally equals zero. The strength of supporting tonus in intact dogs differs in several points from that of decerebellate dogs. When intact and decerebellate dogs are put on a balance in a dorsal position, the head adjusted symmetrically to the trunk, the muzzle pointing vertically upwards, to determine the strength of supporting tonus of the hindlegs (counterpressure of the balance minus bodyweight), the force of supporting tonus in intact dogs, when they are calm and do not make defensive movements, is always very low, sometimes almost equal to zero (less than 1-2 kg). In decerebellate dogs, on the contrary, it is still strong, although somewhat lower than in standing posture of the animals[1] , (Table 4).

---

(1) The maximum strength of supporting tonus is not measured by this method. For the determination of amount of supporting tonus the limbs stretched backwards by static stress have to be moved passively forwards till they stand vertically on the balance. In doing so, a lessening of the strength of supporting tonus occurs. Besides, in the dorsal position of the animals, the strength of supporting tonus of the hindlegs is greater when the mnzzle is dircted ventrally than when it is directed vertically upwards.

TABLE 4

The strength of supporting tonus of hindlegs in decerebellate dogs in standing and in dorsal position(*)

	Standing	Posture	Dorsal	Position
	kgs	kgs	kgs	kgs
decerebellate dog Wolf	+12	—15	+7 —9	—10
decerebellate dog Erik	+ 8	—11	+6	— 7
decerebellate dog Piccolino	+ 4½	— 5	+3	— 4
decerebellate dog Moor	+13	—14	+9	—10

* In this and in the following tables, the numbers denoting the counterpressure well supported by the limbs, are marked with +. The numbers denoting the counterpressure under which the limbs will instantly give way, are marked with —.

In the forelegs, this difference is less constant, as the forelegs of intact dogs, in opposition to the hindlegs, sometimes show a more or less distinct and strong supporting tonus in the dorsal position, but sometimes the limbs are also drawn into a flexed position under static stress.

In decerebellate dogs, the strength of supporting tonus of the forelegs is always strong in the dorsal position.

In summing up it appears that the strength of supporting tonus in decerebellate dogs in the stage of permanent impairment :

1. *is not reduced in the standing posture.*

2. *is usually stronger in the hindlegs in the dorsal position, than it is in intact animals.*

B. STRENGTH OF SUPPORTING TONUS AFTER UNILATERAL EXTIRPATION OF THE CEREBELLUM

The supporting tonus of the hindlegs behaved very differently in different animals in the first stage after extirpation. In the first days after the unilateral extirpation of the cerebellum it was clearly reduced in some animals in the legs on the side of the operation, in others, however, they remained nearly as strong as before extirpation. Thus, the investigation determining the force of supporting tonus in the dog Tip, whose right half of the cerebellum was extirpated, resulted in the values shown in Table 5.

Thus, in this animal, the ipsilateral right hindleg showed no distinct reduction in carrying capacity. When the animal was set down on the extended right hindleg, the paw was able to carry the rearpart of the body. On the contrary, the similarly right-sided decerebellate dog Fox showed a distinct reduction of carrying capacity in the right hindleg. When the animal was set down on this limb, it gave way immediately under the weight of the rearpart of the body (Fig. 105, No. 2). This is all the more remarkable since the legs on the side of operation in this dog as well as in the dog Tip, in the dorsal position on a supporting base as well as in different positions in the air, showed a distinctly *increased* extensor tonus. This abnormal extensor tonus of the right hindleg, however, was lower than the normal force of supporting tonus and, therefore, not sufficient to carry the rearpart of the body. In this stage, the supporting reactions in the ipsilateral legs of both dogs showed impairments; because when the animals were set down on the passively flexed ipsilateral legs, neither in the dog Tip nor in the dog Fox there occurred any extension of the legs or lifting of the trunk, while this happened instantly when the animals were set down on the passively flexed contralateral legs.

Table 5.

Strength of supporting tonus after extirpation of the *right* half of cerebellum in the dog Tip (body weight 9½ kg)

	Standing	Posture	Dorsal	Position
	Left hindleg	Right hindleg	Left hindleg	Right hindleg
	kgs	kgs	kgs	kgs
Prior to operation	9	9	0	0
2 days after operation	8	10	6	9½
3 days after operation	7	8	3—7	8
5 days after operation	5	6		
10 days after operation	8	6	9	7½
15 days after operation	9	8	7	7
19 days after operation	11	11	7	6
42 days after operation	11	9	6	5
180 days after operation (body weight 11 kgs)	15	16	5	5

Pressure pushes on the soles of the feet of the extended ipsilateral paws indeed produced a distinct increase in muscle tension and of resistance against passive flexion.

Gradually the higher extensor tonus decreased and with it, at the same time, also the resistance to pressure on the soles (see Table 5, 5 days after extirpation), to increase again, however, soon afterwards due to the reappearance of magnet- and supporting reactions, in order to not only return to normal, but to become even stronger on both sides (see Tables 5 and 6).

In the early phase after the unilateral extirpation of the cerebellum, the animals cannot stand, but fall down on the side of the operation when they are set up on their legs. It is possible that in the dog

Fox, this was due to the reduced resistance against pressure on the soles of the ipsilateral limbs. However, an explanation is still lacking why, first of all, the dog Tip could not stand unassisted and, in the second place, Tip and Fox still fell down after the reappearance of supporting reactions and after the ipsilateral limbs showed a strength of supporting tonus which amounted to not only half, but more than the whole bodyweight. (The dogs Tip and Fox were able to stand and run freely only in the second month after extirpation).

Table 6

Strength of supporting tonus after extirpation of the *right* half of cerebellum in the dog Fox (body weight 7.3 kg)

	Standing	Posture	Dorsal	Position
	Left hindleg	Right hindleg	Left hindleg	Right hindleg
	kg	kg	kg	kg
1 month after extirpation	$8\frac{1}{2}$	7	$2\frac{1}{2}$—3	7
2 months after extirpation	11	7	4—5	7
8 months after extirpation	$10\frac{1}{2}$	$10\frac{1}{2}$	$1\frac{1}{2}$—2	6

In the stage of permanent impairment, the animals are not only able to stand and run with unloaded, but also with a burdened back. At this stage, the dog Fox was very well able to carry around a sandbag of $5\frac{1}{2}$ kg (Fig. 105). The dog Peter (bodyweight 12 kg), one year after extirpation of the right half of the cerebellum, was capable of standing with a sandbag of 12 kg on his back and to run around with ease with a sandbag of 7 kg. Standing on one foreleg, the right or the left, he could support on his shoulders a bag of $11\frac{1}{2}$ kg without his leg giving way, and when standing on one hindleg, neither the left nor the right leg gave way under a load of 7 kg on the pelvis. Testing on the balance resulted in the following values for each of the animal's legs :

left foreleg   +17   —18,   left hindleg   +13   —15kg,

right foreleg   +17   —18,   right hindleg   +13   —15kg.

We have seen that decerebellate dogs, in distinction from intact and decorticate animals, also show a supporting tonus of considerable force when lying supine. In agreement with this, in unilaterally decerebellate dogs, a distinct supporting tonus could always be established in the dorsal position (Tables 5 and 6). In the dog Fox, among others, it was vigorously present only in the limbs on the side of the operation (Table 6). In some others, however, as for instance in the dog Tip, it appeared forcefully on static stress on the limbs of both sides (associated lesion of the other half of the cerebellum ?).

In the dog Fox, the dorsal position indeed caused an inhibition of supporting reactions in the contralateral limbs, but not in the ipsilateral ones. At the end of the first month after extirpation, the force of supporting tonus in the right hindleg, when he was set down on it, was still distinctly reduced and weaker than in the left leg. Conversely, in the dorsal position with the muzzle pointing vertically upwards, the supporting tonus of the right hindleg was considerably stronger than in the left (Table 7). Thus, the right hindleg showed in a standing posture a reduced strength of supporting tonus and in the dorsal position an increased one.

Table 7

Dog Fox (body weight 7.3 kgs), 1 month after extirpation of right half of cerebellum

| | Strength of supporting tonus of | | | |
| | Left hindleg | | Right hindleg | |
	kg	kg	kg	kg
while standing on one hindleg	$+8\frac{1}{2}$	-9	$+7$	$-7\frac{1}{2}$
in dorsal position	$+2\frac{1}{2}$	-3	$+7$	$-7\frac{1}{2}$

As shown in the values of Table 7, the amount of supporting tonus of the right hindleg was equally strong in standing posture (7 kg) as in the dorsal position (7 kg). The dorsal position did not cause a reduction in the strength of supporting tonus of the ipsilateral right hindleg; thus the leg showed the same behavior shown by both hindlegs after a total extirpation of the cerebellum. The left hindleg, however, behaved like the hindlegs of intact or decorticate dogs : its strength of supporting tonus was slight in the dorsal position.

These observations show that it is not possible to draw any conclusions from the strength of supporting tonus in the dorsal position as to the strength of supporting tonus in the standing posture of the animal and, secondly, that statements concerning the amount of supporting tonus are worthless without precise descriptions of the experimental conditions.

The values in Table 8 were reached only when the hindlegs were alternately placed under static strain. In the dorsal position, the strength of supporting tonus of the ipsilateral right hindleg was considerably lower when the left hindleg was placed under static strain at the same time.

At this stage (one month after extirpation), the strength of supporting tonus in the dog Fox was strongest in the ipsilateral limbs in the dorsal position, and in the contralateral limbs in a standing posture.

At the stage of permanent impairment, the strength of supporting tonus was equal and normally strong on both sides in the standing posture but in the dorsal position, however, it was still stronger in the contralateral right legs than in the left ones (Table 8).

The right legs, however, now showed in the dorsal position a distinctly lesser strength of supporting tonus than in the standing posture. The dorsal position now also exerted a distinctly inhibiting influence on the supporting tonus of the right legs, though not as strong as that exerted on the left ones.

Table 8.

Dog Fox (body weight 8.5 kg) 8 months after extirpation of right half of cerebellum

	Intensity of Supporting Tonus			
	Left hindleg kg	Right hindleg kg	Left foreleg kg	Right foreleg kg
While standing on one leg in dorsal	$+10\frac{1}{2}(-11)$	$+10\frac{1}{2}(-11)$	$+15(\pm16)$	$+15(\pm16)$
position	$\pm 1 (-2)$	$+ 5\frac{1}{2}-6\frac{1}{2}$	$+ 4\frac{1}{2}$	$+ 7\frac{1}{2}$

To sum up, the observations made after unilateral extirpation of cerebellum showed that the legs ipsilateral to extirpation can sometimes carry only a slight load in the initial stage of temporary impairment, but that in the stage of permanent impairment, the supporting tonus appears equal on both sides and in normal strength when the animals are set down alternately on their left and on their right legs, while in the dorsal position with the muzzle pointing upwards, the supporting tonus is strongly evident in some of the animals on both sides, and in others only on the side of the operation.

C. INTENSITY OF SUPPORTING TONUS AFTER TOTAL EXTIRPATION OF THE CEREBRAL CORTEX

After removal of one half of the cerebral cortex as well as after the extirpation of the second half carried out a few weeks later, the dogs are usually able to stand and run again within the first 12 hours. On running around, when they knock against a wall, they will stand up on their hindlegs, scratch the wall alternately with the forepaws and sometimes remain in this posture for a considerable time. The strength of supporting tonus of both hindlegs together only one day after extirpation, in this position, amounts to almost as much as the weight of the animal. Compared with this, the strength of supporting tonus is reduced in the standing animal with a loaded back, since even with a moderate load, first the hindlegs and then the forelegs give way. Gradually, the intensity of supporting tonus will again increase, but in the hindlegs it will usually be distinctly reduced for a considerable time. Thus, the decorticate dogs, Miesel (body weight 6.7 kg.), Robbie (body weight 6½ kg), and Fuchs (body weight 8 kg) after 4 weeks, 5 weeks, and 9 months respectively, were not able to carry on their backs a sandbag of 2 kg (Fig. 78). Investigations on the resistance of supporting tonus by means of the balance at various times after extirpation gave the values given in Table 9.

Table 9.

Intensity of supporting tonus in the limbs of decorticate dogs

decorticate dogs	body-weight	Force of supporting tonus in standing on			
		both forelegs	one foreleg	both hindlegs	one hindleg
	kgs	kgs	kgs	kgs	kgs
Robbie					
32 days after operation	6.5	+ 8 (—10)	+4 (—7)	—4!	
38 days after operation	6.8	+15 (—16)	+7 (—10)	—4!	
Miesel					
24 days after operation	6.75	+8 (—10)		—4!	
32 days after operation			+8	—5	—2.4(+2.2)
35 days after operation		+20	+11	—4!	—1.5
Fuchs					
10 days after operation	6.5			—4!	—3(+2½)
23 days after operation				—3!(+2)	
190 days after operation	6	+18(—20)	+11(—12)	+5(—6)	
240 days after operation	8	+17(—19)		+5(—6)	
Bob ; before operation	8				
6 days after operation		+24(—30)	+15(—17)	+15—20(—21)	—8 (—11)
10 days after operation	6½	+18(—20)		+15(—20)	
12 days after operation		+28(—20)	+8	+20(—21)	
				+20 —25	

As shown in this table, in the first days after extirpation, the resistance of supporting tonus is considerably reduced in the forelegs as well as in the hindlegs; however, it increases again rapidly in the forelegs, while it remains reduced for a long time in the hindlegs of the dogs Robbie, Miesel and Fuchs (Fig. 259).   Contrary to these three dogs, the decorticate dog Bob already showed, 6-10 days after extirpation, a normal intensity of supporting tonus in the hindlegs although neither distinct magnet reactions nor well established reflexes promoting the supporting tonus could be detected.   These were also lacking in the other three dogs.   He showed an excessive hyperactivity, ran around almost the whole day till the soles of the feet were sore, pushed himself constantly and forcefully against the walls of the cage or the animal room so that the skin of the head showed numerous deep injuries from impact, and decubitus.   In investigations on strength of supporting tonus, he made much more defensive movements and attempts to escape than the other animals (1).

There is another question : why is the intensity of supporting touns reduced in the dogs Robbie, Miesel and Fuchs in spite of the presence of distinct and positive and negative supporting reactions ? Perhaps the absence of the exteroceptive component, the magnet reactions, share the responsibility for the reduction of tonus, since the intensity of supporting tonus remains reduced for several weeks   also in dogs whose feet have been anesthetized by the severance of cutaneous nerves. The absence of magnet reactions alone cannot, however, cause the reduction of tonus.   With a gradually increasing loading of the back it could be observed, in the dogs Robbie, Miesel and Fuchs, that the extensors

(1) The dog Bob died of an infection of the cranial cavity due to the tearing open of the suture of the pericranium.   The obduction showed that the cerebrum was totally removed while the thalamus was softened by the infection. It was therefore not possible to cut the brainstem in series in order to investigate the question whether the different behavior of the strength of supporting tonus was perhaps brought about by the fact that in the process of extirpation some of the brain tissue was left more than in the other dogs. In the dogs Miesel and Robbie who died of distemper encephalitis, and also in the dog Fuchs, who lived for eight months without a cerebrum, the cerebrum was totally removed (Figs. 281, 282, 286 and 287).

of knee- and hip joints were passively elongated more and more till these joints became increasingly flexed and finally the legs totally collapse *without being lifted or transferred* (Fig. 78).

Fig. 78. 1. Dog Miesel, 1 month after extirpation of the cerebrum, in a standing posture. 2. When the back is loaded with a sandbag of 2 kgs, the hindlegs are totally flexed without being lifted or transferred.

It also seems that in decorticate animals the increase of muscle stretching in the hindlegs does not cause an increase in reflex contraction, and overstretching does not produce a reflex flexion movement (lifting of the paw).

The contradictory behavior of the strength of supporting tonus in the fore- and hindlegs of decorticate animals is still obscure and needs further investigations. This much however is certain : that the greater strength of supporting tonus in the forelegs is partly caused by the fact that not only the muscles of the upper arm but also the flexors of the wrist joint and digits, (flex. digit. sublimis, flex. digit. profundus, and flex. carpi ulnaris) which have origin on the humerus, participate in the fixation of the elbow joint. In the standing posture, these muscles are passively and strongly extended causing a fixation of the elbow joint. It is remarkable that with an increase in load, the wrist joint is passively overstretched, the tarsal joint, however, is more or less passively flexed (see Fig. 26 and 50.) The overstretching of the wrist joint produces an increased fixation of the elbow joint in the extended position, while with flexion of the tarsal joint, the muscles of the tendon Achilles (in the dog m. gastrocnemius, m. biceps femoris and m. semitendinosus) as well as the m. flex. digit. ped. sublimis, which all originate on the posterior of the femur, tend to pull the knee joint into flexion. Till now it has not yet been ascertained whether the reactions of extensors of the proximal joints are more or less impaired by increased extension in the forelegs. Only this much is certain that, contrary to the hindlegs, the forelegs are *raised* on overloading of the shoulder and are then laid on the ground in a flexed position; thus, it seems that they react in quite a normal way.

Loading of the back with a sandbag of 2 kg often also caused, after the collapse of the hindlegs, a giving way of the forelegs although the strength of the supporting tonus of these legs amounts to several kilograms. On the collapse of the hindlegs, the trunk moves more or less backwards cuasing the forelegs to move forwards at the shoulders. With this change of posture, the strength of supporting tonus diminishes and the forelegs give way.

Investigation of decorticate dogs in a dorsal position is very difficult since in this position they usually make defensive movements and, under the influence of the labyrinthine righting reflexes, try to get up[1] .

Although in this way the values established by means of the balance become inaccurate, one can distinctly perceive that the strength of supporting tonus in the fore- and hindlegs is considerably lower in this position than in the standing posture and sometimes does not amount to 1 kg in the hindlegs. The dog Bob is an exception in which neither the passively stretched foreleg, nor the hindleg could withstand a pressure on the soles of 5 kg on the 10th day after extirpation, while the passively stretched hindlegs on the 12th day, in the same position, did not give way under a pressure of 17 kg.

Thus, also in decorticate animals the dorsal position exerts a distinctly inhibiting influence on the supporting tonus. This influence, however, could not be observed in the dog Bob.

---

(1) In the endeavor, under the influence of labyrinthine righting reflexes to lift the head and move it ventrally, there sometimes occurs a vigorous extension of the hindlegs with a strong resistance to passive flexion.

*In decorticate dogs, the strength of supporting tonus immediately after ex-
tirpation, is distinctly reduced for some time, although strong enough to carry
the trunk. In the forelegs, however, it recovers quickly while in the hindlegs
it remains nearly always reduced.*

*In the dorsal position of the animals, the strength of supporting tonus is
usually, but not regularly, much lower than in the standing posture.*

D. STRENGTH OF SUPPORTING TONUS IN THE DECEREBELLATE DECORTICATE ANIMAL

The dog Robbie, whose cerebellum was removed 40 days after
extirpation of the cerebrum, was at first totally rigid and lay on his
side with extended fore and hindlegs, a dorsiflexed neck and a dorsally
curved back. In the lateral and dorsal position on a supporting surface, the legs were kept in an extended posture (Fig. 79, No. 3), while
the decorticate animals in the same position kept them flexed. When
he was set up on the limbs these were very well able to carry the trunk
(Fig. 79, No. 4). Due to the stiff neck the head was held with the

Fig. 79. Dog Robbie, whose cerebrum was removed first and 6 weeks later also the
cerebellum. 1 and 2. The animal standing freely 3 weeks after extirpation of the
cerebrum. 3 and 4. The animal 6 days after extirpation of the cerebellum, 46
days after ablation of the cerebrum. 3. In a dorsal position on the supporting
surface the animal keeps the limbs extended upwards in the air. 4. Placed down
on its paws the animal keeps its legs excessively extended. They do not give
way under the weight of the trunk. Head and tail are also raised. (In the
first 3 days after the operation, the rigidity was much more pronounced).

muzzle somewhat above the horizontal in the air, the tail lifted dorsal and the back, curved in lateral position, sagged a little but without becoming limp. Thus, the animal showed extensor rigidity; a hypertonia of the extensors of fore- and hindlegs as well as of the muscles of the neck, back and tail.

When, however, the anterior part of the body was lifted from the ground so that the animal would stand on its hindlegs or when on standing on all four legs, a slight pressure was exerted on the pelvis, the hindlegs instantly gave way. In spite of a distinct hypertonia of the extensors the hindlegs, on the balance, could not even support a counterpressure of 2 kg (Fig. 80, No. 1). Set down on its forelegs which, when not under static stress, showed an increased resistance to movement, the animal could carry on its shoulders a sandbag of 5.3 kg (Fig. 80, No. 3). On the balance, the forelegs did not give way even with a counterpressure of 9 kg. and collapsed only under a counterpressure of more than 10 kg. In comparison to the counterpressure of more than 15 kg which could be supported by the forelegs of the thalamus animal prior to the extirpation of cerebellum, the resistance now was indeed distinctly reduced (see Table 9).

Strength of supporting tonus of the forelegs in the dog Robbie :

38 days after extirpation of cerebrum,   46 days after extirpation of cerebrum,

2 days *before* extirpation of cerebellum :   6 days *after* extirpation of cerebellum :

+ 15 kg (—16 kg)   + 9 kg (—10 kg)

Contrary to this, the elbow- and shoulder joints did not go into flexion at a counterpressure of 7 kg (but only at 8 kg) when the forelegs were set on the surface with the dorsum of the paw instead of the sole, while the forelegs of the decorticate animal prior to extirpation of cerebellum would give way under a very slight pressure with the carpus in dorsal position.

At the end of the first week, the extensor rigidity gradually subsided, simultaneously, the supporting tonus increased so that 3 weeks after extirpation of cerebellum the hindlegs no longer gave way when the anterior part of the body was lifted from the ground. Unfortunately the animal died of distemper encephalitis 38 days after extirpation of the cerebellum.

As in the intact animal, extirpation of the cerebellum caused in the decorticate dog Robbie a gradually lessening reduction of the strength of supporting tonus : 38 days after extirpation of the cerebellum, the strength of supporting tonus had not completely recovered in the animal suffering of distemper. In the intact animal, the strength of supporting tonus is always completely recovered after extirpation of the cerebellum. Whether the strength of supporting tonus ultimately regains its normal strength after extirpation of the cerebellum in the decorticate animal, could not be excluded for the time of observation was too short in the dog Robbie.

Fig. 80. Decorticate and decerebellate dog Robbie (body weight 6.1 kgs). 6 days after extirpation of the cerebellum, 46 days after that of the cerebrum. 1. with a counterpressure of less than 2 kgs the hindlegs cannot carry the trunk and give way. 2. The forelegs can withstand a counterpressure of 9 kgs. 3. In agreement with this, the forelegs do not give way when a sandbag of 5.3 kgs is laid on the shoulders.

Extirpation of the cerebellum in decorticate animals causes :

1. *an abnormal tendency to an extended posture of the legs with increased fixation of the joints against passive, not static, flexion;*

2. *an increased fixation of shoulder- and elbow joints when set up with the back of the paw downwards;*

3. *a reduced resistance against pressure on the soles of the feet in standing posture (reduced strength of supporting tonus);*

4. *a gradual decrease of the increased extensor tonus with simultaneous increase of supporting tonus.*

E. STRENGTH OF SUPPORTING TONUS AFTER UNILATERAL DECORTICATION

After unilateral extirpation of cerebrum, dogs are able to run around within the first 12 hours; consequently, the strength of supporting tonus of the legs ipsi- and contralateral to the extirpation must amount at least to about half the bodyweight. The contralateral legs show an increased tendency to a slightly raised extensor tonus. Nevertheless, the strength of supporting tonus appears at first considerably reduced on loading the back as well as on testing with the balance.

Thus, for instance, the dog Wolf (bodyweight 8.5 kg) was not yet able to carry on his back a sandbag of 5.3 kg. three weeks after the extirpation of the left half of the cerebrum; the right hindleg gave way instantly followed by a collapse of the left hindleg. When the animal was set down on one leg only, the right hindleg gave way already on burdening the pelvis with a sandbag of 1.7 kg while the left collapsed only under a load of 5.3 kg. Similarly, the right foreleg would already give way by loading the shoulder with 7 kg, and the left, however, with only a load of 11 kg.

Thus, after the unilateral extirpation of cerebral cortex, the contralateral paws show the repeatedly observed combination of increased extensor tonus with a reduced strength of supporting tonus.

The values of strength of supporting tonus after the unilateral extirpation of the cerebrum, as established on the balance, are shown in Table 10, which shows that the strength of supporting tonus in the contralateral paws is constantly reduced after unilateral extirpation of the cerebrum. Later, it increases again a little, especially in the forelegs but, for a long time, it remains inferior to the strength of supporting tonus of the ipsilateral paws. Whether this difference is equalized after some time, as is the case after unilateral extirpation of the cerebellum, could not yet be decided for want of sufficient material and adequate time for observation.

In the dog Jenny who, in addition, was labyrinthectomized on both sides the difference in strength of supporting tonus between the two sides had almost disappeared 260 days after unilateral decortication, in the forelegs as well as in the hindlegs, i.e., in the forelegs due to the distinct increase of supporting tonus in the contralateral foreleg and in the hindlegs owing to the reduction of the strength of supporting tonus in the ipsilateral hindleg (see Table 10).

TABLE 10

Strength of supporting tonus after unilateral extirpation of the cerebrum
(the animals were alternately set up on one of the forelegs)

	Body weight	Contra-lateral foreleg	Ipsi-lateral foreleg	Contra-lateral hindleg	Ipsi-lateral hindleg
	kgs	kgs	kgs	kgs	kgs
Dog Putter					
before extirpation	6.5	+11, —12	+11, —12	+ 6, — 6	+ 6, — 6½
12 days after extirpation	5.5	+ 7, — 9	+ 9, —10½	+ 3½, — 4	+ 7, — 8
17 days after extirpation		+ 6, — 8	+10, —11	+ 3½, — 4	+ 7, — 8
Dog Bob					
before extirpation	8	+14, —17	+14, —17	+10, —12	+10, —12
9 days after extirpation		+10	+16* —20	+ 6, — 8	+10, —11
22 days after extirpation	7	+ 9, —11	+11* —13	+ 7, — 9	+ 9, —11
Dog Jenny					
before extirpation	6	+10, —11	+10, —11	+ 7, — 8	+ 7, — 8
88 days after extirpation	5.2	+ 7, — 9	+10, —11	+ 5½,— 6	+ 7, — 8
260 days after extirpation	6	+ 9, —10	+10, —11	+ 4, — 5	+ 4, — 5
Dog Schwarzweisser					
12 days after extirpation	5			+ 3 — 3½	+ 5½, — 6½

(*) In this animal the strength of supporting tonus varies in ipsilateral leg according to whether the animal keeps calm or is very agitated and makes continuous attempts to escape.

F. STRENGTH OF SUPPORTING TONUS IN THE UNILATERALLY DECORTICATE TOTALLY DECEREBELLATE DOG

In the dogs Vici and Daumling, the *right* half of the cerebrum and the entire cerebellum were removed, the *left* limbs showed an increased tendency to an extended posture (Fig. 81). In a ventral position in the air, the left limbs are extended and offer a distinct resistance against passive flexion, while the right limbs are flexed and show a much lower response to stretch.

In spite of the increase in response to stretch, the strength of supporting tonus appears greatly reduced in the left legs. This can be most

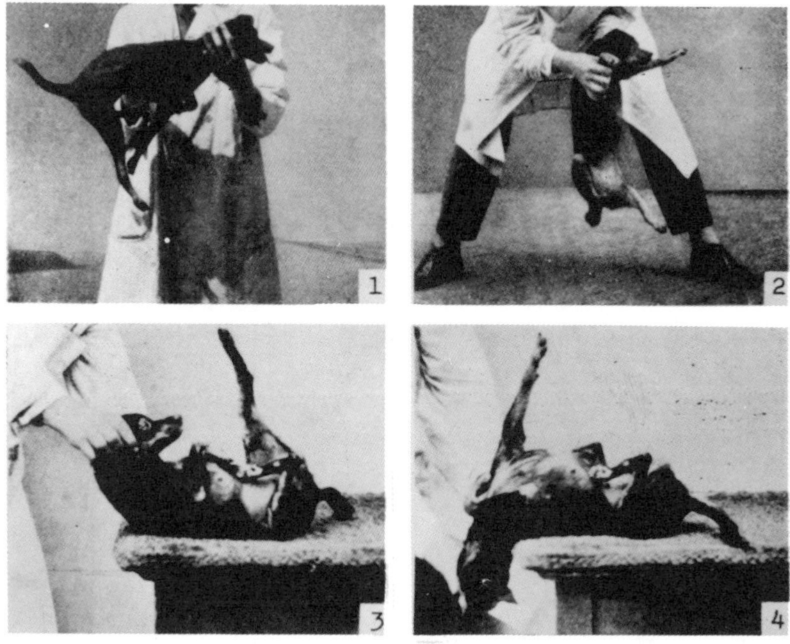

Fig. 81. Decerebellate, right side decorticate dog Vici. 1. In ventral position in the air. The animal keeps the right legs flexed. The left extended with increased extensor tonus. 2. Suspended head upwards in the air. The animal here also has the left limbs extended and the right flexed. (The right hindleg is restrained with one hand to prevent the running movements always performed in this posture, in which the left legs make ample flexion and extension movements, the right, however, perform only slight movement. 3 and 4. In a dorsal position on a supporting surface. With ventrally directed muzzle (3) the left hindlimb is rigidly extended in the air. When the muzzle is held below the horizontal (4) the left foreleg goes into an extended position with increased extensor tonus.

clearly observed when the animal is set up alternately on one of the four limbs followed by the loading of the shoulder or pelvis. In the dog Vici, the left foreleg gives way already with a loading of 1.7 kg on the shoulders, the right foreleg, however, can well bear on its shoulder a load of 5.3 kg (Fig. 82).

Thus, the distribution of supporting tonus shows the same impairments as after the extirpation of the right side of the cerebrum alone. Several distinct differences can, however, be also observed.

In the first place, the increased tendency to an extended posture of the left legs is much more pronounced, secondly, the reduction of the strength of supporting tonus is greater and can be also detected in the left foreleg and, thirdly, the increased response to stretch as well as the reduction of the strength of supporting tonus were still present to the same degree $1\frac{1}{2}$ years after extirpation in the dog Vici. Although the right

Fig. 82. 1. Dog Vici (bodyweight 5.1 kgs). (Simultaneous extirpation of the cerebellum and right half of cerebrum 7 April, 1925). The animal set down on the right forelimb. The leg does not give way under a load of 5 kgs on the shoulder. Film taken 20 May 1926. 2. Dog Daumling (bodyweight 5.3 kgs). (Bilateral extirpation of cerebellum, and right half of cerebrum; extirpation of cerebellum 30 Dec. 1925, of right half of cerebrum 20 Feb. 1926). The animal is set down on the right forefoot and the shoulder loaded with a sandbag of 5 kgs. The limb does not give way. Photo taken 8 May 1926. Dog Vici as well as dog Daumling have the left foreleg extended backwards in the air above the surface (absence of preparation for standing). In both animals this leg gives way already with a shoulder load of 1.7 kgs.

174

legs showed a normal strength of supporting tonus and in these legs the increased extensor tonus was absent, the idea suggests itself that extirpation of the cerebellum plays a role in these differences.

It is remarkable that the left legs in these animals, in addition to proprioceptive supporting reactions, also show facile magnet reactions; the decrease of strength of supporting tonus thus cannot be ascribed to an absence of the exteroceptive component.

We have, till now, no explanation for the cause of impairment in supporting tonus in these and in the unilaterally decorticate animals. One has here to be content with working hypotheses and speculations. Thus, one could think that in the adaptation to the load of the supporting tonus, the contralateral half of the cerebrum normally exerts a promoting influence which is lacking after extirpation and which is partly compensated by the cerebellum in animals with unilateral extirpation of the cerebrum alone. Another interpretation is, however, also possible, i.e., that each half of cerebrum has an inhibiting influence on the adaptation of the ipsilateral legs which is normally compensated by the other half of the cerebrum and the cerebellum.

For further research in these and other related questions, more investigations in totally and unilaterally decerebellate thalamus animals are absolutely necessary.

G. STRENGTH OF EXTENSOR TONUS IN DECEREBRATE DOGS

In the preceding paragraphs, we have seen that the increased extensor tonus[1] was almost always accompanied by a decrease in the strength of supporting tonus. We will now investigate the question as to what pressure on the soles is required in decerebrate animals to make the paws give way or collapse (Table 11).

The investigation on the balance indicated that in dog A (Fig. 6, No. 1) who showed a particularly pronounced rigidity after decere-

---

(1) It should be made clear that Rademaker is referring to intensity of extensor rigidity estimated by passive flexion of each joint, avoiding pressure on the sole of the foot, and concurrent stress on all the joints by which he estimates the intensity of "supporting tonus". (Ed.)

bration, the carrying capacity was much reduced in the forelegs as well as in the hindlegs. In dog B (Fig. 6, No. 2) where the level of decerebration was not so caudal, but transversely through the red nuclei, and who only showed a moderate rigidity, the resistance against pressure on the soles of the feet was only reduced in the hindlegs and the force in the forelegs was unchanged.

Table 11

	Carrying capacity of legs			
	before decere-bration		after decere-bration	
Dog A (bodyweight 5.2 kgs)				
both forelegs	+15	—17	+ 6	— 8
both hindlegs	+12	—14	+ 9	—10
Dog B (bodyweight 6.5 kgs)				
both forelegs	+13	—14	+13	—14
both hindlegs	+10	—11	+ 8*	— 9

(*) With strongly raised trunk +9 (—10) kg (see Fig. 83).

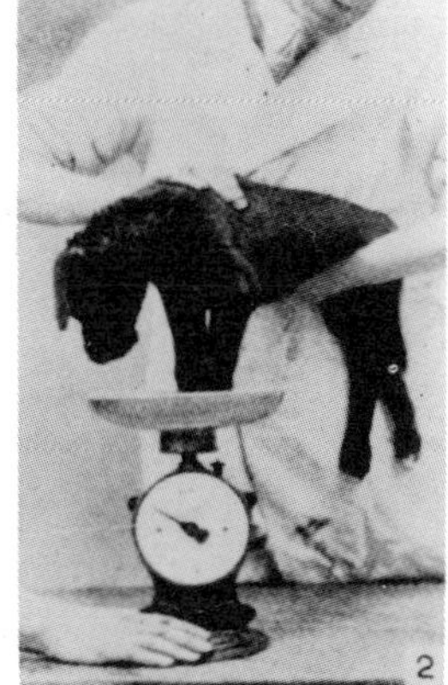

Fig. 83. Decerebrate dog B, set down on a balance with the hindlegs (1) and the forelegs (2). Neither hindlegs nor forelegs give way under a counter-pressure of 9 kgs.

When the forelegs give way, a flexion of the shoulder- and elbow joints first occurs, then suddenly the wrist joint goes into flexion and simultaneously the other joints flex. In the hindlegs, the flexion of hip-, knee- and tarsal joints increases more and more till the paw is completely flexed. *A raising and transferring of the limbs as in intact animals can never be observed in decerebrates.*

Table 12

	Force of extension of paws of decerebrate dog A in			
	standing posture, kg		dorsal position, kg	
both forelegs	$+6$	-8	$+10$	-11
both hindlegs	$+9$	-10	$+9$	-10

In the flexed position of the phalanges the carrying capacity of the paws is less but not abolished (Fig. 83).

Carrying capacity of the forelegs while standing :

on the soles	in dog A $+ 6(-8)$ kg
	in dog B $+14(-15)$ kg
on the dorsum of the foot	in dog A $+ 2(-3)$ kg
	in dog B $+ 6(-7)$ kg

Carrying capacity of the hindlegs while standing :

on the soles	in dog A $+9 (-10)$ kg
	in dog B $+8 (-9)$ kg
on the dorsum of the foot	in dog A $+4 (-5)$ kg
	in dog B $+6 (-7)$ kg

It is worth mentioning that decerebrate dogs can sometimes bear very well a moderate load on the shoulders or the pelvis at first, but after long and differing times, at best after 5 minutes, the limbs gradually begin to give way. This agrees with Liddell and Sherrington's observation that myotatic stretch reflexes usually last for about 6 minutes[1].   Due to the influence of tonic labyrinthine reflexes, the resistance of the limbs to pressure on the soles of the feet is often stronger in the dorsal position of the animal than in the standing posture (Table 12).

According to these observations, flexion of the wrist joint and digits has a much greater influence on the carrying capacity than the flexion of the toes of the hindpaws; the cause is probably that the extensors of the wrist joint are strongly stretched by the flexion of the joint and originate on the anterior aspect of the humerus.

Why the legs of decerebrate dogs in a standing posture can only carry a pressure inferior to that carried by intact dogs in spite of the strong increase of reflex contraction of the extensor muscles on stretching, is still unknown.   Liddell and Sherrington (173) observed in decerebrate animals an increase in tension of the knee extensors of 2 kg on a stretching of 1 mm.   Whether an equal stretching would perhaps cause a higher increase in intact animals has not yet been investigated.

The tonic labyrinthine reflexes probably also play a role in the low resistance to pressure on the soles of the feet, since the head of the animal, when it is set down on its paws, is in the minimum position of these reflexes and, according to Liddell and Sherrington (174), the increase in tension on stretching is much lower in decerebrate animals in this posture of the head than in the dorsal position of the head (maximum position).   Besides, in decerebrate animals the magnet reaction is missing while the effect of static stress on the soles has not yet been thoroughly investigated.*

---

(1) In a personal communication Liddell told me that he could sometimes observe myotatic reflexes lasting for half an hour caused by only a slight stretch; however, the tension of the quadriceps in a decerebrate cat on a strong stretch, for instance of 9 mm, did not last longer than 5-10 minutes.

* Static pressure on the sole of the foot greatly increases the stretch reflex in soleus in the decerebrate cat, Denny - Brown (54).

H. STRENGTH OF SUPPORTING TONUS IN LABYRINTHECTOMIZED DOGS

Ewald showed after extirpation of the labyrinth, a temporary reduction of tonus in the muscles of the extremities on the side of the labyrinthectomy. We investigated the strength of supporting tonus in several dogs labyrinthectomized on both sides by De Kleyn, *but a distinct change in the strength of supporting tonus could not be detected.* For instance, the strength of supporting tonus in the dog Jenny was:

	Immediately before extirpation		18 hours after extirpation	
in both forelegs	+20	—21	+20	—21
in one foreleg	+10	—11	+10	—11
in one hindleg	+ 6	— 7	+ 5.5	— 6.5

and in the dog Max :

	Immediately before extirpation		3 days after extirpation	
in both forelegs	+17		+16	—17
in one foreleg	+ 9	—10	+ 8	— 9
in both hindlegs	+12	—14	+14	—16
in one hindleg	+ 7	— 8	+ 7	— 8

In the labyrinthectomized dog Jenny, an extirpation of the left side of the cerebrum carried out later produced an increased stretch tonus and a reduced strength of supporting tonus in the right leg, just as in dogs with intact labyrinths.

J. ON THE STRENGTH OF SUPPORTING TONUS IN MAN

With reference to the strength of supporting tonus in diseases of the central nervous system there are no definite determinations. It can often be observed that hemiplegics in a standing posture, support themselves on the intact leg and the muscles of this extremity feel harder and more strained than the muscles of the hypertonic paralysed leg. Thévenard reported a case of postencephalitic left-sided hemiparkinsonism in a woman who showed in lying supine an evident hypertonia and a distinctly increased resistance to passive flexion of the left leg. While standing, however, the woman supported herself on the right leg whose muscles felt harder than those of the diseased left side; in this case, the right patella was fixed while the left could easily be moved laterally. The patient could stand alone only on the right leg, and scarcely, if at all, on the left one. ("Standing on one foot on the right leg is good. It is almost impossible on the left leg"). The literature contains numerous similar observations which show that the clinical conception of hypertonia does not comprise an increased strength of supporting tonus, but that in man, too, a hypertonia of the leg muscles can be accompanied by a reduced strength of supporting tonus.

In summary, the determinations of strength of supporting tonus result in the following conclusions :

1. After extirpation of the cerebellum supporting is at first distinctly reduced, sometimes even to zero; later, however, it is not less; in the supine (dorsal) position, it is even nearly always stronger than in intact dogs.

2. After unilateral extirpation of the cerebellum, the amount of supporting tonus in the legs ipsilateral to the extirpation is sometimes greatly reduced at first and thus weaker than in the contralateral legs, but it gradually increases and finally is of equal strength on both sides while standing alternately on the left and on the right legs, and thus is as strong as in intact dogs. In the supine (dorsal) position, it is usually stronger in the ipsilateral than in the contralateral legs.

3. After extirpation of the cerebrum, the strength of supporting tonus in the legs is considerably reduced; nevertheless, within the first 24 hours after extirpation, it is already so strong that it can carry the body. The decrease in the forelegs soon disappears again, but in the hindlegs it remains present for a certain time (not regularly) but sometimes even constantly.

4. In decerebellate and decorticate dog, the strength of supporting tonus remained reduced during the 6 weeks the animal stayed alive, and weaker than after an exclusive extirpation of the cerebrum, but here also a gradual increase was apparent.

5. After the unilateral extirpation of the cerebrum, the animals showed a considerable reduction of strength of supporting tonus in the contralateral legs, which was still distinctly apparent several weeks later, although the animals were almost always able to stand and run within the first 12-24 hours.

6. In two unilaterally decorticate and totally decerebellate dogs, there was a strong reduction in the strength of supporting tonus in the legs contralateral to the extirpation of cerebrum, which could still be shown in one of the animals $1\frac{1}{2}$ years after extirpation.

7. The legs of decerebrate animals in the standing posture do not show an increased resistance but usually a much reduced one against pressure on the sole of the foot. In the dorsal (supine) position, however, this resistance is indeed sometimes stronger than in the standing posture and even stronger than in intact dogs in the same position.

8. Bilateral extirpation of the labyrinth does not cause any distinct reduction in strength of the supporting tonus.

9. Often, a combination of reduced strength of the supporting tonus and an abnormally strong tendency of the limbs to an extended posture with increased extensor tonus can be demonstrated.

10. Intact dogs in the dorsal position usually have a much lower strength of supporting tonus than in the standing posture. The same manifestation is found, but not constantly, in the thalamus (decorticate) dogs and also, but to a much slighter degree, in decerebellate animals. On the contrary, in decerebrate animals the resistance to pressure on the soles is usually stronger in the dorsal position than in a standing posture.

Thus, the dorsal position exerts an inhibiting influence on the supporting tonus in intact and decorticate dogs, to a lesser degree in decerebellate animals, and usually causes a stronger reduction in the contralateral, than in the ipsilateral legs in unilaterally decerebellate dogs. This influence is absent in decerebellate animals.

Table 13

Weight of the load which can be carried on the back by animals in normal standing posture on all four legs (— sign = fails at given value).

	Body weight kg	Load on the back kg
Black normal dog	7.6	+ 5.3(—8.8)
Erik, 2½ years after extirpation of cerebellum	7	+ 9
Moor, 1 yr. after extirpation of cerebellum	9	+12—14(—15.5)
Peter, 1 yr. after extirpation of right half of cerebellum	12	+12
Fox, 3 mo. after extirpation of right half of cerebellum	7.8	+ 8
Robbie, 1 mo. after extirpation of cerebrum	6.5	— 1.7
Miesel, 1 mo. after extirpation of cerebrum	6.7	— 1.7
Bob, 6 days after extirpation of cerebrum	6.5	+ 5.3
Fuchs, 10 mo. after extirpation of cerebrum	8	— 2

Table 14

Strength of supporting tonus of the 4 legs with simultaneous pressure on shoulders and pelvis

	Body weight kgs	strength of supporting tonus of the 4 legs kgs	
Black normal dog	8	+30 —35	(—40)
Decerebellate dogs :			
Moor	9	+40 —45	(—50)
Erik	7	+30	(—32)
Piccolino	4.6	+20	(—22)

Table 15

Strength of supporting tonus in standing on two legs

	Body weight	Strength of supporting tonus of	
		2 forelegs	2 hindlegs
	kgs	kgs	kgs
Dogs :			
Black normal dog	8	+24	+15—18
Wolf, decerebellate	9½	+35	+27
Moor, decerebellate	9	+30	+20
Erik, decerebellate	7	+23	+15
Piccolino, decerebellate	4.6	+18	+ 9
Peter, unilaterally decerebellate	12	+35	+25
Fox, unilaterally decerebellate			
Miesel, decorticate after 4 weeks	6.7	+16	+ 3½ (—4½)
decorticate after 5 weeks		+20	+ 4 (—6)
Robbie, decorticate after 4 weeks	6.5	+ 9	
decorticate after 5 weeks	6.8	+15	+ 3 (—4)
3 weeks after extirpation of cerebellum		+ 9	— 2
Bob, decorticate after 10 days	6.5		+20
after 12 days			+20—25
Fuchs, decorticate after 10 months	8.5	+18	— 3
Cats :			
Carolus, decerebellate after 2 years	4.35	+11	+10
Pierrette, decerebellate after 2 years	4	+10	+ 7.5
Nikker, decerebellate after 2 months	2.1	+ 4.5	+ 3.8

Table 16

Strength of supporting tonus in standing on one leg

	Body weight	Strength of supporting tonus of the			
		Right foreleg	Left foreleg	Right hindleg	Left hindleg
	kgs	kgs	kgs	kgs	kgs
Black normal dog	8	+11	+11	+ 9	+ 9
Wolf, decerebellate	9.5	+17	+17	+12	+12
Moor, decerebellate	9	+13—15	+13—15	+10—13	+10—13
Erik, decerebellate	7	+13	+13	+ 8	+ 8
Piccolino, decerebellate	4.6	+ 9	+ 9	+ 4½	+ 4½
Miesel, decorticate	5.3	+11	+11	— 2	— 2
Robbie, decorticate	6.8	+ 7	+ 7	— 2	— 2
Bob, decorticate	7.5	+11—13	+11—13	+ 9	+ 9
Fuchs, decorticate	8	+14	+11	— 2	— 2
Peter, right sided decerebellate	12	+17	+17	+13	+13
Fox, right sided decerebellate	8.5	+15	+15	+10.5	+10.5
Bob, left sided decorticate	7.5	+10	+16	+ 6	+10
Putter, left sided decorticate	6.5	+6—7	+ 9—10	+ 3½	+ 7
Schwarzweiss, left sided decorticate	5			+ 3	+ 5½
Jenny, left sided decorticate	5.2	+ 7	+10	+ 5½	+ 7

VII. ALTERATION OF SUPPORTING TONUS BY RAISING AND LOWERING OF THE HEAD. THE SUPPORTING TONUS AND THE VERTEBRA PROMINENS REFLEX

The strength of supporting tonus can be influenced by alterations of static conditions, for instance by alterations in the position of the supporting surface for the animal. The adaptation of supporting tonus to static conditions is produced by the coordination if numerous reflexes emanating from different parts of the body. It will now be our task in the following chapters to discuss these reflexes in detail.

Fig. 84.   Normal cat.   1. The animal raises the head to look at a piece of meat held high in the air.   Due to the effect of tonic neck reflexes, a vigorous extension of the forelegs with raising of the anterior part of the body takes place.   2. The meat is held at ground level.   The animal looks down and holds the neck ventrally flexed.   Consequently, the forelegs are in a flexed posture and the anterior part of the body is lowered (from R. Magnus : Körperstellung.   Figs. 32 and 33).

Magnus and De Kleyn (139) have shown that in decerebrate animals, leg position and muscle tonus in the extremities are regularly altered by raising and lowering of the head (tonic neck and labyrinthine reflexes).   In decerebrate dogs and cats, a raising of the head causes in the forelegs an increase and in the hindlegs a decrease of extensor posture and tonus.   Conversely, a lowering of the head produces a decrease of extensor posture in the forelegs and an increase in the hindlegs.   These reactions are present in decerebrate animals in the ventral, lateral and dorsal position.   Following Magnus and De Kleyn, the influence of tonic neck and labyrinthine reflexes can also be detected in the daily life of intact animals (Fig. 84).

On the other hand, neither intact nor decorticate or decerebellate dogs or cats, on lifting and lowering of the head in lateral or dorsal position, show any distinct change of muscle tonus and position of the extremities (Fig. 85).

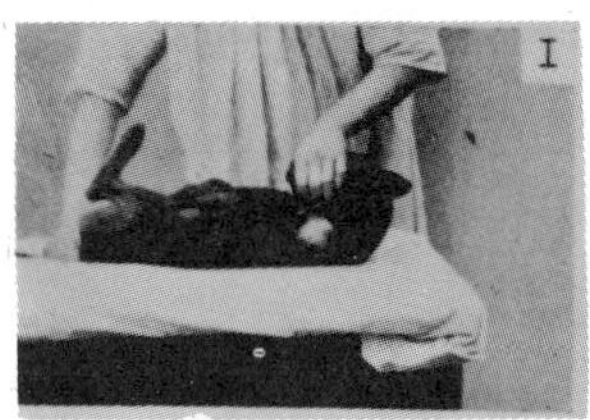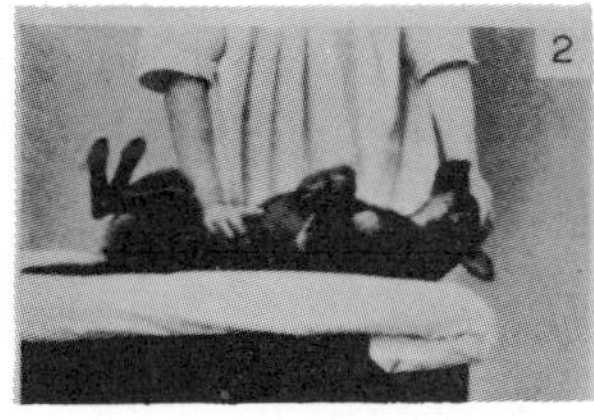

Fig. 85. Decerebellate dog Moor in dorsal position. 1. Muzzle in ventral direction. 2. Muzzle kept vertically upwards. 3. Muzzle lowered to below the horizontal level in dorsal direction. The head posture does not exert a distinct influence on the position of the limbs.

The investigations on supporting tonus succeeded in clearing up this contradiction. When a decerebellate dog is laid on its back, the muzzle directed vertically upwards or ventrally, we have seen that hindlegs on static stress go instantly into an extended posture and can only be flexed with difficulty by a pressure on the soles. When the head is slowly moved backwards, one can feel that the resistance diminishes and in lowering the head below the horizontal, after a short interval it will suddenly cease and simultaneously the hindlegs are moved forwards (Fig. 86). Conversely, in the ventral movement of the head, the hindlegs on static stress are extended again as soon as the muzzle stands above the horizontal and with this movement the hand placed on the paws is pushed away with considerable force.

Thus, in a dorsal position, the supporting tonus is maximally strong when the muzzle is directed vertically upwards or ventrally, in return it is equal to zero when the muzzle is held 45-60° below the horizontal.

In the first mentioned position of the head, the magnet reaction is clearly present while it is absent in the latter; neither on touching the soles nor on static stress an extension of the hindlegs occurs (Fig. 87).

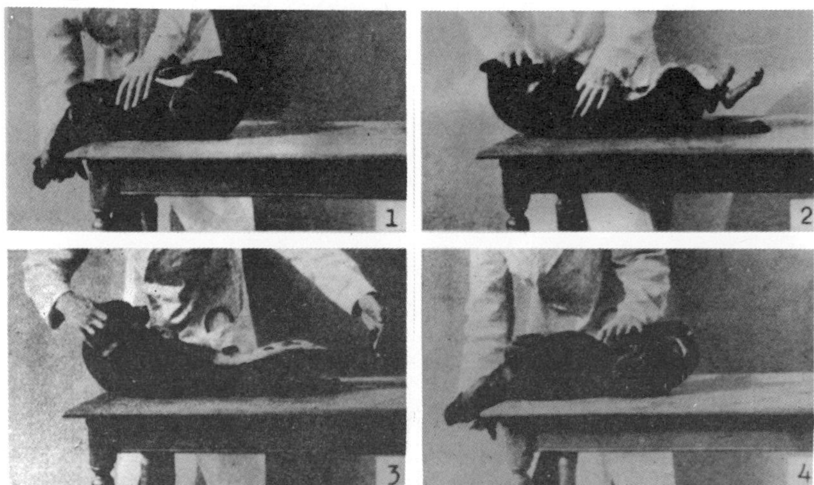

Fig. 86. Influence of lifting and lowering the head on the posture and the supporting tonus of the hindlegs in the decerebellate dog Moor in a dorsal position. 1 and 2. The hindlegs are in a posture of flexion with dorsally extended as well as with ventrally flexed head.   3. With a ventrally directed muzzle the hindlegs on static stress are extended backwards and the hand laid on the soles is forcibly pushed away.   4. On dorsal movement of the head to below the horizontal level the extension and fixation of the hindlegs gradually cease and they are flexed and moved forwards.

Lifting and lowering of the head in the dorsal position also exerts a distinct, even if less strong, influence on the supporting tonus of the forelegs. In opposition to the hindlegs, an extended posture and a resistance against flexion of the forelegs in static stress are at a maximum when the muzzle lies below the horizontal, it is much lower when the muzzle is directed vertically upwards and lower still when directed ventrally, but even here the supporting tonus is clearly present.

The same influence of head posture can be observed in the lateral and in the standing positions. When in a standing dog the head is moved downwards, the hindlegs are strongly extended backwards while the forelegs are flexed in shoulder - and elbow joints (Fig. 88, No. 1).

In the lifting of the head, the supporting tonus of the hindlegs decreases and disappears when the muzzle is moved dorsally to or beyond the vertical. In this case, the hindlegs usually flex by themselves and are at the same time moved forwards. With a lifted head, the forelegs are totally extended and offer maximum resistance against pressure on the shoulders (Fig. 88, No. 2).

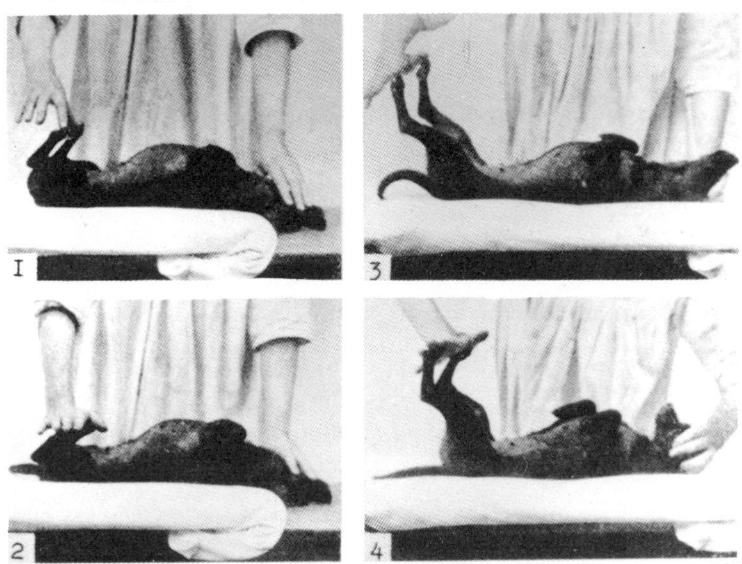

Fig. 87. Influence of head posture on the magnet and supporting reactions in the decerebellate dog Piccolino in dorsal position. 1 and 2. In dorsal position with the muzzle below the horizontal level neither touching (1) nor static strain (2) causes an extension of the hindlegs. 3 and 4. Conversely, a brisk magnet reaction on touching the sole appears in dorsal position with the muzzle pointing upwards (3), pressure on the soles encounters considerable resistance (4).

Fig. 88. Influence of lifting and lowering the head on the supporting tonus of the legs in the decerebellate dog Piccolino in a standing posture. 1. With passively depressed head the hindlegs are extended and show a strong supporting tonus while the forelegs are somewhat flexed at the shoulder and elbow joints. 2. On passive lifting of the head, the hindlegs flex and the forelegs are maximally extended.

The reactions on lifting and lowering of the head are produced, above all, by alterations in posture of the neck and by the tonic neck reflexes. The fact that tonic labyrinthine reflexes play only an inferior role is proved by the following observations: Firstly, dorsal and ventral movements of the head produce similar alterations of supporting tonus in standing and in dorsal positions, although the changes in the position of the labyrinth are not the same. Secondly, lifting and lowering of the head as well as the tonic neck reflexes exert an opposite influence on the tonus of fore- and hindlegs. However, the tonic labyrinthine reflexes are directed similarly. Thirdly, the legs also show typical alterations in their supporting tonus when the position of the neck with regard to the trunk, and not the position of the head in space, is alone altered (Fig. 89). In the fourth place, changes in posture and supporting tonus also occur distinctly in labyrinthectomized dogs. To produce these alterations, it is not necessary to even move the head and neck and they also appear when, with head and neck in an unchanged position, the trunk is moved dorsally and ventrally (Fig. 90, No. 3 and 4). The position of the head in relation to the neck in the atlanto-occipital joint (Fig. 90, No. 1 and 2) as well as that of the neck in relation to the trunk (Fig. 90, No. 3 and 4) exert an influence.

Fig. 89. In the decerebellate dog Piccolino, the head is moved dorsally (1) and ventrally (2) to the trunk while the head is held with the muzzle at an angle of 45° above the horizontal. Although the position of the head in space is not altered, the limbs show a distinct supporting tonus that diminishes with the dorsal movement of the neck (1).

Ventral movement of the head as well as ventral movement of the neck exert a facilitating influence on magnet and supporting reactions of the hindlegs while they are inhibited by a dorsally directed change in position. The influence of change in the head position in the atlanto-occipital joint can be observed particularly clearly in the dorsal position when the neck is directed dorsal to the trunk (Fig. 91).

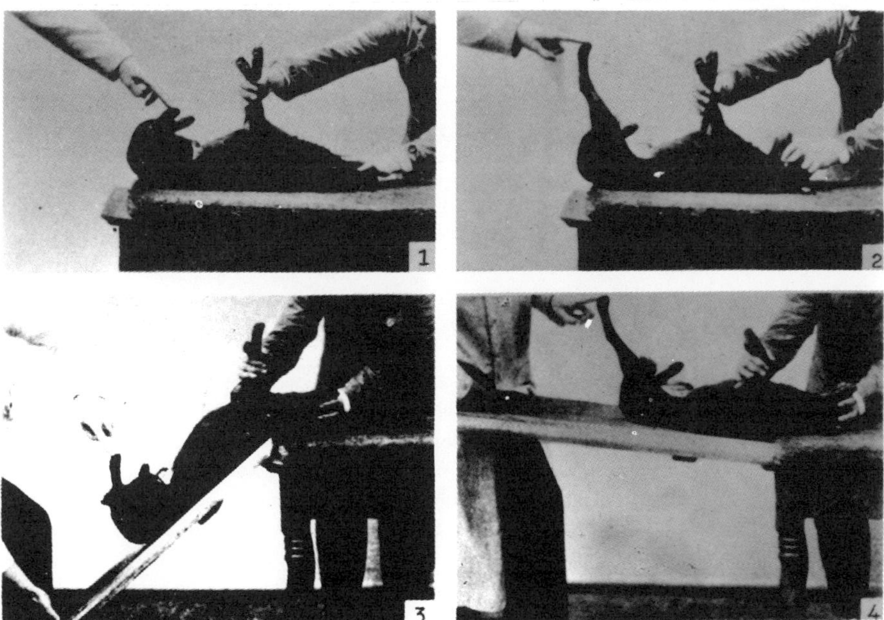

Fig. 90. 1. Decerebellate dog Erik in a dorsal position with the neck flat on the supporting surface, the muzzle horizontal. On touching the sole, the hindleg shows no distinct magnet reaction. 2. Without changing the posture of the neck, the head is ventrally flexed in the atlanto - occipital joint till the muzzle points vertically upwards. On touching the sole, the hindleg instantly goes into an extended posture and is fixed in this position. 3. Decerebellate dog Erik in dorsal position; the neck lies on a table with the muzzle pointing upwards at an angle of 45°, while the trunk lies on a board which is moved up and down from the caudal end. When the board is moved below the plane of the table, touching of the sole does not produce any magnet reaction in the hindlegs. 4. This reaction, however, is produced immediately when the end of the board is raised above the plane of the table.

In decerebrate cats, Magnus and De Kleyn could not detect any distinct influence on the extensor rigidity of changes in position of the atlanto-occipital joint. Magnus and Storm van Leeuwen (190) have proved that stimulations for tonic neck reflexes in cats emanate from the most rostral parts of the neck, since these reflexes are abolished by severing the posterior roots of C I, C II, and C III. It appears from our observations that tonic neck reflexes in dogs can also be produced by stimulations from other parts of the neck. The neck reflexes have a tonic effect on the supporting tonus, for it persists in the hindlegs as long as the head and neck remain ventrally flexed (up to 45 minutes and more).

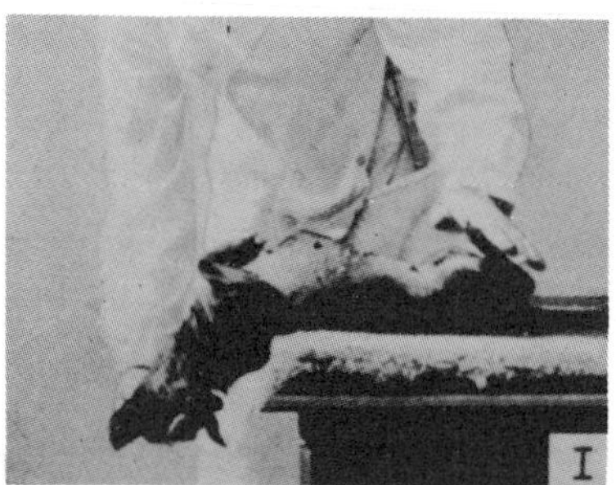
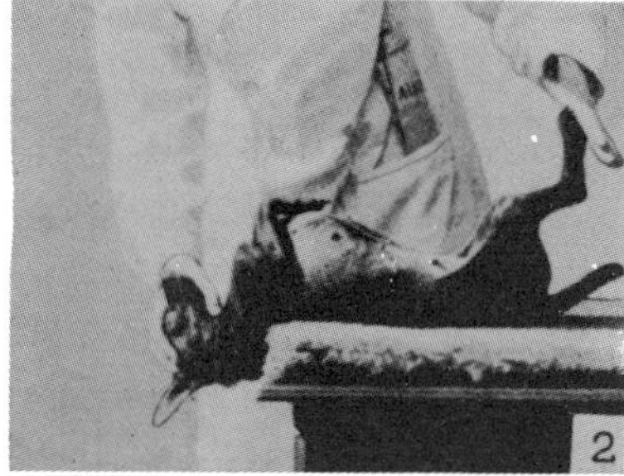

Fig. 91. Decerebellate dog Piccolino in a dorsal position table: the neck is hanging dorsally over the edge of the table and the head is moved dorsally and ventrally to the neck. In a dorsal flexion, the supporting tonus of the hindlegs disappears (1). With a ventral flexion it returns together with the magnet reaction. Thus the influence of the position of the head in relation to the neck takes precedence over the influence of the position of neck to trunk. (These alterations also occur when the anterior part of the neck is fixed and the change of position occurs only in the atlanto-occipital joint).

Thus, in the regulation of supporting tonus, the reflexes produced by the alteration in position of the different parts of the neck play an important role, though not all the observed alterations have to be regarded as direct consequences of these reflexes. For instance, the posture of the pelvis is changed by lifting and lowering of the head in the dorsal position of the animal, thus causing a secondary alteration of supporting tonus (see Chapter X). However, the limbs also show the typical changes when the pelvis is fixed to the supporting surface. When the head of the standing animal is raised, first the trunk is displaced caudally, causing the legs to move forwards in hip- and shoulder joints, secondly, there occurs a change in the curvature of the spinal column and, thirdly, the posture of the pelvis is altered (Fig. 88 and 142); all these three facts influence the supporting tonus as we shall see later.

After elimination of these three factors, however, there still remains the distinct and strong influence of the tonic neck reflexes on the supporting tonus produced by the lifting of the head.

In *intact dogs*, the influence of the tonic neck reflexes on the supporting tonus can be clearly observed *in the standing posture* : with the raising of the head, the support of the hindlegs decreases, while it increases in the lowering of the head. However, the influence of the reflexes in the *dorsal position* is variable and less distinct. The supporting tonus is always completely absent in the dorsal position when the muzzle lies below the horizontal, but sometimes also when the muzzle is directed upwards or ventrally. Some of the animals in the dorsal position show a distinct extension of the statically stressed hindlegs when the head is moved ventrally. In most of them, however, this extension does not appear. (In this, in addition to cerebral conditioned inhibitions, unconditioned inhibitions also play a role, as we shall see in Chapter XIV).

In *decerebellate dogs*, the alterations of supporting tonus on lifting and lowering of the head always appear conspicuously strong and abrupt. In the dorsal position, the hindlegs always totally collapse under static stress when the muzzle is moved dorsally below the horizontal, even when no strong pressure is applied and the soles are simply touched; conversely, in the ventral flexion of the head, the hand pressing on the soles is pushed away with considerable force and a maximum extension occurs. With a vertically or ventrally directed muzzle, the extended hindlegs are able to lift a board loaded with several kilograms, laid on the soles, and to keep it raised as long as the head remains in this posture (Fig. 92). The alterations appear also remarkably strong and automatic in the standing position. In the passive lowering of the head, the extension of the hindlegs is sometimes so strong that the rearpart of the body jumps into the air (Fig. 93).

In the daily life of decerebellate animals, abnormal reactions can be observed in the *active* lifting and lowering of the head, i.e., particularly distinct in the stage when the animals again begin to stand and to run. When the animals (cats and dogs) lower the head in order to drink in a standing position, the hindlegs are usually so excessively strongly extended that the muzzle is plunged into the milk bowl. Frightened, the animals withdraw the head quickly backwards and, with this movement, the hindlegs bend in and the forelegs are extended so strongly that the anterior part of the body is lifted from the ground and the animals almost turn into a backwards somersault. Later, the animals avoid excessive movements of the head and gradually lower their shaking head and almost always lie down on their bellies for the purpose of drinking. However, the strong reactions with active abrupt head movements, still continue to appear when the animal is frightened.

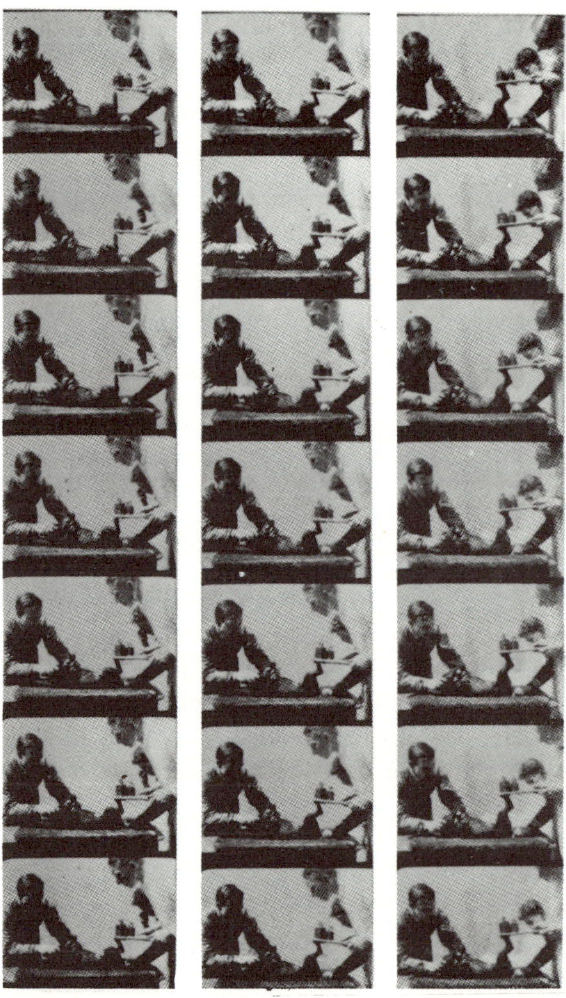

Fig. 92. Decerebellate dog Piccolino (bodyweight 4½ kgs) in dorsal position with ventrally directed muzzle. A board loaded with 3 kgs is laid on the soles of the flexed hindlimbs (3). Thereupon the hindlegs are extended, lifting the board, and keep it elevated as long as the position of the head remains unchanged.

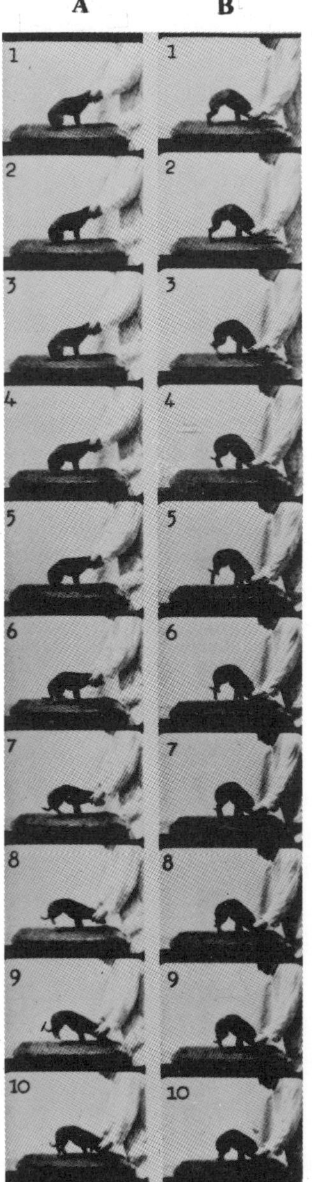

Fig. 93. Decerebellate dog Piccolino. In passive lowering of the head the posterior part of the body is sometimes raised so high that the hindlegs leave the ground. This is brought about in different ways. A. By the strong fixation of the cervical and dorsal spine the posterior part of the body is sometimes lifted as if with a lever. A 1–5 : Passive lowering of the head of the animal sitting on the posterior part of the body. A 6 : When the neck is lowered to the horizontal the posterior and hindlegs are raised from the ground. The hindlegs are still flexed. A 7–9 : With further lowering of the head, the posterior is raised further. The back is still fixed and almost straight (A 9). A 9–10 : Finally the hindlegs are extended and the posterior moves downwards. B. The raising of the posterior part of the body is sometimes brought about by a particularly strong and sudden extension of the hindlegs so that the posterior jumps upwards. B 1: The head of the animal is slowly lowered till the muzzle touches the ground. With this the posterior has not lifted itself from the ground but the back is curved. (Compare the curve of the back in the same position of the head in A 9). B 2–5 : The passive lowering of the head, however, produced such a strong extension of the hindlegs that by it the posterior jumps upwards.

Thus, on lifting and lowering the head in a standing posture as well as in a dorsal position, decerebellate dogs, as soon as they can stand and run again, show a particularly strong, abrupt and automatic increase and decrease of the extensor posture and supporting tonus of the legs under static stress, while in the legs not under static stress no distinct alterations of position and tonus can be observed.

In the first few days after extirpation of the cerebellum, the limbs are sometimes limp, and supporting and magnet reactions are absent. Similarly, the tonic neck reflexes cannot be demonstrated. Other animals following extirpation, become rigid and the stiffly extended legs show tonic cervical reflexes like those in decerebrate animals even when the legs are not under static stress.

In the early phase after extirpation in *unilaterally decerebellate dogs*, tonic neck reflexes can distinctly be observed in the limbs without static stress, first in the limbs ipsilateral to the extirpation, just as with increased extensor tonus. The contralateral limbs then usually react only on dorsal and ventral movements of the head when they are under static stress.

Later, the tonic neck reflexes in limbs not statically stressed disappear simultaneously with the dying away of the increased extensor tonus. The alterations of head position then produce only changes in position and tonus of the limbs when the soles are touched or under static stress. In the stage of permanent impairment, the strength of supporting tonus of the hindlegs is only very weak in the standing posture of the animal with the head raised high; however, with a ventrally bent head, it is considerable and equally strong on both sides. In the dorsal position, when the head is held down dorsally below the horizontal, the statically stressed hindlegs are flexed and their force of supporting tonus is nil; with the muzzle directed vertically upwards or ventrally, both hindlegs are extended, but the increase in supporting tonus in this case is sometimes stronger in the ipsilateral leg than in the contralateral one (as for instance in the dog Fox).

Thus, the alteration of head position in a ventral direction produces an increase in supporting tonus which, in the standing posture is equally strong in both hindlegs. In the dorsal position, however, it is of different strength on the two sides and, especially in the contralateral limb, it is usually considerably lower than that in a standing posture.

For instance, when the dog Fox was set down alternately on one or other of the hindlimbs, the strength of supporting tonus with a raised head was about zero in each of the hindlegs and with a ventrally flexed head it was 10 kg. Thus, the alteration of the head position caused in each of the hindlegs an increase of $\pm$ 10 kg.

In the dorsal position with the muzzle below the horizontal, the strength of supporting tonus was also equal to zero, however, with the muzzle pointing vertically upwards, in the right hindleg, ipsilateral to the extirpation, the supporting tonus was 6 kg and in the left it was 1 kg. Thus, the increase here was lower by 4 kg in the ipsilateral limb and in the contralateral limb even lower by 9 kg than in the standing posture. On raising the head in the standing animal a distinct increase of the extended posture and strength of supporting tonus can usually be observed in the two forelegs (tonic neck reflexes) but sometimes there occurs a flexion of the ipsilateral limb. This phenomenon is caused by the stretch myotatic reflex of the sternocleidomastoid muscle, which terminates in dogs on the distal portion of the humerus. The flexion appears when the animal leans heavily on the contralateral limb, although the soles of both forelimbs touch the ground.

Sometimes in decorticate dogs in the first days after extirpation, weak tonic neck reflexes can be observed in limbs not under static stress. These, however, disappear later. Alterations of the head position do not then produce reactions in the extremities when they are under static stress. The strength of supporting tonus of the hindlegs of decorticate dogs in the dorsal position is almost zero in all positions of the head. The decorticate dog Bob was an exception; its hindlegs in the dorsal position with an upwards or ventrally directed muzzle showed a much stronger supporting tonus.

In standing decorticate dogs, however, lifting and lowering of the head exerts an evident influence. With a raised head, a distinct supporting tonus is lacking in the hindlegs and with a lowered head, it is clearly present but less strong than in intact dogs (again with the exception of the decorticate dog Bob).

In unilaterally decorticate dogs in the dorsal position, on lifting and on lowering the head, the legs contralateral to the extirpation usually show distinct, though not strong, tonic neck reflexes (Fig. 94) at first, which later gradually grow weaker and usually disappear completely after a certain time. Only minor alterations can be observed in this case in the paws under static stress. In the standing posture, however, a lowering of the raised head causes the appearance of supporting tonus in the hindlegs of both sides, but the force of supporting tonus is less strong in the hindleg contralateral to the extirpation than in the ipsilateral one.

Thus, for instance, the dog Jenny, 3 months after extirpation of the left half of cerebrum, when alternately set down on one or other of the two hindlegs, showed with the head raised a strength of supporting tonus equal to zero, in both legs with a lowered head, however, a strength of supporting tonus of 5 kg in the right hindleg and 7 kg in the left one. The increase in the left hindleg amounts to just as much as that before the unilateral extirpation of the cerebrum. In the dorsal position, the strength of supporting tonus in the right as well as in the left hindleg is equal to zero when the head is moved dorsally with the muzzle below the horizontal, while with a ventrally flexed head it amounts to some-what less than 1 kg in the contralateral right leg, and to a little more than 1 kg in the left one.

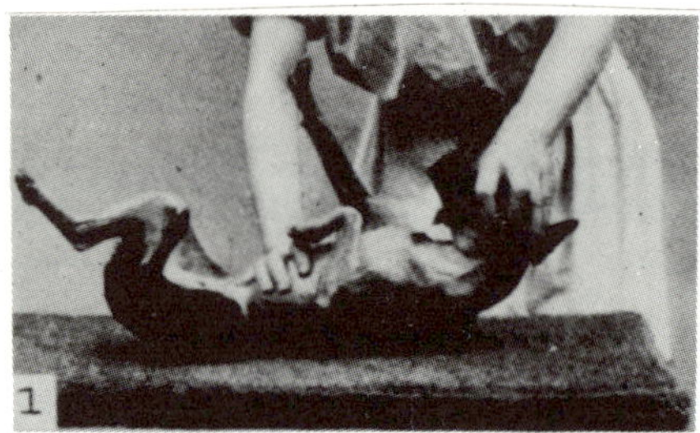

Fig. 94. Tonic neck reflexes on lifting and lowering of the head in a unilaterally decorticate dog. Labyrinthectomized dog Jenny after extirpation of the *left* cortex in dorsal position. 1. With ventrally directed muzzle, the left legs are flexed, the *right* extended with increased static tonus. 2. On lowering the muzzle below the horizontal, the extended posture of the right hindleg disappears while that of the right foreleg increases. The left legs do not show any distinct alterations in position.

In the unilaterally decorticate, totally decerebellate dog Vici, the reactions on the dorsal and ventral movements of the head show several remark-able differences. In this *rightsided* decorticate animal the *left* legs, not statically stressed in the dorsal position of the animal, showed strong tonic neck reflexes which could not be observed in the right legs (Fig. 95, No. 1 and 2). This agrees with our findings after the unilateral extirpation of the cerebrum alone with the only difference that here the tonic neck reflexes in the contralateral legs were usually clearly present for only a short time, but in the dog Vici they remained remarkably strong 1½ years after extirpation.

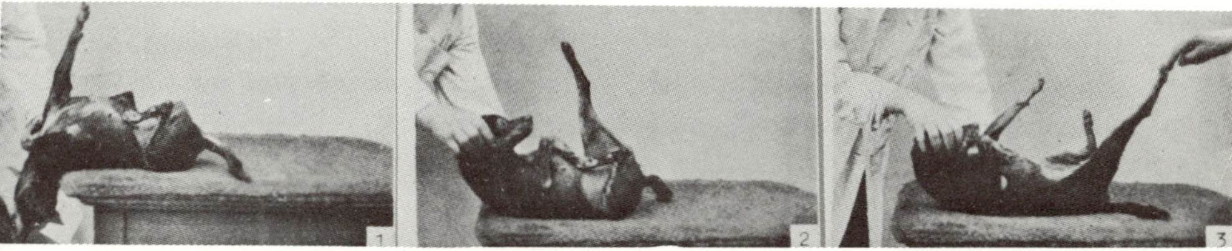

Fig. 95. Tonic neck reflexes in the decerebellate, rightsided decorticate dog   Vici in
dorsal position.   1. Where the muzzle lies below the horizontal, the left foreleg
goes into maximum extended posture and shows a strong extensor tonus,
while the other three legs are kept flexed.   2. On ventral flexion of the
head the extensor tonus of the left foreleg diminishes, the paw usually (not
always) falls down into flexion, while the left hindleg is extended and shows
a strong resistance against passive flexion.   The right limbs remain flexed
even now.   On lifting and lowering of the head, the limbs show no tonic
neck reflexes , while the left paws show lively one.   3. When, in the dorsal
position with the muzzle below the horizontal, the sole of the left hindleg
is touched or statically stressed and the head flexed ventrally, this leg then
goes into a maximum extended posture while no extension of the right leg
occurs.

The reactions of the left hindleg disappeared on touching or static
stress on the sole of the right hindleg.   In this case, however, the leg
showed distinct alterations in supporting tonus on lifting and lowering
of the head, so that now on the ventral movement of the head, the right
leg instead of the left one went into an extended posture (Fig. 95, No. 3).

The diminution of the tonic neck reflexes on static stress on the
opposite leg can sometimes also be observed in unilateral decorticate
or unilaterally decerebellate dogs.   In the first phase after the unilateral extirpation of the cerebellum, the hindleg on the side of extirpation without static stress in the dorsal position of the animal, as we have
seen, shows distinct tonic neck reflexes.   These also fail to appear when
the sole of the contralateral paw is touched or statically stressed. Thus,
the static stress on the paws contralateral to the extirpation in unilaterally decerebellate animals produces not only a decrease of extensor
supporting tonus in the opposite paw, but also an inhibition of the tonic
neck reflexes in these legs.

After unilateral decortication alone, static stress on the hindleg ipsilateral to the extirpation has sometimes the same effect as in the dog Vici. Sometimes, static strain in the dorsal position with an upward directed muzzle produces no supporting tonus in the ipsilateral leg and, in this case, the leg also shows no distinct changes in supporting tonus on alteration of the head posture and the tonic neck reflex effect on the contralateral leg is less inhibited. In the dog Vici, however, the stressed right hindleg shows a particularly lively change of supporting tonus on alterations of head position, probably as a result of the extirpation of the cerebellum, while at the same time the tonic neck reflexes of the left hindleg are completely inhibited.

With a simultaneous stress on the limbs of both sides, strong bilateral alterations in position and tonus occur with ventral and dorsal movements of the head as well as after the extirpation of the cerebellum alone (Fig. 96). On the ventral flexion of the head, both hindlegs are extended and show a vigorous resistance to passive flexion which is stronger in the right hindleg.

Standing alternately on the left and the right legs, the supporting tonus distinctly changes on lifting and lowering of the head. With a raised head, the strength of supporting tonus is zero in both hindlegs, and on lowering a vigorous supporting tonus of the hindlegs appears, the right side being stronger than the left.

Thus, similarly to exclusively decerebellate animals, the animal Vici with right decortication shows in a dorsal position neither an inhibition of magnet reactions and supporting reactions, nor the reactions on lifting and lowering of the head. Similarly to exclusively unilateral decorticate animals, the increase in supporting tonus on ventral movement of the head is much lower in the hindleg contralateral to the extirpation of cortex than in the ipsilateral one, although the contralateral leg reacts to the changes in head position with a strong extensor tonus when the legs are not under static stress, a reaction which is absent in the ipsilateral hindleg.

In the decorticate and decerebellate dog Robbie, the unstressed legs showed lively tonic neck reflexes on the lifting and lowering of the head. Unfortunately, the animal died before the cessation of rigidity so that investigations on the influence of the support alone could not be carried out.

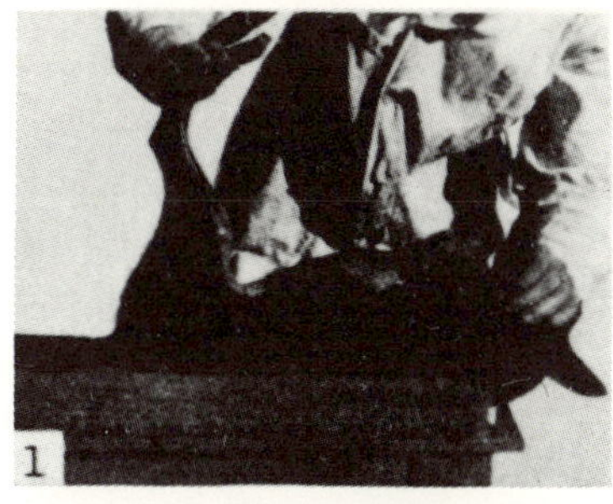
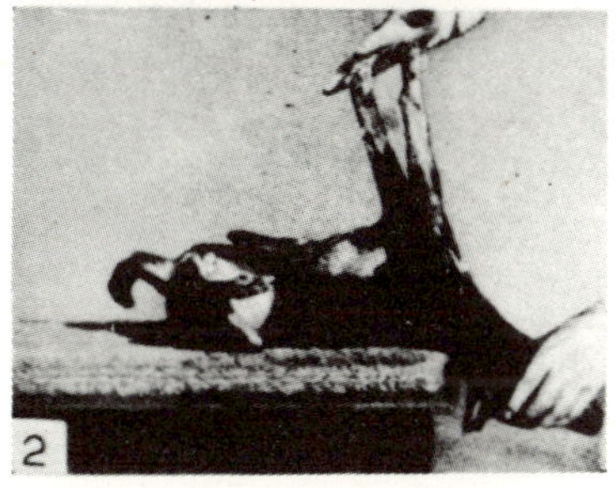

Fig. 96. Decerebellate, rightsided decorticate dog Vici in dorsal position.
1. When both hindlegs are under static stress, both go into extended
posture on the ventral flexion of the head, which they maintain with
a strong supporting tonus as long as the muzzle is ventrally directed.
2. On lowering the head below the horizontal, both forelegs under
static stress show a maximum extension posture with strong fixation.

Labyrinthectomized dogs, in their behavior on lifting and lowering of
the head, behave essentially like intact dogs.

In decrebrate animals, the rigid extremities show lively tonic neck
reflexes. As mentioned before, the tonic neck reflexes were observed
in decerebrate animals for the first time by Magnus and De Kleyn.

In man, also under pathological conditions , see, for example,
Tonische Hals- und Labyrinthreflexe beim Menschen [Tonic neck and
labyrinthine reflexes in man in R. Magnus, *Körperstellung*, p. 113], and
in normal children in the first half year of life (Landau, Schaltenbrand),
alterations in position and tonus in the extremities not under static
stress on lifting and lowering of the head can be observed.   Lowering
of the head in some cases causes an extension and in others a flexion
of the arms.

Whether these alterations of the head position under pathological
conditions (for instance in cerebellar diseases) could also change the
supporting tonus of the legs has not yet been investigated.   In normal
adults, they do not exert any distinct influence on the supporting tonus.

When a man in a standing posture, without moving the anterior part of the body, alternatively moves his head strongly forwards and backwards, one can usually feel that with backward flexion of the head the psoas and the muscles of the back of the thigh feel somewhat lax, while the muscles of the anterior aspect are particularly strained. With the head bent forwards, however, the psoas and the muscles of the back of the thigh are more strained and those of the front somewhat relaxed. The impression is gained that in normal human beings, under certain conditions, the posture of the head exerts some, though little, influence on the distribution of tonus in the supporting legs.

In summing up, our observations lead to the following conclusions :

1. Lifting and lowering of the head in dogs and cats produces distinct alterations in supporting tonus and posture of the statically stressed legs, owing to stimulations emanating from both rostral and caudal parts of the neck.

2. Lifting and lowering of the head influence the supporting tonus and the posture of fore- and hindlegs in an opposite sense; the alterations are more considerable in the hindlegs than in the forelegs.

3. In a standing posture with head flexed strongly backwards, the strength of supporting tonus in the hindlegs is slight, so that the legs sometimes give way; with a ventrally flexed head, the legs are extended and show a vigorous supporting tonus.

4. The alterations in supporting tonus also appear when the unstressed legs do not show any tonic neck reflexes.
   Generally, the increase in supporting tonus on ventral flexion of the head is slighter in the legs which show strong tonic neck reflexes without being under static stress than in those legs in which these reflexes are absent.

5. In intact dogs, the alterations in supporting tonus and posture on lifting and lowering of the head are more prominent in the standing posture. In the dorsal (supine) position, they are weaker and inconstant.

6. In decerebellate dogs, the alterations in supporting tonus and posture are very vigorous some time after extirpation in the standing posture as well as in the dorsal position. The abrupt extension of the statically stressed hindlegs on the ventral move-

ment of the head is particularly remarkable.

7. In a late phase after extirpation, unilaterally decerebellate dogs in a standing posture show on both sides equally strong changes of supporting tonus; however, in the dorsal position, the increase of supporting tonus on ventral movement of the head is usually stronger in the ipsilateral hindlegs than in the contralateral ones. The increase in the ipsilateral hindleg is sometimes almost as strong as the increase in standing posture while the increase in supporting tonus in the contralateral leg is considerably lower.

8. Decerebellate dogs, in the standing posture (with the exception of dog Bob), have less increase in supporting tonus of the hindlegs on ventral flexion of the head than in intact dogs. In the dorsal position, it is totally absent.

9. Unilaterally decerebellate dogs show, in standing posture as well as in the dorsal position, a less strong increase in supporting tonus in the contralateral legs than in the ipsilateral ones and in the dorsal position this increase is very slight on both sides.

THE IMPORTANCE OF ALTERATIONS IN SUPPORTING TONUS ON LIFTING AND LOWERING OF THE HEAD FOR ADAPTATION TO STATIC CONDITIONS

When an animal stands on a supporting surface which is raised by the tail end (Fig. 97, No. 2) and lowered by the head end (Fig. 98, No. 1), the head is also passively moved forwards. By altering the position of the head in space, the labyrinthine righting reflexes take effect and lead the head back into its initial position in space, i.e., they move the head so far dorsally to the trunk, that it maintains its position in space. The alteration of position of the head in relation to the trunk will produce a decrease in supporting tonus and of the extension posture of the hindlegs, and an increase in the forelegs, and thus participate in the adaptation of the limb posture to the position of the supporting surface.

Conversely, on lowering the tail end (Fig. 97, No. 1) or raising the head end of the board (Fig. 98, No. 2), the head is moved ventrally to the trunk by the labyrinthine righting reflex through which supporting tonus and extensor posture of the hindlegs are increased.

Thus, the labyrinthine righting reflexes together with the tonic neck

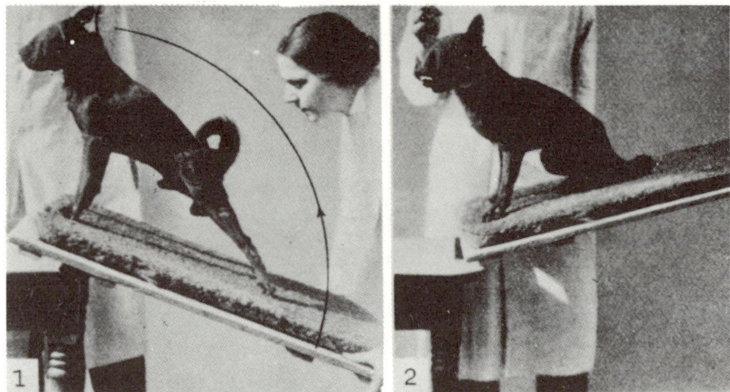

Fig. 97. Decerebellate dog Erik with blindfold on a support which can be raised at the tail end.   1. The tail end of the board is lower than the head end. Head in normal posture, i.e., the muzzle at an angle of 45° below the horizontal: neck and spine form almost a straight line. The hindlegs maximally extended, the muscles of the hindlegs in strong contraction. 2. The tail end of the board is higher than the head end. Due to the influence of labyrinthine righting reflexes, the head was moved dorsally to the trunk, so that the muzzle is now also situated 45° below the horizontal. Neck and dorsal spine form an upward obtuse angle. In agreement with the alteration of head posture and the trunk, the supporting tonus of the hindlegs has diminished and the legs have gone into a flexed position.

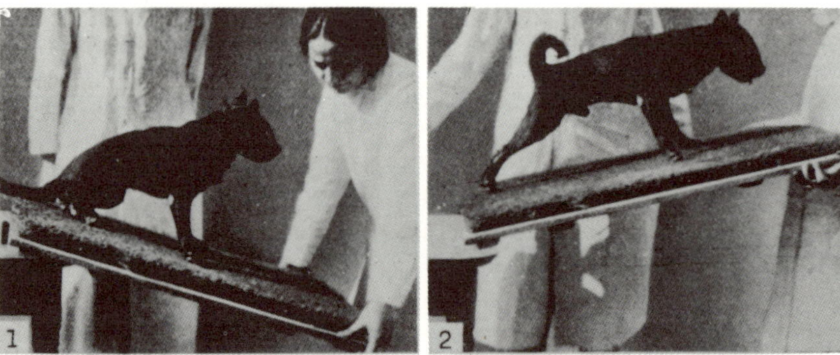

Fig. 98. Decerebellate dog Erik, with blindfold, on a board which is moved up and down at the head end. 1. Head end of the board is lower than the tail end. The dog keeps its muzzle $\pm 45°$ below the horizontal, neck and dorsal spine form an upward opened obtuse angle, hindlegs are flexed so that the posterior part of the body is in a sitting posture. 2. Head end of the board is above the horizontal. The muzzle now also stands at an angle of $\pm 45°$ below the horizontal due to a ventral flexion of the head, neck and dorsal spine form an almost straight line.   The hindlegs are extended with a strong supporting tonus.

reflexes are able to produce the alterations in supporting tonus and position of the extremities necessary to the adaptation to static conditions. However, it became evident through observations on labyrinthectomized dogs that the alterations of the limb posture on lifting and lowering the head or tail end of the supporting surface were not produced exclusively by the labyrinthine righting reflexes and the tonic neck reflexes, because on slowly shifting the supporting base, labyrinthectomized dogs with blindfolded eyes show the same alterations in supporting tonus and in position of the extremities (Fig. 99). In this case the alterations occur in spite of the fact that the head does not maintain its position in space. A flexion of the hindlegs with a ventrally bent head, i.e., against the effect of the tonic neck reflexes, can often be observed (Fig. 99).

Fig. 99. Labyrinthectomized dog with blindfold on a board the tail end of which is moved up and down. 1. Tail end of the board is lowered. The dog has its hindlegs maximally extended. 2. Tail end above the horizontal. The hindlegs are in flexed posture in spite of the ventral flexion of the head.

THE VERTEBRA - PROMINENS - REFLEX
AND SUPPORTING TONUS

Magnus and De Kleyn observed in decerebrate animals a decrease in extensor tonus in fore- and hindlegs by the vertebra-prominens-reflex which is produced either by a ventral displacement of the entire neck in the lowest cervical joints or by pressure on the spinous processes of the last cervical and the first thoracic vertebrae. This reflex can sometimes also be observed in the daily life of animals, as, for instance, when a cat creeps under a cupboard (Magnus). In a standing dog, when the neck is ventrally displaced by pressure on the cervical vertebra, an increase in supporting tonus first of all appears. The four legs prop up, sometimes the hindlegs are maximally extended and the back curves. With further pressure, the supporting tonus suddenly vanishes and the animals lie flat on the supporting surface on their chest and belly (Fig. 100). This reaction can be observed in intact, decerebellate and decorticate dogs.

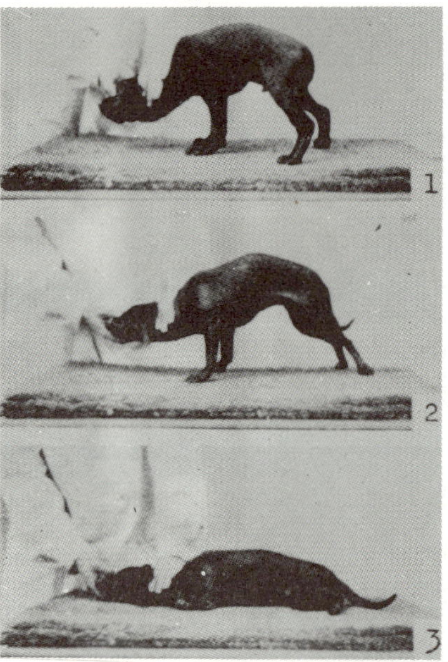

Fig. 100. Vertebra-prominens-reflex and abolition of supporting tonus. 1. Normal young dog in standing posture. By a pressure on the cervical vertebra the neck is moved downwards, the legs prop up and the back is curved. 2. Similar reactions under similar conditions are also shown by the decerebellate dog Piccolino. 3. With a stronger pressure the supporting tonus suddenly vanishes and the dog lies down flat on the ground.

VIII. ALTERATION OF SUPPORTING TONUS BY ROTATING AND TURNING THE HEAD
A. THE CHANGES OF SUPPORTING TONUS UPON TURNING THE HEAD

Magnus and De Kleyn (139) observed, on rotation of the head of the decerebrate animal, a decrease in extensor rigidity in the so-called "cranial legs" (the extremities of the side towards which the vertex of the cranium is turned) and an increase of extensor tonus in the "mandibular legs" (the extremities of the side towards which the lower jaw is turned). [1] The "cranial legs" are more or less flexed while

(1) Magnus and De Kleyn's term rotation of the head is a movement round the

the "mandibular legs" show an increase in extensor posture. In the daily life of animals, rotation of the head is also sometimes accompanied by the same alterations in tonus and posture of the extremities, for instance when lying down on their side. When an animal standing on its four legs rotates its head to the right, lying down on the right side lateral position will not, however, always follow. In this case, the decrease in tonus is only slight or absent and sometimes one even sees that the right limbs are more extended and braced laterally, in opposition to the influence of the tonic neck reflexes. This opposite reaction can be particularly and distinctly observed when the head of a standing animal is rotated; in this case the mandibular legs show a flexion while the cranial legs extend and brace outwards to prevent a fall towards the cranial side (Fig. 101).

When the back of a standing animal is loaded with a sandbag in a weight equal to the animal's bodyweight and the head is then rotated,

---

muzzle-occipital foramen axis, or a rotation in the atlanto-epistropheal joint round the process of epistropheus as an axis. By turning the head, they mean a movement round a dorsoventral axis. These definitions, however, are not quite correct, because when the animal holds the neck vertically upwards and the muzzle horizontally, then a movement of the head round a dorsoventral axis comes about by a rotation in the atlanto-epistropheal joint with the process of epistropheus as an axis. The reaction on rotating and turning of the head also appear in labyrinthectomized animals; therefore, the tonic neck reflexes must play a role. The alterations are always the same. When the animal holds its head horizontally and it is turned round the muzzle-occipital foramen axis, a spiral torsion is produced. This is also the case when the muzzle and neck are directed upwards. When, however, the neck is directed upwards and the muzzle is horizontal, the movement of the head round the muzzle-occipital foramen axis, for instance towards the right, will produce a concave curvature of the neck towards this side. Also, in the dorsal position when the neck is flat on the ground and the muzzle is directed upwards, this movement of the head will cause a curvature of the neck. Thus, the movement of the head round the muzzle-occipital foramen axis will sometimes produce a spiral torsion and sometimes a lateral curvature of the neck. Likewise, a movement of the head round the dorsoventral axis will sometimes produce a curvature and sometimes a spiral torsion of the neck, according to muzzle and neck forming a more or less straight line or standing almost vertically to one another. From this we see that it is absolutely necessary, for avoiding any misunderstanding, to state exactly each time what is meant by rotation and turning and from which initial posture of the head and neck is the alteration of head posture carried out. When in the following a rotating and turning of the head is mentioned, we mean the rotation of the head by a movement in the atlanto-epistropheal joint with the process of epistropheus as axis, followed by a spiral torsion of neck, while with turning, we mean a movement of the head round a dorsoventral axis *of the neck* followed by a lateral curvature of the neck.

206

Fig. 101. Decerebellate dog Moor with a sandbag on its back. The head is passively rotated by 90° to the left, i.e., with the vertex of the cranium towards the left shoulder. The left hindleg (cranial leg) is strongly extended, the right (mandibular leg) is slightly flexed.

for instance to the left, the posterior part of the body will fall down, but in this case *not onto a left side, but onto a right lateral position*. Thus, the strength of supporting tonus of the right hindleg (mandibular leg) decreases. The decrease is not, however, a direct consequence of the rotation of the head; it is not produced by tonic cervical and labyrinthine reflexes. As we shall see later, the alterations in tonus and posture here are caused by a change of position of the legs relative to the trunk caused by an associated rotation of the thorax and pelvis (see Chapter XI and XIII).

When an intact dog is set down on a balance with only one hindleg and the other three limbs are kept raised, there appears no change in strength of the supporting tonus in this leg when the head is rotated 90° from the intermediate position to the right or to the left (see Table 18). Likewise, in the rotation of the head from a dorsal position, the typical alteration of tonus and position is absent when the trunk is fixed by hand on the supporting surface to avoid an associated rotation movement. The supporting tonus of the statically stressed legs, which is already low in this posture, will nearly always disappear completely on both sides on rotation of the head.

Thus, a typical influence of tonic neck and labyrinthine reflexes on the force of supporting tonus produced by rotation of the head cannot be observed *in intact dogs*, either when standing on one leg or in the dorsal position, while standing on all four legs opposite reactions sometimes appear on rotation of the head.

In decerebellate dogs, immediately after extirpation, the legs act very differently to rotation of the head, depending on whether or not the animal has become rigid after extirpation. In the first case, the legs behave like those of decerebrate animals, while in animals that have become limp after extirpation, no reactions whatever on rotation of the head can be observed in the legs. After a lapse of time after extirpation, decerebellate dogs in the dorsal position keep their limbs flexed and this posture is not altered on the passive rotation of the head; distinct tonic neck reflexes are not seen. When, however, the hindlegs are extended by static stress, a rotation of the head from the intermediate position (muzzle pointing vertically upwards) to a lateral position usually causes a decrease of supporting tonus in all four legs. In the mandibular legs, this decrease indeed is not constant and is usually lower than in the cranial legs (see Table 17). When the head is rotated by $\pm 180°$, the supporting tonus totally disappears in all four legs. Thus, decerebellate dogs in a dorsal position show a distinct alteration in supporting tonus on rotation of the head as well as on lifting and lowering which agree with changes of tonus caused by tonic neck and labyrinthine reflexes observed by Magnus and De Kleyn in decerebrate animals.

Table 17

Influence of rotation of head on the strength of supporting tonus of the hindlegs in decerebrate dogs in dorsal position.

	Intermediate position of head, muzzle pointing vertically upwards	Head in lateral position cranial leg	mandibular leg
Erik : 1. investigation	$+6$	$+4$	$+6$
2. investigation	$+5$	$+2$	$+4$
3. investigation	$+6\frac{1}{2}$	$+4\frac{1}{2}$	$+5$
4. investigation	$+7$	$+3\frac{1}{2}$	$+5$
Piccolino : 1. investigation	$+3$	$+2\frac{1}{2}$	$+2\frac{1}{2}$
2. investigation	$+3$	$+2$	$+3$
3. investigation	$+2\frac{1}{2}$	$+2$	$+2$
4. investigation	$+2\frac{1}{2}$	$+1\frac{3}{4}$	$+1\frac{3}{4}$

Since the supporting tonus decreases on both sides, this increase is probably caused by a change of position of the labyrinths. In a dorsal position with the muzzle pointing vertically upwards, the head is approximately in the maximum posture, with a rotation of 180° nearly in the minimum position of the tonic labyrinthine reflexes.

When the head is rotated only by 90°, i.e., in the lateral position, the influence of tonic labyrinthine reflexes on extensor tonus also decreases but to a slighter degree. Through rotation, the extremities of one side are now considered as "cranial" and those of the other side as "mandibular" legs. As shown by Magnus and De Kleyn, the tonic neck reflexes exert on the mandibular legs a facilitation of extensor tonus and in the cranial legs a decrease.

The decrease in extensor tonus will therefore be considerable in the cranial legs (decrease of facilitation by tonic labyrinthine reflexes + reduced effect of tonic neck reflexes), and in the mandibular legs it will be lower or absent (decrease of facilitation by tonic labyrinthine reflexes + facilitation by tonic neck reflexes). This behavior could be observed in the dog Erik but conversely in the dog Piccolino both sides showed an almost equal decrease. Thus, in the dog Erik, the combined influence of tonic neck and labyrinthine reflexes could be observed, while in the dog Piccolino only the influence of tonic labyrinthine reflexes was apparent.

In a decerebellate dog set down on its fore- and hindlegs, no distinct alterations of strength of supporting tonus can be detected on rotation of the head (Table 18).

These values show that a rotation of the head sometimes causes an alteration in the strength of supporting tonus which, however, always remains low and, in addition, inconstant. It is probably produced by the change in position of the observed hindleg in relation to the trunk, which is caused by the rotation of the head. It is nearly impossible to preserve unchanged the relative position to the trunk of the investigated leg while rotating the head. In a second investigation in which the trunk was kept somewhat elevated, the observation resulted in the following values for the dog Erik :

Table 18

Influence of rotation of the head on the strength of supporting tonus of the right hindleg in intact and decerebellate dogs. (The dogs are standing on this leg only.)

Position of Head	Normal young dog (5 kgs), right hindleg	Decerebellate dog Erik, right hindleg	Decerebellate dog Wolf, right hindleg	Decerebellate dog Piccolino, right hindleg
Intermediate position	+4 kg (—5)	+9 kg (—10)	+12 kg (—10)	+5½ kg (—6)
Right lateral position (head rotated 90° to the right)	+3½ kg (—5) cranial leg	+8 kg (—10) cranial leg	+13 kg (—12) cranial leg	+4½ kg (—5½) cranial leg
Left lateral position (head rotated 90° to the left)	+4 kg (—5) mandibular leg	+8 kg (—10) mandibular leg	+11 kg (—14) mandibular leg	+4½ kg (—5½) mandibular leg

strength of supporting			**tonus of the hindleg**	
in intermediate position of head	+	10 kg		(−11)
in lateral position of head:				
cranial leg	+	12 kg		(−13)
mandibular leg	+	10 kg		(−11)

On comparing these values with those of Table 13, the strong influence of even a slight alteration in position of the hindleg in relation to the trunk becomes evident.

While standing on all fours and rotating the head, decerebellate dogs usually show particularly distinct alterations in posture and supporting tonus (Fig. 101) which, however, are contrary to the influences of tonic neck and labyrinthine reflexes (increase of supporting tonus and lateral bracing of the cranial legs, decrease of supporting tonus and more or less flexion of the mandibular legs) and are only produced secondarily by the position of the head.

Thus, decerebellate dogs show typical alterations of supporting tonus on rotating the head when lying in a dorsal (supine) position, and in a standing posture on one leg show no distinct alterations of supporting tonus. Standing on all four limbs, however, they show distinct alterations which are contrary to the influence of tonic neck and labyrinthine reflexes.

The behavior while standing on one leg and standing on four limbs agrees with that of intact animals, but the behavior in a dorsal position is different. In this position, intact animals do not show any distinct changes of supporting tonus on rotation of the head. The finding that on standing on one leg, a distinct alteration of force of supporting tonus on rotation of the head does not appear is not yet fully understood. The first change in position of the labyrinths reduces the extensor tonus; the second increases it. The facilitating effect does not become as clearly apparent in the already strong supporting tonus of the standing decerebellate animal, as does the reducing influence on the already weakened and labile supporting tonus in the dorsal position.

In the earliest days after the unilateral extirpation vf the cerebellum[1] the

---

(1) The reactions on rotating and turning of the head behave very differently in unilaterally decerebellate animals particularly in the first stage after extirpation. In some animals there appears increased extensor tonus and tonic neck reflexes not only in the ipsilateral but also in the contralateral legs. We report here only the observations made on the dog Fox.

ipsilateral limbs, in the dorsal position of the animal or in the ventral position in the air (thus in the absence of static stress), show lively tonic reflexes on rotating the head. These are usually absent in the contralateral legs or can only be weakly detected (Fig. 102, No. 1).

Thus, for instance, the dog Fox in a dorsal position, immediately after extirpation of the right half of the cerebellum, showed in the right legs :

in intermediate position of the head (maximum position of tonic labyrinthine reflexes)	a strong extensor tonus (Fig. 102, No. 1)
in rotating the head 90° to the right (cranial leg position)	± no extensor tonus (Fig. 102, No. 2)
in rotating the head 90° to the left (mandibular leg position)	a distinct extensor tonus (Fig. 102, No. 3)
in rotating the head 180° to the right or left (minimum position of tonic labyrinthine reflexes)	no extensor tonus

while the left legs remain flexed in all four postures of the head and show nearly no resistance against passive, non-static flexion.

In a 90° rotation of the head from a dorsal position with the muzzle pointing vertically upwards, the influence of tonic neck reflexes becomes evident. On rotating to one side, the increased extensor tonus of the ipsilateral legs is maintained (mandibular leg position) and on rotating to the other side, it disappears (cranial leg position). All four legs are flexed in this case (Fig. 102, No. 2).

With a rotation of 180°, the increased extensor tonus always disappears in the ipsilateral legs when the head is rotated to the right as well as to the left. Since with a rotation of 180° the head is almost brought into the minimum position of the tonic labyrinthine reflexes, these probably participate in the disappearance of the increased extensor tonus.

In this stage, the ipsilateral paws do not yet show a distinct (at least not a lasting) alteration of tension on static stress, while in the contralateral limb a distinct supporting tonus is produced only in intermediate and in mandibular leg positions. This is the cause of the fact that a pressure on the soles encounters the strongest resistance in the

212

ipsilateral legs with one lateral position of the head, and in the contralateral legs with the other lateral position.

For instance, in the dog Fox (right half of cerebellum extirpated) in the dorsal position, when the head is rotated to a lateral position to the left, all four legs are flexed and a distinct supporting tonus appears on static stress on the soles only in the left paws (mandibular legs) (Fig. 103, No. 3). The limbs, in this case, offer the strongest resistance to pressure on the soles. In a lateral position to the right, however, static stress on the left legs (cranial legs) does not produce any supporting tonus, while the right legs show increased extensor tonus and thus offer the strongest resistance to pressure on the soles of the feet (Fig. 103, No. 4).

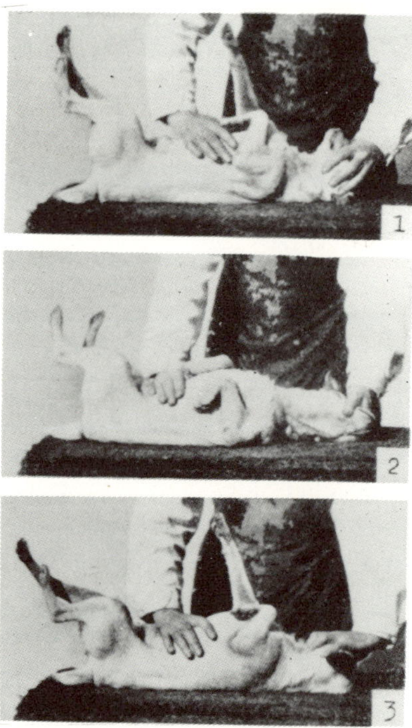

Fig. 102. Influence of head rotation on the muscle tonus of the extremities after the unilateral extirpation of the cerebellum. 1. The dog Fox (right half of cerebellum extirpated) in a dorsal position with the head symmetrical to the trunk. Right legs extended in the air, left legs flexed. 2. Head rotated by 90° to the right, i.e., into left lateral position. Position of left legs unchanged flexion of the right (cranial legs). Leg position equal on both sides. The head is rotated into a right lateral position. Right legs (mandibular legs) again in extended posture, left ones flexed.

The difference in resistance in a lateral position of the head to the left is intensified by the fact that simultaneously with the appearance of supporting tonus in the left limbs due to static stress, the increased extensor tonus in the right limbs is reduced. This reduction does not appear when static stress does not cause the occurrence of supporting tonus, just as is the case in a lateral position of the head to the right.

In the second week after unilateral extirpation, with the animal lying in dorsal position, static stress produces a distinct supporting tonus also in the ipsilateral legs, but not in cranial leg position. In the dorsal position of the animals, a distinct supporting tonus on static stress is now seen in the left legs only in one lateral position and in the right legs only in the other lateral position, and the resistance to pressure on the soles of the feet appears stronger sometimes in the legs on the side of extirpation and sometimes in the contralateral ones, according to the lateral position of the head.

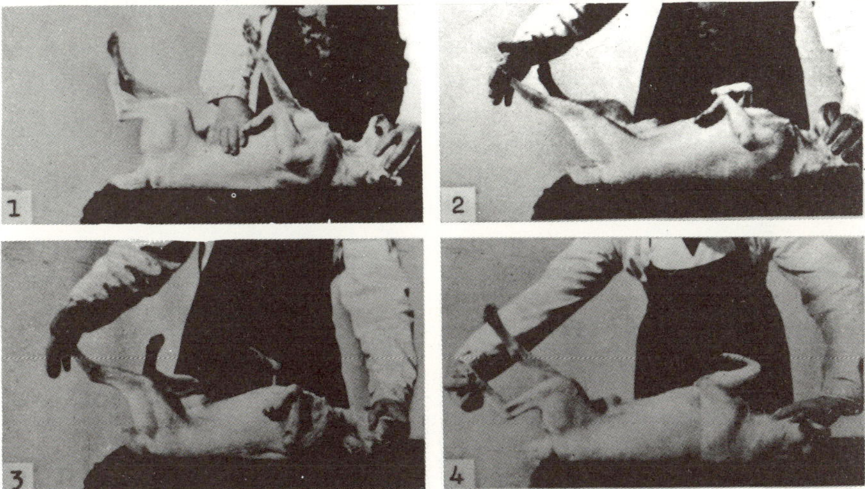

Fig. 103. Decerebellate dog Fox after extirpation of the right half of the cerebellum, in dorsal position, the muzzle pointing vertically upwards. The animal keeps the right hindleg extended, the left one flexed. 2. On static stress, the left hindleg goes into an extended position and shows a distinct supporting tonus. At the same time, a decrease in supporting tonus appeared in the right hindleg. 3. The head is passively rotated into a left lateral position. The left hindleg (mandibular leg) still shows a distinct, though weaker, supporting tonus; the extensor tonus of the right hindleg (cranial leg) has decreased still further. 4. The head is passively rotated into a right lateral position. The left hindleg scarcely shows any supporting tonus; the right, however, shows again a strong extensor tonus.

On setting down the animals with the left or the right legs alternately on a supporting surface, and also in the ventral position in the air, quite different conditions appear. In both lateral postures of the head, the contralateral legs show the stronger supporting tonus. When the animal is set down alternately with the left and with the right legs on a supporting surface, the ipsilateral legs then give way in both lateral positions of the head, but not the contralateral legs. The latter are able to carry the trunk in both lateral postures of the head; the force of supporting tonus of the ipsilateral legs is not able to do so in both lateral positions. This is, therefore, unlike the case in the dorsal position when sometimes the ipsilateral or the contralateral legs have the stronger supporting tonus, here the contralateral legs always show the stronger carrying ability.

This is brought about by the fact that the supporting tonus appears more evident in these experiments, in contrast to the investigation in the dorsal position. In the following days, the supporting tonus in the ipsilateral legs increases steadily. One then sees that the ipsilateral hindleg is able to support the trunk in the mandibular leg position, but not yet in the other positions of the head (Fig. 104).

Four weeks after extirpation, supporting reaction in the ipsilateral limbs is usually so strong that the hindleg of this side is able to carry the posterior part of the body in both lateral positions, as well as in the intermediate position of the head. Determination of the strength of supporting tonus on the balance in the intermediate position as well as in both lateral positions of the head resulted in the remarkable values given in Table 19.

Rotation of the head from one lateral position to the other now produces an alteration in the strength of supporting tonus not only in the dorsal position of the animal, but also while standing on one leg. Owing to this, the strongest supporting tonus is shown by the ipsilateral legs in one lateral position of the head and in the contralateral legs in the other lateral position, both in the dorsal position as well as in the standing posture on one leg. The alterations are distinctly stronger in the dorsal position than in the standing posture, where they are only slight. Although the strength of supporting tonus of both hindlegs (8–8½ kg) amounts to twice the bodyweight, yet the animal is not yet able to stand unassisted.

Fig. 104. The dog Fox (3 weeks after extirpation of the right half of cerebellum) set down on the right hindleg. 1. When the head is passively rotated with the vertex of the cranium towards the left shoulder, the right hindleg (mandibular leg) has a supporting tonus sufficient to carry the trunk. 2. Conversely, when the head is held in an intermediate position (or in right lateral position), the leg instantly gives way under the weight of the posterior part of the body.

Two months after extirpation the animal is able to carry on its back a load of 7 kg and to run around with it. On rotating the head into the left or right lateral position, the legs no longer give way even when the back is loaded with a sandbag (Fig. 105).

Table 19

Dog Fox, bodyweight 7 kgs, 4 weeks after extirpation of *right* half of cerebellum.

	Alterations of supporting tonus of hind-leg on rotation of the head in standing posture and dorsal position	
	left hindleg kgs	right hindleg kgs
In standing posture:		
head in intermediate position	+8 (—9)	+6½ (—7½)
head rotated 90° to the left	+6 (—6½) cranial leg	+6½ (—7) mandibular leg
head rotated 90° to the right	+7½ (—8) mandibular leg	+6 (—6½) cranial leg
In dorsal position:		
head in intermediate position	+2½ (—3)	+7 (—7½)
head rotated 90° to the left	+2½ (—1) cranial leg	+7½ (—8) mandibular leg
head rotated 90° to the right	+4 (—4½) mandibular leg	+2½ (—3) cranial leg

In an animal set down on one leg, rotation of the head no longer causes any distinct alterations of supporting tonus (see Table 20: 27th of April). The ipsilateral legs, however, show a lower strength of supporting tonus than the contralateral ones, in intermediate as well as in both of the lateral positions.

In the dorsal position, however, rotation of the head still produces a considerable decrease in supporting tonus in the cranial leg, so that the ipsilateral leg shows a stronger supporting tonus in one lateral position of the head and the contralateral one in the other.

In *median* position of the head in relation to the investigated leg the strongest supporting tonus is now always found in a standing posture in the contralateral leg and in a dorsal position in the ipsilateral leg.

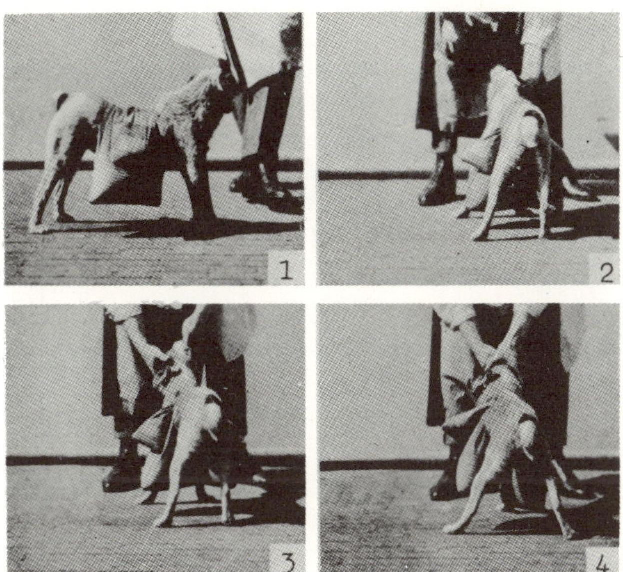

Fig. 105. The dog Fox, bodyweight 7.8 kg, 2 months after extirpation of right half of the cerebellum. 1. Standing on all four legs, the animal is able to carry on its back a sandbag of 7 kg. 2. On the passive rotation of the head into a right lateral position, the legs do not give way under the load. 3 and 4. Just as they do not give way in a left lateral position of the head, neither in the vertical posture of the legs (3), nor in a broad based posture (4).

Standing posture

| mandibular leg : left hindleg | $+10$ kg | right hindleg | 8 kg |
| cranial leg : left hindleg | 10 kg | right hindleg | 7 kg |

Dorsal position

| mandibular leg : left hindleg | 2 kg | right hindleg | 7 kg |
| cranial leg : left hindleg | ± 2 kg | right hindleg | 2 kg |

In the stage of permanent impairment, when the animal is set down on a balance alternately with one or other of the four legs and when the strength of supporting tonus of the individual legs is determined in an intermediate position and in the two lateral positions of the head, it becomes evident that, firstly, the rotation of the head is not able under these circumstances to produce any distinct alterations in supporting tonus and that, secondly, the strength of supporting tonus in the ipsi- and

Table 20

Alterations of supporting tonus on rotating of the head in the contra- and ipsilateral hindlegs in the dog Fox at different times after extirpation (16. II. 1927) of the right half of cerebellum.

	left hindleg (contralateral)				right hindleg (ipsilateral)			
	14 March	27 April	17 May	29 Oct.	14 March	27 April	17 May	29 Oct.
In standing on one hindleg								
Head in intermediate position	+8	+10	+10	+10	+6½	+7	+10	+10
Mandibular leg	+7½	+10	+10	+ 9	+6½	+8	+10	+ 9
Cranial leg	+6	+10	+ 9	+ 9	+6	+7	+ 9	+ 9
In dorsal position								
Head in intermediate position	+2½	+4	+4	+4	+7	+7	+5	+6
Mandibular leg	+4	+2	+2	+3	+7½	+7	+4	+5
Cranial leg	+½	±2	±1	+1	+2½	+2	+1½	+3½

contralateral legs is equally strong in all three head postures. In the dog Fox, the strength of supporting tonus of the ipsi- and the contralateral hindleg in the intermediate position and in the lateral positions of the head, amounts at this stage to ±9 kg (Table 20: 29th October). With the animal in the dorsal position testing static stress on the individual legs, however, rotation of the head still exerts a distinct influence on the strength of supporting tonus. On rotating the head from the intermediate position into a lateral position, the cranial leg shows a considerable decrease (Table 20: 29 October) and the mandibular leg no essential change or only a slight decrease. As in the intermediate position of the head, the strength of supporting tonus in the ipsilateral hindleg is considerably stronger than in the contralateral one, the rotation of the head into a lateral position to the left produces an almost equally strong strength of supporting tonus on both sides due to the decrease in the cranial leg (ipsilateral hindleg), while in the lateral position to the right, as well as in the intermediate position of the head, that of ipsilateral right hindleg is distinctly stronger than that of the left hindleg.

The dog Fox	left hindleg	right hindleg
head in intermediate position	+4 kg	+6 kg
head in lateral position to the left	+3 (mandibular leg)	+3½ (cranial leg)
head in lateral position to the right	+1 kg (cranial leg)	+5 kg (mandibular leg)

In the dorsal position, a rotation of the head does not produce tonic neck and labyrinthine reflexes when the legs are not under static stress. In the stage of permanent impairment, the unilaterally decerebellate animal usually keeps all four limbs flexed and this position is not altered on rotating the head.

We have dis ussed in detail these findings in unilaterally decerebellate animals for the reason that the conception of the cerebellum being a tonus-promoting organ is based mainly on observations after unilateral extirpation of the cerebellum. Luciani and many others after him came to the conclusion that the cerebellum has a facilitating effect on muscle tonus and that each half of the cerebellum exerts this effect only on the tonus of the muscles of the ipsilateral half of the body. According to Luciani, the phenomena shown by unilaterally decerebellate animals in crawling, standing and running are the consequences

220

of an atonia of the muscles of the body ipsilateral to the extirpation
(see p. 538). Luciani, however, did not perform any investigation in
the dorsal (supine) postures, in the different positions of the head, nor
did he make any estimations on the balance.

The impairments in the distribution of muscle tonus, according to
our investigations and findings, are not as simple as Luciani thought.
At any rate, there is *no question of a decrease in muscle tonus on the side of
extirpation* after unilateral extirpation of the cerebellum. As shown by
the investigations described in the three preceding chapters, the muscle
tonus of the extremities in unilaterally decerebellate animals:

according to the time elapsed since extirpation,

according to the fact whether the paws are under static stress or
not;

according to whether the investigations are carried out in the dorsal
position,   ventral position or standing posture, and

according to whether the head is lifted or lowered, rotated to one
or the other side or kept in an intermediate position,

is sometimes stronger on one side, sometimes on the other and sometimes equally strong on both sides.

Decerebellate dogs, in the dorsal position, when they are calm,
keep their four legs flexed and do not alter this posture on rotating the
head. Tonic neck and labyrinthine reflexes of the limbs not under
static strain cannot be observed. Investigations of alterations of supporting tonus on rotating the head in the dorsal position of the animals
encounter great difficulties because, firstly, they constantly try to get
up when they are laid on a hard supporting surface in the dorsal position, and alternatively extend and flex their limbs and, secondly, the
static stress in this posture is usually not followed by a distinct supporting tonus of the hindlegs (with the exception of the dog Bob)
(Fig. 259, No. 1).

Also, on standing on all four legs, an investigation of the influence
of head rotation on strength of supporting tonus is almost impossible, since
the four legs are braced forwards as soon as the head is touched and
the animal tries to liberate itself by pulling backwards (Fig. 261, No. 2).

The investigation, however, is more successful in the standing position on one foreleg. In this type of experiment, rotation of the head does not cause any strong or typical alterations in the strength of supporting tonus. For instance in the dog Fuchs, investigations resulted in the values shown in Table 21 for the strength of supporting tonus in the forelegs in both lateral positions and in the intermediate position of the head. As shown by these values, a rotation of the head into a lateral position produces in this animal a slight decrease in strength of supporting tonus in the cranial leg as well as in the mandibular leg.

In animals set up on one hindleg, a distinct influence of rotation of the head was also absent; in the intermediate position as well as in both lateral positions of the head, the leg gave way under the weight of the posterior part of the body (in the dogs Robbie, Miesel and Fuchs).

Table 21

Influence of rotating the head on strength of supporting tonus in the forelegs of the decorticate dog Fuchs, standing on one foreleg.

	3 months after extirpation of cortex, right foreleg	9 months after extirpation of cortex, right foreleg
head in intermediate position	+11 kg (—12)	+11 kg (—13)
head in lateral position to the right	+ 9 kg (—10) cranial leg	+8½ kg (—9½) cranial leg
head in lateral position to the left	+ 9 kg (—10) mandibular leg	+9½ kg (—10½) mandibular leg

After the unilateral extirpation of the cerebral cortex, animals sometimes on rotation of the head and in the dorsal position show tonic neck and labyrinthine reflexes in the contralateral legs which, however, are not strong and later on decrease further. When the jaw was rotated towards the side of the operation, the contralateral limbs were flexed; conversely they were more or less extended in an intermediate position and in the other lateral position of the head, while the ipsilateral limbs remained in the flexed position in all three postures of the head. In a dorsal position, the animals show in the intermediate position as well as in the lateral position of the head only a very slight supporting tonus of the hindlegs on static stress, so that it was impossible to measure, on the balance, the small alterations caused by rotation of the head. While standing on one leg, the legs on the side of extirpation show no distinct alterations; on the other hand, the reduced strength of supporting tonus of the contralateral legs in cranial leg position is considerably weaker than in the intermediate position and in the other lateral position of the head. Thus, for instance, the investigations on the balance 12 days after extirpation of the left half of the cerebral cortex shows for the dog Schwarzweisser the following values (Table 22).

Table 22

Dog Schwarzweisser, bodyweight 5 kg, 12 days after extirpation of the left half of cerebrum. Influence of head rotation of strength of supporting tonus on standing on one hindleg.

	left hindleg	right hindleg
head in intermediate position	$+5\frac{1}{2}$ kg (—$6\frac{1}{2}$)	$+3$ kg (—$3\frac{1}{2}$)
head rotated 90° to the right	$+5$ kg (—6) mandibular leg	$+1\frac{1}{2}$ kg (—2) cranial leg
head rotated 90° to the left	$+5$ kg (—6) cranial leg	$+3$ kg (—0) mandibular leg

At later stages neither the contralateral nor the ipsilateral legs show any essential alterations of strength of supporting tonus on rotation of the head.

In the decorticate and decerebellate dog Robbie, a rotation of the head produced distinct alterations of extensor tonus and of the extended posture of the extremities due, as it appears, predominantly to the influence of the tonic labyrinthine reflexes.

In adult human beings under normal conditions, a rotation of the head does not cause any distinct alterations of tension in the muscular system of the extremities, but under certain pathological conditions (see reference 196) and in normal infants in the first months of life, it causes tonic neck and labyrinthine reflexes in the extremities not under static strain. Unfortunately, we do not yet possess any observations on the influence of rotation of the head on the supporting tonus.

Summing up, we find :

A. *Standing on one leg*, rotation of the head in intact, in totally or unilaterally decerebellate animals in a late phase after extirpation, and in totally or unilaterally decorticate dogs causes no distinct alterations of supporting tonus (sometimes rotation of the head causes a slight decrease in strength of supporting tonus in the mandibular leg as well as in the cranial leg); in the same standing posture, however, a rotation of the head produces distinct alteration in supporting tonus:

    1. in decerebellate animals when they are beginning to stand again;

    2. in unilaterally decerebellate animals in the limbs on the side of extirpation early after extirpation;

    3. in unilaterally decorticate animals in the limbs opposite the side of extirpation early after the operation.

B. *In the dorsal position*, totally and unilaterally decerebellate dogs on rotation of the head usually show distinct alterations of strength of supporting tonus of the hindlegs, even late after extirpation; intact and totally or unilaterally decorticate dogs, however, show none.

C. Rotation of the head produces alterations in tonus and position of the extremities when they are *not* under static stress:

    1. early after extirpation of cerebellum in dogs which became rigid after extirpation;

    2. early after unilateral extirpation of the cerebellum in the legs on the side of extirpation;

    3. sometimes but not constantly, early after unilateral decortication, in the legs opposite to the side of the operation;

    4. in a decorticate and decerebellate dog;

    5. in decerebrate dogs.

In intact and in decorticate dogs, these alterations are totally absent, as in animals after total or unilateral extirpation of the cerebellum and the unilateral extirpation of the cerebrum, after some time has elapsed since extirpation.

B. THE IMPORTANCE OF ALTERATIONS OF SUPPORTING TONUS ON ROTATION OF THE HEAD FOR THE ADAPTATION TO STATIC CONDITIONS

When an intact or decerebellate dog with blindfold is set down on a supporting surface which is raised on the left side of the animal or lowered on the right side so that the animal is passively rotated to the right, round the longitudinal axis, the head of the animal is rotated to the left, the left legs are simultaneously flexed and the right ones are extended (Fig. 112).

The rotation of the head by which the position in space remains unchanged is probably brought about by the labyrinthine righting reflexes. By this rotation of the head, the left legs become "cranial legs" and the right "mandibular legs" and, accordingly, the extended position of the first decreases and of the latter increases. Thus, one has the impression that the labyrinthine righting reflexes with associated tonic neck reflexes participate in the adaptation of posture of the extremities to the position of the supporting surface when it is tilted to the right or left side.

These mechanisms, however, are not absolutely necessary for adaptation, as proved by observations on labyrinthectomized dogs (Fig. 106 and 107).

When a labyrinthectomized dog with blindfold stands on a surface which is tilted from one side (Fig. 106, No. 2 and 3), the head also rotates towards the raised side. In this case the rotation is not produced by the labyrinthine righting reflexes, and the alterations in the position of the legs, which are nearly the same in labyrinthectomized and in intact dogs (Fig. 107), are, therefore, not secondary results of labyrinthine righting reflexes.

The question remains open whether they are caused by tonic neck reflexes or whether the rotation of the head is primary and the position of the legs produced by it secondary. In adapting to the position of the surface, intact and labyrinthectomized dogs not only rotate the head but also the thorax and pelvis towards the raised side (Fig. 106, No. 2, and Fig. 107). By the rotation of the trunk, the head is, or remains, symmetrically related to the trunk; the tonic neck reflexes, therefore, are not activated; nevertheless, the legs show the typical posture of adaptation.

Fig. 106. Labyrinthectomized dog Moritz, with blindfold, on a supporting surface which is raised on the left side of the animal.   1. The animal on a nearly horizontal supporting surface; the bitemporal axis of the head is horizontal.   2. The supporting surface is raised on the left side of the animal; by a rotation of the head to the left, the bitemporal axis has maintained its horizontal position.   3. By a further raising of the supporting surface the animal is brought into a left lateral position; the head, too, is brought into this position.

Fig. 107. Labyrinthectomized dog Moritz, with blindfold, on a supporting surface which is lowered on the right side (which, for adaptation, is equal to a raising of the left side). The hindlegs show a distinct adaptation to static conditions; the hindleg of the lowered side is strongly extended and abducted, that of the right side flexed and adducted. In addition, the thorax and pelvis have turned away so far from the lowered side that the bitemporal axis remained horizontal.

On precise observation, one sometimes can see that the limbs have already changed their position and, consequently, the thorax and pelvis have been turned towards the raised side before a rotation of the head took place (Fig. 108, No. 1).

When the head is passively turned away from the raised side, the position of the legs in intact and labyrinthectomized dogs still adapts to static conditions (Fig. 108, No. 2) even against the influence of tonic neck reflexes.

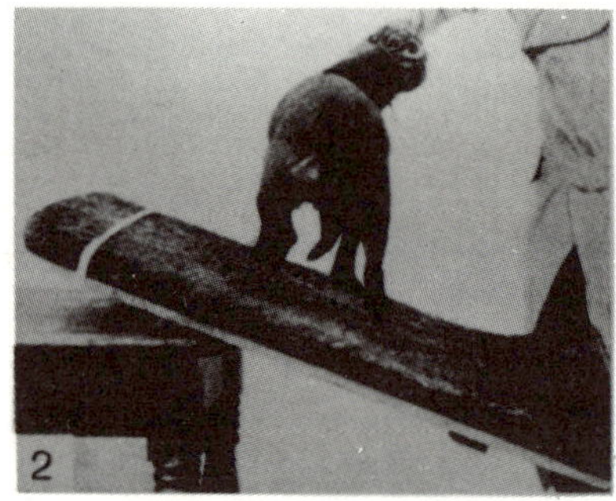

Fig. 108. 1. Labyrinthectomized dog Moritz, with blindfold, on a supporting surface which is lowered on the left side. The animal's legs have adapted themselves to the static conditions. Thorax and pelvis have turned away from the lowered side so that the bitemporal axes of the trunk are horizontal, while the bitemporal axis of the head has remained parallel to the base. Although the head is rotated towards the trunk to the left, the left legs (cranial legs) show an extended posture and the right ones (mandibular legs) a flexed posture. Thus, the alteration of leg position took place contrary to the effect of tonic neck reflexes, the rotation of the thorax contrary to the effect of cervical righting reflexes. 2. Normal dog. The head of the animal is passively rotated to the right lateral position, the supporting surface on the right side. Leg position is adapted to the position of the surface and, contrary to the influence of tonic neck reflexes, the left legs (mandibular legs) show a decrease and the right legs (cranial legs) an increase in extension.

C. ALTERATION OF SUPPORTING TONUS ON TURNING THE HEAD

Magnus and De Kleyn found in decerebrate animals that a turning of the head produces alterations in tonus in the muscles of the extremities. On turning the head to the right, i.e., by approaching the right ear to the right shoulder, the right legs (mandibular legs) show an increase in extensor tonus and the left (cranial legs) a decrease of extended posture. In a certain position of the head and neck, i.e., in the vertical position of the dorsoventral axis, the alterations in tonus are exclusively produced by the tonic neck reflexes; in another position of the axes, for instance a horizontal, tonic labyrinthine reflexes can also be produced. In opposition to decerebrate animals, intact, decorticate and decerebellate dogs in the dorsal position do not show any distinct alterations of posture of the extremities on turning the head; the legs remain flexed. On standing on all fours, however, a turning of the head exerts a distinct influence. Magnus (196) already recognized this influence and observed "that when a cat standing in normal posture turns its head, for instance to the right and shifts its center of gravity to the right, the extension of the right foreleg becomes evident and counteracts this shifting of the center of gravity, while the left foreleg shows a decrease in extensor tonus." Thus, one gets the impression that on

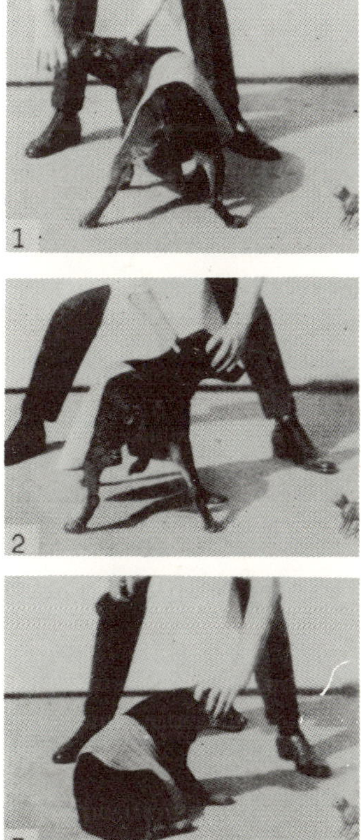

Fig. 109. The dog Moor (16 months after extirpation of the cerebellum), body-weight $9\frac{1}{2}$ kg, with a sandbag of 5.3 kg on its back. 1. Head turned passively by 90° to the left. Left hindleg vigorously extended and abducted and the right one somewhat flexed and adducted. 2. Head passively turned to the right. Right hindleg abducted and strongly extended, the left one less abducted and slightly flexed. 3. Head passively turned to the right by 135°. Left hindleg collapsed so that the posterior part of the body falls to the left side.

standing on all four legs, the tonic reflexes caused by the turning of the head and which affect the extremities are of great importance for the adaptation to static conditions. In a standing posture, intact, decorticate and decerebellate dogs show particularly distinct alterations on passive and active turning of the head, when the back is loaded with a heavy sandbag (Fig. 109). When in this case the head is passively turned, for instance to the right, the right hindleg (mandibular leg) is very vigorously extended, while the left (cranial leg) is increasingly flexed and finally gives way so that the posterior part of the body goes into a lateral position to the left (Fig. 109, No. 2 and 3).

These alterations of position and tonus can be observed particularly clearly in decerebellate dogs. When the head of a unilaterally decerebellate dog is alternatively turned to the right and to the left with equal force, the hindleg ipsilateral to the side of extirpation will give way sooner than the contralateral, especially in the early phase after extirpation (Fig. 110). After some time, however, this difference is no longer evident. Conversely, after unilateral extirpation of the cerebral cortex, the hindleg contralateral to the extirpation gives way sooner than the ipsilateral one.

As we have seen, in the animal standing on all four legs, a turning of the head produces alterations of tonus (increase of supporting tonus in the mandibular legs, decrease in the cranial legs) which agree with those observed in the decerebrate animals of Magnus and De Kleyn. It

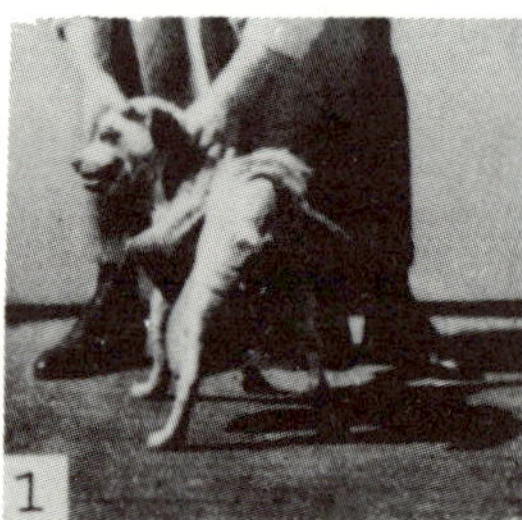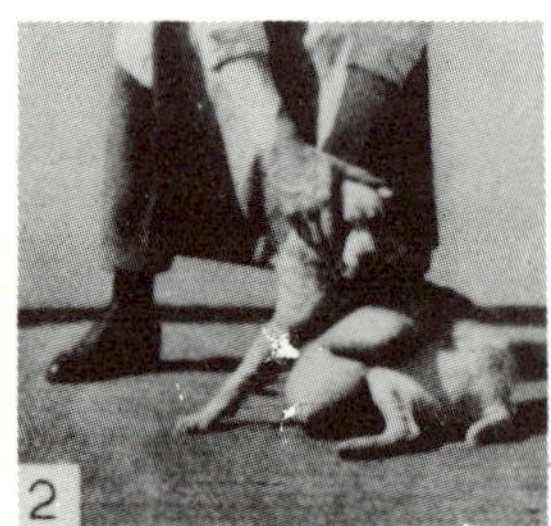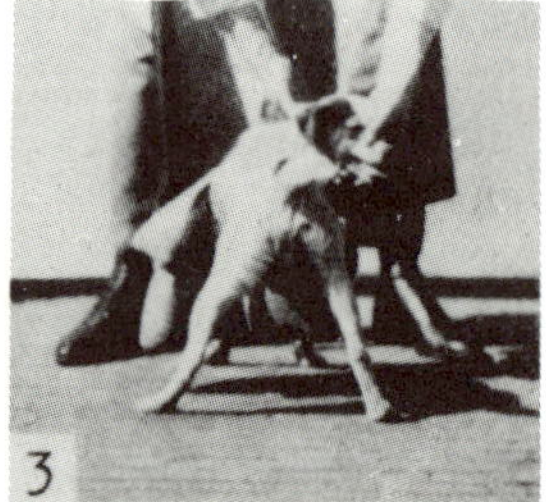

Fig. 110. The dog Fox (3 months after extirpation of the *right* half of the cerebellum), bodyweight 7.8 kg, the back laden with two sandbags of a total weight of 7 kg, 1. Head is passively turned to the left by 90°, the legs do not give way under the load. 2. Head passively ,turned 135° to the left, the *right* hindleg gives way and the posterior part of the body falls to the right. 3. Head passively turned 135° to the right. The left hindleg does not give way.

might be supposed that just as with alterations of extensor tonus in the decerebrate animal, the alterations of supporting tonus in the intact animal standing on all four legs are brought about by tonic neck reflexes. Investigation showed, however, that these reflexes do not play any role at all, or only a minor one. In the animal standing on all four legs, a turning of the head not only causes a lateral curvature of the neck, but also of the back, by which the position of the legs in relation to the trunk is altered in the hip and shoulder joint. When the lateral curvature of the back is passively prevented, no distinct alterations of position and supporting tonus of the legs occur (Fig. 111).

When the animals are set down on one paw or when they are examined in a dorsal position, a turning of the head will likewise not cause any alteration in the strength of supporting tonus as long as the position of the leg in relation to the trunk remains unchanged. Thus, the alteration of supporting tonus on turning the head of an animal standing on all fours is not caused by the tonic neck reflexes but by the secondary change of position of the legs in relation to the trunk (see also Chapter XIII).

In intact dogs in a dorsal position, a turning of the head does not cause any tonic neck reflexes and the legs remain flexed. On legs under static stress, a turning of the head does not produce any distinct alterations in strength of supporting tonus either in the dorsal position or on standing on one leg (Table 25).

In decerebellate dogs in a dorsal position, the effects of turning the head are different depending on whether a short or long time has elapsed since extirpation. Dogs which remained flaccid after extirpation are not influenced in any way; rigid animals, however, react in a manner similar to decerebrate ones. In the stage of permanent impairment decerebellate animals keep their legs flexed in the dorsal position and do not change this posture on turning the head. Similarly, only a little distinct alteration in tonus can be observed in the static stress on the paws, in the dorsal position of the animal or on standing on one leg (Table 23).

On standing on one leg, the strength of supporting tonus is sometimes less strong in the cranial leg position than in the mandibular leg position, but not always. For instance, a second investigation of the strength of supporting tonus of the hindlegs in the dog Erik showed the values of

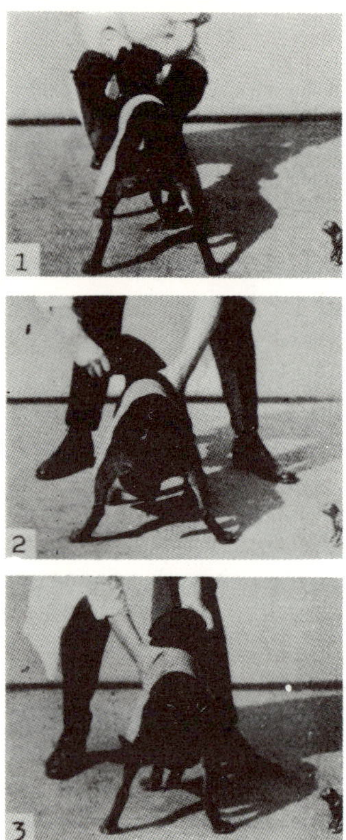

Fig. 111. Decerebellate dog Moor, its back loaded with a sandbag of 5.3 kg.
1. With the head held passively in a symmetrical position. 2 and 3.
While the hand prevents a lateral curvature of the vertebral column,
the head is passively turned 90° to the left (2) and to the right (3). The
hindlegs do not show any distinct alterations in position and supporting
tonus (compare Fig. 110, No. 2).

+ 10 kg in intermediate position, +10½ kg in mandibular leg position
and 9½ kg in cranial leg position, but the method does not allow one
to attach importance to a difference of 10 per cent.

In unilaterally decerebellate dogs in the dorsal position, the ipsilateral
limbs at first show distinct tonic neck reflexes on turning the head.
When the muzzle is turned towards the side of the operation, the legs
on this side are considerably more extended and offer a stronger resistance against static flexion than when the head is turned towards the
other side.   Gradually, the alterations in position and tonus grow less
and disappear some time after extirpation.   In this case, there is no
distinct influence no turning the head even under static stress, whether
in the dorsal position or in the standing posture (Table 24).

Table 23

Influence of turning the head on the strength of supporting tonus in the hindlegs of decerebellate dogs in dorsal position and in standing on one leg.

	strength of supporting tonus in the right hindleg			
	dog Piccolino kg	dog Erik kg	dog Wolf kg	dog Moor kg
In dorsal position				
Intermediate position of head	+3½ (—4)	+5 (—6)		
Head turned to the right (mandibular leg)	+2½ (—3½)	+5 (—6)		
Head turned to the left (cranial leg)	+2½ (—3½)	+5 (—6) +4½ (—5½)		
Standing on one hindleg				
Intermediate position of head	+5½ (—6)	+11 (—12)	+12 (—13)	+12 (—14)
Head turned to the right (mandibular leg)	+6 (—6½)	+10 (—12)	+12 (—13)	+13 (—14)
Head turned to the left (cranial leg)	+5 (—6)	+10 (—12)	+12 (—13)	+12 (—14)

Table 24

Influence of turning the head on the strength of supporting tonus in the hindleg in dog Fuchs, 3 and 9 months after extirpation of right half of cerebellum.

	Strength of supporting tonus of	
	right hindleg kg	left hindleg kg
Standing posture on one hindleg 3 months after extirpation:		
Head in intermediate position	+8 (— 9)	+10 (—11)
Head turned to the right	+8 (— 9) mandibular leg	+ 9 (—10) cranial leg
Head turned to the left	+9 —9½ cranial leg	+9½ —9½ mandibular leg
9 months after extirpation:		
Head in intermediate position	+9 (—10)	+ 9 (—10)
Head turned to the right	+8 (— 9) mandibular leg	+7½ (—8½) cranial leg
Head turned to the left	±8 (— 9) cranial leg	+8½ (—9½) mandibular leg
Dorsal position 9 months after extirpation:		
Head in intermediate position	+5 (— 6)	+4 (— 5)
Head turned to the right	+4 (— 5) mandibular leg	+1 (—2) cranial leg
Head turned to the left	+4 (— 5) cranial leg	+3½ (—4½) mandibular leg

In thalamus (decorticate) animals, in the dorsal position, a turning of the head does not cause distinct tonic neck reflexes in the legs without static stress. The strength of supporting tonus of the hindleg is usually as with a turned head as in the intermediate position. In the standing posture on one leg also no influence or only a slight one can be traced (Table 25).

Table 25

Influence of turning the head on the strength of supporting tonus of the foreleg in a normal and in a decorticate dog (3 months after extirpation of cerebrum). The animal stood on one leg at a time.

	Strength of supporting tonus in the left foreleg	
	normal dog (6 ½ kg) kg	decorticate dog Fuchs (6 kg) kg
Head in intermediate position	+6½　(—7)	+11　(—12)
Head turned towards the right	+6½　(—7) cranial leg	+10　(—11) cranial leg
Head turned towards the left	+6½　(—7) mandibular leg	+11　(—12) mandibular leg

In unilaterally decorticate dogs, it was found that when the animal lay on its back the limb opposite the ablation showed at first only a weak tonic neck reflex on head turning. In these animals, neither in dorsal position nor on standing on a limb was there any distinct influence of head turning on supporting tonus. The response remained just as slight in later stages of recovery as it was soon after ablation, when the supporting tonus of the contralateral limb was distinctly lessened (Table 26).

Table 26

Influence of head turning on supporting tonus of the hindlimbs in the dog Schwarzweisser (5 kg) 12 days after ablation of the *left* cerebral hemisphere. The animal was made to stand alternately on one and then the other leg.

	Supporting tonus of	
	right hindleg kg	left hindleg kg
With head in median position	+3　(—4)	+5　(—6)
Head turned to right	±2　(—3) mandibular leg	+5　(—6) cranial leg
Head turned to left	+3　(—4) cranial leg	+5　(—6) mandibular leg

In a standing intact *man*, no alterations in tension of the leg muscles can be observed on turning (inclination) of the head, but usually an alteration in tonus of the sacrolumbal muscles can be distinctly established. Unfortunately, till now, investigations are lacking whether a turning of the head can eventually also cause alterations in supporting tonus under pathological conditions, in agreement with the observed alterations in tonus and position of the extremities not under static stress (see reference 196).

Summing up, we find that the tonic neck reflexes in the unstressed legs caused by a turning of the head usually occur :

1. in decerebellate dogs early after extirpation, but only when the legs are rigid;

2. in unilaterally decerebellate dogs early after extirpation, i.e., usually only in the legs ipsilateral to extirpation;

3. in unilaterally decorticate dogs early after extirpation in the contralateral legs (in the dog Vici whose cerebellum was also removed, they were still particularly evident $1\frac{1}{2}$ years after extirpation);

4. in decerebrate dogs.

In the dorsal position and in standing on one leg, however, a turning of the head never produces any constant, measurable alterations in supporting tonus.

The influence of a head turning on the strength of supporting tonus is, therefore, less strong by far than the influence of a rotation of the head which, on its part, causes much slighter alterations than the lifting and lowering of the head.

D. THE ROLE OF TURNING OF THE HEAD IN ADAPTATION TO STATIC CONDITIONS

When the supporting surface of a dog standing on all four legs is raised or lowered from one side, the head is often turned towards the lowered side and away from the raised one (Fig. 112; Fig. 249, No. 4), even when its eyes are blindfolded.[1] When the surface is raised, for instance on the left side of the animal, usually in addition to a rota-

---

(1) What produces this turning of the head will be discussed in Chapter XIII.

tion to the left, a turning of the head to the right takes place. Simultaneously, the left legs flex and are adducted, while the right ones show a strong extension (Fig. 112) and are brought into a position of abduction at the same time the whole spinal column is concavely curved to the right.

Fig. 112. Decerebellate dog Moor, with blindfold, standing on a supporting surface which is lifted on the left side of the animal. It has turned its head to the right (in addition to a left rotation); the right legs (mandibular legs) are vigorously extended and abducted, the left (cranial legs) are flexed. The entire vertebral column is curved with concavity to the right.

Fig. 113. The head of an intact dog is passively turned to the left while the supporting surface is raised on the left side of the animal. With this posture of the head, the legs also adapted themselves instantly to the altered position of the supporting surface.

Although the position of the legs corresponds to the influence of the tonic neck reflexes on turning the head, the above mentioned observations show that the adaptation of leg posture to the position of the supporting surface is probably not produced by these reflexes. Our supposition agrees with the findings that the alterations in position of the legs still occur distinctly when the head is passively turned towards the raised side (Fig. 113).

The tonic neck reflexes on turning the head are perhaps able to promote the adaptation of leg posture to static conditions, but they surely do not play an important role.

IX. SUPPORTING TONUS AFFECTED BY LABYRINTHINE REFLEXES

Under certain circumstances supporting tonus can be altered :

A. by labyrinthine righting reflexes;

B. by tonic labyrinthine reflexes;

C. by labyrinthine reflexes on movements of the head in a straight line in dorsoventral direction (lifting reaction); movements of the head (the labyrinths) in a straight line in occipito-nasal or bitemporal direction do not produce any distinct alterations in supporting tonus;

D. by labyrinthine reflexes on rotating round the bitemporal axis of the head;

E. by labyrinthine reflexes on rotating movements round the longitudinal axis of the head; rotating movements round the dorsoventral axis of the head do not cause any distinct alterations in support tonus.

A. SUPPORTING TONUS AFFECTED BY LABYRINTHINE RIGHTING REFLEXES

The labyrinthine righting reflexes indirectly influence the supporting tonus. By a slow leaning forward movement of the trunk in space, the head, under the influence of the labyrinthine righting reflexes, moves dorsally to the thorax; the tonic neck reflexes produced by this movement cause a reduction of supporting tonus in the hindlegs and an increase in the forelegs (Fig. 114).

Conversely, when the thorax is leaned backwards the labyrinthine righting reflexes indirectly produce an increase in supporting tonus in the hindlegs and a decrease in the forelegs. The importance of the indirect influence of the labyrinthine righting reflexes on the supporting tonus to the adaptation to static conditions in standing on a slanting surface has already been discussed in Chapter VII.

When an animal tries to raise its head from lying in a lateral position, the legs go simultaneously into flexion even when the movement of the head is prevented. Based on these and other observations one gets the impression that the coming into effect of the labyrinthine righting reflexes could directly affect the muscle tonus of the extremities.

B. SUPPORTING TONUS AFFECTED BY TONIC LABYRINTHINE REFLEXES

The influence of tonic labyrinthine reflexes on the supporting tonus can be better observed in decerebellate dogs than in intact animals. Nevertheless, it is also advisable in such dogs to exactly adhere to the experimental conditions, as otherwise the alterations produced by the tonic labyrinthine reflexes could easily be masked by other influences. The position of the head has to be exactly fixed in relation to the trunk as well as the position of the entire spinal column and of the pelvis, and the position of the legs has to remain unchanged as much as possible in the hip and shoulder joints. It is most appropriate to place the dogs with the back on a board and there to fasten the head, neck and thorax with straps and to hold the muzzle with the hand (Fig. 115) and to fix the pelvis by the tail pulled through a cleft in the board. Fixed in this way, the buccal cleft, cervical and thoracic spine form an almost straight line. The animal is now held in the air in various positions, static stress is applied to the hindlegs with one hand and the supporting tonus is investigated.

Fig. 114. 1. Decerebellate dog Piccolino, with blindfold, suspended in the air, head upwards. The animal keeps its muzzle at an angle of 45-60° below the horizontal ventrally flexed towards the trunk; the hindlegs show a strong supporting tonus. 2. The animal has been rotated head downwards so that now the head is hanging downwards. In doing so, the head moved dorsally to the trunk, due to the influence of labyrinthine righting reflexes, so that the muzzle is again at an angle of 45-60° below the horizontal. The hindleg now does not show any distinct supporting tonus on static stress.

With this mode of investigation, it appears that *in the ventral position with the buccal cleft at an angle of about 45° above the horizontal, the hindlegs show a strong supporting tonus* (Fig. 115, No. 1). However, the supporting tonus is weak with a horizontal buccal cleft and, with the buccal cleft directed 45° downwards, it is totally absent. Neither touching nor static stress on the sole can then produce an extension and fixation in the extended posture of the hindlegs '(Fig. 115, No. 2).

Furthermore, *in dorsal position*, too, a distinct supporting tonus appears in the hindlegs on static stress when the muzzle is directed upwards (Fig. 116, No. 1) while with the horizontal cleft it is weak, and it is totally absent when the buccal cleft is directed downwards (Fig. 116, No. 2).

When the muzzle points upwards, the hindlegs under static stress show a distinct supporting tonus even with a *vertically placed* thoracic longitudinal axis and buccal cleft; this tonus fails to appear when the head is directed downwards.

Fig. 115. Decerebellate dog Piccolino; neck and trunk are fixed with straps to a board, the pelvis is fixed by the root of the tail, while the head is held with one hand in such a way that the buccal cleft falls into the extension of the longitudinal axis of the trunk. 1. In a ventral position with the buccal cleft at an angle of 30-45° above the horizontal, the hindlegs show a distinct supporting tonus. 2. In a ventral position with the buccal cleft at an angle of 30-45° below the horizontal, no supporting tonus appears in the hindlegs on static stress.

Thus in the hindlegs there is a distinct supporting tonus in the ventral position, dorsal position and vertical posture when the buccal cleft is pointing upwards; however, it is absent in all postures with downward directed buccal cleft and can neither be produced by touching nor by static stress on the soles of the hindpaws.

The "neutral positions" are positions with a horizontal buccal cleft. In them, the supporting tonus is weak or only just present and slight alterations of position produce strong alterations in supporting tonus. From a ventral as well as from a dorsal position with a horizontal buccal cleft, a slight alteration of position of the head upwards (over 10-15°) causes a strong increase in supporting tonus and a slight downwards movement causes a total disappearance in supporting tonus (Fig. 117).

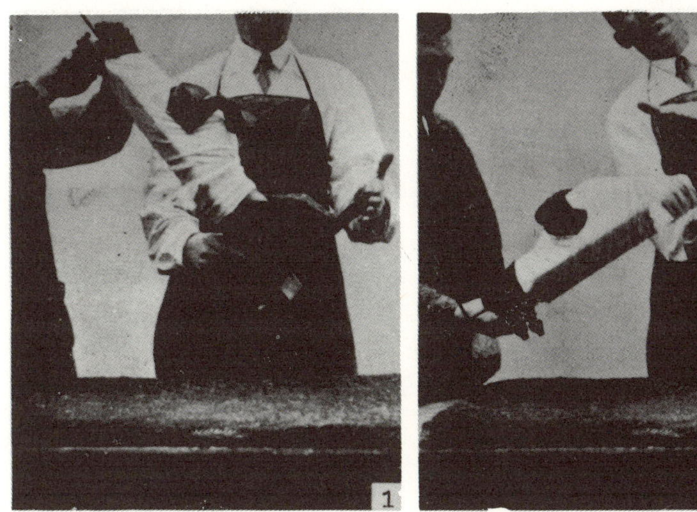

Fig. 116. Decerebellate dog Piccolino, fixed on a board in the same way as in Fig. 115. 1. In a dorsal position with the muzzle pointing upwards, the buccal cleft at an angle of about 45° above the horizontal, a distinct supporting tonus appears in the hindlegs on static stress. 2. In a dorsal position with the muzzle pointing downwards, the buccal cleft at an angle of about 45° below the horizontal, static stress produces neither extension nor a supporting tonus in the hindlegs.

In opposition to these strong alterations of supporting tonus on slight changes of position, alterations of position of the muzzle to the vertical when the animal is suspended head downwards do not influence

Fig. 117. Decerebellate dog Piccolino fixed in a dorsal position on a board. 1. When the buccal cleft stands at a moderate angle above the horizontal, the hindlegs show a distinct supporting tonus on static stress. 2. Conversely, supporting and magnet reactions are totally absent, when the buccal cleft stands at a very moderate angle below the horizontal.

the supporting tonus in any way. On static stress the supporting tonus fails to appear in any position with the muzzle pointing downwards. Conversely, with suspension head upwards, any changes in position in which the plane of the buccal cleft goes through the vertical produce distinct increase and decrease but not a disappearance of supporting tonus. For instance, when animals are changed from a dorsal position through a suspended position head upwards, then into a ventral position, usually after passing through the vertical posture, an increase takes place which lasts as long as the new position is maintained (Fig. 118) and which disappears in going back into the initial posture.

Thus, these alterations in tonus produced by position show the same characteristics as tonic labyrinthine reflexes, especially as they, too, continue tonically. Nevertheless, they are not necessarily labyrinthine, i.e. produced by the position of the head in space, because in moving from the dorsal into a ventral position and vice versa, the relationship of the position of the thorax to the supporting surface is also changed, and simultaneously with it the plane of application of counter-

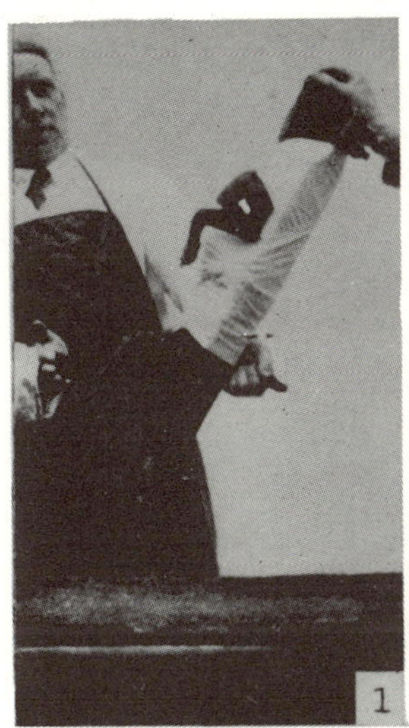

Fig. 118. Decerebellate dog Piccolino fixed on a board.   1. In a dorsal position with the muzzle directed obliquely upwards, the buccal cleft at an angle of 75° above the horizontal, the hindlegs show a moderately strong supporting tonus.   2. On a slow head forwards rotation of the animal beyond the vertical position, an increase in supporting tonus suddenly appears which continues as long as the animal is held in a ventral position with a strongly upwards directed buccal cleft.   Even with a strong pressure on the soles of the feet, the legs now do not give way.

pressure. This is exerted in the first position by the board on the back, and in the second by the straps on the ventral side of the thorax. As we shall see later, the counterpressure on the back reduces the supporting tonus. At any rate, the supporting tonus is strongest in the ventral position of the animals with the buccal cleft upwards at an angle of 45-75° to the horizontal. In this position, the hindlegs offer a stronger resistance to pressure than in the dorsal position with the buccal cleft directed upwards at an angle of 45-75°. This finding is all the more surprising as Magnus and De Kleyn, in *decerebrate* animals, i.e., in rabbits, cats and dogs (see Figs. 7 and 8) and monkeys found the maximum position for tonic labyrinthine reflexes on the extremities to be in the dorsal position with the buccal cleft at an angle of 45° above the horizontal.

In the determination of maxima by Magnus and De Kleyn, it turned out that considerable individual variation may happen in the different experimental animals. *The maximum is always found in a dorsal position,* only the angle of buccal cleft to the horizontal differs in the maximum position of the various animals. Thus, Magnus and De Kleyn usually found the strongest extensor rigidity in decerebrate dogs in a dorsal position, the buccal cleft at an angle of 45° above the horizontal, but also sometimes with the buccal cleft pointing vertically upwards and once with a horizontal buccal cleft. Magnus and De Kleyn attributed the cause of these differences to the anatomical variations of the vestibular apparatus.

The difference between these findings and ours is probably based on the fact that the counterpressure exerted on the back produces in intact, decorticate and decerebellate dogs a decrease in supporting tonus and in decerebrate animals, however, a decrease in extensor rigidity (see Chapter XIV). Thus, in decerebrate animals in a dorsal position, the tonic labyrinthine reflexes exert their influence unimpeded, while in the other animals the labyrinthine effects are more or less counteracted by the counterpressure on the back. I consider this observation important for the reason that in human beings with hypertonia of the muscles, tonic labyrinthine reflexes in the extremities can be observed which have their maximum position sometimes in the dorsal, sometimes in the ventral position.

Thus, Magnus and De Kleyn, in a child of 16 months with amaurotic idiocy, suffering from spasms and epileptiform attacks, found the maximum in the dorsal position when the oral half of the thoracic longitudinal axis stood at an angle of 45° below the horizontal. Pette, on the other hand, in three cases of chronic spastic hemiparesis or hemiplegia, observed that the stretch tonus was strongest in the ventral position, i.e., in one case in the ventral position with the rostral half of the thoracic longitudinal axis 45° below the horizontal and in the other two cases in the ventral position with the rostral half of the thoracic longitudinal axis 0-45° *above* the horizontal.

Magnus and De Kleyn as well as Pette in their investigations always fixed the head towards the thorax in such a way that the plane of the buccal cleft stood vertically to the thoracic longitudinal axis. Thus, in all dorsal positions, the buccal cleft was directed upwards and in all ventral positions, downwards.

Magnus considers the different maximum positions found in man as expressions of a reversal of reflexe; causeed by cerebral lesions. He observed that in animals cerebral lesions can cause a reflex reversal of tonic labyrinthine reflexes. To our knowledge, however, it has not yet been excluded that perhaps also the counterpressure of the supporting surface may participate in the fact that in man the maximum position of these reflexes is sometimes found in the dorsal sometimes in the ventral position. It seems that, just as in animals, in the normal man, too, the dorsal (supine) position exerts an inhibiting influence on different reactions, while under pathological conditions (as in the decerebrate or decerebellate animal) this influence is absent or reduced.

When the head is kept in a different position to the trunk, similar alterations of supporting tonus in the hindlegs appear on changing the position of the head in space. For instance, when the head is fixed in such a way that the buccal cleft and the ventral side of the thoracic longitudinal axis for an angle of $45°$, then, too, there is a strong supporting tonus in all positions with the muzzle directed upwards, while in the ventral (Fig. 119, No. 1) and dorsal position with a horizontal buccal cleft, the hindlegs show a weak supporting tonus which, however, owing to tonic neck reflexes is somewhat stronger than when the buccal cleft is held in the same plane as the thoracic longitudinal axis.

In all positions in which the muzzle is directed downwards, the supporting tonus is only just present or totally absent.

When dogs are held in a ventral position with a horizontal thoracic longitudinal axis, i.e., with the buccal cleft at an angle of $45°$ below the horizontal (Fig. 119, No. 2), the hanging hindlegs can be flexed almost without resistance by a pressure on the soles. This is all the more remarkable as the longitudinal axis of the trunk is also horizontal in normal standing and the buccal cleft is also below the horizontal at an angle of $20\text{-}45°$. Since a freely standing animal also holds the neck somewhat raised, one should expect in the standing dog a weaker, or at most a similar supporting tonus of the hindlegs, and one is all the more astonished to find, in reality, a very strong supporting tonus, while in an animal kept fixed to a board in a ventral position, only a slight supporting tonus is evident.

It is natural to ascribe this difference to inhibiting influences, for instance the pressure of the straps, physical stress, etc. These, however, cannot be exclusively the cause, for there is a strong supporting tonus in positions in which the buccal cleft is directed upwards. The difference is brought about mainly by the fact that by raising and fixation of the pelvis by the root of the tail, the dorsal side of the pelvis, lumbar and thoracic spine form a straight line (see Chapter X). When the pelvis is lowered, a strong supporting tonus of the statically stressed hindlegs also occurs in dogs fixed to a board in a ventral position. In these investigations the hindlegs must stand vertical to the rabbit board as closely as possible.[1] When they are held directed forwards at the hip joint, the supporting tonus is considerably stronger in all positions with the muzzle pointing upwards and only disappears when the buccal cleft stands at a larger angle below the horizontal.

The supporting tonus of the forelegs shows the same differences as the hindlegs in the various positions. It is also very strong in all positions with the muzzle directed upwards and weaker in positions with the muzzle directed downwards, but is still clearly evident. The forelegs here still offer a considerable resistance to pressure on the soles.

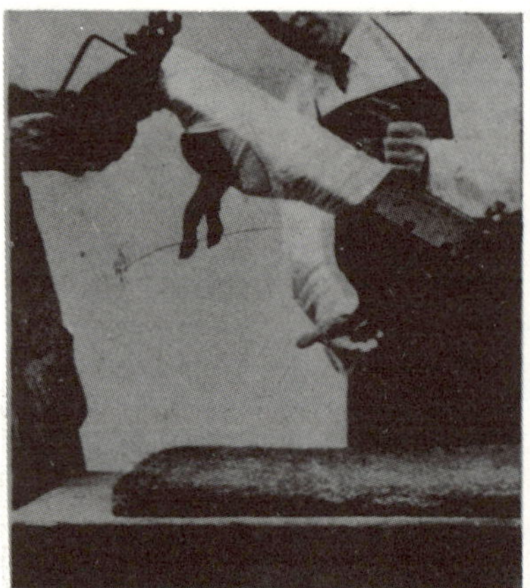

Fig. 119. Decerebellate dog Piccolino, fixed on a board. The head is held fast in such a way that the plane of the buccal cleft forms an angle of 45° with the longitudinal axis of the trunk on the ventral side. 1. The animal in a ventral position with horizontal buccal cleft. On static stress, the hindlegs show a distinct, though not strong supporting tonus. 2. The animal in a ventral position with a horizontal longitudinal axis of the trunk. On static stress, the supporting tonus of the hindlegs is just present. (This position is nearly the same as in a normal standing position.)

---

(1) It is impossible to keep the legs in exactly the same position in relation to the trunk without a disturbing influence on the reactions, because simultaneously with the appearance of supporting tonus they are moved backwards at the hip joint and with the disappearance of supporting tonus they are moved forwards.

For the strength of supporting tonus in fore- and hindlegs, it is of no consequence from which initial position and in which way (by rotation forwards or backwards or by rotation round the longitudinal axis) the animal had been brought into a certain position. In this certain position, after a distinct latency, a certain equally strong supporting tonus always appears which lasts as long as the animal remains in this position. The differences in supporting tonus in the various ventral and dorsal positions, therefore, must be produced by tonic labyrinthine reflexes, i.e., by the position of the animal in space.

In *decerebellate* dogs, the alterations in supporting tonus produced by the changes of position appear distinctly and quite automatically, far more regularly than in intact dogs since these usually show no distinct supporting tonus in the various dorsal positions. Possibly, intact dogs are more inhibited by experimental conditions (dorsal position!) than decerebellate animals. In decorticate dogs, when they are fixed on a board, investigations on alteration of supporting tonus are utterly impossible because of the continual adverse movements and attempts to free themselves.

The relations of the *unstressed* leg to alterations of position of the head in space, which can be particularly well observed in decerebrate animals, are totally absent in the intact as well as in the docorticate and decerebellate dogs in the later stages of recovery. During the early phase after extirpation, they are indeed present, but only in dogs which became completely rigid after extirpation. They usually also appear in the legs on the side of extirpation during the time immediately after the unilateral extirpation of the cerebellum and in the opposite legs after the unilateral extirpation of cerebrum.

Since, in a change of position of the supporting surface, the head retains its position in space due to the influence of the labyrinthine righting reflexes, the tonic labyrinthine reflexes do not play any role in the adaptation of limb posture to the position of the supporting surface.

C. SUPPORTING TONUS AFFECTED BY LABYRINTHINE REFLEXES PRODUCED BY MOVEMENTS IN STRAIGHT LINE

(The lifting reactions and preparation for jump)

When a dog is set down on a horizontal board which is moved up and down in a vertical direction, all four legs are flexed at the start of the lifting and extended very strongly when the movement ceases; conversely, an increase in extension occurs at the start of lowering and a flexion of the legs at its cessation. In this case, the supporting tonus sometimes disappears completely so that the trunk comes to lie on the board (*lifting reaction* of Magnus and De Kleyn). After cessation of the board movement and the fading away of the reactions, the animal usually returns to its initial position.

At the start of lifting, the head is ventrally flexed and, at cessation, it is dorsally flexed while a lowering of the board causes a lifting of the head at the start and a lowering at the cessation. Moreover, there appears a dorsal curvature of the spinal column simultaneously with the extension of the legs and a lordosis of the back together with the flexion. The reactions of the extremities still appear when the head is fixed to the trunk during the lifting movement and are thus not primarily caused by the head reactions.

By the ventral and dorsal movement of the head, tonic neck reflexes, tonic labyrinthine reflexes and reflexes on movement of rotation round the bitemporal axis can be produced. The reflexes partly exert a facilitating and partly an opposite influence on the supporting tonus. For instance, at the cessation of the lifting movement there occurs a raising of the head which can produce the following reflexes of the fore - and hindlegs.

	Effect of reflexes	
	on the forelegs	on the hindlegs
1. tonic neck reflexes:	increase in supp. ton.	decrease in supp. ton.
2. tonic labyrinthine reflexes:	increase in supp. ton.	decrease in supp. ton.
3. reflexes on rotating movements round the bitemporal axis:	decrease in supp. ton.	increase in supp. ton.

While 4 : the lifting movement exerts a facilitation of supporting tonus in all four legs. It is not quite certain whether the different influences which partly counteract one another really all go into effect at the cessation of the upwards movement; at any rate, as a total result an extension with increase of supporting tonus can always be observed in all four legs.

At the start of lowering as well as at the cessation of lifting, a dorsal movement of the head, extension of the four legs and a curvature of the back take place and, at the cessation of lowering as well as at the start of lifting, a ventral movement of the head, flexion of the legs and a lordosis of the back occur.

The lifting reaction is absent after a bilateral extirpation of the labyrinths, but it is distinctly present in intact, decorticate and decerebellate animals. In the latter, there is usually a strikingly brusque and strong increase in extensor posture and dorsal curvature of the spinal column. Usually, the lifting reaction is also present in decerebrate animals, but the downwards movement, when there is pronounced rigidity, does not cause a disappearance but only a reduction in extensor tonus and extensor posture. In man also, particularly in children, Freeman and Morin (81, 82), Schaltenbrand (265) and others observed reactions which correspond to lifting reactions and preparation for jump. Gamper (91) observed them in a case of arhinencephaly.

In vertical lifting and lowering movements, the same reactions are not always perceptible in all limbs; for instance, when a dog is held suspended head downwards (with or without blindfold), the head occupies its normal posture in space (with the muzzle 45° below the horizontal) and thus is directed dorsal to the trunk. When the animal is now moved up and down with static stress on the paws, only the forelegs show a distinct alteration in supporting tonus although head and labyrinths are passively moved in the same way as in standing on the board. Even when the paws are not under static stress, the downward movement produces only an extension of the forelegs (preparation for jump of Magnus and De Kleyn), the reactions of the hindlegs are inhibited by the dorsal position of the head in relation to the trunk (Fig. 114). When the animal is held suspended head upwards in the air by the skin of the neck, only the hindlegs show alterations in supporting and extensor tonus due to the ventral flexion of the neck (Fig. 233, B) and the reactions of the forelegs are inhibited.

The role of lifting reactions in the maintenance of balance in standing on a moving supporting surface

When a dog is set down on a seesaw with the forepaws on the outer rim and the board is rapidly moved upwards, the animal will perform a rotating movement round the bitemporal axis and a vertical movement simultaneously (Fig. 120).

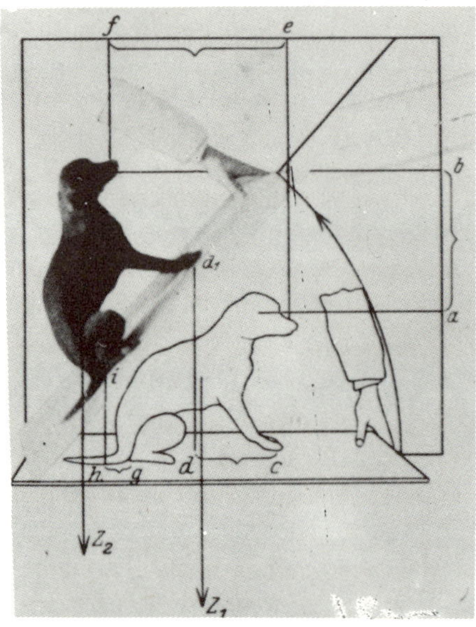

Fig. 120. When the horizontal supporting surface of a dog is raised at the head end till it forms an angle of 50° with the initial position, then, with unchanged situation of the animal to the surface, the following would happen :   1. The animal is rotated backwards by 50°.   2. The head is moved upwards vertically over the distance a - b; the soles of the forelegs over the distance d-d^1; the tarsal joints over the distance h - i.   3. The head is caudally displaced in space over the distance e - f; the soles of the forelegs over the distance c - d; the tarsal joints over the distance g - h.   By this alteration in position, the line of the center of gravity falls beyond the supporting surface and the animal would turn a backward somersault if no reaction would occur.   With a rapid raising of the head end, the occurrence of a somersault is even accelerated by the fact that the head is moved dorsally and upwards with a higher speed (ef = ± 10 X gh) than the posterior part of the body; head and anterior part of the body acquire a greater inertia so that on cessation of movement of the board they are flung dorsally and upwards with a higher speed.

With a *rapid lifting* of the head and of the board, intact and decerebellate animals will flex the paws at the start of lifting in agreement with the movement in the vertical (dorsoventral) direction and move the head ventrally and thus avoid making a somersault backwards like labyrinthectomized dogs who lack these reactions.

With a *rapid downwards movement* of the seesaw, the animal tends to fall downwards and the four legs are extended. With a cessation of the movement, however, the animal again flexes the four legs and thus avoids falling head foremost on the downward slanting board. Thus the lifting reaction, under certain circumstances, can participate in the maintenance of balance, i.e., when the animal is standing far from the axis of rotation on a supporting surface which is moved up and down by the head or tail end (see further, also p. 194.

Reactions on movements in oral, caudal and bitemporal directions

When a dog in a ventral position in the air is moved forwards and backwards or to the right and left instead of a dorsoventral direction, no distinct reactions can be observed in the legs under static stress, just as little as when the animals perform these movements in the dorsal position. Conversely, when the animal is set down on a board which is shifted forwards and backwards, to the right and to the left on a horizontal plane, distinct alterations in posture and muscle tension appear in the extremities.

In a sudden forward displacement of the board, the legs are at first passively moved forwards towards the trunk and in so doing the hindlegs usually flex a little. The passive movement is followed immediately by a distinct muscular contraction which moves the legs backwards in relation to the trunk and, by this, moves the trunk forwards. The head, too, shows reactions : at first it moves passively backwards and then actively forwards. The leg reactions also occur when the head is fixed in relation to the trunk, thus they are not produced by the movements of the head.

When the movement of the board ceases, the forward movement of the trunk continues owing to inertia, the legs are passively moved backwards. Muscle contraction also appears instantly which now moves the trunk backwards, the legs are propped forwards, to return finally to the intermediate position. The legs are braced forwards more strongly when the forward movement not only ceases but is transformed into a backward movement. The reactions can best be observed in the hindlegs which are strongly extended backwards on a forward movement of the supporting surface and with a backward movement are protracted forwards.

Similar phenomena also appear in shifting the board to the right and to the left. For instance, when the board is suddenly moved to the right, the legs are pulled passively to the right of the body, i.e., the right legs go into abduction, the left into adduction. Then muscle contraction appears immediately which adduct the right legs and abduct the left. These reactions, however, are not produced by the labyrinths because they also promptly occur in labyrinthectomized dogs standing on a board with blindfolded eyes. We mention them in this place for the reason that they were hitherto often ascribed to labyrinthine reactions.

Tait and McNally (279) observed similar reactions in intact and decerebrate frogs. After destruction of the ampullae of the semicircular canals they were absent or greatly impaired (after cessation of movement the animals returned to their initial position). Unfortunately, the authors did not state how long after extirpation the investigations were carried out, because in dogs these reactions, which later appear promptly, are usually greatly impaired in the first days after a bilateral extirpation of the labyrinths.

When the board is suddenly shifted very quickly, the above-mentioned reactions do not occur, but the paws are raised and set down afresh. With a rapid shift of the surface forwards, they hop backwards, by a shift backwards they hop forwards and by a shift to the right they hop to the left. These movements of the paws appear promptly in labyrinthectomized dogs.

(On the mechanism of these different reactions in intact, decorticate and decerebellate dogs, see Chapter XIII.)

D. SUPPORTING TONUS AFFECTED BY LABYRINTHINE REFLEXES PRODUCED BY ROTATION OF THE HEAD ROUND THE BITEMPORAL AXIS

In the investigations into alterations of supporting tonus on rotation of the head round the bitemporal axis, the position of the head in relation to the trunk must not be changed. It was found most practicable to attach the dogs with the ventral side of the trunk on a wooden beam padded with cotton-wool and to fix head, neck, thorax, pelvis and tail with strapping to this beam (Fig. 121). The straps must be attached in such a way that the legs remain actively and passively freely movable. Cervical and thoracic spine then form an almost straight line, continued in the plane of the buccal cleft which lies parallel with the wooden beam. After horizontally adjusting the bitemporal axes, the wooden beam is fixed laterally just below the level of the auditory meatus in such a way that this part represents the axis of rotation and the end of the beam is then moved to and fro vertically and the influence of these movements on the supporting tonus of the hind- and forelegs is determined under static stress.

It now appears that sometimes an increase in supporting tonus of the hind-legs occurs at the start of the backwards rotation, and a decrease in supporting tonus at the start of the forwards head rotation; secondly, that the alterations in supporting tonus continue and disappear when the rotation ceases or are transformed into opposite reactions, and, thirdly, that one and the same animal shows alterations in tonus on rotations beyond a certain angle which are absent in rotations over another certain angle of the same value.

For instance, when an animal is successively rotated by 90° till it returns to its initial position, i.e., has performed a total rotation of 360°, it appears that every 90° rotation can produce different alterations in supporting tonus. It became clear that the following factors participate in the "total reaction" produced by a certain rotation :

1. the initial position which produces a certain starting supporting tonus under the influence of labyrinthine reflexes;

2. the start of the rotating movement causes an alteration of supporting tonus which continues during the rotation;

3. the end of the rotating movement in which these alterations of supporting tonus disappear or the opposite rotating reaction occurs;

4. the final position can produce a certain supporting tonus, different from that of the initial position.

For instance, when an animal is rotated backwards by 90° from a ventral position with the buccal cleft directed upwards at an angle of 45° (Fig. 121), so that the animal through a vertical position turns into a dorsal position with the buccal cleft directed upwards by 45° (Fig. 121: caudal end of the beam moves a ➔ b), the vigorous supporting tonus of the hindlegs produced by the initial position shows an increase at the start of the movement; the legs become more extended and the hand exerting the static stress feels a stronger resistance.

At the end of the movement, the resistance felt at first distinctly decreases and the hindlegs become more or less flexed (after-reaction), but in the final position, an extension and a vigorous supporting tonus soon appear again.

The total reaction in this rotating movement (Fig. 121: the buccal cleft from c —▸— d) is composed of the following components:

1. strong supporting tonus (owing to the initial position with the muzzle directed upwards);
2. increase of supporting tonus (owing to backwards rotation);
3. decrease of supporting tonus (flexion due to after-reaction);
4. return of a strong supporting tonus and extension caused by the final position with the muzzle directed upwards.

Sometimes, the flexion and decrease in supporting tonus due to the cessation of the rotating movement are instantly compensated by the tonic labyrinthine reflexes of the final position and are therefore not distinctly evident. Similarly, the increase at the start of the movement is sometimes insignificant due to a vigorous extensor position with strong supporting tonus in the initial position. In this case, the legs are kept extended by the strong supporting tonus before, during and after the movement, which produces no distinct alterations.

When the head is moved further by $90°$, i.e., into a dorsal position with the buccal cleft directed downwards by $45°$ (Fig. 121: buccal cleft from d —▸— a, tail end from b —▸— c), the vigorous supporting tonus undergoes at the start of the movement, in most cases, a sudden increase and at the end of the movement a sudden decrease which gradually leads to the total disappearance of supporting tonus. With a very vigorous initial supporting tonus the increase cannot be distinctly perceived.

Total reaction:

1. vigorous supporting tonus;
2. more or less distinct increase;
3 and 4. decrease and subsequent disappearance of supporting tonus.

With the next $90°$ rotation (Fig. 121: buccal cleft from a —▸— b, tail end from c —▸— d) the animal changes from a dorsal position with the muzzle directed downwards by 45°, into a ventral posture with a similar position of the muzzle; at the start of rotation a strong extension of the flexed hindlegs occurs and with cessation of the rotating movement the extension ceases, and the legs return into a flexed posture and remain flexed.

Total reaction :

1. flexed posture without supporting tonus;
2. vigorous extension;
3. flexion;
4. flexed posture without supporting tonus.

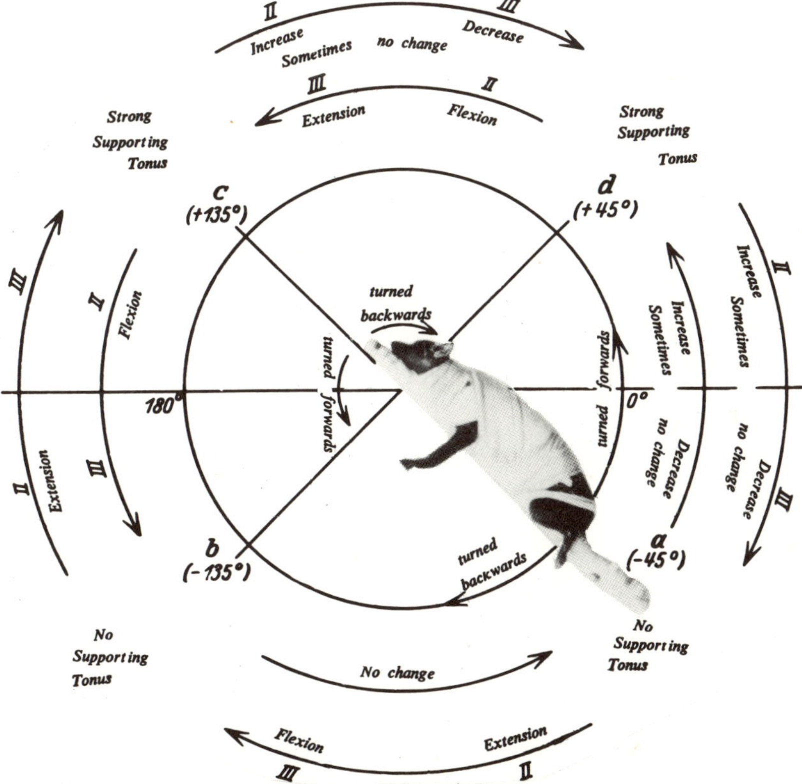

Fig. 121. Reaction of the hindlegs under static stress on rotating movement around the bitemporal axis of the head by 90° in a dog riding on a padded wooden beam on which it is fixed with strapping. The reactions produced at the start of the rotating movement are noted in number II, the reactions occurring at the termination of movement in number III. (The numerals between brackets indicate the different position of the head in space according to rotations suggested by Magnus.)

When with a further 90° rotation the animal returns to its initial position, i.e., into a ventral position with the buccal cleft directed upwards by 45°, the flexed hindlegs show a lively extension at the start of rotation which sometimes relaxes all at once at the end of the movement and suddenly changes into a flexion which is again followed by an extension caused by tonic labyrinthine reflexes. Sometimes, the flexion is compensated instantly by the tonic labyrinthine reflexes and the extended posture continues from the start of the rotation.

Total reaction :

   1. flexed posture without supporting tonus;

   2. strong extension;

   3. temporary flexion;

   4. extended posture with vigorous supporting tonus.

In a forward head rotation, the total reactions also behave differently according to the initial and final positions. When the animal, from a ventral position with the buccal cleft directed upwards by 45°, is rotated forwards by 90° (Fig. 121: buccal cleft from c ⟶— b, tail end from a ⟶—d), the supporting tonus decreases at the start of the rotation, at the termination of the movement a slight extension movement sometimes occurs; however, the limbs usually remain flexed due to tonic labyrinthine reflexes.

Total reaction :

   1. extended posture with vigorous supporting tonus;

   2. flexion;

   3. temporary extension;

   4. flexed position without supporting tonus.

With the next 90° rotation (Fig. 121: buccal cleft from b ⟶— a, tail end from d ⟶— c) no decrease in supporting tonus can take place at the start of the movement as the totally flexed limbs have no supporting tonus at all; a distinct after-reaction is also absent and the final position, too, produces a flexed posture. Thus, in this rotation movement, reactions of the hindlegs under static stress are absent and they remain flexed before, during and after the rotation.

Total reaction:

   1. flexed posture without supporting tonus;

2. flexed posture without supporting tonus;

3. (———————————————————————)

4. ,, ,,   ,, ,,   ,, ,,   ,, ,,   ,, ,,

With the following forward head rotation by 90°, i.e., from a dorsal position with the buccal cleft directed downwards by 45° to a dorsal position with the buccal cleft directed upwards by 45° (Fig. 121: buccal cleft from a —→— d, tail end from c —→— b) no decrease in supporting tonus can occur at the start of the movement; at the end, however, extension with vigorous supporting tonus usua'ly appears after a rather long latency.

Total reaction:

1. flexed posture without supporting tonus;

2. flexed posture without supporting tonus;

3. (———————————————————————)

4. extended posture with vigorous supporting tonus.

Finally, with the rotating movement into the initial position, i.e., a ventral position with the buccal cleft directed upwards by 45° (Fig. 121 : buccal cleft from d —→— c, tail end from b —→— a), at the start of the movement, there is a sudden flexion of the extended hindlegs with decrease in supporting tonus, at the end of the movement a sudden extension occurs which continues after cessation of the movement.

Total reaction:

1. extended posture with vigorous supporting tonus;

2. flexion;

3. sudden extension;

4. extended posture with vigorous supporting tonus.

Thus, the backwards rotation in a vertical plane produces at the start an increase in supporting tonus in the hindlegs, when in the initial position the supporting tonus is weak, while at the start of the forward head rotation a decrease in supporting tonus occurs when there is a distinct support in the initial position.

Since in daily life the rotating movement of the head round the bitemporal axis nearly always takes place in the vertical plane, the reactions on rotation in this plane have been discussed in detail. For the study of labyrinthine reflexes which are produced by a rotation round the bitemporal axis, however, a rotation of the animal in a vertical plane is less suited.

For this, a rotation in the horizontal plane is more appropriate, the head in a lateral position, by which the influence of the initial and terminal position, i.e., the influence of tonic labyrinthine reflexes are eliminated. It is best to set down the animal on a turntable and to fix the head in a lateral position above the axis of rotation (Fig. 122).

For the occurrence of reactions, it is unimportant whether the trunk is in a ventral (Fig. 122) or in a dorsal position (Fig. 123, No. 1 and 2) or how the head is kept in a lateral position (Fig. 123, No. 3 and 4).

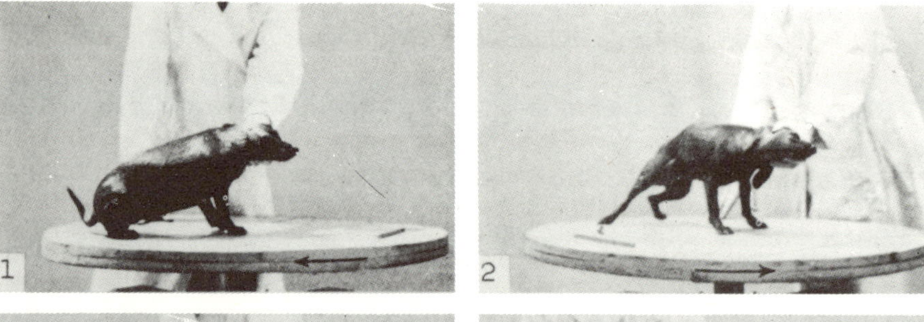

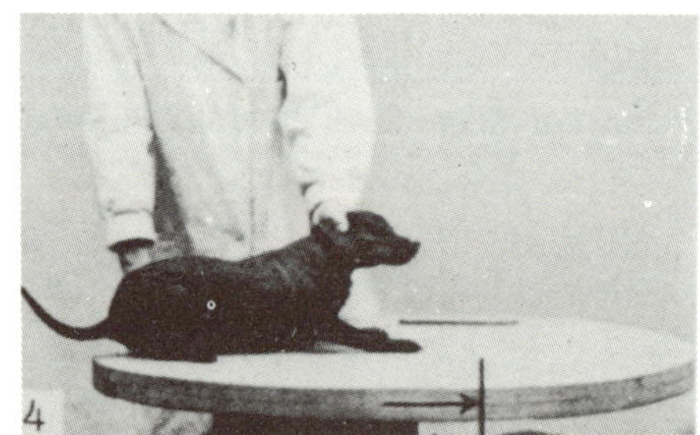

Fig. 122. Decerebellate dog Piccolino in a standing posture on a turntable. Head is fixed above the axis of rotation in a left lateral position. 1. During the clockwise rotation of the disk (head forwards rotation), the animal props the forelegs forwards, the pelvis moves ventrad, the hindlegs also go forwards and flex, the animal tries to move the head dorsally. 2. and 3. In an anti-clockwise rotation (backwards rotation) the four legs are moved caudad, thus pressing the trunk forwards, the hindlegs are strongly extended backwards, the pelvis moves dorsad, the head ventrad. 4. On cessation of the backwards rotation, the four legs are propped forwards as in a head forwards rotation (1) through which the trunk is pulled caudad, the hindlegs collapse, the pelvis moves ventrad, the head is dorsally flexed (after reaction on backwards rotation).

In this experimental arrangement, the backwards rotation always produces a distinct increase in supporting tonus in the hindlegs under static stress and, in addition, a dorsal movement of the pelvis by which the hindlegs are maximally extended and moved backwards. The legs remain in this posture during rotation, and when rotation ceases opposite reactions appear and the legs finally return to their original position.

Conversely, a forward head rotation in the lateral position of the head constantly produces a decrease in supporting tonus and flexion of the hindlegs which, just as the pelvis, are moved ventrally.

In a forward head rotation, it can sometimes be seen that the hindlegs are indeed at first flexed but are then extended again. In opposition to the extension in the backwards rotation, the paws here are braced *forwards*. The forward extension of the hindleg occurs particularly when the forward rotation is carried out very rapidly; in this case,

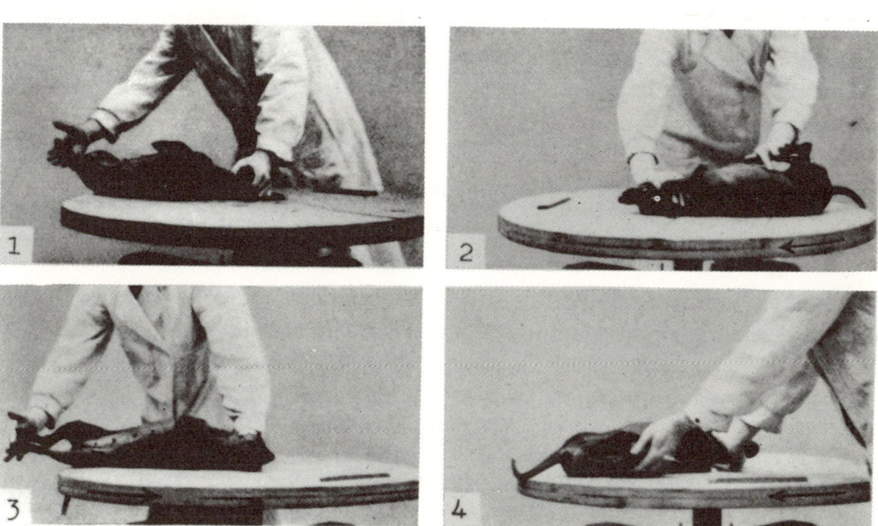

Fig. 123. Decerebellate dog Piccolino on a turntable, the head fixed in left lateral position. (In 1. the muzzle is clasped between thumb and index.) 1 and 2. Head in left lateral position, trunk in dorsal position. 1. In anticlockwise rotation (backwards rotation), the statically stressed hindlegs are extended and moved caudally, together with the pelvis. 2. In a clockwise rotation (forwards rotation), the hindlegs under static stress as well as the pelvis are ventrally moved and flex. 3 and 4. Head and trunk both in left lateral position. 3. In an anti-clockwise rotation (backwards rotation), the hindlegs are caudally extended, while 4. in clockwise (forwards rotation) they retract and fold ventrally.

the pelvis and lumbar parts of the spine are strongly curved ventrad and this alteration of the spine's posture causes an extension of the hindlegs (see Chapter X).

Rotating round the bitemporal axis, the forelegs, too, show reactions which proceed differently according to the initial posture of the legs. When the forelegs under static stress are kept somewhat passively flexed in the elbow and shoulder joints, the resistance against passive flexion usually distinctly decreases at the start of the backwards rotation, the flexed posture of elbow and shoulder joints grows stronger under the influence of the hand pressing against the soles, with the cessation of rotation the resistance again increases so that the hand is pushed away. The increase in resistance and extension becomes distinctly stronger when the backwards rotation is suddenly transformed into a head forwards rotation. *Thus, the backwards rotation produces a decrease in supporting tonus and extended position in the forelegs and the forwards head rotation an increase*; consequently, the reactions are just the opposite of those of the hindlegs. The alterations in supporting tonus are less strong in the forelegs than in the hindlegs; that is especially in the case when the forelegs under static stress have shown a strong extension and supporting tonus prior to rotation. The backwards rotation, in this case, causes only a very slight, sometimes almost imperceptible flexion of the elbow joint. Under these conditions, however, there appear distinct changes in the shoulders. In the forward head rotation, the extended forelegs protract at the shoulders and, in doing so, the scapula distinctly revolves and the ventral part of the scapula is strongly moved forwards. In the backwards rotation, the scapula moves back and the forelegs are retracted at the shoulders. These movements can also be clearly observed when the animals are rotated in a vertical plane. Thus, the reactions of the forelegs are to some extent similarly directed with those of the hindlegs : *in the backwards rotation all four legs move backwards, in the forward head rotation, forwards.* In these investigations, the head is immovably fixed to the trunk. Distinct alterations of posture appear when the head and neck remain freely movable. In the forward head rotation of the animal, the head moves dorsad at the atlanto-occipital joint and the neck curves concavely backwards; in the backwards rotation, the head goes ventrally at the atlanto-occipital joint and the neck has a convex backward curve.

Alterations in head position will secondarily influence the supporting tonus of the legs, [1] but they are not able to abolish the reactions to rotation which become apparent with similar strength whether the head is freely movable or fixed. The reactions of the head and neck can be observed most distinctly when the animal, in the course of rotation, is held in a lateral position and in the case of rotation in a vertical plane, they can easily be impaired by the labyrinthine righting reflexes.

When the legs are not under static stress and hang down freely, they also show distinct reactions, the most lively being during rotation of the animal in a vertical plane from a ventral position. In the forward head rotation, the forelegs are extended forwards while the hindlegs remain flexed. Reversing the direction of rotation, an extension of the hindlegs takes place and the forelegs return to a flexed posture. [2]

The reactions produced by the rotation round a bitemporal axis continue during rotation, at the end of which, however, the various parts of the body return to their initial positions. In more rapid rotation over greater angles (180° and more) distinct after-reactions to rotation can be observed; the parts of the body at first show opposite reactions and then occupy an intermediate posture. Reactions to rotation round a bitemporal axis appear abrupt, strong and regular in decerebellate animals, so that these animals are particularly suited for a systematic investigation of these reflexes. Although the reflexes are equally present in intact dogs, a systematic investigation is much more difficult because they are much more inhibited by the experimental conditions (especially by the attachment to the padded wooden beam). In decorticate, thalamus dogs, the reactions can also be easily demonstrated; but a systematic investigation does not usually succeed because of the constant attempts to free themselves. In labyrinthectomized dogs the reactions are totally absent.

We may sum up as follows :

A. *A backwards rotation round the bitemporal axis* produces at the start of the rotation :

    1. a ventral flexion of the head at the atlanto-occipital joint;

---

(1) As we have seen, a ventral flexion in the atlanto-occipital joint produces an extension and increase in supporting tonus in the hindlegs, a decrease in supporting tonus in the forelegs, while the dorsal curvature of the neck which occurs simultaneously causes just the opposite alterations in supporting tonus.

(2) Schaltenbrand (265) observed similar reactions in infants.

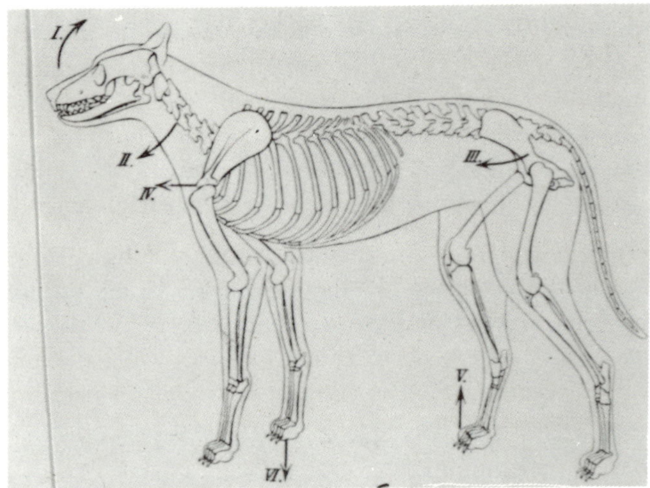

Fig. 124. Reactions on rotation of the head in space around the bitemporal axis. I. Rotation of the head dorsally at the atlanto-occipital joint. II. Concave curvature of the neck and the upper part of the thoracic vertebral column dorsally. III. Ventral movement of pelvis and *the forward movement of the hindlegs.* IV. *Forwards movement of the scapulae and forelegs.* V. *Decrease in supporting tonus and flexion of the hindlegs.* VI. Increase in supporting tonus in the forelegs.

2. a convex dorsal curvature of the neck and of the upper part of the thoracic spinal column (this curvature occurs also when the head is fixed. When the head is not fixed, the neck also moves ventrally. The curvature of the upper part of the thoracic spinal column appears particularly strong when the forelegs are under static stress and move backwards in the course of the rotation);

3. a dorsal movement of the pelvis;

4. an extension with a strong supporting tonus of the hindlegs which move caudad;

5. a rotation of the scapulae so that the shoulder joint is moved backwards;

6. a decrease in extension and supporting tonus in the forelegs with retraction at the shoulders;

7. a vertical deviation of the eyeballs downwards with vertical nystagmus of the eyes and vertical nystagmus of the head with the rapid component upwards.

B. A forward rotation of the head round the bitemporal axis at the start of rotation :

1. a dorsal movement of the head in the atlanto-occipital joint;

2. a concave dorsal curvature of the neck and of the upper part of the thoracic spinal column (with non-fixed head also a dorsal movement of the neck in relation to the trunk);

3. a ventral movement of the pelvis;

4. a flexion of the hindlegs with disappearance of supporting tonus and movement forwards (with a very strong curvature of the back, followed by an extension forwards);

5. a rotation of the scapulae so that the shoulder joints are directed forwards;

6. extension and protraction of the forelegs with increase in supporting tonus;

7. a vertical deviation of the eyeballs upwards, vertical eye and head nystagmus with the rapid component downwards.

The reactions to rotation round the bitemporal axis and its importance for the adaptation to static conditions and for the maintenance of balance

When a dog is set down with the forelegs on the axis of rotation of a hinged board in such a way that this runs parallel to the bitemporal axis of the head and when then the head end of the board is rapidly moved downwards or the tail end upwards, the animal will make a passive rotation forwards. The four legs move forwards in relation to the trunk which, since the legs are supporting themselves on the board, moves backwards in relation to the supporting surface. The forelegs are braced forwards and vigorously extended in the elbow and shoulder joints, the hindlegs also go forwards, but are flexed in the knee and tarsal joints (Fig. 125). Thus, sometimes a sitting posture is reached; moreover, the head moves backwards.

With the cessation of the movement, these reactions do not lessen, or only very little. Under the influence of the labyrinthine righting reflexes, the head and neck remain directed dorsally in relation to the trunk and the legs retain their posture due to this position of the head and the changed static conditions.

Conversely, with a rapid lifting of the head end or lowering of the tail end, the animal is passively rotated backwards, with this the trunk moves forwards towards the board, i.e., the four legs are moved backwards in relation to the trunk. The forelegs are somewhat flexed in the elbow joints, the hindlegs strongly extended in knee and tarsal joints while head and neck move forwards in relation to the trunk (Fig. 125, No. 1).

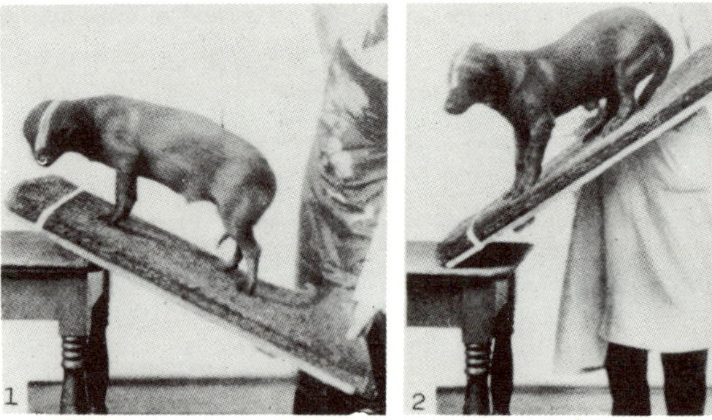

Fig. 125. Normal dog with blindfold on a board which is moved upwards (1) and downwards (2) at the tailend. The forelegs stand almost vertically above the axis of rotation (1). 1. In a rapid lowering of the tailend, the four legs go backwards and the trunk forwards towards the board. The forelegs are slightly bent in the elbow joints, the hindlegs strongly extended in knee and tarsal joints. Head and neck are ventrally flexed in space and towards the trunk. 2. Sudden and rapid lifting of the tail end. The legs are braced forwards, the trunk moves backwards towards the board. The elbow joints are strongly extended, the knee and tarsal joints somewhat flexed. Head and neck are dorsally directed towards the trunk. (See also alterations in position of the spinal column, pelvis and scapulae.)

With the help of these labyrinthine reactions, the leg positions are capable of adapting themselves to the respective positions of the supporting surface. The *line of the center of gravity* of the body always remains within the supporting surface so that balance is not lost. The reactions occur also if the head is held fixed in relation to the trunk or to the supporting surface.

When a dog is set down on the end of a seesaw with the bitemporal axis of the head parallel to the axis of rotation, on moving the board it will show different reactions, depending on the distance of the two axes from one another. The farther from the axis of rotation the animal stands, the smaller is the angle of rotation with a similar movement in the vertical direction or, in other words, the larger is the up and down movement with similar rotation (Fig. 126).

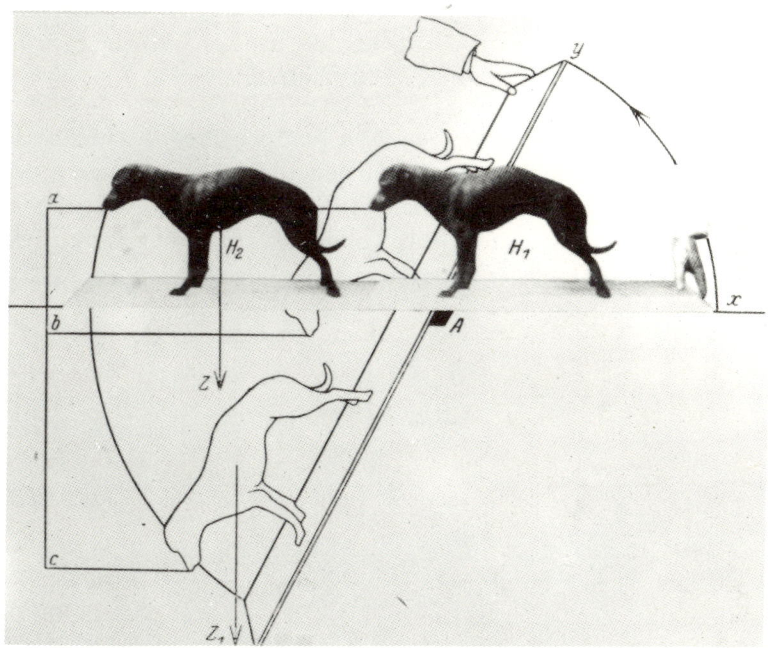

Fig. 126. Two dogs, H_1 and H_2, in standing posture one behind the other on a horizontal swinging board. Dog H_1 with the forelegs on the axis of rotation (A) dog H_2 with the forelegs on the end of the board. An upwards movement of the tailend of the board from X to Y, if there were no positive reactions of the animals, would cause the following: 1. both animals make an equal head forwards rotation; 2. both animals move the head downwards but over different distances, dog H_1 over the distance a-b, dog H_2 over the much larger distance a - c; 3. the heads are shifted in a horizontal direction, by the dog H_1 forwards, by the dog H_2 backwards. By the forward rotation of the animals the line of center of gravity is shifted orally to the supporting surface in such a way that both animals would fall forwards.

When the animal stands far from the axis of rotation, a change in position of the swinging board over a small angle will result in that only slight reactions to rotation are produced or none at all, whereas there will be a distinct lifting movement and the animal will only show a lifting reaction. On raising the head end of the swinging board over a large angle, an animal standing above the axis of rotation will show distinct rotating reactions, and it is to be expected that the animal standing at the end of the board will show a combination of lifting and rotating reactions since it has to perform a large lifting movement, in addition to the same rotation. In the rapid lowering of the board, however, only a lifting reaction will appear probably because the animal standing at the end will nearly or totally lose contact with the board and because of this will fall almost in a straight line.

When a dog stands at the end of a swinging board with the head towards the outer end, the animal will make a lifting movement upwards and a backwards rotation on raising the board, and on lowering it, a lifting movement downwards with a more or less strong forward head rotation.

These movements produce:

	lifting movement upwards	*backwards rotation*
at the start:	flexion of the hindlegs	extension of hindlegs
	flexion of the forelegs	flexion of forelegs
		movement of the four legs backwards
at the end of movement:	extension of hindlegs	(flexion of hindlegs)
	extension of forelegs	(extension of forelegs)
at the start:	lifting movement downwards	head forwards rotation
	extension of hindlegs	flexion of hindlegs
	extension of forelegs	extension of forelegs
		movement of the four legs forwards
at the end of movement :	flexion of hindlegs	(extension of hindlegs)
	flexion of forelegs	(flexion of forelegs)

Our hypothesis was found to be correct in actual investigation. A dog standing at the end of a swinging board with the head towards the outer rim shows at the start of downwards movement an extension of the four legs, at its termination it lies flat on its chest and belly (lifting movement on a downward movement). When the head end is moved upwards, the legs go backwards and are extended so that at the termination of the movement, the animal stands on its four extended and caudally directed legs (combination of lifting and rotating reactions).

These reactions produced by rapid rocking movements can be observed not only in intact dogs but also in decorticate and decerebellate dogs, though in decorticate dogs they sometimes appear with a slight delay. In decerebellate animals, they are particularly lively. In decerebrate and labyrinthectomized dogs, they are totally absent. A labyrinthectomized dog, therefore, falls head first when the tail end of the board is rapidly moved upwards, when he is standing above the axis of rotation as well as when he is standing at the end and turns a somersault at a rapid upward movement of the head end.

As we have seen, in a slow change of position of the supporting surface, labyrinthectomized dogs with blindfold show almost as prompt an adaptation to static conditions as intact ones; conversely, after a bilateral extirpation of labyrinths any adaptation to rapid swinging movements of the supporting base round an axis running in bitemporal direction is absent even months or years later.

Thus, adaptation to static conditions and the maintenance of balance in standing on a rapidly up and down swinging board are principally produced by labyrinthine reflexes, while these reflexes only have minimal importance for adaptation to slow changes in position of the supporting base.

Tait and McNally (296) observed similar reactions in frogs when they set down the animals on a swinging board and moved one end rapidly up and down. After destruction of the vertical semicircular canals these reactions failed to appear. Tait observed corresponding reactions in intact human beings which were set down with closed eyes in a knee-elbow position on a rapidly up and down moving swinging board, with the longitudinal axis of the body vertical to the axis of rotation. According to Tait, these reactions also fail to appear in human beings when the labyrinthine functions are absent. Unfortunately, the authors did not state how the labyrinth patients reacted on a slow lifting or lowering.

According to v. Stein (291), De Haan (114) and others, patients with impaired labyrinthine functions, when they stood upright, also showed impairments of adaptation to a slow inclination of the supporting base. The patients were tested in a military posture with feet close together and with v. Stein they even had to maintain themselves quite rigid and were not permitted to bend the knees. (The subject must maintain himself upright, vertically very straight, avoid bending the knees and without inclining the body backwards so as to make the inclination more rapid.) Thus, the patients were compelled to more or less suppress the adapting reactions.

E. SUPPORTING TONUS AFFECTED BY LABYRINTHINE REFLEXES PRODUCED BY ROTATION OF THE HEAD AROUND THE LONGITUDINAL AXIS OF THE BODY

For the investigation of these reactions the animals were attached to a wooden beam and alternately rotated from a ventral into a left or right lateral position. Simultaneously, the supporting tonus of the hind- and forelegs, put under static stress by means of the examiner's is investigated.

On rotation from a ventral position into a left-sided lateral position (Fig. 127 from A – – – B), the supporting tonus of the right legs decreases

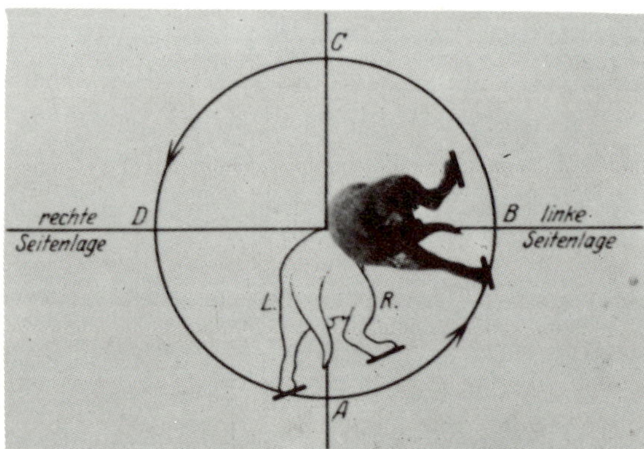

Fig. 127. Reactions of the limbs under static stress on rotation around the longitudinal axis of the animal. At the start and during a rotation to the left, i.e., seen from the tail in an anti-clockwise direction, the right legs show a flexion and decrease in supporting tonus, the left an extension with increase in supporting tonus and abduction.

at the beginning of rotation or disappears completely so that the legs go into flexion and are adducted, while the left legs are extended and abducted and show an increase in supporting tonus. At the termination of the rotation, opposite reactions occur.

Similar reactions appear when the animal is rotated from the left-sided lateral into a dorsal position (Fig. 127 from B → C), from a dorsal into a right-sided lateral position (Fig. 127 from C → D) or from a right-sided lateral into a ventral position (Fig. 127 from D → A).

Conversely, in a rotation towards the right,[1] i.e., viewed from behind in a clockwise direction, an extension with increase of supporting tonus and adduction appears in the right legs and a flexion, decrease of supporting tonus and abduction in the left ones. Thus, the legs of the two sides show opposite reactions.

It is of no importance for these reactions whether the position of the head is fixed in relation to the trunk during rotation in such a way that the plane of the buccal cleft stands parallel to the longitudinal axis of the body or forms an angle of $\pm$ 30° ventral to the latter (the usual position of the head). When the head is not fixed, the legs also show similar reactions, besides, in this case, a rotation of the head towards the trunk occurs; for instance, in a rotation to the left, the vertex of the cranium goes towards the right shoulder.

When the longitudinal axis of the animal is placed vertical instead

(1) A rotation of the head in which the vertex of the cranium goes to the right shoulder, i.e., the head into a right lateral position, is termed by Magnus as a dextro-rotation. In agreement with this convention, here also a rotation in which the animal passes over from a ventral into a lateral position to the right is referred to as a rotation to the right, although some would rather call it a rotation towards the left. Movements in the sequence: ventral position → right-sided lateral position → dorsal position → left-sided lateral position are termed rolling movements towards the right. When, on a turntable which turns in a clockwise direction, an animal is brought into :

> ventral position (buccal cleft horizontal or $\pm$ 30° downwards directed) it turns to the right;
>
> dorsal position (buccal cleft horizontal or $\pm$ 30° upwards directed) it turns to the left;
>
> suspended head downwards (buccal cleft vertical or $\pm$ 30° ventral) it turns to the right;
>
> suspended head upwards (buccal cleft vertical or $\pm$ 30° ventral) it turns to the left.

of horizontal, the reactions of the legs can also be observed. For instance, when the animal is held suspended head upwards, with the buccal cleft directed vertically or at an angle of 30° ventrally to the vertical, a flexion and a decrease in supporting tonus are evident in the right hindleg, in a clockwise rotation, i.e., to the right, an extension and increase in supporting tonus in the left (Fig. 128). These reactions occur also when the head is immovably fixed. Legs which are not under static stress show no distinct reactions.

In this investigation, the experimental conditions have to be adhered to very strictly. It is most successful when one sits down in the center of the turntable, the buccal cleft of the animal held in the vertical axis of rotation or at an angle of ± 30° ventrally to the latter and tests each paw under static stress alternately for its reactions. The position of the trunk is of no importance for the reactions on rotation, but it is the most comfortable to hold the trunk in a suspended or dorsal position (Fig. 128).

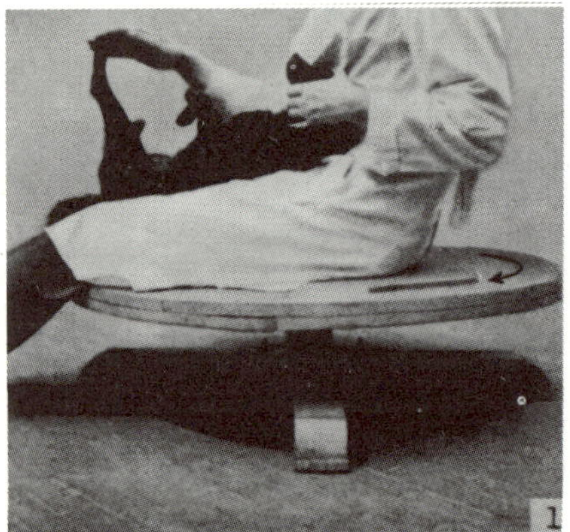 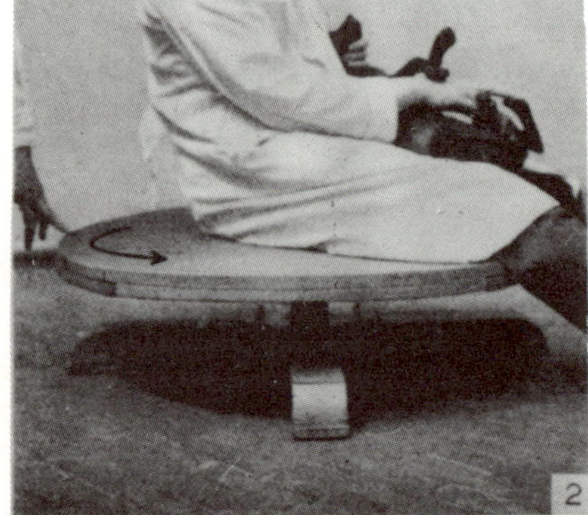

Fig. 128. Cerebellumless dog Erik in a dorsal position on a turntable. The upward pointing muzzle is ventrally directed at an angle of ± 30° to the vertical. The left hindleg is under static strain.   1. By rotating the animal to the right, i.e., clockwise, the leg under static strain is extended and shows a strong supporting tonus. 2. On rotating the animal to the left, i.e., anti-clockwise, the statically strained left hindleg goes into flexion and the supporting tonus disappears.

The investigator must keep the arm of the investigating hand strongly against the body so that the hand is not abducted or adducted by the rotating movement and does not pull the limb with it. It is not admissible either to put both hind- or forelegs simultaneously with the hand under static stress because in an animal suspended head upwards, if both fore-legs are put under static strain with one hand and the head is rotated passively (for instance, with the vertex of the cranium towards the right shoulder), the thorax is found also to rotate. By this movement, the right foreleg is abducted and the left adducted to the thorax since the soles are fixed by the hand. One then gets the impression that both forelegs are moving to the right. Since, in the investigation on the turntable, the thorax rotates too, these alterations in position of the legs in relation to the trunk also occur and not only simulate the only apparent labyrinthine rotating reactions of the extremities, but also impair the true labyrinthine reactions of the legs. In order to establish the latter incontestably, the position of head and trunk should be fixed as much as possible.

In addition to the reactions of the extremities and the head, a rotation of the pelvis in relation to the anterior part of the body also takes place. When the animal suspended head upwards, with the muzzle pointing upwards, is rotated in a clockwise direction (to the right), the dorsal side of the pelvis moves towards the right side of the trunk. The rotating reaction of the head and of the anterior part of the body also rotates to the left so that the whole trunk rotates contrary to the turntable.

Moreover, a concavity of the spinal column to the left side with a raising of the left iliac crest occurs.

Sometimes, the posterior part of the body and the tail also show peculiar rotating movements. In a clockwise rotation, the tip of the tail makes great circles in the same direction round the root of the tail and after cessation of the movement, several circles in the opposite direction occur, i.e., anti-clockwise.

In decerebellate dogs, the reactions on rotation round the longitudinal axis are particularly abrupt and regular, in standing as well as suspended head upwards or downwards and also in the dorsal position of the trunk with the muzzle pointing upwards (Fig. 129). In investigations on suspended and dorsal positions, they are usually inhibited in intact dogs. In labyrinthectomized dogs, they are totally absent.

According to Dusser de Barenne (60), frogs show corresponding reactions. In a frog suspended vertically by a thread running through the maxilla, Dusser de Barenne observed that, on rotation of the animal in a clockwise direction, the symmetrical, slightly flexed posture of the hind-legs was altered as follows: "the preceding left hindleg was suddenly extended stiffly in the knee joint and at the same time abducted by a lively flexion in the hip joint and raised somewhat forwards and the toes were spread apart. The right hindleg was flexed a little more in the hip and knee joints and at the same time directed dorsally, i.e., towards the back."[1] Occasionally, extension and adduction of the left forelegs together with flexion also appeared.

Dusser de Barenne found that these alterations of position are still present after the transverse section of the midbrain behind the optic lobes and after severance of the posterior roots of C II and C III but that they are absent in the spinal frog and after bilateral extirpation of labyrinths.

Thus, in summing up, a rotation round the longitudinal axis towards the right (seen from behind in a clockwise direction) produces in dogs the following reactions:

1. extension, increase in supporting tonus and abduction of the right legs;

2. flexion, decrease in supporting tonus and adduction of the left legs;

3. rotation of the head to the trunk towards the left, i.e., with the vertex of the cranium towards the left shoulder;

4. rotation of the anterior part of the body and the pelvis towards the left side and a concavity of the spinal column towards the right side;

5. rotary deviation of the eyeballs and rotary nystagmus.

The importance of reactions on rotation round the longitudinal axis for the adaptation of leg position to the position of the supporting surface and for the maintenance of balance

When an intact dog is set down blindfolded on a hinged board with the longitudinal axis of the animal parallel and directly above the axis

---

(1) The description is quoted verbatim as it does not seem clear to me how the left hindleg is lifted forwards by a flexion in the hip joint and, conversely, the right also flexed in the hip joint, is directed backwards. In my opinion, a flexion at the hip joint must always produce a movement forwards. At any rate, frogs as well as dogs, on rotation in a clockwise direction show an abduction of the left hindleg with extension in the knee joint and, occasionally, an extension and abduction of the left foreleg while the legs of the right side are flexed at the knee and elbow joints.

of rotation and the right or left side of the board is rapidly moved up and down, the animal is passively rotated round the longitudinal axis, in which case alterations occur in posture and supporting tonus which correspond to the reactions on rotation just described. When, for instance, the board is moved rapidly upwards by 30-45° on the right side of the animal, from a horizontal position, the animal is passively rotated to the left, the left legs are instantly extended and abducted, the right ones are flexed and adducted and the head and thorax rotate towards the raised side, i.e., towards the right. With a faster rotation at a larger angle, the animal, due to these reactions, goes into a *right lateral position*. Under the same conditions, alterations in position of the legs fail to appear in labyrinthectomized dogs and they thus fall to the *left* side.

When an intact dog is blindfolded and placed on its belly on the axis of rotation, the animal, on the rapid lifting of the board from the right side, will also actively rotate to the right and go into a right lateral position, in which case the left legs are more or less extended and abducted, while the labyrinthectomized animals here, too, roll to the left side.

At the termination of lifting the board, the reactions usually only lessen partially since the changes in position produced by the reactions are maintained by the altered static conditions.

When an animal is set down at the end of a board with the longitudinal axis parallel to the axis of rotation, it passively performs a rotating movement round the longitudinal axis and, in addition, a vertical movement. Then, on lowering the board, distinct lifting reactions can be observed. The four legs are extended at the start and flexed at the termination of the movement so that the animal will lie flat on the thorax and belly.

When the board is raised, again an extension of all four legs takes place at the end of the lifting so that the animal is again in a standing posture; however, the legs are abducted on one side and extended more than the legs of the other side, which go into adduction. Thus, a combination of lifting and rotating reactions takes place.

After the bilateral extirpation of the labyrinths, the different reactions for adaptation to rapid lifting and lowering of the sides of the supporting base are absent. Even after several months, the animals instantly fall on their sides. Thus, the adapting reactions are produced by the labyrinths. It is remarkable that, on slow lifting and lowering, labyrinthectomized dogs show some adaptation which is, therefore, produced by non-labyrinthine influences.

On standing or lying on a supporting surface which is rapidly moved up and down on the left or right side of the animal, labyrinthine reflexes produced by rotation round the longitudinal axis play an important role in the adaptation of the position of legs and trunk to the position of the supporting surface and in the maintenance of balance.

Strikingly sharp reactions are shown by decerebellate dogs when they are set down on a rocking board and the sides of the board are rapidly raised and lowered. In the rapid raising over larger angles, for instance of the right side, the animals fall to a right lateral position like intact dogs and not to the left like labyrinthectomized animals. Also, in decorticate animals, distinct adapting reactions can be detected, though they sometimes appear somewhat late, particularly in the hindlegs. In decerebrate animals, these reactions are totally absent, with each movement of the rocking board they instantly fall down as if dead.

Whether in decerebrate animals a rotation round the longitudinal axis produces an influence on the extensor rigidity has not yet been precisely investigated. According to Magnus and De Kleyn, this rotation produces compensatory rotating reactions of the head.

Frogs, too, when set down on a rocking board with the longitudinal axis parallel with and above the axis of rotation, show corresponding reactions on the rocking of the board. After the destruction of the vertical semicircular canals, Tait and McNally (215, 297) could no longer elicit these reactions.

McNally destroyed one, several and all semicircular canals in frogs. The animals were then set down on a swinging or rocking board and the impairment of reactions to rotation round the bitemporal, the longitudinal and both diagonal axes were investigated. Based on these findings, Tait and McNally came to the following conclusions :

1. When the longitudinal axis of the frog stands vertical to the axis of rotation, the movement of the board from the horizontal with the head end pointing downwards produces a stimulation of the two anterior vertical semicircular canals similarly to the downward movement of the two posterior vertical semicircular canals.

2. When the longitudinal axis of the animal stands parallel to and above the axis of rotation, a lowering of the right side of the board produces a stimulation of the right vertical and a lowering of the left side produces a stimulation of the left vertical semicircular canals.

3. When the diagonal axis left-forward right-backward lies above and parallel to the axis of rotation, a downward movement of the right side of the board produces a stimulation of the right anterior vertical semicircular canal and an upward movement produces a stimulation of the left posterior vertical semicircular canal.

4. When the diagonal axis right-forward left-backward stands parallel to the axis of rotation, in a downward movement of the right side of the board, a stimulation of the right posterior vertical semicircular canal takes place and, in a downward movement of the left side, that of the left anterior vertical semicircular canal.

5. Stimulation of :

the left anterior vertical semicircular canal produces an extension forwards and downwards of the left foreleg;

the right anterior vertical semicircular canal produces a similar movement of the right foreleg;

the left posterior vertical semicircular canal produces an extension backwards and outwards of the left hindleg;

the right posterior vertical semicircular canal produces a similar movement of the right hindleg.

According to Tait and McNally, a rotation round the longitudinal axis towards the right thus produces in the frog a stimulation of the right vertical semicircular canals and causes an extension and abduction of the right legs. Their findings do not correspond with those of Dusser de Barenne, who observed on rotation round the longitudinal axis in the frog reactions in all four legs, in agreement with our observations in dogs.

Similar reactions to rapid up and down rocking movements are shown also by intact human beings when they are placed on a swinging board in a knee-elbow position, with closed eyes and with the longitudinal axis of the trunk parallel to the axis of rotation of the board. When labyrinthine function had been lost, Tait and McNally could not detect these reactions. Whether such patients in knee-elbow positions show an adaptation to the position of the board on slow lateral rocking movements round the longitudinal axis has, unfortunately, not been investigated. When such a patient is placed upright on a swinging board with closed eyes so that he looks in the direction of the axis of rotation, with his legs a little apart on both sides of this axis, with the slow lifting of one side of the board, he will fall to the other side, at an inclination of the board to which an intact man is very well able to adapt himself (Tait). Thus, adaptation appears impaired in the upright position in a slow rotation round the *dorsoventral axis*.

SUPPORTING TONUS AFFECTED BY A ROTATION ROUND THE DORSOVENTRAL AXIS

When an intact dog, with or without blindfold, is set down on a turntable moved in a clockwise direction,[1] the head turns to the left shoulder, the spinal column curves concavely to the left, the dorsal side of the pelvis usually rotates slightly to the right side and, in addition, the right legs go into flexion and adduction and the left ones into an extension and abduction (Figs. 129 and 130). The alterations are usually constant and more distinctly observed in the hindlegs than in the forelegs. The same reactions are present in decorticate and decerebellate dogs, but not in labyrinthectomized animals. Thus, one gets the impression that all these reactions are of a labyrinthine origin. When, however, intact dogs in a ventral position in the air are rotated round the dorsoventral axis, reactions of the head and a curvature of the spinal column indeed occur, but distinct and constant alterations in position and supporting tonus of the legs under static stress are absent. When, during rotation, the head is also fixed to the trunk, with the buccal cleft horizontal or directed downwards at an angle of 0-45°, the rotation round the dorsoventral axis produces no reactions in the extremities nor when they are hanging down freely or when under static stress. When the animals are held in the air in a dorsal position or lying in a dorsal position on a turntable, with the buccal cleft horizontal or directed upwards at an angle of 0-45°, a rotation round the dorsoventral

---

(1) A rotation round the dorsoventral axis seen dorsally in a clockwise direction and seen ventrally in an anticlockwise direction, is called a rotation to the right.

axis also causes a turning of the head;[1] the reaction of the legs, however, fails to appear. When the vertex of the cranium stands directly above the axis of the rotation as well as when it stands on the outer edge of the rotating table, whether the head is fixed to the trunk or is freely movable to the right and left, the legs do not show any distinct reactions. The rotation round the dorsoventral axis, under these circumstances, does not produce reactions of the legs, either in intact or in decorticate and decerebellate dogs, but a turning of the head always appears. It even appears in decerebrate animals, but in labyrinthectomized animals it is absent.

Besides the horizontal deviation of the eyeballs and the horizontal nystagmus of the eyes, turning of the head is the only constant labyrinthine reaction which appears on rotation round the dorsoventral axis. No specific alterations in posture or supporting tonus can be established as a direct result of labyrinthine stimulations by rotation round the dorsoventral axis, either in intact or in decorticate and decerebellate animals.

It is the general opinion that the labyrinthine reflexes of rotation round the dorsoventral axis are produced by stimulations of the horizontal semicircular canals. Grahe (108) used a unilateral caloric stimulation of the horizontal semicircular canal in rabbits, i.e., rabbits in a dorsal position with the buccal cleft horizontal or 0-45° directed upwards. He rinsed one ear with cold or hot water and observed that alterations in tonus and position occurred in the forelegs. One foreleg showed extension and abduction, the other flexion and adduction. With a horizontal buccal cleft, the reactions were opposite to those with the buccal cleft pointing upwards; with a right-sided stimulating rinse with a horizontal buccal cleft, he observed extension and abduction of the left foreleg, flexion and adduction of the right foreleg; with the buccal cleft pointing upwards by ± 50°, however, gave flexion and adduction of the left foreleg and extension and abduction of the right foreleg.

Grahe put the rabbits in a dorsal position on a turntable with the buccal cleft horizontal or pointing upwards by 50°. Rotation also produced alterations in tonus and position in the forelegs which, however,

---

(1) In the dorsal position of the animal, the rotation round the dorsoventral axis in a clockwise direction (rotation towards the right) causes a turning of the head towards the left shoulder.

appeared far less regularly. They were frequently absent and when positive they were slight, rarely distinct and consistent. . Besides, in the majority of observations, the results of test rotations were opposite to those of calorical stimulation.

How does the different behavior of intact and labyrinthectomized dogs on the turntable come about ? Probably in this case the turning of the head with subsequent curvature of the spinal column plays an important role. As we have seen in Chapter VIII, a passive turning of the head in a dorsal position and in a ventral position in the air does not produce any alterations of supporting tonus of the extremities; when the animal is standing on all four legs, however, very distinct ones are produced (Fig. 109). Rotation of the turntable to the right causes standing animals to turn their heads towards the left shoulder followed by a concavity of the spinal column to the left, rotation of the dorsal side of the pelvis towards the right, adduction with flexion and decrease in supporting tonus in the right legs, abduction and extension in the left legs, i.e., exactly the same reactions which appear on the passive turning of the head in the animal standing on all four legs (Fig. 109). Since, in rotation, turning of the head is caused by a stimulation of the labyrinths, the manifestations produced by turning the head are secondary effects of labyrinth stimulation and thus are the explanation of the different behavior of intact and labyrinthectomized dogs on the turntable. There must be factors other than labyrinthine rotation participating in the limb reactions. An animal standing on a turntable turning towards the right is always passively rotated round the dorsoventral axis towards the right regardless of whether the muzzle is directed towards the center or towards the outer rim, into the direction of rotation or against it. The rotating reactions always have to be the same. The response relative to turning the head with a rotation of the table to the right, clockwise a turning of the head towards the left shoulder, always appears. The legs, however, do not always show the same reaction; they differ according to the position of the animal on the table; sometimes the position of the legs is even altered during rotation (Fig. 129).

By rotation of the turntable, the centrifugal force comes into effect and pulls the animal's body forwards or backwards, to the right or to the left according to its position (Fig. 129), by which the position of the leg to trunk is actively changed. The passive alteration of position is able to produce a reflex reaction in the muscles (see Chapter XIII) which actively change the position of the legs to the trunk. The al-

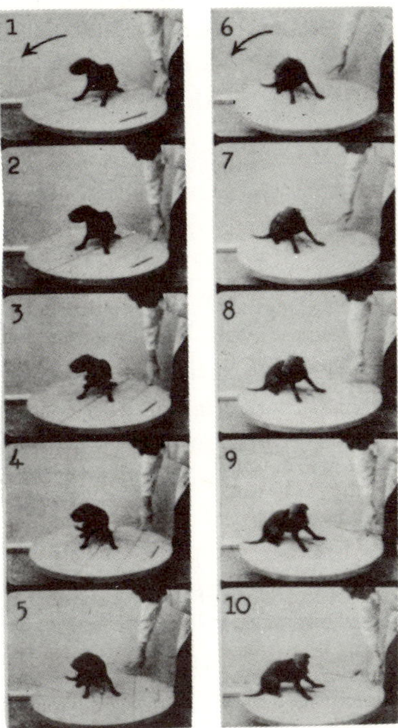

Fig. 129. Decerebellate dog Piccolino, with blindfold, on a turntable which is rotated towards the left. Typical reaction. The head is turned towards the right, the spinal column concavely curved to the right, the right foreleg extended and abducted, the left adducted and slightly flexed in the elbow joint (1 and 2). During rotation, however, the leg position is altered in such a way that both forelegs show at first almost the same posture (4), then the left foreleg is extended and abducted, the right slightly flexed and adducted (5-7).

terations caused by the centrifugal force can perhaps explain the different behavior of the extremities in the various postures of the animals on the turntable and the changes in leg position during rotation, but they are not a key to the different behavior of intact and labyrinthectomized dogs under the same experimental conditions, since labyrinthectomized dogs, too, show these proprioceptive reactions. When the head of the intact dog does not stand exactly above the axis of rotation, the centrifugal force will probably affect and stimulate the labyrinths. Wittmaack has shown that in guinea pigs the otoliths were dislocated by rapid centrifuging (2000 rotations per minute for 1/2-3/4 minutes).

278

Thus, centrifugal force causes strong alterations in the intralabyrinthine distribution of pressure and a pulling of the otoliths on the nerve endings. What reflexes are produced by this and whether a less rapid rotation is generally sufficiently strong for the production of reflexes have not yet been investigated.

McNally (215) reported that frogs set down on the turntable showed no reactions to rotation and, among others, no turning of the head when the speed of rotation was gradually increased. In high speed rotation, however, different reactions appear which vary according to the position of the animal on the turntable. When the animal is situated exactly in the center of the disk, no reactions appear, but when it sits with the longitudinal axis of the body parallel with the median line of the disk, the head directed towards the outer rim, the head and the anterior part of the body are raised and the forelegs are strongly braced forwards. When the head is directed towards the center, it is ventrally flexed till the mandible touches the table , the caudal part of the back is curved (the lumbar depression disappears), and the forelegs are extended laterally while the animals with increased velocities, creep towards the center. When the animal is placed with one side, for instance the left, parallel to the outer rim, then with a high speed of rotation *to the left as well as to the right*, the head and the anterior part of the body move downwards on the right side so that the animal hangs over to the right. With a still higher speed, the right foreleg flexes, while the left is laterally extended and with a further increase in speed, similar reactions appear on the hindlegs. The animals in this case occupy an almost lateral position to the right, the right legs are flexed and the left extended. When the right side stands parallel to the outer rim, however, the animal lies in a lateral position to the left with flexed left and laterally extended right forelegs.

These reactions also appear in decerebrate frogs and in frogs in which all six ampullae of the semicircular canals have been destroyed, but they are absent after a bilateral utricular lesion. (Unfortunately it was not stated how long after destruction of the utricular otoliths the experiments were carried out, because labyrinthectomized dogs, after having recovered from shock some time after the operation, again show reactions which counteract the centrifugal force.)

Finally, it seems that probably the linear displacement of the supporting surface which occurs, with the same direction of rotation, forwards, backwards, to the right or left, depending on the position of the animal, also exerts an influence on the posture of the limbs. When the animal sits or stands on the turntable with the axis of the body parallel to the median line with the head directed outwards, then on rotating for instance to the left, it is not only pulled forwards by the centrifugal force but also passively to the right. When the head is directed towards the center, the animal, with a similar rotation, will be pulled to the right; with the longitudinal axis vertical to the median line, it will be passively pulled forwards or backwards. The displacement of the supporting surface to the left, right, forwards or backwards produces passive alterations of the position of the legs in relation to the trunk, which, in their turn, cause reflex contractions of the muscles, thus producing active changes of limb posture. As we have already discussed (p.273), however, reactions on linear displacement on a horizontal plane are not labyrinthine reactions since they also occur in labyrinthectomized dogs. These reactions, too, can contribute to the explanation of the different behavior of limb posture in the altered position of the animals on the turntable and on the change in posture of the legs during rotation, but they cannot be the cause of the difference in behavior of intact and labyrinthectomized dogs.

These reactions can be particularly well observed in the hindlegs, when a labyrinthectomized dog is placed on the turntable with the head directed towards the center and the rear part of the body towards the rim and when the table is then moved alternatively to the right and left over small distances, for instance by 15-30°. The hindlegs then show the same reactions they show on the lateral displacement of the supporting surface in a straight line.

Thus, the different behavior of intact and labyrinthectomized dogs on the turntable probably depends exclusively on the labyrinthine turning of the head with consequent alterations in position of the spinal column and the extremities. The eventual participation of reactions on labyrinthine stimulations by centrifugal force has not yet been established with certainty.

Summing up :

1. In dogs, on rotation of the labyrinths round the dorsoventral axis, apart from reactions of the eyeballs, only a turning of the head, eventually with head nystagmus, can be observed as the only direct effect of labyrinthine stimulation; turning of the head

280

occurs in intact, decorticate, decerebellateless and decerebrate dogs.

2. Intact, decorticate and decerebellate dogs, on standing or in a ventral position on the turntable, show, on rotation round the dorsoventral axis, numerous other reactions. Among others, those in the legs are more or less absent in labyrinthless dogs; these reactions, at least partly, are only secondary effects of labyrinthine stimulations and are caused by turning of the head.

3. Other factors, too, participate in the occurrence of reactions of the extremities while standing on the turntable. Among others are the centrifugal force and the linear displacement of the supporting surface; it is, however, not established whether these factors also cause labyrinthine reflexes and can thus be made responsible for the different behavior of intact and labyrinthless dogs on the turntable.

The importance of labyrinthine reflexes by rotation round the dorsoventral axis for the maintenance of balance on the turntable

Although labyrinths are considered the organs of balance par excellence, labyrinthectomized animals, with or without blindfold, in rotation round the dorsoventral axis on a turntable are capable of maintaining their balance under various circumstances as well as and even better than intact dogs on the turntable.

A labyrinthectomized dog stands quite calmly on the turntable and is able to run around on the table without falling. Even with some lesion or injury to a hindleg which causes it to be held flexed, the animal can stand well and securely on the other three legs. In an increase or decrease in the speed of rotation or a change in direction, labyrinthectomized dogs show distinct adapting reactions of the extremities and of the spinal column which seem absolutely necessary for keeping balance under all circumstances.[1] These reactions also appear in intact dogs; however, labyrinthine reactions then are added which, according to circumstances, strengthen the reactions so that they become too strong, or counteract them. This can be observed particularly clearly in decerebellate animals which show a very abrupt turning of the head followed by especially strong secondary reactions of the spinal

---

(1) The causes and mechanism of these reactions will be discussed later.

column and of the legs. According to circumstances, these secondary reactions will either help or impede the maintenance of balance.

When a dog with intact labyrinths is placed on a turntable rotating in a clockwise direction towards the right, the labyrinthine reactions produce a turning of the head towards the left followed by a concavity of the spinal column towards the left, an abduction of the left legs and an adduction of the right ones (Fig. 130), and this happens not only when the left side (Fig. 130, No.1) but also when the right one (Fig. 130, No. 2) lies towards the outer rim of the table and when the head is directed towards the center (Fig. 130, No. 3) or to the outer edge (Fig. 130, No.4). When the left side of the animal lies towards the outer rim (Fig.130, No. 1),the outward bracing of the left legs will counteract the centrifugal force and impede the falling over or the throwing of the animal from the table. When it lies with the right side towards the outer rim (Fig.130, No.2), however, the abduction of the left paired with the adduction of the right leg will promote the falling over and throwing off by centrifugal force. When a labyrinthectomized dog is placed with the left side towards the outer rim, the left legs also go into abduction and are propped outwards, the right ones are adducted and the spinal column shows a slight concavity to the left. Thus, alterations in posture are similar to those of a dog with intact labyrinths and counteract the centrifugal force.

When the labyrinthectomized animal stands with the right side towards the outer rim, however, the right legs are braced outwards, the left are adducted and a slight concavity to the right appears in the spinal column (see Chapter XIII). Thus, the reactions are contrary to those of intact animals and in this case, also, counteract the centrifugal force.

Thus, labyrinthine reactions with secondary reactions sometimes produce in intact animals, depending on the position of the animals, abduction of the outward directed legs with an outwardly directed concavity of the spinal column (Fig. 130, No. 1) and sometimes an abduction of the inward directed legs with an inwardly directed concavity of the spinal column (Fig. 130, No. 2). In the first case, the centrifugal force is counteracted and in the second case the tendency to fall over and be thrown off by centrifugal force is enhanced by the reactions. In labyrinthectomized animals standing with one side towards the outer rim, an abduction of the outwards directed leg with an outwards directed

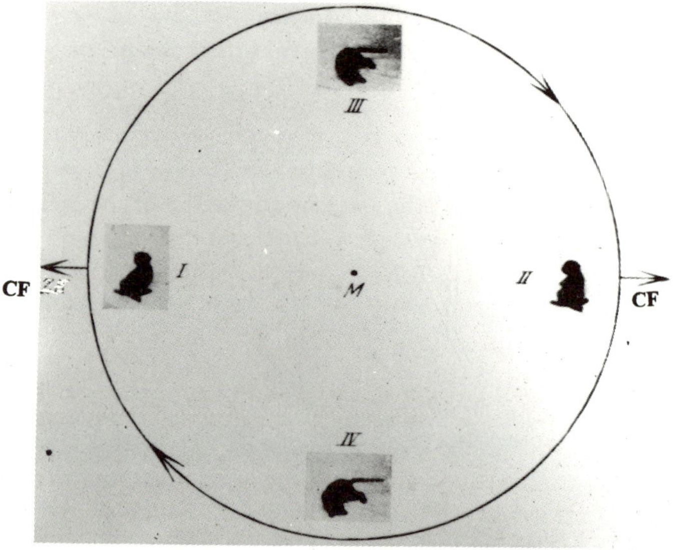

Fig. 130. On a turntable rotating towards the right are two dogs (I and II) and two cats (III and IV), all blindfolded. The rotation produces in all four the same typical reactions : turning of the head towards the left, convex curvature of the spinal column towards the right, adduction and flexion of the right legs, abduction and a more or less extension of the left legs. In dog I, which sits with the head in the direction of rotation, the abduction of the left legs and the curvature of the spinal column will counteract the fall and the thrust of the animal caused by the centrifugal force; in the dog II, sitting with its head against the direction of rotation, however, it will promote them. When the table is suddenly rotated more rapidly, in animal III, which stands in the direction of the radius with the head towards the center, the legs, already directed towards the left are once more passively drawn towards the left and the animal will fall on the right side. In animal IV, which stands also in the direction of the radius but with the head outwards, this will not be the case as the left directed legs are passively drawn to the right. (This animal, however, will fall more easily into a right lateral position than animal III, at the sudden cessation of rotation or a change in its direction.)

concavity of the spinal column always occurs; thus the influence of centrifugal force here is always counteracted. The tendency to be thrown off by centrifugal force, in any specific direction of rotation, is in intact animals dependent on whether they stand parallel to the outer rim of the disk with the right or with the left side, which is completely unimportant with labyrinthectomized dogs.

When an intact animal stands in the direction of the radius (Fig. 130, Nos. II and IV) the four legs are directed towards the left in a clockwise rotation, when the head points towards the center as well as when it points outwards. In the first case, the turning of the table will pull the legs passively towards the left of the trunk and cause a shifting of the center of gravity to the supporting surface towards the right. In the second case, the legs are passively pulled to the right and the center of gravity is shifted to the left. It is now clear that the labyrinthine alterations of position in the first case will accelerate the loss of balance and, in the second, promote the maintenance of balance. In the labyrinthectomized animal, however, the reactions always facilitate the preservation of balance. When a labyrinthectomized animal stands on a turntable rotating in a clockwise direction with the head towards the center, the right legs are braced outwards and when the head is directed outwards, the left legs do so.

Moreover, intact animals are often forced by labyrinthine reactions to run around in circles during rotation as well as afterwards, in which case they often stumble. Labyrinthectomized dogs, on the contrary, are able to stand quite still.

After cessation of, for instance, a rotation to the right, a turning of the head towards the right with a concavity of the spinal column towards this side appears, caused by the after-reaction to rotation. In this case, the animals often run around in circles to the right. In cats, particularly in decerebellate cats, these movements often appear quite violent.

On the turntable, when the speed and direction of rotation are alternately changed, labyrinthectomized dogs are more capable of maintaining their balance than intact animals and particularly more than decerebellate ones where balance is very difficult to maintain because of the abrupt occurrence of labyrinthine and secondary reactions.

In the last three chapters, we have seen that the supporting tonus is influenced by the posture as well as by movement of the head and that reflexes from different parts of the neck (tonic neck reflexes) and from the labyrinths play a role therein. Distinct alterations of support tonus are produced :

1. by alterations in position of head in relation to neck;

2. by alterations in position of neck in relation to trunk;

3. by alteration in position of the head in space (tonic labyrinthine reflexes);

4. by the linear movement of the head (the labyrinthins) in dorsoventral direction;

5. by rotating movements of the head round the bitemporal axis;

6. by rotating movements of the head round the longitudinal axis of the body.

Alterations in supporting tonus as a direct effect of labyrinthine stimulations have not been observed either in the linear movement in rostro-caudal or in the bitemporal direction, or in the rotating movement round the dorsoventral axis.

In the following chapter, we will see that a change of posture of the head can also influence the supporting tonus in another way.

X. SUPPORTING TONUS AFFECTED BY THE POSTURE OF THE SPINAL COLUMN AND PELVIS

Just as the position of the neck and the reactions on static stress exert a reciprocal influence, interrelations also exist between these reactions and the posture of the spinal column. We have already discussed several changes in tension of the vertebral muscles caused by static stress on the feet, for example, and how in this way a hollow, sagging back was straightened and fixed, even dorsally curved in decerebellate dogs. In what follows, we shall see how, on the contrary, the posture of the spinal column influences both the reactions on static stress and the strength of supporting tonus.

We will successively discuss :

1. the factors which can produce changes in posture of the spinal column and the pelvis;

2. the reactions of static stress to the posture of the spinal column;

3. the reactions of static stress to the posture of the pelvis.

A. THE POSTURE OF THE SPINAL COLUMN

Several of the factors exerting an influence on the posture of the spinal column, such as contact with and static stress on the soles of the feet, rotating movements round the bitemporal axis and others have already been discussed in the preceding chapters. Another factor is the position of the legs in relation to the trunk which exerts a certain influence on the posture of the spinal column. When the hindlegs of a standing dog are shifted caudally, the back forms a concave hollow in intact and decorticate animals while, in decerebellate animals, the dorsal convexity decreases. The greatest change of posture occurs at the transition between the thoracic and the lumbar parts of the spine (Fig. 131).

Fig. 131. Decerebellate dog Moor. 1. With the hindlegs stretched backwards, the spinal column forms an almost straight line. 2. With a vertical posture of the hindlegs, the spinal column forms a strongly convex upward curve.

With the caudal shifting of the hindlegs, a stretching of the m. quadratus lumborum and m. ileopsoas takes place which probably causes the lordosis of the vertebral column.

The m. quadratus lumborum in the dog originates on the ventral surface of the last two thoracic vertebrae, the last rib and on the transverse processes of the lumbar vertebrae and inserts on the ventral surface of the upper portion of the ilium, the sacrum and on the anterior iliac spine.

The m. ileopsoas originates on the ventral surface of the vertebral body and the transverse processes of the last two thoracic and lumbar vertebrae and ends on the ileo-pectineal crest and on the trochanter.

In shifting forwards on the hindlegs, opposite changes occur. In intact dogs, the hollowed back is straightened or somewhat convexly curved dorsally, while in decerebellate dogs the dorsal curvature greatly increases (Fig.131 , No. 2). The change in posture of the spinal column is probably brought about by an extension of the ms. latissiumus dorsi, longissimus dorsi and spinalis dorsi. It is not only produced passively by the extension, but evidently *reflex* alterations n contraction of these muscles and of the extensors of the hip joint also participate.

For the contraction of vertebral muscles, a static stress on the protracted hindlegs is not necessary. When a dog is lifted from the ground by the thigh (Fig. 133, No. 1), the vertebral column of the vertically suspended trunk is slightly fixed and the anterior part of the body more or less swings in space in a to and fro movement. When the thigh is passively moved forwards at the hip joints, however, the vertebral muscles are suddenly stressed and pull the anterior part of the body upwards (Fig. 133, No. 2). The tension is tonic; with the thigh directed forwards, the anterior part of the body remains actively suspended in a nearly horizontal position. In this case, the cervical muscles also go into contraction so that the animal keeps its muzzle directed upwards in the air (the longissimus dorsi terminates on the 2nd-7th cervical vertebra, Fig. 132). The entire vertebral column is strongly and elastically fixed as can be seen clearly when the animal, without changing the relative positions of the thighs and trunk, is moved up and down in a vertical direction.

The tension of the vertebral muscles is strongest *during* the forward movement of the thighs; on cessation of the movement, the anterior part of the body relaxes a little and a distinct "clicking" of the joints can be felt. The phenomenon supports the view of a participation of reflex contraction and also of the fact that the contractions change when the animals are moved up and down in a vertical direction, since the tension decreases in downwards movement, probably under the influence of labyrinthine reflexes, and increases in an upwards movement. When the hindlegs are retracted backwards, the anterior part of the body instantly falls again vertically downwards and the vertebral muscles become limp.

In decerebellate dogs, the cotnraction of the vertebral muscles on the ventral movement of the thighs appears particularly strong and abrupt. With a forward movement of the thighs of a decerebellate animal held suspended in the air, the anterior part of the body is some-

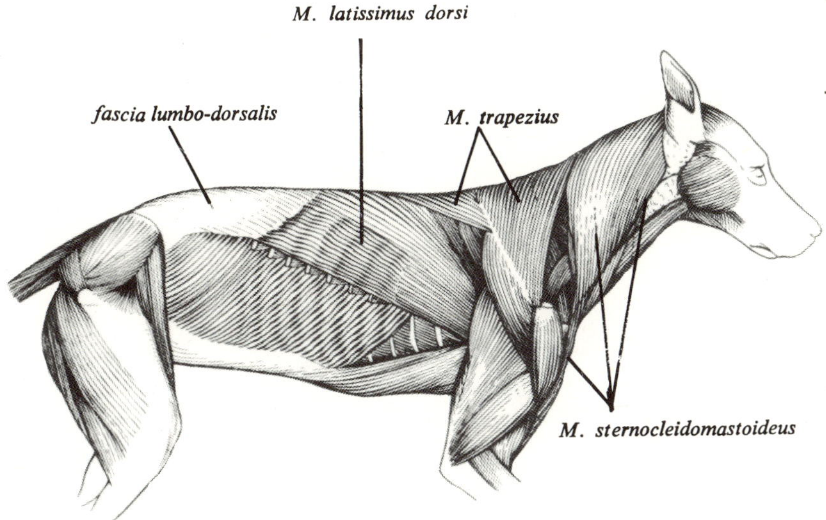

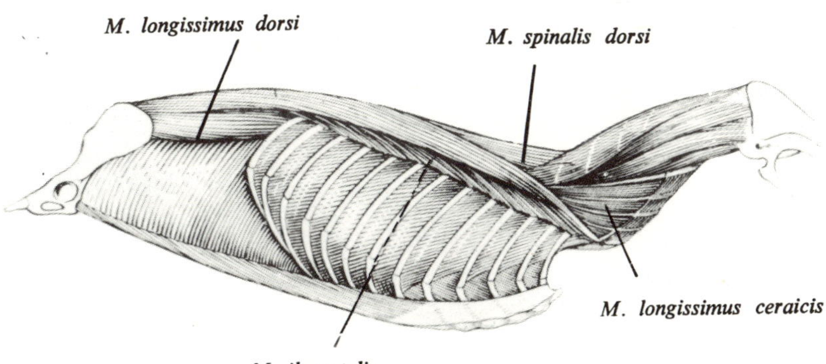

Fig. 132. Back muscles of the dog, superficial layer. *M. latissimus dorsi* : originates on the fascia lumbodorsalis, on the processi spinosi of the lumbar and the last seven thoracic vertebrae and on the last 2-3 ribs and terminates with a tendon on the linea tuberculi minoris humeris. *M. trapezius*: originates on the median aponeurosis of the back from the 3rd cervical to the 10th thoracic vertebra or on the fascia lumbodorsalis and terminates on the free edge of the spine of the scapula. 2. Back muscles of the dog, deep layer. M. erector trunci (*M. sacrospinalis*) : *M. longissimus dorsi* : originates on the crista and on the median surface of the ilium inserts on a spur of the 7th cervical vertebra; due to its fusion with the m. longissimus cervicis and the m. spinalis, it reaches up to the 2nd cervical vertebra. *M. spinalis dorsi*: originates from the longissimus dorsi, inserts by aponeuroses on the spinous processes of the 2nd-7th cervical vertebra. *M. ilecostalis* (sacro-lumbalis) : originates from the longissimus dorsi and terminates on the costal margins. (Figures and nomenclature after W. Ellenberger and H. Baum : Anatomie des Hundes, Berlin, 1891.)

times lifted above the vertical and falls backwards. Similarly, an exaggerated strong contraction with a lifting of the anterior part of the body can be seen when the hindlegs are placed too far forwards or slip forwards during running. In the early post-operative stage when the animals just start to stand and run, sometimes, when both hindlegs are set on the ground in a forwards direction, the animals will turn a somersault caused by the exaggerated strong and sudden contraction.

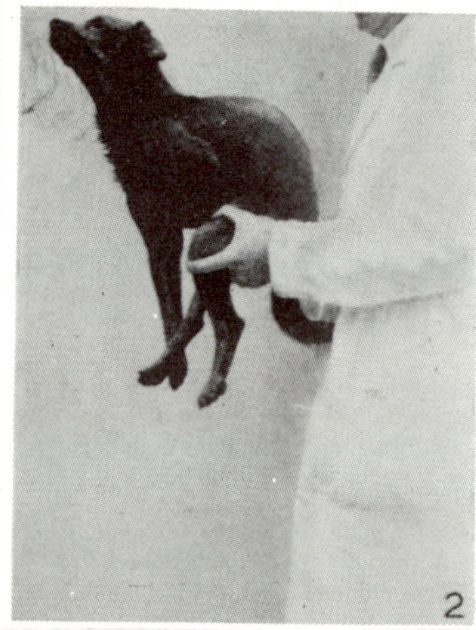

Fig. 133. Decerebellate dog Moor. 1. The animal is lifted from the ground by the thigh. The trunk hangs down vertically. Vertebral column mobile. 2. Thigh is passively ventrally flexed causing a tension in the vertebral muscles and raising of the anterior part of the body which remains almost horizontal, freely supported in the air as long as the thigh is directed forwards.

We have already seen that the levators of the anterior part of the body in decerebellate animals are also particularly vigorously contracted by touching and static stress on the soles of the hindfeet (Fig. 32); when the hindlegs are directed forwards, this is also the case, but in a much greater degree.

Tension in the vertebral muscles also occurs in decorticate and in labyrinthectomized dogs; it is, however, absent in deep narcosis and in new-born dogs during the first weeks of life. Although the contraction is not produced by optical impulses, it can yet be counteracted by them. When animals (without blindfold) are slowly moved downwards towards a supporting surface, the tension in the vertebral muscles suddenly ceases when they reach within a small distance of the surface and the forelegs are set down on it (Fig. 134) decrease in tension of the vertebral muscles due to optical preparation for standing (visual placing).

Fig. 134. Decerebellate dog Piccolino.　1. The animal is held in the air by the forward directed hindlegs.　By means of contractions of the vertebral muscles the animal keeps the anterior part of the body raised. 2. By bringing the head to a table, the tension is reduced although the hindlegs are still kept in a forwards direction and the animal sets down its forelegs on the table.

When the eyes are blindfolded, this relaxation only occurs when the forelegs touch the surface.　When, in a dog hanging in the air, only one, for instance the left thigh, is ventrally moved, the anterior part of the body is still raised and fixed in this position.　Simultaneously, a concavity of the spinal column and a turning of the head towards the left occurs (Fig. 135).

The curvature of the back can also be influenced by the position of the *forelegs;* thus backward movement of the forelegs in a standing dog (Fig. 136) causes a dorsal convexity of the thoracic spinal column, while in the upper part of the shoulder girdle an indentation appears (extension of m. rhomboideus and of the pars thoracalis of m. serratus anticus ? see Fig. 137).

On moving the forelegs forwards, however, the back of intact dogs sinks in and in decerebellate animals the dorsal convexity decreases

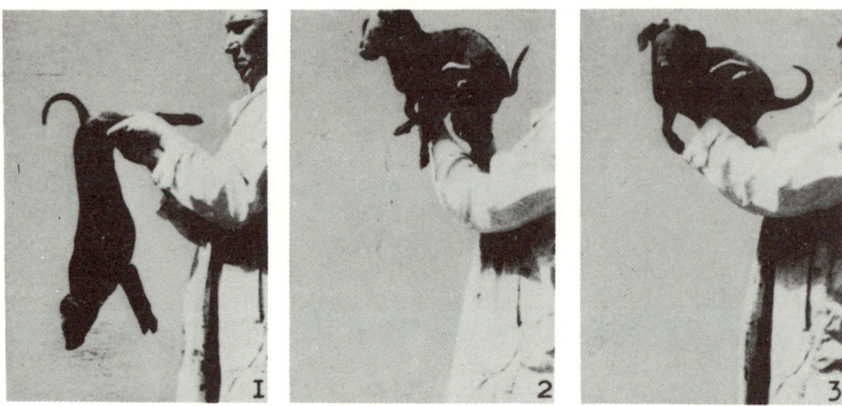

Fig.135. Decerebellate dog Piccolino with blindfold. 1. The animal is held by the thighs in a hanging position, head downwards. 2 and 3. On the ventral flexion of the left thigh alone, raising of the anterior part of the body in addition to a turning of the head and a concavity of the vertebral column to the left.

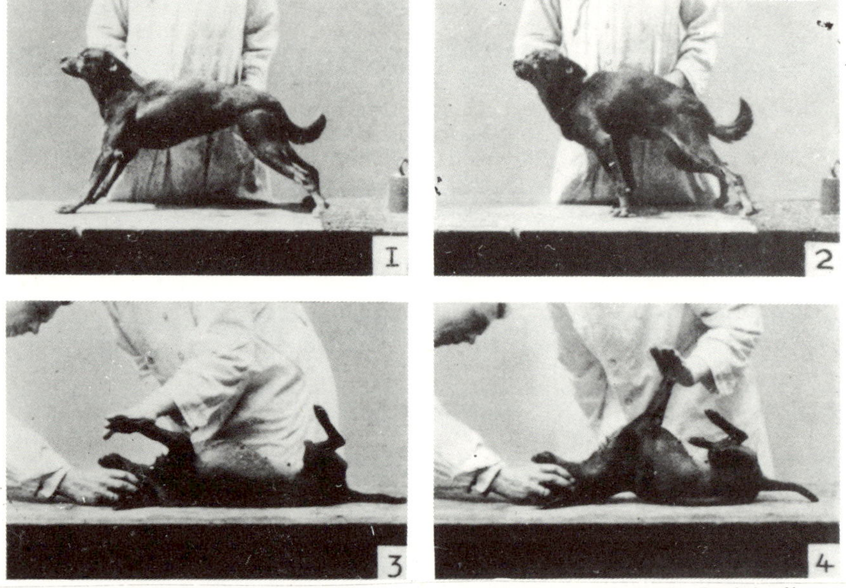

Fig. 136. Decerebellate dog Moor in a standing posture. 1. With the forelegs directed forwards, the spinal column forms an almost straight line. 2. When the forelegs are propped backwards, the thoracic spinal column is convexly curved upwards, while an indentation appears in the shoulder girdle. 3 and 4. Decerebellate dog Piccolino in a dorsal position with the muzzle at an angle of 45° above the horizontal, the forelegs under static strain. 3. With ventrally extended forelegs, the spinal column lies flat on the supporting surface. 4. By a caudal movement of the forelegs, the shoulder girdle is lifted from the surface, the thoracic spine curves and the pelvis is ventrally moved.

(extension of m. latissimus dorsi and of the m. trapezius ?). In these alterations in position of the vertebral column, reflex processes also play a role, as is particularly seen in the dorsal position (Fig. 136, No. 3 and 4).

Alterations in position of the spinal column are also produced when the legs of one side are passively abducted or adducted; for instance with an abduction of both right legs, a concavity of the vertebral column to the right appears; similarly, the animals actively prop their right legs outwards as, for instance, when standing on an inclined plane (Fig. 207).

Under certain conditions, the position of the vertebral column can also *be influenced by the position of the head in relation to the neck*. This influence can be observed most clearly in decerebellate animals in a dorsal position, with the forelegs under static stress. When the muzzle is pointing vertically upwards, the vertebral column lies flat on the

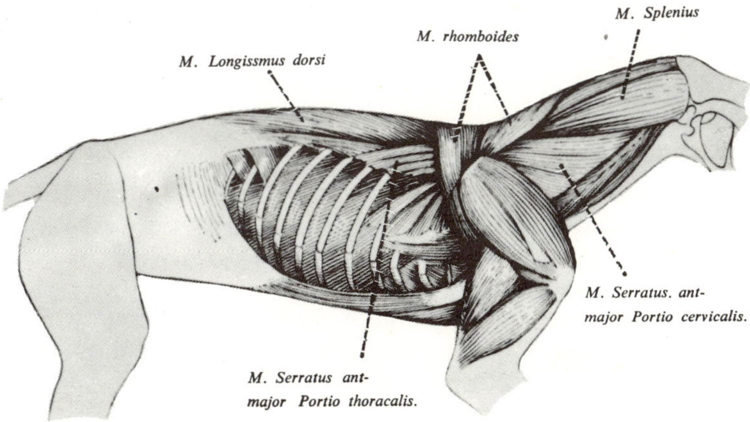

Fig. 137. Muscles of the shoulder girdle of the dog : intermediate layer (for the superficial and deep layers see Fig.132). *Mm. rhomboides* originate on the median aponeurosis of the neck and on the spinous processes of the six first thoracic very tebrae and terminate on the median surface of the scapula. *M. serratus anticus major* spreads from the scapula in the shape of a fan so that, on the one handl it reaches to the 7th rib (portio thoracalis), on the other hanp to the transverse processes of the 3rd cervical vertebrae (portio cervicalis).

supporting surface, while with the muzzle already directed upwards to 30-45° the shoulder girdle is lifted from the surface (Fig. 138).

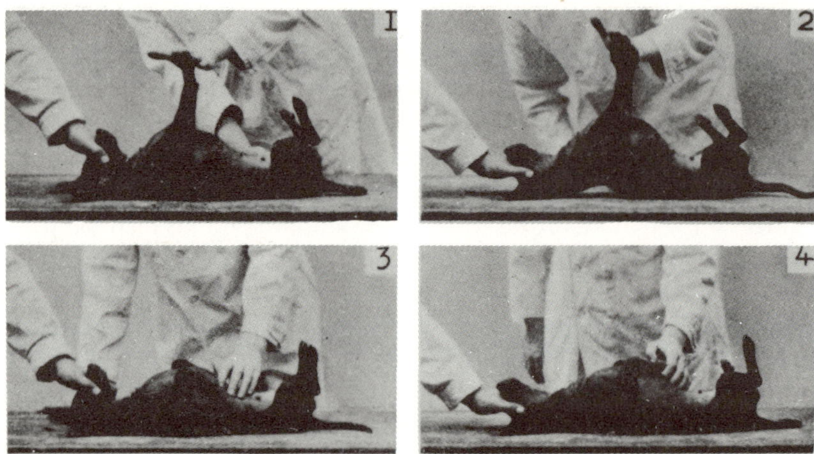

Fig. 138. Decerebellate dog Piccolino. 1. In a dorsal position with the muzzle pointing vertically upwards and the forelegs under static stress, the entire vertebral column lies flat on the supporting surface. 2. On the dorsal movement of the head in the atlanto-occipital joint, the shoulder girdle is lifted from the surface. 3 and 4. These alterations of position of the spinal column fail to appear when the forelegs are not under static stress.

This concave curvature of the spinal column is produced by strong reflex contraction of the muscles of the shoulder girdle (Fig. 137) and of the muscles which lead from the vertebrae of the cervical vertebral column to the thoracic vertebrae (Fig. 139).

On simultaneous alterations of posture of head and neck in relation to the thoracic spinal column, the position of back and pelvis also changes; thus, in intact and decorticate dogs in a standing position, the back curves convexly upwards when the head is ventrally flexed (Fig. 140, No. 1, 3); moreover, a ventral movement of the pelvis takes place. In decerebellate dogs, the curvature of the back increases (Fig. 140, No. 4 and 5). In them, the alterations of posture of the spinal column and pelvis are also evident in the dorsal position (Fig. 140, No. 6 and 7).

When the dogs are held in a dorsal position in the air by their hindlegs and pelvis, a ventral flexion of head and neck is brought about by labyrinthine righting reflexes which results in a dorsal directed convexity of the back and a ventral flexion, of the pelvis and the hindlegs. In this case, the anterior part of the body is sometimes raised into a

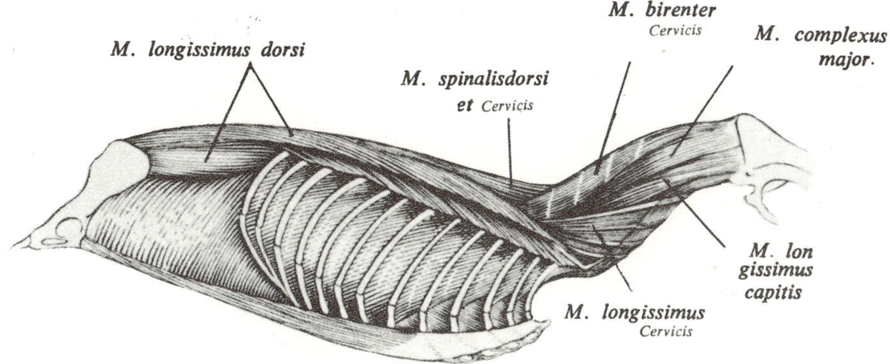

Fig. 139. The neck muscles of the dog, *M. longissimus dorsi* from the pelvis to the 7th cervical vertebra. *M. longissimus cervici* originates from the longissimus dorsi and from the spinous processes of the lst-5th thoracic vertebrae and terminates on the proc. transversi of the 2nd to 7th cervical vertebrae. *M.spinalis dorsi et cervicis* originates from the longissimus dorsi in the height of the 7th - 12th thoracic vertebrae and terminates in the articular and spinous processes of the 2nd - 7th cervical vertebrae. *M. longissimus capitis (m. complexus minor)* originates on the proc. spinosi and proc. transversi of the lst - 4th thoracic vertebrae and on the proc. obligui of the 5th-7th cervical vertebra and terminates on the proc. mastoideus ossis temporalis. *M. semispinalis cervicis et capitis*: *m. biventer cervicis* + *m. complexus major*. *M. biventer cervicis* originate on the proc. transversi of the 5th -6th and the proc. spinosi of the 2nd-5th thoracic vertebra, on the nuchal ligament and median strip of the neck. *M. complexus major* originates on the proc. transversi of the lst - 4th thoracic vertebra and on the articular processes of the 3rd - 7th cervical vertebra. Both parts fuse at the level of the atlas and terminate with a strong tendon near the protuberance and on the linea nuchalis of the occiput. *M. Plenius* originates from an aponeurosis which attaches itself to the spinous processes of the lst - 5th thoracic vertebra and on the tendinous median strip of the neck and terminates on the nuchal line of the occiput and mastoid process. (See Fig. 135.)

vertical position and fixed in this posture by a sustained contraction of the muscles (Fig. 141). In this experimental arrangement, there occurs a distinct contraction of the quadratus lumborum, ileopsoas and of the abdominal wall muscles which all participate in the alteration of posture of the pelvis and the vertebral column.

When, in a standing posture of the animal, the head and neck are *moved dorsad* till the buccal cleft is directed upwards 60-90°, the back

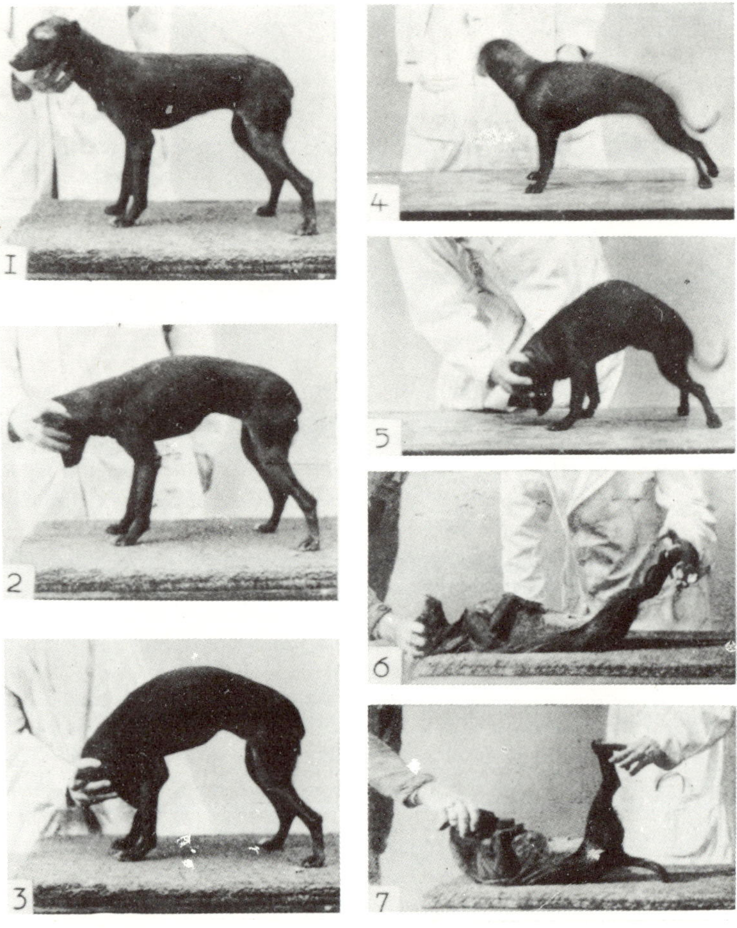

Fig. 140. Influence of the posture of head and neck on the curvature of the back. 1-3. Intact dog in a standing posture. 1. The head is held in a normal posture, the back is almost straight. 2 and 3. On a ventral flexion of head and neck, a dorsally directed convexity of the back appears by which the pelvis is slightly ventrally flexed. 4-7. Decerebellate dog Piccolino, 4 and 5, in a standing posture, 6 and 7, in a dorsal position. In 4 the head is held in a normal posture. Pelvis and hindlegs are strongly caudally extended . 5. On ventral flexion of head and neck, the dorsal convexity has strongly increased at the border between thoracic and lumbar spine. Pelvis and thigh are ventrally flexed and the right hindleg is displaced forwards. 6. In a dorsal position with the muzzle pointing vertically upwards, neck, back and dorsal side of the pelvis are flat on the supporting base, while the statically stressed hindlegs are caudally extended. 7. On ventral flexion of the head and neck a curvature of the spinal column occurs by which the pelvis and lumbar spine are lifted from the surface and the statically stressed hindlegs move forwards.

sinks in a little at the anterior part of the thoracic spinal column, the muscles of the shoulder girdle are vigorously extended and sometimes protrude in the form of a clearly visible, hard bulge (Fig. 142). Strange to say that this alteration in position of head and neck is also accompanied by a ventral flexion of the pelvis (in addition, the hindlegs more or less bend and sometimes altogether give way under the weight of the posterior part of the body).

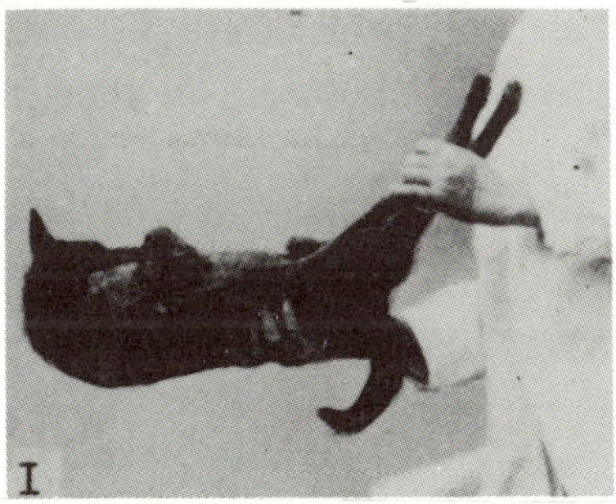

Fig. 141. Decerebellate dog Erik held in a dorsal position in the air by the hindlegs and pelvis. 1. Under the influence of labyrinthine and optical righting reflexes, the animal has flexed its head ventrally so far that the buccal cleft stands directed ventrally. At the same time, the downward hanging anterior part of the body has raised itself to the horizontal and hollow back is straightened. The spinal column is strongly and elastically fixed. 2. The animal has flexed its head even more ventrally so that the muzzle is now directed downwards at an angle of 50°. At the border of the thoracic and lumbar spine, the vertebral column shows a strong convexity while the pelvis and hindlegs are ventrally directed. By means of the various muscle contractions, the anterior part of the body is lifted to an almost vertical position.

These alterations in posture associated with the dorsal movement of head and neck are not only found in intact but also in decorticate and labyrinthectomized dogs. These synergies are particularly vigorous in decerebellate dogs, both in the standing posture as well as in the dorsal position (Figs. 142 and 143).

When the head is dorsally moved *beyond the vertical*, the pelvis moves even more ventrally, both in the standing posture as well as in the dorsal position of the animal (Fig. 143), and with it the entire lumbar part of the vertebral column, especially when the animal tries to liberate itself. In this case, as in the ventral flexion of the head, a convex curvature of the back appears at the point of adjunction of the thoracic

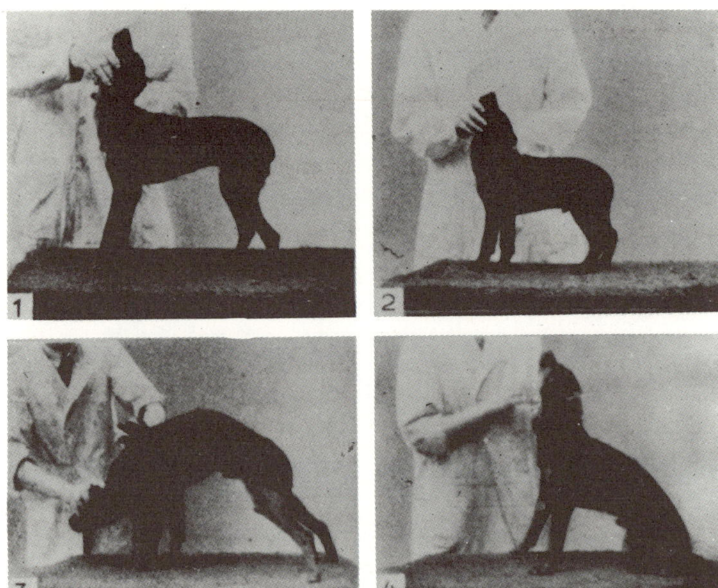

Fig. 142. 1 and 2. In two intact dogs, the muzzle is held vertically upwards. In so doing, the anterior part of the thoracic spine sinks in a little, and the muscles of the shoulder girdle can distinctly be felt as a bulge. Moreover, the pelvis is ventrally flexed and the hindlegs are flexed. 3 and 4. Decerebellate dog Moor in a standing posture. 3. Head is passively ventrally flexed. In so doing, the back is strongly and convexly curved upwards and the pelvis moves dorsally (see the concavity of the caudal part of the lumbar spine). Hindlegs caudally directed and strongly extended. 4. Head is passively lifted vertically. The curvature has decreased at the border between the thoracic and lumbar spine, the pelvis is ventrally flexed, and the hindlegs collapsed.

and lumbar parts of the spine, as well as a forward extension of the hindlegs. The defensive movements as well as all other phenomena also occur in decorticate dogs. They are caused by reflexes.

As we shall see later, a strong curvature of the back will produce a forward extension of the hindlegs.

Thus, the vertebral column and the pelvis show the following alterations in position on lifting and lowering the head and neck. *With a moderately strong ventral flexion*: (a) very strong convex curvature of the back especially at the point of junction of the thoracic and lumbar parts of the spinal column; (b) dorsal movement of the pelvis. *With a very strong ventral flexion* : (a) very strong convex curvature of the back; (b) sometimes a slight ventral movement of the pelvis. *With a moderately strong dorsal movement* : (a) lordosis of the anterior part of the thoracic spinal column; (b) straightening or hollowing of the back at the point of junction of the thoracic and lumbar parts of the spinal column; (c) ventral flexion of the pelvis. *With a very strong dorsal movement*: (a) flipping back of the lumbar part of the spine so that the back is again convexly curved; (b) strong ventral flexion of the pelvis.

The alterations in position of the thoracic and lumbar parts of the vertebral column do not always occur with equal strength but show large differences depending on the initial distribution of muscle tonus of the trunk, on the initial positions of spinal column and pelvis and on the centers of support of the animal.

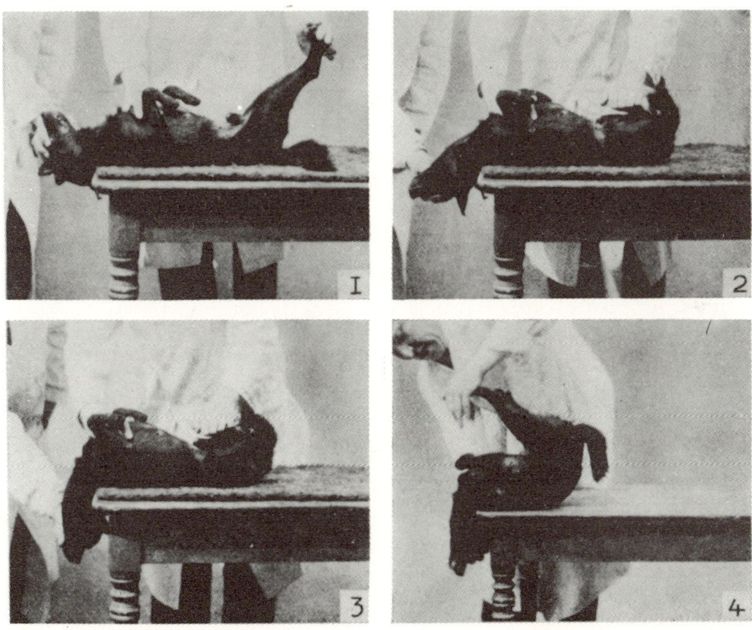

Fig. 143. Decerebellate dog Erik in a dorsal position.   1. With the muzzle directed vertically upwards the hindlegs, extended under static stress, as well as the pelvis show a dorsally directed posture.   2 and 3. On the dorsal movement of the head till the muzzle stands below the horizontal at an angle of 90°, the pelvis moves ventrally (see the insertion of the root of the tail) and the supporting tonus of the hindlegs diminishes. 4. With a very vigorous dorsal movement of the head defensive movements sometimes appear in the course of which the pelvis goes strongly ventrad, the entire lumbar spine is lifted from the surface and the hindlegs are extended forwards.

Since the position of the spinal column and pelvis influence the supporting tonus, the positions of head and neck exert an indirect influence on the supporting tonus. Not only dorsal and ventral flexion but also rotating and turning of the head produce alterations in posture of the spinal column and pelvis. A rotation of the head is accompanied by a helical rotation of the entire vertebral column; for instance, a turning of the head to the right is followed by a concavity of the spinal column towards that side.

B. THE SUPPORTING TONUS AFFECTED BY THE POSITION OF THE BACK

The influence of the position of the back on the supporting tonus becomes distinctly evident when a dog is set down with the hindpaws on a hand which is moved up and down while the head and shoulder girdle are held fast (Fig. 144). With the downward movement, the convex curvature of the back increases at the point of junction of the thoracic and lumbar part of the spinal column, the hindlegs are strongly extended and show a vigorous supporting tonus; with the upward movement, the back becomes hollowed, the hindlegs are flexed and the supporting tonus is reduced.

Decorticate dogs show the same phenomena. In decerebellate dogs the supporting tonus usually subsides as soon as the back becomes

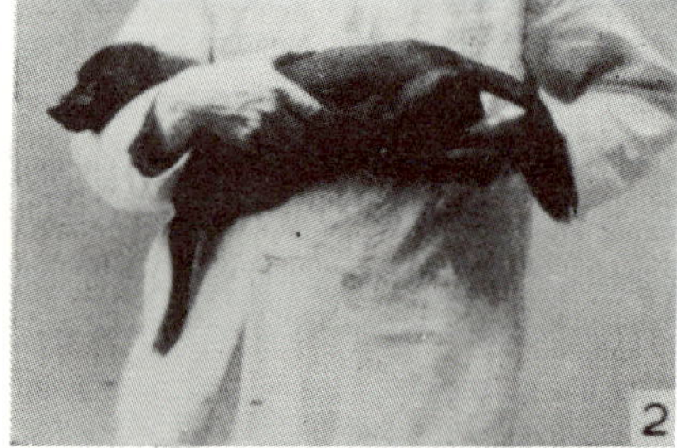

Fig. 144. Decerebellate dog Piccolino. The animal stands with the hindlegs on the left hand of the investigator while the head of the animal rests on the right forearm and the anterior part of the thoracic spine is held fast with the right hand. 1. The left hand is moved downwards. The vertebral column of the dog is strongly curved convexly at the border between the thoracic and the lumbar spine, the hindlegs are extended with a strong supporting tonus. 2. The left hand is moved upwards. The convexity of the spinal column has disappeared, the supporting tonus has decreased and the hindlegs have become flexed.

straight (Fig. 144).   In the dorsal position, too, the influence of curvature of the back on the supporting tonus of the hindlegs can be observed, particularly in decerebellate dogs since in them the supporting reactions are less inhibited in the dorsal position (Fig. 145).

The influence of the position of the spinal column on the supporting tonus is the cause of the fact that the legs show less strength of supporting tonus by loading the back than by burdening the shoulders and pelvis (see Chapter VI). Loading the back causes a lordosis and by that a reduction of supporting reactions and of strength of supporting tonus.

Fig. 145.   Decerebellate dog Erik in a dorsal position : head, neck and anterior part of thoracic spine on a table, the lumbar spine, however, on a board which can be moved up and down.   The anterior part of the body is immovably fixed by the forelegs, the muzzle pointing upwards at an angle of $\pm 45°$.   1. The board is moved upwards till the back is convexly curved at the transition point between the thoracic and lumbar spine.   On static stress as well as on touching the sole, the hindleg instantly extends and a strong supporting tonus appears.   2. On a downwards movement of the board, a dorsally directed concavity is produced at the transition point between the thoracic and the lumbar spine. Neither touching of the sole nor static strain can now bring about an extension and supporting tonus in the hindleg.

With an intact dog which is able to carry a sandbag of 6 kg on its pelvis without its hindlegs giving way, these collapse instantly on burdening the back with 8 kg although the load is distributed on the fore- and hindlegs.   One hindleg of the decerebellate dog Moor was well able to carry on the pelvis a load of 9 kg, but both legs gave way with a load of 14 kg on the back.

On the helical rotation and lateral curvature of the vertebral column no distinct primary alterations of supporting tonus appear; in a ventral position in the air and in the dorsal position, they are not able to exert any influence on the supporting tonus, or at most, only a slight one. The alterations in supporting tonus of the hindlegs which appear when the anterior part of the body of a standing animal is lifted from the ground and at the same time head and shoulder girdle are rotated into a lateral position are caused only secondarily, as we shall see later, by a change in the posture of the back. Similarly, alterations in supporting tonus can be observed in the forelegs when the posterior part of the body is lifted from the ground and rotated to a lateral position or turned sideways.

C. INFLUENCE OF THE BACK CURVATURE ON ADAPTATION OF LIMB POSITION TO STATIC CONDITIONS

The influence of the back curvature on the adaptation of limb posture can be observed most distinctly when the fore- and hindlegs stand on two different supporting surfaces whose position to each other is changed (Fig. 146).

These adapting reactions occur in intact, decorticate, decerebellate and labyrinthectomized dogs.

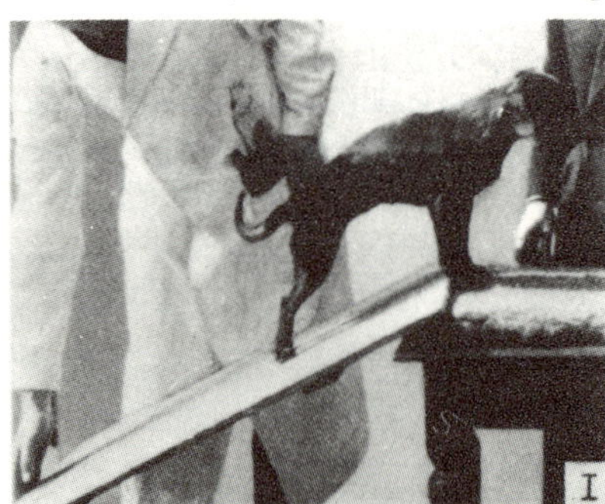
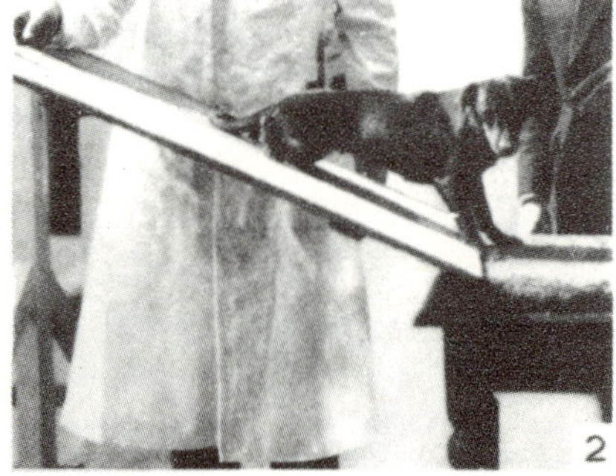

Fig. 146.   Decerebellate dog Piccolino. The animal stands with the forelegs on a table, with the hindlegs on a board.   On altering the position of the board,   the hindlegs show   distinct   adjusting   reactions. 1. Board is moved downwards causing a strong curvature of the back and a vigorous extensor position of the hindlegs. (Please note the angle formed by the thoracic and the lumbar spine.) 2. On raising the board, the back is straightened and the hindlegs flexed. (These reactions also appear in the hindlegs if the board is simply raised vertically.)

D. INFLUENCE OF THE POSITION OF THE PELVIS ON THE SUPPORTING TONUS OF THE HINDLEGS

When the pelvis is moved ventrally or dorsally at the intervertebral joints of the caudal part of the lumbar spine, alterations in supporting tonus in the hindlegs also take place. For instance, a dog is placed on a board in a ventral position in such a way that the pelvis and hindlegs hang down over the edge while the thorax and the anterior part of the lumbar spine are fixed; when the pelvis is now moved ventrally and dorsally by the tail with a dorsal movement, a vigorous supporting tonus appears in the hindlegs under static stress which weakens considerably with a ventral movement. In decerebellate dogs, this influence also appears in the dorsal position. When a decerebellate dog in a dorsal position with the muzzle directed upwards at $\pm$ 45°is laid flat and motionless with the cervical, thoracic and oral part of the lumbar spine on a table and with the dorsal part of the pelvis on a board that is moved up and down by one end (Fig. 147), a distinct increase and decrease in supporting tonus of the hindlegs can be observed on movements of the board.

On a downward movement of the board and a dorsal movement of the pelvis caused by it, a strong extension and a strong supporting tonus with distinct resistance will appear in the hindlegs under static stress. When the board is now moved up and the pelvis is moved ventrally with it, the resistance subsides, the legs are flexed, and touching the soles does not cause any magnet reaction. On static stress, too, extension and supporting tonus fail to appear. In intact, decorticate and labyrinthectomized dogs, this influence of pelvis position on supporting tonus can usually be observed only in standing postures.

In opposition to the effect of ventral and dorsal movements of the pelvis, no distinct alterations in supporting tonus appear on rotating or turning of the pelvis at the caudal intervertebral joints of the lumbar part of spine.

The influence of alteration of the position of the pelvis on the supporting tonus of the hindlegs is seen when the animal stands with the forelegs on a supporting surface which is moved up and down in relation to the surface for the hindlegs. A passive alteration in position of the pelvis to the vertebral column and a change in supporting tonus are brought about which, on their part, cause an adaptation of paw position to static conditions. The alteration of the position of the pelvis, however, does not play a principal role in adaptation which, above all, is produced by a passive flexion and extension of the thigh in the hip joints, as we have already seen (Fig. 55).

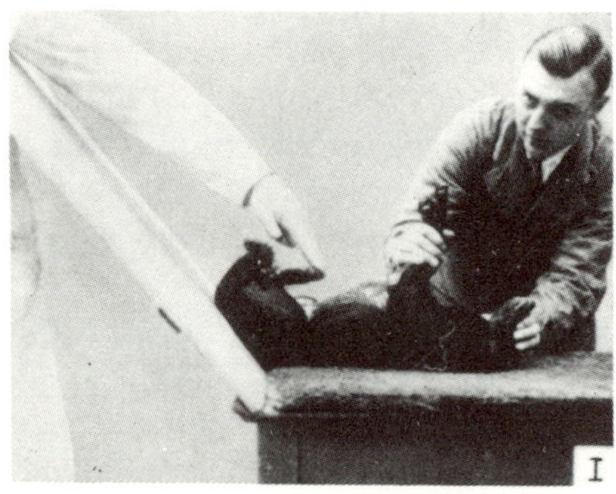

Fig. 147. Decerebellate dog Erik in a dorsal position, the muzzle pointing upwards at an angle of $\pm 45°$C the cervical, thoracic and oral part of the lumbar spine lie flat on a table, the dorsal part of the pelvis on a board moved up and down from one side. 1. Board is moved upwards thus moving the pelvis ventrally. No extension of the flexed hindleg on touching the sole or static stress. 2. Board is moved downwards thus moving the pelvis dorsally. On touching the sole instant extension of the hindleg and strong resistance against the pressing hand.

When one examines the results of investigations discussed in this chapter together with those of Chapter VII, alterations in supporting tonus are found in changing the position of head in relation to neck, of neck to trunk, of thoracic to lumbar part of spine and of lumbar spine to pelvis (Fig. 148).

Thus :

the dorsal movement of the head in the atlanto-occipital joint produces : a decrease in supporting tonus of the hindlegs, increase of supporting tonus in the forelegs;

the dorsal movement of neck relative to thorax produces : a decrease of supporting tonus in the hindlegs, increase of supporting tonus in the forelegs;

the dorso-convex curvature of the back produces : an increase of supporting tonus in the hindlegs;

the dorsal movement of the pelvis produces: an increase of supporting tonus in the hindlegs;

while alterations of position in a reverse direction produce opposite changes in supporting tonus.

In man, the influence of spinal curvature and pelvic position on the supporting tonus of the legs has not yet been closely investigated, but there exist some observations which indicate a conformity with conditions in animals. Thus, every wrestler knows that the legs of the opponent are easily brought into flexion by a pressure under the lower jaw from behind the shoulder as soon as he succeeds in bending the spinal column at the point of junction of the thoracic and lumbar part of the spine.

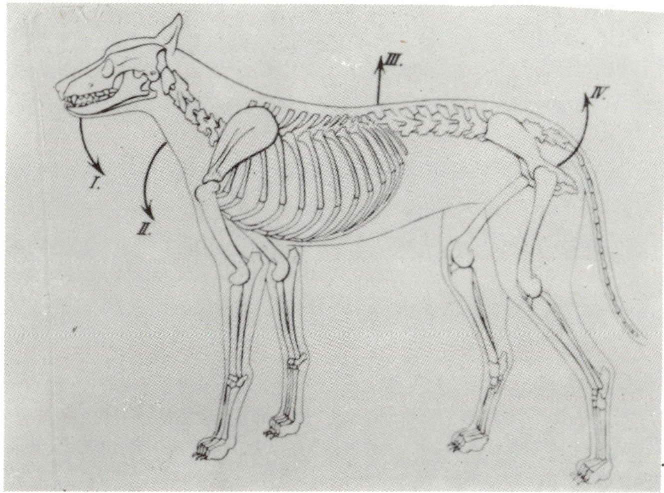

Fig. 148. Skeleton of a dog. The arrows indicate the alterations in position of head, neck, spinal column and pelvis which are followed by an increase in supporting tonus in the hindlegs.

In a standing position with extended legs, a backwards movement of the upper part of the body causes a flexion of the knee joints which, conversely, are strongly extended by forward flexion of the neck. In a maximum displacement of the head forward, however, with a strong ventral flexion of the pelvis on the spine, flexion of the limbs again appears.

It thus seems that in man, similar to animals, lordosis of the back and ventral flexion of the pelvis cause a decrease in supporting tonus.

In standing posture with bent knees, however, the flexion usually increases with a forward flexion of the thorax and decreases again in the movement backwards.[1] Thus, the effects are different, depending on the initial posture of the legs. Whether the differing behavior in different initial postures is brought about by coordination or by other factors is still unknown. It has also to be taken into consideration that, in movements of the anterior part of the body, not only the position of the thorax in relation to lumbar part of spine and of lumbar part of spine to pelvis is changed, but also that of the thigh to hip joint. According to circumstances, either one or the other change in position becomes more prominent and exerts a stronger influence. Thus, observations in man are not clear and further investigations are needed.

In the clinical literature, several cases of disorder of the brain are mentioned in which abnormal synergies on the forwards and backwards movement of the anterior part of the body have been observed. I consider it important to compare these observations with findings in animals where certain parts of the central nervous system have been extirpated. Babinski (3) observed in patients with cerebellar diseases the absence of flexion in the knee joints when the anterior part of the body was moved backwards in a standing posture; he considers this synergy as more or less characteristic of cerebellar lesions. This concept, to attribute the absence of normal synergies to cerebellar dysfunction, by no means corresponds to the results of our experiments with animals because, as we have seen, in decerebellate animals the hindlegs not only show typical but even particularly exaggerated synergies on altering the posture of the vertebral column. In the case observed by Babinski, a pontine-angle tumor was found which, however, had also gravely injured the pons. Patients of André Thomas (303, 310) with pure cerebellar affections did not show these "asynergies"

---

(1) Conversely, in a strong movement backwards, the flexion again increases. Thus, hollowing of the back is constantly accompanied by an increase in flexion. It is remarkable that the extensor of the knee joint, m. quadriceps, remains stressed in the course of this flexion, while, on stretching the knee joint on a head forward bending of the thorax, the extensors usually relax, as is shown by the mobility of the patellae. The flexors, however, are strongly stretched, sometimes even to the point of being painful.

and patients with a unilateral lesion of the cerebellum flexed both tarsal and knee joints with equal force on the backwards movement of the anterior part of the body (Fig. 149, 4).

André Thomas, Levy-Valenski and Besson (310) also observed the same synergy to be unaffected in lesions of other parts of the brain;

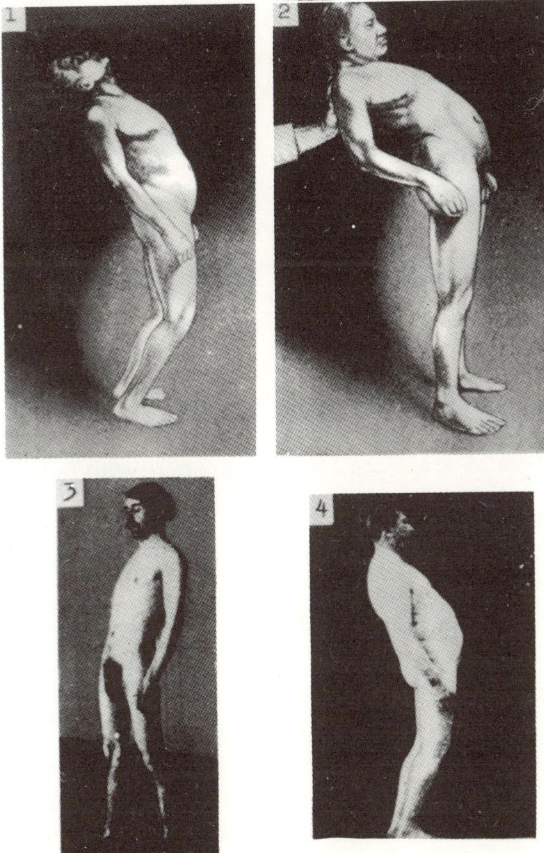

Fig. 149. 1. Normal synergy on backwards movement of the trunk : knee and ankle joint are flexed. 2. Asynergia in a case of pontine-angle tumor. Flexion of knee and ankle joints is absent. From Babinski: De l'asynergie cerebelleuse. *Revue Neurologique*. 3. The same asynergy in a case of tuberculoma of the left lobus temporalis (mesencephalon and pons of the left side strongly compressed and displaced). From André Thomas, Lévy Valensi et Benon. 4. Man with gunshot injury at the right cerebellar hemisphere. On backward movement of the trunk both ankle and knee joints show an equally strong flexion. From André Thomas : *Etude sur les Blessures du Cervelet*. Paris, Vigor Freres. 1918.

thus a patient with a tuberculoma of the temporal lobe also showed symptoms of a lesion of the midbrain (among others, a partial oculomotor paresis). Claude and Lhermitte (43) observed these synergies in a man with a bilateral gunshot wound of the paracentral lobule. When one considers further that, in the patient observed by Babinski, the pons was also severely affected, the assertion that this synergy is caused by an absence of cerebellar functions loses its validity. Babinski also described another synergy which, according to his opinion, is more or less typical for cerebellar lesions. When a patient with cerebellar disease is asked to raise himself from a dorsal position with his arms crossed on his chest, the extended legs are instantly lifted high as soon as the anterior part of the body is raised, which makes it impossible for the patient to sit up. This observation corresponds to findings in decerebellate animals in the fact that in these animals, too, the hindlegs under static stress are forcefully moved ventrally on ventral flexion of head and thorax (Fig. 140, No. 6 and 7). This symptom described by Babinski was also distinctly present in those of André Thomas (Fig. 150). André Thomas, in addition, observed that both legs in patients with bilateral cerebellar disease and only the ipsilateral leg in the case of a unilateral disease are lifted abnormally high. It seems that this symptom really has a certain relation to cerebellar lesion.

On the ventral flexion of the head and thorax in intact dogs, too, a ventral movement of the hindlegs occurs which, however, is less pronounced. Similarly, in intact persons, too, who wish to raise themselves from a dorsal position with their arms crossed over their chests, show some flexion of the hip joint during which the heels are more or less lifted from the ground. At a certain age (neither in young children nor in older adults), it is possible to keep the heels on the ground but always with a more or less strong effort. The abnormally high lifting of the legs in cerebellar patients is thus not an expression of an asynergy but, on the contrary, of a particularly strong, unrestrained synergy. One could rather call it a hyper-synergy. As a result of these considerations, it is very doubtful whether the absence of cerebellar function in man can actually produce asynergy, contrary to manifestations in animals.

In diseases of the cerebrum, too, impairments of synergies on forward and backward movement of the thorax have been described. We already mentioned the absence of flexion in the knee joints on the backward movement of the thorax in a case of tuberculoma of the temporal lobe and in a case of the bilateral destruction of the paracentral lobula by a gunshot injury. Levi found that hemiplegics in a standing

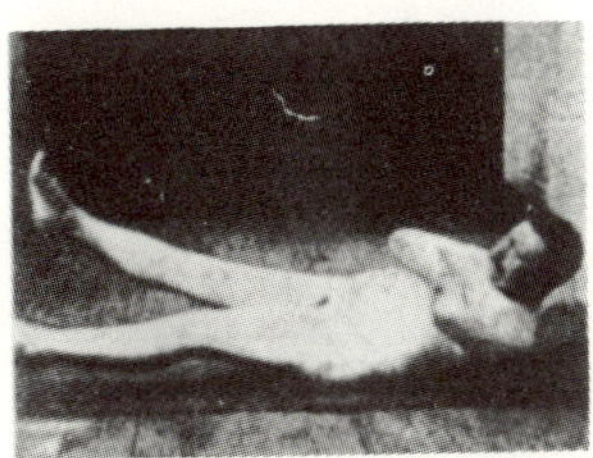

Fig. 150. Man with gunshot injury of the right cerebellar hemisphere. In the attempt to get up, both legs are lifted from the ground, the right more than the left. The right shoulder is also raised more than the left; this movement causes the trunk to turn towards the left side. From André Thomas: *Pathologie du Cervelet Nouveau Traité de Médecine.* G.H. Roger, F. Widal et P. J. Tessier, 19, 755 Masson et Cie, Editeurs Paris, 1925.

posture on bending the head forward on the thorax, showed an extension of the knee joint in the healthy leg and a flexion of the joint on the hemiplegic side (Fig. 151). Levi considered this phenomenon as an important differential diagnostic sign between organic and hysterical hemiplegia.

Babinski observed in cases of hemiplegia that in an attempt to raise oneself from a dorsal position with the arms crossed over the chest, the affected leg was lifted abnormally high. Thus, in cerebral lesions, synergy can be absent (absence of knee flexion by bending the anterior part of the body backwards) or proceed abnormally (flexion instead of extension of the knee joint in flexing the head on the thorax) or it may be excessive. These findings do not entirely correspond to the results of animal experiments since in decorticate animals a change in position of the vertebral column produces typical synergies of the hindlegs, even if the increase in supporting tonus caused by alterations in position does not occur as strongly as in intact dogs. However, find-

308

ings in both man and animals till today, have been analysed too little to allow positive conclusions. In addition to Babinski's observation that hemiplegics in raising the thorax from a dorsal position lift the hemiplegic leg higher than the intact one, we wish to report the following observations in animals : when unilaterally decerebellate dogs are placed in a dorsal position and the head and anterior part of

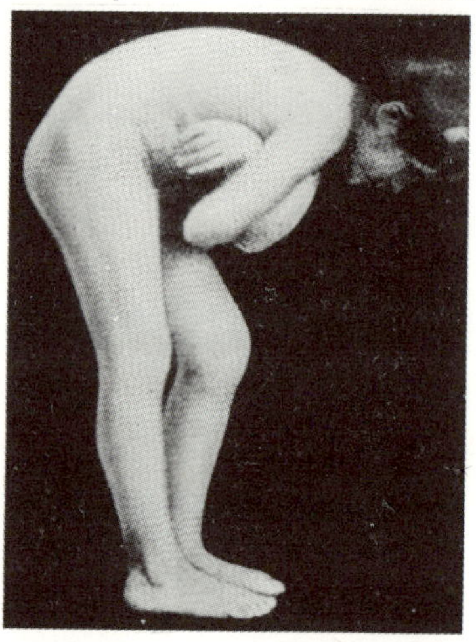

Fig. 151. Léri's knee-bending phenomenon. Girl with left-sided hemiplegia. On flexion of the trunk, the diseased hypertonic leg is flexed at the knee joint while the healthy leg remains in extended posture. From André Léri : *Hemiplegie nerveuse*. Ch. Archard, A. Badoein, Laignel Lavastine, André Leri and Leopold Levi, Fig. 93, p. 484, Paris, J. B. Bailliere et fils, 1925.

the body are then moved ventrally, occasionally but usually only early after the extirpation, the contralateral hindleg is extended while the ipsilateral one remains flexed. In the dog Vici, whose entire cerebellum was removed in addition to the right half of the cerebrum, the left hindleg was extended on the ventral flexion of the head and thorax even 1½ years after extirpation (Fig. 152).

Just as in left-sided hemiplegics, in this animal, too, the heel of the left extremity was lifted higher than the other side on a ventral movement of the head and thorax. However, while in human patients the higher lifting was produced by a strong flexion of the hip joint, in the dog Vici it was caused by a stronger extension of the knee joint.

In intact dogs in the dorsal position, the hindlegs not under static stress are not extended on the ventral movement of head and the anterior part of body, just as in decorticate and decerebellate animals, but are quite extended in decerebrate dogs. In unilaterally decerebellate dogs, there is a temporary extension of the ipsilateral hindleg and in unilaterally decorticates of the contralateral hindleg.

Fig. 152. Dog Vici, right-sided decorticate, totally decerebellate, in a dorsal position. On a passive ventral movement of the head, neck and anterior part of the body, the pelvis moved ventrally and the *left* hindleg became extended.

In summarizing the results of these observations, we find the following : in man, too, alterations in position of the spinal column are accompanied by changes in position and muscle tonus in the lower extremities, which correspond in certain respect with those in animals and disagree in others.   It is not known whether these differences are to be attributed to the different initial positions (in dogs, in a dorsal position or in a standing posture, the hip joints are flexed or at most extended at $\pm$ 90°, while in man, in a dorsal position as well as in a standing posture, the hip joints are almost maximally extended) or to a varying initial distribution of tonus or to other causes.

XI. SUPPORTING TONUS OF ONE LEG AFFECTED BY THE POSTURE OF THE OPPOSITE LEG

A. THE REACTIONS ON ABDUCTION AND ADDUCTION OF THE OPPOSITE LEG, THE SO - CALLED ROCKING (HOPPING*) REACTIONS

When a foreleg, for instance the right one, of a standing dog is passively lifted, one can feel on passive displacement of the trunk to the right a vigorous extension and abduction of the lifted leg. The leg offers a strong resistance to passive flexion which continues as long as the anterior part of the body remains inclined to the right, but subsides instantly on a return movement to the left. These reactions are produced neither by optical stimulation nor by labyrinthine reflexes, since they also occur with closed eyes and similarly when the head is fixed in space and only the trunk is moved, as they are also distinctly present in labyrinthectomized dogs with blindfolded eyes. They are also not produced by tonic neck reflexes because they are not absent when the head is fixed to the thorax. They still appear when the posterior part of the body is lifted from the ground and the animal stands on only one leg (Fig. 153).

The reactions of the lifted leg (leg taking no weight) are produced by the alterations in position of the opposite leg (supporting leg), which stands on the ground. When the animal is set down on an even surface with one foreleg and this leg is alternatively pushed into a posture of abduction and adduction, or when the animal is set down with one foreleg on a board and this board is moved to the right and to the left while the trunk remains motionless, the same reactions appear in the leg taking no weight. These observations thus show that the reactions in the leg taking no weight are produced by the passive abduc-

---

* Originally called "Schunkelreaktion" (rocking or see-saw reaction) but obviously an essential part of what is now generally called the "hopping" reactions (following Bard. *Arch. Neurol. Psychiat* Vol. 30, p. 40, 1933) or in a broader sense "displacement" reactions. (Ed.).

tion and adduction of the supporting leg. Probably, the stretching and relaxation of the abductors and adductors of the supporting leg play a role in these reactions.

The passive abduction of the supporting leg, which is accompanied by a stretching of adductors (ms. pectoralis major and minor, Fig. 154) and relaxation of the abductors, causes an active extension and abduction of the leg taking no weight, with a strong increase of the stretch or supporting tonus, while the passive adduction of the supporting leg, which is associated with a relaxation of the ms. pectoralis and a stretch of the abductors of the upper arm (among others m. deltoideus) and certain scapular muscles (among others m. serratus ant.) causes a de-

Fig. 153. Rocking (displacement) reactions of the forelegs in the decerebellate dog Moor. Right foreleg and posterior part of body lifted from the ground so that the animal stands only on the left foreleg. On passive movement of the trunk to the right, by which the left foreleg is passively pulled into abduction posture, the right foreleg was actively and vigorously extended and abducted and offers strong resistance against passive, static and non-static flexion. By a movement of the trunk towards the left, the left foreleg is passively adducted and the support tonus of the right leg has relaxed and can now easily be flexed, returning it to position 1.

crease in supporting tonus and a pulling of the leg taking no weight into a flexed position. Sometimes a slight alteration in position such as 10° angle in the supporting leg is sufficient to produce an appearance or a decrease in supporting tonus in the leg taking no weight.

Phenomena similar to those of the forelegs can also be observed in the hindlegs. When a dog stands on only one hindleg, the passive abduction of this leg (in which the adductors m. gracilis, ms. adductor longus, magnus and brevis are stretched) causes an extension and abduction movement with a strong stretch reflex and supporting tonus in the contralateral hindleg.

Conversely, an adduction of the supporting leg not only produces a decrease in supporting tonus but even a flexor tonus in the opposite leg which is pulled into a flexed position. When a dog is set down on

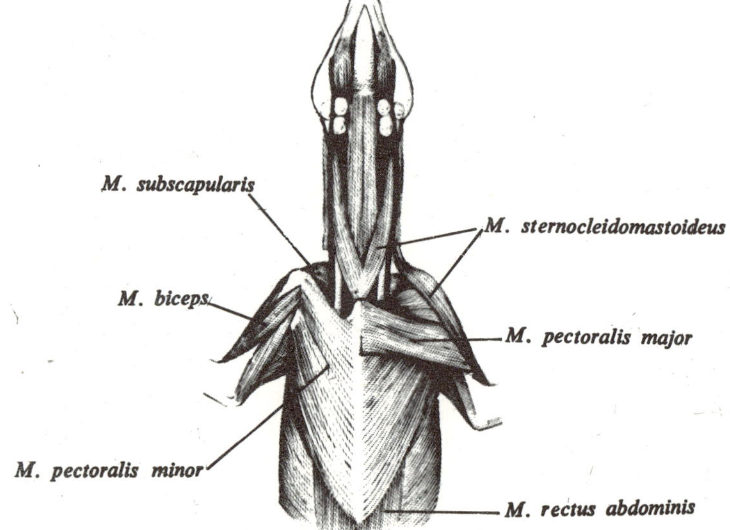

Fig. 154. The thoracic muscles of the dog (ventral view). Adductors of the upper arm : m. *pectoralis major* originates on the outer surface of the sternum from manubrium sterni to the cartilage of the 3rd rib and terminates on the entire linea tuberculi majoris humeri. M. *pectoralis minor* originates on the sternum and terminates with a tendon on the tuberculus minus. This tendon sends a fibrous plate to the tuberculum majus; besides, fro mthe middle of the muscle there usually proceeds a bundle of fibers which directly reaches the middle of the humerus. From W. Ellenberger and H. Baum : *Anatomie des Hunds.*

its hindlegs (Fig. 155, No. 2) and the trunk is displaced alternatively to the right (Fig. 155, No. 1) and to the left (Fig. 155, No. 3 and 4), the right and the left hindlegs when rocked to the right and to the left, respectively, are passively brought into an adduction posture by the weight of the trunk, and in doing so the left and the right hindlegs are alternatively lifted from the ground and flexed.

The reactions of the leg taking no weight only occur in the case of static stress on the supporting leg. On a passive abduction and adduction of a leg not under static stress no distinct alterations in position and tonus can be observed in the opposite leg and this probably for the reason that on static stress the adductors and abductors of the supporting leg are in reflex contraction thus having a strong initial tonus. In agreement with this supposition, a passive abduction and adduction of a leg under static stress does not usually produce distinct reactions in the opposite legs of intact dogs in a dorsal position, but it does so in decerebellate animals where in the dorsal position strong reflex contractions on static stress can always be detected.

After operations on the central nervous system, the rocking reactions are sometimes unilaterally or bilaterally impaired. In the following, we shall speak of an absence of rocking reactions in a leg, for instance the right foreleg, when it does not show any reactions while taking no weight, in response to passive abduction or adduction of the stressed left foreleg. This definition is necessary because the absence of rocking reactions in the leg taking no weight can be produced by three different means: firstly, the stimulations from the muscles of the supporting leg may fail to appear or cannot reach their centers in the central nervous system; secondly, the centers of the rocking reactions can possibly lose their ability to function or were destroyed and, thirdly, the muscles of the leg taking no weight may be unable to react or the path from the centers to the muscles may be obstructed.

Thus, the cause of impairment may be found in the muscles of the supporting leg, in the sensory nerves of these muscles, in the centers or in the motor nerves of the muscles of the leg taking no weight. Consequently, it is possible that after a unilateral extirpation or lesion of the central nervous system, the affected leg may promptly show rocking reactions, even on absence of reactions in the apparently intact legs of the other side. For instance, when the posterior roots belonging to the

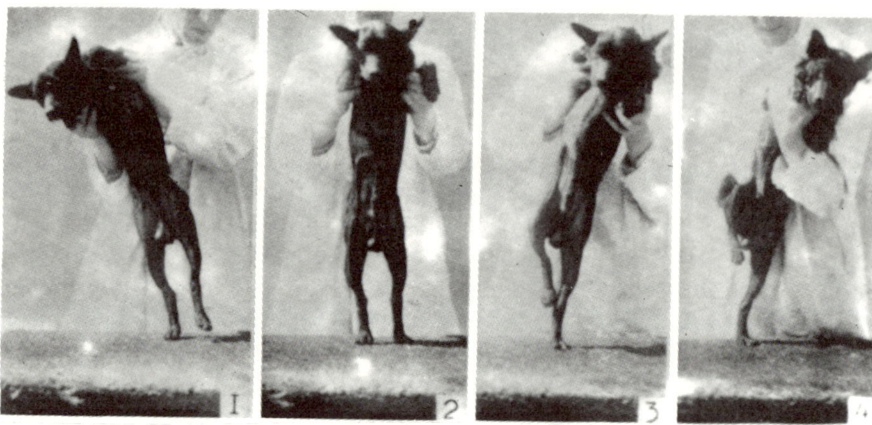

Fig. 155. Rocking reaction of the hindlegs in the decerebellate dog Erik.
1. The trunk of the animal standing on the hindlegs is moved to the
right : in so doing the left hindleg is raised from the ground and
flexed in the knee and tarsal joint. 2. Trunk is brought back
into an intermediate position : left leg again extended and set down
on the supporting base : the leg shows a strong supporting tonus.
3. Trunk is moved to the left. The right hindleg is now raised from
the ground and flexed in the knee and tarsal joint. 4. On a further
movement of the trunk to the left the flexion of the right leg
increases.

right hindleg of a dog are severed, the right hindleg will show lively
extension and flexion movements when the animal, standing on the
left hindleg, is rocked to the right and to the left; when, however, the
animal is set down on its right hindleg, the rocking movements will
produce no reactions in the intact left leg. This observation shows
that one has to be extremely prudent after a unilateral extirpation or
lesion of the central nervous system, in drawing any conclusions based
on the absence of reaction as to the mechanism of these reactions, because it is possible after surgical intervention that reactions fail to appear without the center of these reactions or the tracts leading from the
center to the muscles being injured.

When the posterior roots of the hindlegs are severed on both sides,
the rocking reactions of the hindlegs totally disappear.

After extirpation of the cerebellum, the rocking reactions are at first
totally absent on both sides in rigid as well as in flaccid animals, but
usually soon reappear, sometimes even within the first weeks. Just

as in intact dogs, very slight alterations of position in the supporting leg (about 10°) are sufficient to eliminate or to enhance the supporting tonus in the leg taking no weight. One has the impression that the supporting tonus in decerebellate animals decreases with less adduction posture of the supporting leg than in intact animals and, on the other hand, that it needs a stronger abduction posture in order to reappear. In addition, the leg taking no weight is lifted higher on adduction of the supporting leg (Fig. 155, No. 4). In distinction from intact and decorticate animals, decerebellate dogs further show, in dorsal position, a distinct decrease and increase in supporting tonus of the opposite leg on passive abduction and adduction of the legs under static stress even if, in the dorsal position, they occur after a longer latency period than in a standing posture.

After unilateral extirpation of the cerebellum, the rocking reactions are at first totally absent on both sides, but they later return on both sides just as in totally decerebellate dogs. They usually first reappear in the legs on the side of extirpation, but in the hindlegs an opposite sequence of return of reflexes has also been observed. After a certain time they are briskly present on both sides.

In this case, they often still show a distinct asymmetry. When the leg contralateral to extirpation is the supporting leg, an adduction to about the intermediate position (vertical position) in most cases already produces a total decrease of supporting tonus in the opposite leg taking no weight. A slight abduction from this posture, however, produces the instant appearance of a strong supporting tonus.

When, on the other hand, the leg ipsilateral to extirpation is the supporting leg, a considerable abduction from the intermediate position is needed for the appearance of supporting tonus in the leg taking no weight.

For reactions of the leg contralateral to the extirpation, greater alterations in position of the supporting leg are needed than for reactions of the ipsilateral leg when this latter is the leg taking no weight.

In decorticate dogs, the rocking reactions are again already present on the first or second day after extirpation. In all decorticate dogs they appear distinctly and very obviously in the forelegs, but are conspicuously weak in the hindlegs. This probably is due to the reduction of the force in supporting tonus.

The presence of rocking reactions in decerebellate as well as in decorticate animals proves that the reactions do not require reflex arcs through the cerebellum or cerebral cortex.

Fig. 156. Rocking reactions of the hindlegs in the dog Fox after extirpation of the right half of the cerebellum.

In the *totally decerebellate and decorticate dog Robbie*, no rocking reactionswere present, but the observation time of 5½ weeks was not sufficient to draw conclusions as to the central mechanism.

After the unilateral extirpation of the cerebrum, the legs contralateral to extirpation show marked rocking reactions, just as in the ipsilateral foreleg. The reactions of the ipsilateral hindleg (leg taking no weight), however, are usually only just present, probably due to the reduced strength of supporting tonus of the supporting ipsilateral hindleg.

In the dog Vici in which *the cerebellum and the right half of the cerebrum* were removed, the *left* legs show exaggerated rocking reactions. On rocking the right hindleg, the left one is raised abnormally high (Fig. 157, No.1), and it is then extended again (Fig.157, No. 2) but this time it is not set down on the supporting surface. This extension movement combined with an evident abduction is appropriate as such, but it is not able to prevent falling to the left in spite of the abduction movement. (The suspension in air is probably due to the absence of preparation for standing in the extended left hindleg on contact of the opposite leg with the supporting surface.) In static stress on the left legs during rocking, it is clearly evident that the supporting tonus does not react nearly as strongly to stretching as in intact and decerebellate dogs. In the dorsal position of the animal, the left hindleg also shows distinct flexion and extension movements on passive abduction and adduction of the right hindleg under static stress.

In the *right* legs, the rocking reactions are present in the standing posture only with very strong alterations in position of the supporting leg (left fore- or hindleg). The alterations in tonus are clearly evident in the right foreleg, but they are only weakly detectable in the right hindleg. In the dorsal position, there appears in the right legs no, or at best only weakly indicated, alterations in tonus even on very strong abduction and adduction of the left legs.

Thus, the reactions of these animals correspond more or less to those which appear after the extirpation of the cerebellum alone and after the extirpation of the right half of the cerebrum alone. The exaggerated extension and flexion movements in the air carried out by the left hindleg in a standing posture as well as the occurrence of these movements in a dorsal position correspond to observations on decerebellate dogs, while the lesser strength in supporting tonus of the left legs in an extended position (and the absence of preparation for standing on extension) as well as the lesser alterations in position and tonus in the right hindleg on strong rocking movements are manifestations which appear after extirpation of the right half of the cerebrum alone.

After decerebration, the rocking reactions are absent on both sides, but sometimes a temporary, short-lived increase of extensor rigidity appears in the opposite leg on abduction as well as on adduction of one hindleg. (One gets the impression that the direction of the alteration in position here does not play any role.)

After the transverse section of the spinal cord, too, rocking reactions fail to appear and do not return later, even if the animal can be kept alive for several months.

From findings in decerebrate and spinal animals one should not draw the conclusion that rocking reactions do not represent spinal reflexes and that their center must be situated orally from Sherrington's intercollicular section. One has to take into consideration that in the spinal animal the initial tonus of the supporting leg is different from that of the intact animal. The spinal animal is not even capable of standing on one hindleg; the supporting leg usually gives way under the weight of the trunk and, in addition, static stress does not produce

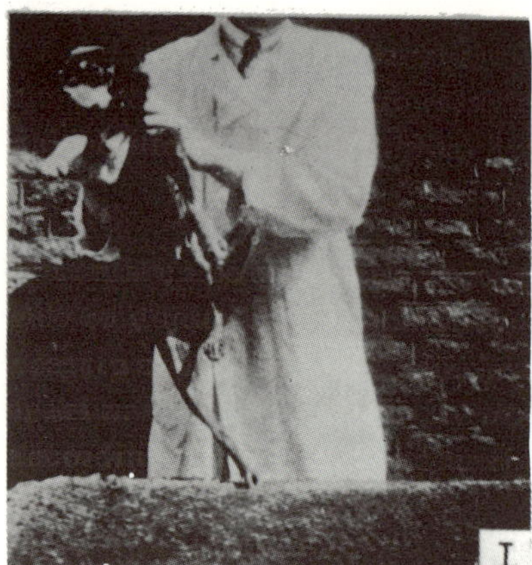 

Fig. 157.   Rocking reactions of the left hindleg in the dog Vici after extirpation of the entire cerebellum and of the right side of the cerebrum. 1. On movement of the anterior part of the body to the right maximal flexion of the left hindleg.   2. On movement of the anterior part of the body to the left extension of the left hindleg and abduction of the hip joint.   The leg is not set down on the supporting base but remains hanging in the air, directed forwards (the left hindleg shows no preparation for standing on contact of the sole of the right hindleg with the surface.)

a tonic contraction of abductors and adductors.   The non-appearance of rocking reactions after the transverse section of the spinal cord must not be caused absolutely by the absence of centers but may perhaps be produced by the inferior initial tension of abductors and adductors of the supporting leg.   Similarly, it would be in itself conceivable that in the decerebrate animal the non-appearance of flexion is caused by the exaggerated extensor tonus and the abnormal tendency to extension postures.

One could imagine the following : that the supporting leg indeed exerts a certain influence which, however, is not sufficient to counteract the increased extensor tonus of the leg taking no weight; or that the increased tendency to extension causes a situation which, by the adduction movement of the supporting leg, produces an extension instead of a flexion of the leg taking no weight.

The abnormal distribution of muscle tonus in the supporting leg possibly also plays a role in the increase of extension.   Contrary to intact animals, the decerebrate animal does not show any myotatic reflexes to the stretch of flexors and the reactions to the stretch of abductors and adductors may also be weakened. As results from the preceding observations, the question of a localization of the central mechanism of rocking reactions cannot be probably solved before it can be recorded by means of the Sherrington method. Firstly, the muscles are detached from their insertions, then the question where and by what stimulations are these reactions produced and whether they are really based on the stretch of abductors and adductors; secondly, which reflex muscle contractions are normally brought about by such stimulation; and thirdly, what alterations are shown by these tensions after decerebration or transverse section of the spinal cord.

The rocking (hopping) reactions are absent not only after decerebration, transverse section of the spinal cord, bilateral severance of posterior roots and temporarily after extirpation of the cerebellum but also in *new-born dogs*. In the third week after birth, however, they are usually present, although still weak.

In *man*, rocking (hopping) reactions are absent in the first months of life. They only appear when the children start to stand and run. They are particularly evident when infants, in their first attempts to walk, are supported below the armpits by their mothers and alternately rocked to the right and to the left. On rocking to the right, the right leg is extended and has enough supporting tonus to carry the trunk, while, on rocking to the left, it is clearly lifted from the ground and raised.

When an adult is grasped from both sides in the lumbar region and the trunk is passively moved to the right and to the left, the legs are alternately lifted when the subject of the experiment does not resist. The raised leg can be passively flexed in the knee joint with ease. André Thomas (307) observed that a patient with a gunshot injury in the *right* cerebellar hemisphere would lift the right, but not the left, leg from the ground on passive lateral displacement of the trunk. Thus, the right ipsilateral leg shows rocking reactions, the left one does not (Fig. 158). André Thomas considers the lifting of the right as an expression of "anisosthenia" of an abnormal innervation of muscles. As we have seen, the lifting is a physiological manifestation; the non-appearance of lifting in the left leg, however, a pathological one.

On the behavior of rocking reactions in other disorders of the brain, I was unable to find any detailed data in the literature. I would like to mention that as with intact dogs, in intact men in the dorsal position, flexor and extensor movements or an increase and decrease in extensor tonus in the opposite leg can be seen as a response to an abduction or adduction of one leg.

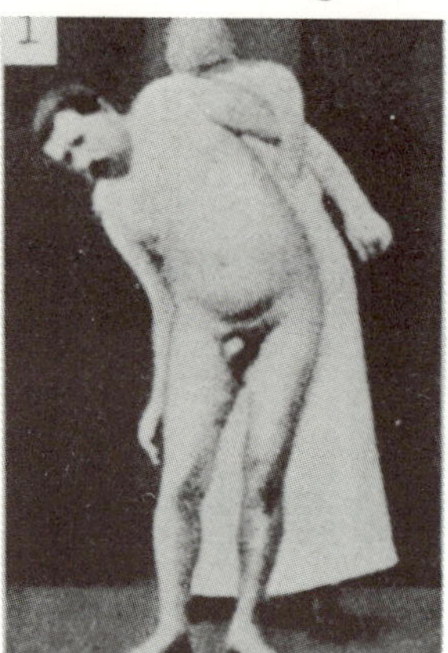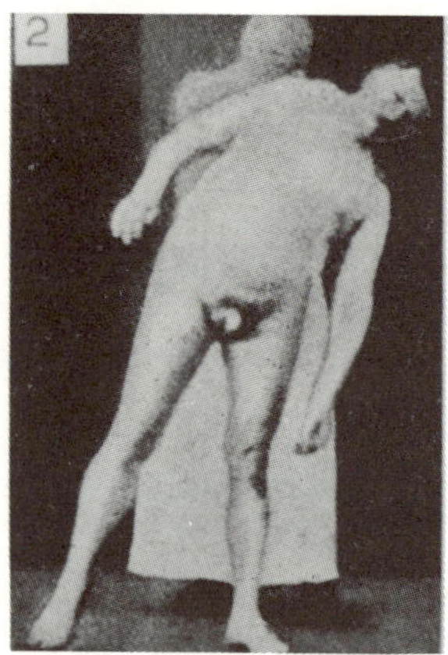

Fig. 158. Man with gunshot injury in the *right* cerebellar hemisphere. Trunk is clasped with two hands laterally in the lumbar region and displaced to the right and left.   1. On passive movement of the trunk towards the right, the *left* leg remains in contact with the ground while the right one is flexed in the knee joint. 2. On passive movement of the trunk towards the left, the *right* leg is instantly raised from the ground, the left remains extended. From André Thomas: *Etude sur less blessures du cervelet*. Vigot Frères, Paris, 1918.

THE IMPORTANCE OF ROCKING (HOPPING) REACTIONS FOR THE MAINTENANCE OF BALANCE AND FOR THE ADAPTATION OF LEG POSTURE TO THE POSITION OF THE SUPPORTING SURFACE

When a dog is pushed to the left while running, for instance while its left fore- or hindleg is raised, the passive abduction of the right leg will produce a rocking reaction in the raised left leg.   The latter instantly makes a vigorous extension and abduction movement and thus prevents the falling to the left side.   When the animal is standing on a board which is moved downwards on the left side while one of the left legs is raised, the force of gravity will pull the trunk towards the left and thus bring the right supporting leg into an abduction posture. The latter produces an active extension and abduction of the raised left leg, so that that animal does not fall to the left and the leg occupies the position necessary for adaptation to the change in static conditions.

When a supporting surface on which an animal is standing on all four legs is moved downwards on the right side, the animal, too, is pulled to the right by the force of gravity and, theoretically, a passive abduction of the left legs followed by a reflex extension and abduction of the right legs should take place.   The alterations in position of the right legs indeed take place (Fig. 159), but the abduction of the right legs is replaced — at least in the case of a slow downwards movement — by adduction and flexion; therefore, the extension and abduction of the right legs cannot be the result of a rocking reaction.   Furthermore, on the abduction of the right legs (Fig. 159) the left leg should be extended with a strong supporting tonus, but actually these legs show a flexed posture and flex more and more on a further downwards movement in spite of the increasing abduction of the right legs and finally collapse entirely.   Thus, in the adaptation to the position of the supporting surface, other factors must participate which interfere with the rocking reaction.

When we now ask ourselves by what means the influence of the rocking reactions is abolished, i.e., why the abduction of the right legs does not produce an extension and abduction of the left legs, the supposition suggests itself that the *active* abduction of the right leg in the adaptation (in opposition to that in passive rocking) is responsible.

In a passive abduction caused by passive rocking, the abductors are relaxed, while in an active abduction they are actively stressed.

Fig. 159.   The supporting platform of a blindfolded, labyrinthectomized dog standing on all four legs is lowered from the right side.   In this case the leg posture is adapting itself to the position of the base : the extended right legs are actively strongly abducted.   The abduction does not produce rocking reactions in the left legs which, on the contrary, go into a flexed posture.

This supposition, however, has proved to be wrong, since we have seen elephants and bears in zoological gardens actively rocking to and fro for a considerable time while showing typical rocking reactions. When a person also raises, for instance, the left leg and actively abducts the right, the left is extended and set down on a supporting surface. When, however, a supporting surface, for instance a chair, is set under the raised leg, an increase in flexion will take place. Similar conditions are also found in animals where the active as well as the passive abduction of a leg can only produce an extension of the flexed opposite leg when the latter does not support itself on an immovable supporting surface. How the influence of a fixed supporting surface is brought about will be discussed in Chapter XIII.

In a dog with a raised leg, the rocking reaction will also appear when the animal is pushed to the side, pulled towards one side or even when the supporting surface of the animal is shifted. The abduction and extension of the leg taking no weight will in this case prevent its falling down and thus represents a balance reaction. The rocking reactions produced by a lateral shifting of the supporting surface can be clearly observed when a dog is set down with the hindlegs on a turntable while the anterior part of the body is held fast (Fig. 160).

Fig. 160. Decerebellate dog Piccolino with the hindlegs on a turntable rotating in clockwise direction. The animal is held fast by the forelegs and neck in such a way that the bitemporal axis of the trunk stands vertical on a radius. 1. The animal raises the right hindleg, the left stands on the table. 2 and 3. By the rotation of the table the left hindleg, the supporting leg, is passively abducted; in so doing, the right hindleg (leg taking no weight) is extended. 4. On further rotation the right hindleg is passively adducted, and the left hindleg is drawn in to a flexed posture.

As a result of the preceding observations, the rocking reactions thus represent a typical balance reaction. Under certain circumstances, it may even play an inferior role in the adaptation of leg posture to the position of the supporting surface which, however, is usually brought about just contrary to the effect of the rocking reaction.

B. REACTIONS TO A STRONG ADDUCTION OF THE OPPOSITE LEG

When a dog standing on its hindlegs is slowly displaced towards the right while the toes of the right hindleg are fixed to the supporting surface with one hand, the left hindleg is raised from the ground, flexed more and more and then suddenly extended (Fig. 161). In this process, it is set down on the surface next the inner side of the right leg.

We thus have here a third example for our supposition that a *strong* stretch of muscles produces a reflex reaction contrary to that produced by a moderate or slight stretch. A slight adduction of the supporting leg accompanied by a slight stretch of the abductors can thus produce a flexion of the leg taking no weight, while a strong adduction accompanied by a strong passive stretch of the abductors, on the other hand, produces an extension of the leg taking no weight.

In this investigation, the supporting leg must be passively fixed to the surface, otherwise it will be raised and simultaneously shifted laterally in a hopping movement with the setting down of the leg taking no weight. The fixed supporting leg, in this case, usually attempts to lift itself from the base several times and, in so doing, one feels how the leg is alternatively flexed and extended while the other leg simultaneously makes alternating extension and flexing movements. These reactions are also very lively in decorticate dogs and are thus caused by reflexes. Here, too, one has the impression that the strong passive stretch of muscles may produce an alternating contraction and relaxation.

The extension of the leg taking no weight also occurs when, with a fixation of the trunk, the supporting leg is passively shifted in strong adduction, or when the adduction is brought about by a shifting of the supporting surface, as on a turntable (Fig. 162).

The forelegs show corresponding reactions; strong adductions of the supporting leg also cause an extension of the flexed leg taking no weight and a setting down of the latter just medial to the supporting leg.

Just as in flexion, the extension of the leg taking no weight also fails to appear when the opposite leg is not under static stress during adduction; blindfolding has no influence whatsoever in this case.

Fig. 161. Dog Vici, decerebellate, right-sided decorticate, set down on the hindlegs. As always the animal supports itself on the right hindleg while the left is held in the air. 1. Trunk is inclined to the left, left hindleg extended. 2. Trunk moved to the right thus leading the right hindleg passively to a slight adduction. Left hindleg actively drawn in to a maximum flexed posture (rocking reaction). 3. Trunk moved even more to the right, in so doing the right hindleg is passively strongly adducted. Left hindleg now actively extended (while the right hindleg gives way).

The extension of the opposite leg on strong adduction of the supporting leg is absent in new-born dogs and in decerebrate and spinal dogs, but it is evident not only in intact but also in labyrinthectomized, decerebellate and decorticate dogs. In decorticate (thalamus) dogs, it appears promptly in the forelegs, but is greatly delayed in the hindlegs, i.e., an extension occurs only on an abnormally strong adduction of the supporting leg, especially when the strength of supporting tonus in the hindleg is greatly reduced. Moreover, the supporting tonus produced by the extension of the hindleg is only very slight.

After extirpation of, for instance, the left half of the cerebrum, an extension of the right legs with a strong adduction of the left ones promptly occurs; however, an extension of the left leg on a strong adduction of the right one usually takes place with considerable delay, i.e., only after an abnormally strong adduction of the latter. The result is that the left leg crosses the right one while extending and is set down

Fig. 162. Decerebellate dog Piccolino (A) and decerebellate, right-sided, decorticate dog Vici (B) with the hindlegs on a turntable rotating in a clockwise direction. A. Dog Piccolino. A 1. The right hindleg is passively adducted by the rotation of the table. Accordingly, the left hindleg is raised from the table (rocking reaction). A 2. On further adduction of the right hindleg, the left one is extended and set down on the table while the right, at the same time, is raised from the table. A 4-6. The same reactions. B. Dog Vici. B 1 and 2. Right hindleg (supporting leg) passively adducted due to the rotation of the table, the left (leg taking no weight) drawn in a flexed posture (rocking reaction). B 3. On a stronger passive adduction of the right hindleg, the left is suddenly extended in the air, the right one, at the same time, is raised. Due to the absence of preparation for standing, the left hindleg is not set down on the table. B 4. The right hindleg is again extended and set down on the table. (Since the left leg does not support itself on the table but remains hanging

Fig. 162. (cont.) in the air, the extension of the right leg is not caused by the passive abduction of the left one. As in B 3 and 4, conditioned strong adduction of the right leg thus shows an extension of the left and a flexion extension movement of the right leg, see also B 8 and 9.) B 5-7. The left hindleg goes again into flexion on a passive slight adduction of the right one. B 8. On strong passive adduction, sudden extension of the left leg in the air while the right one makes a flexion extension movement. Note the more exaggerated reactions of the right hindleg in the dog Vici than those in the dog Piccolino. The reactions of the right hindleg are appropriate in the dog Piccolino; in the dog Vici, however, they are inappropriate, though adequate to the purpose (compare, for instance, A 1-2 with B 7-8).

on the surface beyond the right leg. The delay of extension is stronger and more constant in the hindlegs than in the forelegs on the side of extirpation.

In this investigation, one obtains the impression that the delay was caused by the reduced strength in supporting tonus of the supporting leg. After a total or unilateral extirpation of the cerebellum, extension is absent at first on both sides; it appears again later with a great delay but promptly towards the end. In decerebellate dogs, extension can usually be seen clearly only in the case when the supporting leg is fixed to the surface because when a decerebellate dog is set down on its hindlegs and strongly displaced to the right, the left leg is indeed at first raised, but with further adduction, the supporting leg will be lifted from the surface *before* the extension of the other leg has taken place.

Owing to the irregular behavior of reaction on strong adduction of the supporting leg after extirpation of the cerebrum or cerebellum, there appear typical impairments when the animals are pulled sideways.

When an intact dog standing on all four legs is pulled to the right, the following four reactions appear successively : 1. raising of the left legs; 2. extension and setting down of these legs medial to the right ones; 3. raising of the right legs; 4. setting down of these legs on a spot situated more to the right. Each of these four reactions can be separately and clearly observed.

When a left-sided decorticate dog is pulled to the left, i.e., towards the side of extirpation, the four reactions seemingly appear as in a normal animal. The succession of reactions is also normal on pulling to the right side, i.e., to the intact side, but the extension of the left legs

(2) occurs only on a very strong adduction of the right legs so that the left ones in the course of extension sometimes cross the right ones and are set down lateral to them. Thereupon the strongly adducted right legs make particularly large steps outwards (3 and 4). Thus, on pulling towards the intact side, unilaterally decorticate dogs hop sideways with larger steps than in pulling towards the other side and the legs of the side of extirpation cross the contralateral legs. On a *vigorous* pulling towards the intact side, the reactions 2, 3 and 4 sometimes appear so much delayed, particularly in the hindlegs, that the body falls to the surface.

After bilateral extirpation of the cerebrum similar impairments occur on pulling to the left or to the right. In these animals, too, the reactions of the hindlegs are delayed and since, in addition, the extensor tonus produced by the extension of the hindlegs is only weak, the posterior part of the body easily falls down on a pulling to the right or to the left.

Totally and unilaterally decerebellate dogs sometimes show similar impairments in the crawling stage (when they are not yet able to run around freely) and later, in the stage of permanent impairment, the reactions are not normal and show typically irregular behavior when the animal is pulled sideways.

When, for instance, a decerebellate dog is pulled to the right, the following reactions appear :

1. Increased raising of the left legs; 2. after a long delay and after overcoming a distinct resistance, sudden raising of the right legs with almost simultaneous extension and setting down of the left ones; 3. setting down of the right legs further to the right.

Thus, in decerebellate dogs the interval between the first and the second reaction is abnormally long while the interval shown by intact dogs between the second and third reactions is eliminated. During the long interval between 1 and 2 the left legs are usually raised abnormally high. The second reaction in decerebellate dogs further differs from that of intact ones by the fact that in intact animals the reaction consists of an extension of the left legs, while in decerebellate dogs a raising of the right legs is accompanied by an extension of the left ones.

In decerebellate animals, the raising of the right legs is clearly the preliminary reaction.[1]   Moreover, when decerebellate dogs are pulled to the right, the right legs are raised abnormally high and are transferred to the right over an abnormally large distance. Thus, in a sideways movement over a certain distance the number of steps is less (usually by half) in decerebellate dogs than in intact ones. In pulling these animals sideways, one encounters a distinct resistance. It is difficult to ascertain whether it is increased since the resistance in intact dogs is very varied and sometimes, in obstinate dogs, very strong.

The same difference appears between intact and decerebellate dogs when the trunk of the animals is fixed and the supporting surface is shifted sideways or the animals are set down on a turntable with a fixed trunk.

Unilaterally decerebellate dogs show impairments when they are pulled towards the side of extirpation as well as towards the intact side. In the movement towards the side of extirpation, the succession of reactions is the same as in the intact animal but the steps are abnormally large and the resistance to the movement is slight; in a movement towards the intact side, however, the succession of reactions is irregular as in totally decerebellate dogs and in this case the steps are of normal size and the resistance is rather strong.

As already mentioned, it is very difficult to ascertain whether the resistance is increased or reduced, owing to the greatly varying resistance in intact animals. Still one has the impression that in unilaterally decerebellate dogs the resistance to movement towards the side of extirpation is reduced and increased to movement towards the intact side. Unilaterally decerebellate dogs *constantly* offer a strong resistacne to a movement towards the intact side; totally decerebellate animals to movements towards both sides (for details see Chapter XIII).

The preceding observations show that the extension of the leg taking no weight on strong adduction of the supporting leg represents a component of a complicated balance reaction which prevents the animal pulled or pushed sideways from falling and enables the animal to run sideways.

---

(1) When intact dogs strongly resist pulling to the right, occasionally a raising of the right legs is shown as a second reaction. Conversely, in decerebellate animals a normal succession of reactions can also be observed now and then.

In man, too, the strong adduction of one leg produces an extension of the other leg, for instance, when a person on raising the left leg is suddenly pushed or pulled to the right. In patients with brain diseases these reactions have not yet been investigated.

Moderate adduction of a leg under static stress produces a decrease in supporting tonus of the opposite leg.

Strong adduction produces a reappearance of supporting tonus.

Moderate abduction produces an increase in supporting tonus and extension in the opposite leg.

Strong abduction, under certain circumstances, as we shall see later (P. 381),can produce in the opposite leg a reduction of supporting tonus with raising and tension of the adductors.

C. ALTERATIONS IN SUPPORTING TONUS OF ONE LEG ON THE FORWARD AND BACKWARD MOVEMENT OF THE OPPOSITE LEG

When a dog is set down on a supporting surface in an exactly symmetrical standing position, the opposite leg when lifted from the ground can easily be flexed and does not make any attempt to extend when the sole is resting on the supporting hand (Fig. 163, No. 1). On a forwards movement of the trunk, however, there occurs a vigorous extension movement in which the supporting hand is pushed away (Fig. 163, No. 2). Pressure on the sole encounters a strong resistance and even when the leg is grasped laterally it is difficult to flex it. Similar reactions are shown by the raised leg when, instead of a forwards movement of the trunk, the supporting leg is shifted backwards. Thus, the determining factor is the alteration in a caudal direction of the position of the supporting leg in relation to the trunk.

Extended posture and supporting tonus of the opposite leg continue as long as the supporting leg remains directed backwards and disappear when the trunk and supporting leg are moved forwards into their former positions.

When the trunk is moved backwards from the intermediate position or the supporting leg is shifted forwards, the passively raised contralateral leg is also extended and shows a distinct supporting tonus

330

(Fig. 163, No. 3 and 4). Usually the extension is less strong and abrupt and sometimes only occurs when the supporting leg gives way in the elbow joint.

Fig. 163.   Decerebellate dog Erik with blindfold. 1. The animal stands with the right foreleg exactly in an intermediate position on a supporting surface. The raised left leg which rests with the sole of the foot on a hand can easily be flexed. 2. On passive movement forwards of the trunk the left foreleg is vigorously extended and shows a strong supporting tonus. The limb has pushed away the supporting hand. 3. On moving back the trunk decrease in supporting tonus. 4. On moving the trunk further backwards, the left foreleg is extended and also shows a distinct supporting tonus; simultaneously the right one is flexed at the elbow joint.

These reactions are present in intact, labyrinthectomized, decorticate and decerebellate dogs; they are absent in decerebrate dogs. They appear only when the supporting leg is under static stress. They are promoted by static stress or simple touching of the sole of the foot taking no weight, but even when this leg is only grasped from the side, distinct alterations in tonus occur. These reactions appear also particularly clearly in decorticate dogs when the leg taking no weight is not supported and similarly when in unilaterally decorticate dogs the foreleg contralateral to the extirpation acts as the leg taking no weight (Fig. 164).   In these animals, the reactions of the leg taking no weight are not impaired by the preparation for standing. When, for instance, a right-sided decorticate dog is set down on the edge of a table with the right foreleg, and the trunk is moved caudally so that the supporting leg is directed forwards, the left foreleg goes into a maximum extended posture and remains extended and suspended in the air (Fig. 164, No. 1) while intact dogs under the same circumstances would set down the left foreleg beside the right one on the edge of the table under the influence of preparation for standing. On a forward movement of the trunk, the extensor tonus of the left foreleg in the unilaterally decerebellate dog decreases gradually and the left is flexed. In an intermediate posture of the supporting leg it remains suspended in the air in a flexed position as long as the posture of the animal remains unchanged (Fig. 163, No. 3); due to the absence of preparation for standing, even now the leg is not set down on the supporting base. When the trunk is moved as far forwards as to direct the supporting leg backwards, the left foreleg is again extended in which case the sole of the foot makes contact with a spot on the table situated in front of the right leg (Fig. 164, No. 4). On further forward movements of the trunk, corresponding reactions also occur in the right foreleg in which case the leg is raised at first from the supporting surface, then extended and set down on the table in front of the left one. The reactions of the right leg, however, take place only with a continued forward movement of the trunk; as soon as the movement is interrupted, the leg, due to its preparation for standing, is instantly set down on the table behind, beside or in front of the left one, depending on the position of the animal.

The influence of the forward or backward directed position of the supporting leg on the position and tonus of the opposite leg became particularly evident in the *unilaterally decorticate and totally decerebellate* dogs Vici and Däumling, since in these animals the reactions of the foreleg contralateral to cortical extirpation, on the one hand, were not impaired by preparation for standing and, on the other hand, occurred with exaggerated force due to extirpation of the cerebellum.

As a result of these observations, we conclude that these reactions must play a role in running movements, owing to the forward and backward directed position of the supporting leg. When a decerebrate dog or a decerebrate cat is set down on a table with one foreleg directed forwards, the opposite leg, on the forward movement of the trunk, shows neither flexion nor an evident decrease in extensor tonus. This is all the more remarkable as decerebrate animals show bouts of running movements, in which the left and right forelegs go alternatively forwards and backwards at the shoulder joint and, in so doing, are more or less flexed. Similar conditions can also be observed in spinal animals; alterations in position of the supporting leg to the trunk produce no distinct alterations in tonus of the opposite leg in spite of the fact that they show bouts of "spontaneous" running movements of the hind-

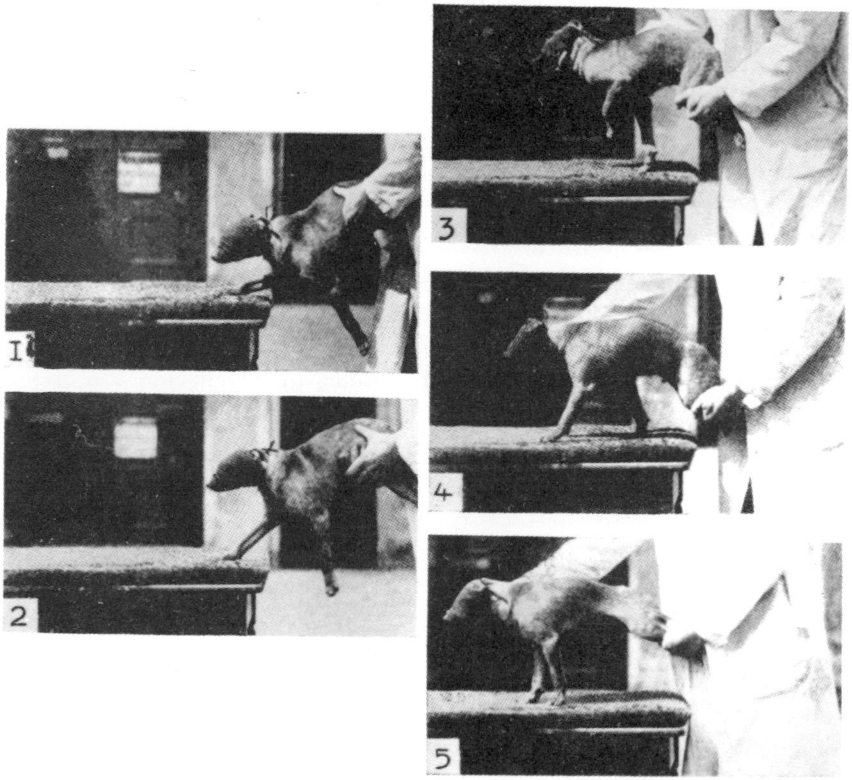

Fig. 164. The dog Däumling, decerebellate, right-sided decorticate, set down with the right foreleg on the edge of a table. Eyes are closed by a blindfold.   1. Supporting leg strongly directed forwards, left leg extended in the air, is not set down on the table because of the absence of preparation for standing.   2. Trunk is slightly moved forwards. Extended position of the left limb diminishes. Leg is somewhat flexed at the elbow joint.   3. On a further movement forwards which brings the supporting leg almost into an intermediate position, the left limb is drawn into a flexed posture.   During this position the animal's leg remains hanging in the air and is not set down on the supporting base. 4. The animal is moved forwards even more. The left foreleg is again extended and is set down on the table in front of the right one and stands extended forwards. 5. On continued forward movement similar reactions occur in the right foreleg; at first it is flexed, then extended and set down on the table in front of the left one.

legs. Worth mentioning in this respect are the observations made
on the decerebellate, right-sided, decorticate dog Vici. When this animal was set down with the right foreleg on a turntable in such a way
that on rotation the leg went backwards, the left foreleg sometimes
showed strong exaggerated reactions carried out in the air. These,
however, would sometimes fail to occur (Fig. 165). Thus, in this case,

Fig. 165. A. Decerebellate, right-sided, decorticate
dog Vici with the right foreleg on
a turntable. 2-4. By the clockwise rotation the right foreleg is moved backwards; in so doing, the left is extended
forwards more and more. 5-6. On further rotation of the table the right limb
is raised and transferred forwards while
the left one goes backwards. (The paw
of this leg can just be seen below the
belly, see also A 1.) 7-8. In setting down
the right leg, the left is moved forwards so that the tip of the toes appears at the anterior side of the chest.
9. On passive movement backwards
of the right leg an extension of the
left one occurs again. B. The animal
with the right foreleg on the table.
1-9. Although the right foreleg is again
moved backwards by the rotation of
the table (B1-3) and then again makes
steps forwards (B 4-6) the left shows
no reactions but remains extended backwards in the air.

the posture of the supporting leg did not produce any reactions in the opposite leg, although the centers of these reactions were surely preserved and able to function as is shown by the occasional appearance of these reactions. In decerebrate animals, the conditions are probably similar.

The hindlegs also show reactions resembling those of the forelegs. With the supporting leg in an intermediate posture, the passively raised opposite leg can easily be flexed, while a strong supporting tonus and extension occur when the supporting leg is directed forwards or backwards. On the extension of the hindleg of intact, labyrinthectomized and decerebellate dogs a stronger supporting tonus usually appears, more than in decorticate animals. In decerebrate dogs this influence is totally absent. When the animal is moved back so that the supporting leg comes into an intermediate position, a change in supporting tonus fails to appear in the leg taking no weight. Alterations in position and supporting tonus continue in decorticate dogs due to the absence of preparation for standing in the leg taking no weight, just as in the unilateral extirpation of the cerebrum when the hindleg contralateral to extirpation functions as the leg taking no wieght (Fig. 166).

When instead of moving the trunk, the supporting leg is shifted forwards or backwards or is moved forwards and backwards by a shifting of the supporting surface or the rotation of a turntable, these reactions can occur in the opposite hindleg (Fig. 167).

In the dorsal position neither intact, labyrinthectomized, nor decorticate dogs show distinct alterations in position or tonus of the opposite leg on a passive forward or backward movement of an extended hindleg. Conversely, decerebellate dogs in the dorsal position usually show most clearly alterations in posture. However, in the decerebellate, *right-sided*, decorticate dog Vici, distinct reactions to the forward and backward movement of the opposite leg could always be observed in the *left* fore- and hindlimbs (Fig. 168).

The reactions of one leg produced by the forward and backward movement of the opposite leg play a role firstly in running forwards and backwards and secondly in participating in the maintenance of balance, as for instance, when an animal is suddenly pushed forwards or backwards while one or two legs are raised. When the front end of

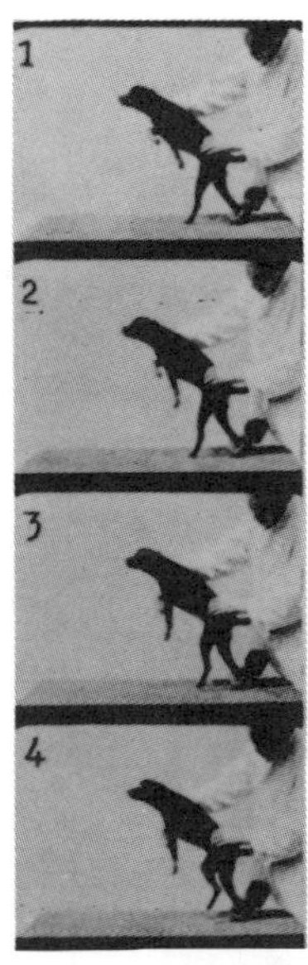

Fig. 166. Decerebellate, right-sided, decorticate dog Vici. The animal stands on the right hindleg which is passively fixed on the supporting surface. 1 and 2. Trunk of the animal is moved forwards so that the right hindleg stands strongly directed backwards. The left hindleg shows an extended posture. 3-7. Trunk is passively moved backwards. By this the left hindleg is raised from the ground and flexed more and more. 8. Supporting leg in an intermediate position; the left leg is drawn into a flexed posture.

the supporting surface is suddenly lowered or shifted while the legs are raised and the animal is drawn forwards or backwards by the force of gravity, these reactions will in a forwards movement also cause a bracing forwards and in a backwards movement a bracing backwards in order to prevent the animal from falling.

Similar mechanisms probably come into effect in man also when he is suddenly pushed forwards or backwards and, in a reflex movement, extends the raised leg.

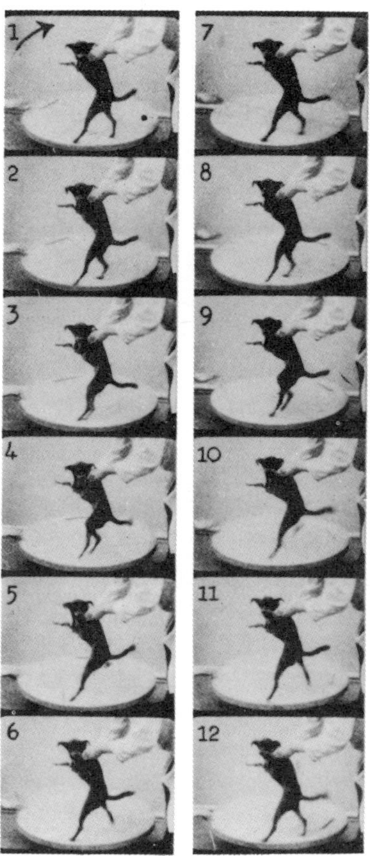

Fig. 167. Decerebellate, right-sided decorticate dog Vici with the hindlegs on a turntable which draws the limbs passively forwards. 1. Right hindleg (supporting leg) directed backwards; left hindleg totally extended. 2-4. Right hindleg is passively moved forwards into an intermediate position; by this the left hindleg is drawn into a flexed posture. 5-7. The right hindleg is suddenly raised abnormally high and makes a large step backwards. At the same time the left hindleg makes an extensor movement and is set down inappropriately (directed forwards!) on the supporting surface. 8-13. The same reactions repeated.

Fig. 168. Decerebellate, right-sided, decorticate dog Vici in a dorsal position. 1. The right hindleg under static stress is held passively forwards, the left fully extended. 2-9. The right hindleg under static stress is passively moved backwards; the left hindleg goes into flexion. 9-16 The right hindleg under static stress is moved backwards even more; the left hindleg is again extended.

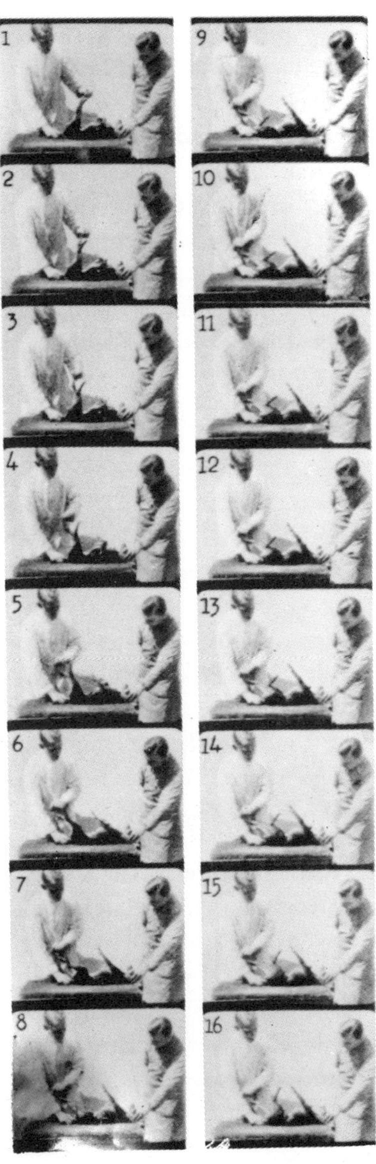

XII. THE SUPPORTING TONUS OF THE HINDLEGS IS AFFECTED BY THE POSITION OF THE FORELEGS

As we have already seen, an alteration in position of the forelegs brings about a curvature of the spinal column which is followed by an alteration of supporting tonus in the hindlegs. Primary alterations of supporting tonus in the hindlegs, too, can be produced by alterations in position of the forelegs, that is, when an intact, decerebrate or decerebellate dog is set down on a solid supporting surface on the forelegs, the soles of the hindfeet are placed and at the same time the trunk is moved caudally so that the forelegs move forwards at the shoulders. One will feel that the hindlegs suddenly extend backwards, push the hand vigorously away and offer a strong resistance against pressure on the soles of the feet. If the trunk is moved forwards, the hindlegs move forwards, the resistance against pressure on the soles decreases, the supporting tonus disappears and finally the hindlegs are drawn into a flexed position (Fig. 169).

In decerebellate dogs, the influence of position of the forelegs appears particularly distinct in the dorsal, supine position. When a decerebellate dog is laid down on its back with the muzzle 0-90° above the horizontal then, on static stress on the fore- and hindlegs, the latter extend vigorously backwards on a forward movement of the forelegs and show supporting tonus as long as the forelegs are held forwards; while on a backward movement of the forelegs the hindlegs go forwards and give way to the slightest pressure and even without any pressure at all. The supporting tonus is absent as long as the forelegs remain directed backwards (Fig. 170). Thus, the reactions are tonic and caused by the position of the forelegs.

When the hindlegs are not under static stress, flexion and extension movements occur less promptly. When only the hindlegs are under static stress and the forelegs are not, distinct but sometimes less strong alterations in supporting tonus appear in the hindlegs on a forward and backward movement of the extended forelegs which, however, are almost absent if the forelegs are held in a flexed posture during the passive forward and backward movement.

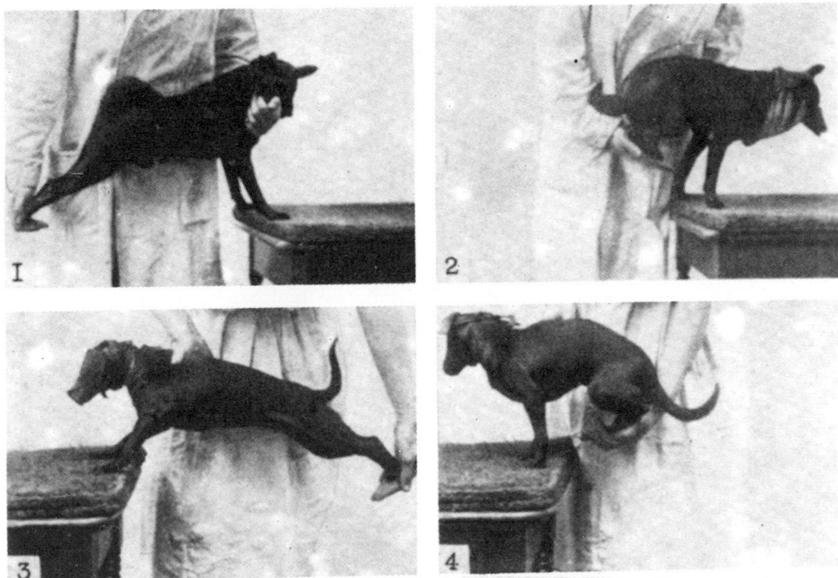

Fig. 169. Decerebellate dogs Erik (1 and 2) and Piccolino (3 and 4), blindfolded; the forelegs on the edge of a table, the hindlegs with soles of
the feet resting on a hand. 1 and 3. The trunk of the animals is passively moved caudally and the forelegs are directed forwards in the
shoulder joint.   The hindlegs are extended pushing the hand away.
The extended posture of the hindlegs and the strong resistance
against pressure on the feet continue as long as the position of the
animals remains unchanged. 2 and 4. On forward movement the
resistance decreases, the supporting tonus disappears and the hindlegs
are gradually moved forwards and flexed as soon as the forelegs are
directed backwards at the shoulders. This position, too, continues
as long as the posture of the animals is unchanged. As shown in
Figs. 2 and 4, the supporting tonus decreases although on the forward
movement the spinal column is curved dorsally more or less strongly.

Intact and decerebellate dogs, when set down on their forelegs,
will place the non-supported hindlegs beside the forelegs on the table
by a forwards movement of the trunk. Thus, in this case, the reactions
of the hindlegs are altered by the preparation for standing. The reactions of the hindlegs appear particularly strong and brusque in decerebellate dogs in a standing posture as well as in a dorsal position. Thus,
on a forwards movement of the forelegs, the hindlegs are strongly
directed backwards and maximally extended. In intact, decerebellate
or labyrinthectomized dogs they can only be well observed in a standing posture and in decerebrate dogs they are totally absent.   In the

latter case a slow forward and backward movement of the forelegs causes no alterations in stretch reflexes in the hindlegs but a rapid movement, like any other strong stimulation, will produce a distinct increase in extensor rigidity which is usually of short duration.

The reactions of the hindlegs also occur, though with less force, when the animals are set down on the edge of the table on *only one foreleg* and the trunk is then moved backwards and forwards; in this case the extensor tonus will increase or decrease in both hindlegs. When in the to and fro movement each hindleg is alternatively put under

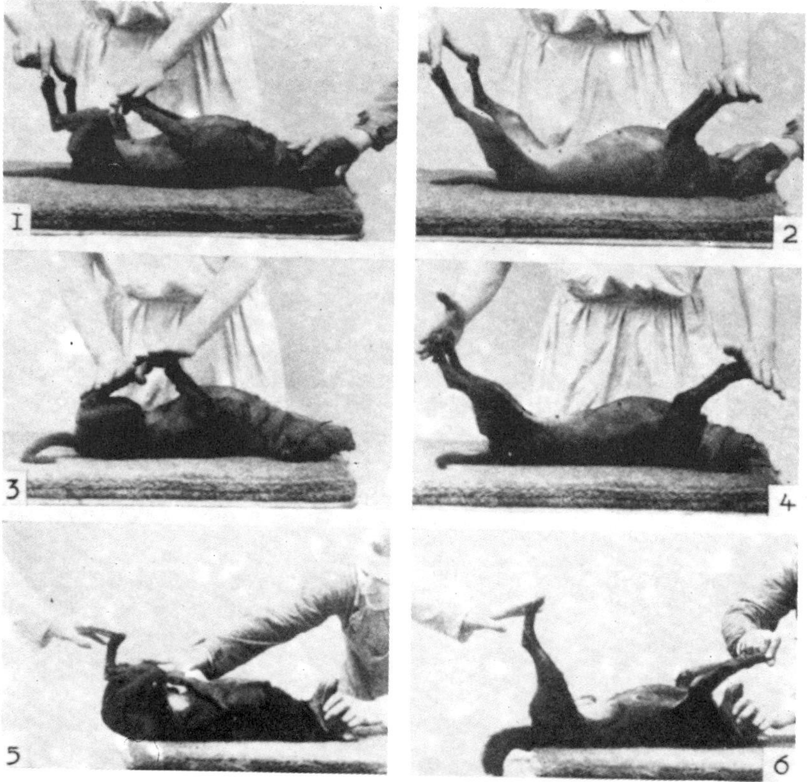

Fig. 170.   The decerebellate dogs Piccolino (1-4) and Erik (5-6) in dorsal position.   1. The forelegs under static stress are held caudally; no magnet reaction of the hindlegs on touching of the soles.   2. The forelegs under static stress are held forwards; hindlegs are instantly extended on touching of the soles.   3. With the forelegs directed backwards there is no extension of the hindlegs on static stress. 4. Conversely, when the forelegs are directed forwards, the hand exerting pressure is instantly pushed upwards and backwards by the hindlegs. 5 and 6. The same reactions in the dog Erik.

static stress one can clearly see that alteration in position of each of the two forelegs exerts a corresponding influence on the supporting tonus of both hindlegs (Fig. 171). Conversely, alterations in position of the hindlegs under static stress bring about alterations in supporting tonus of the forelegs which, however, are inferior and cannot always be clearly observed. When the hindlegs are moved caudally, the supporting tonus increases slightly and on a forward movement it decreases, but it never disappears completely.

When the animal stands on its hindlegs, alterations in posture sometimes appear, but not constantly, in the forelegs not under static stress on forward or backward movements of the trunk (Fig. 172). In the totally decerebellate, *right-sided*, decorticate dog Vici, passive alterations in position of the hindlegs regularly caused strong alterations in

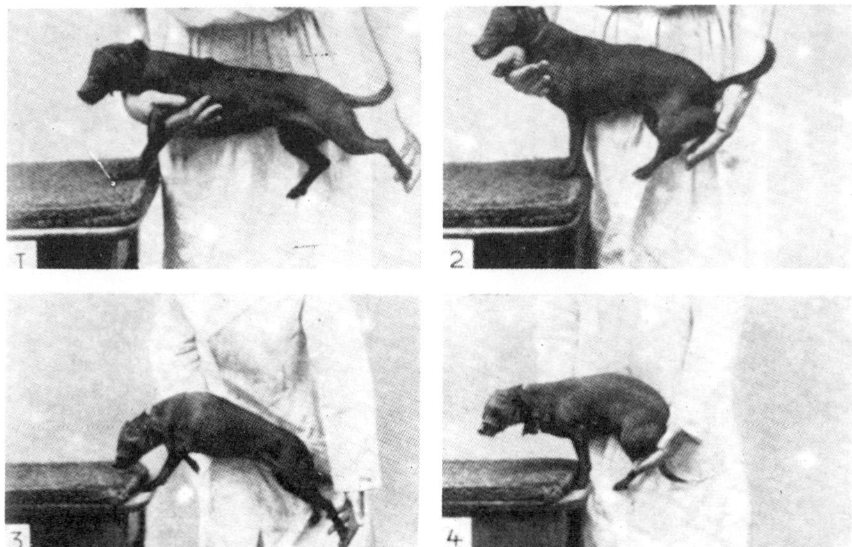

Fig. 171. Decerebellate dog Piccolino with blindfold. 1. Animal with the right foreleg on the edge of a table, the left one passively raised. On caudally directed movement of the trunk the opposite left hindleg, under static stress, is vigorously extended and shows a strong supporting tonus. 2. The tonus again decreases when the trunk is moved forwards again. 3. The animal with only the left foreleg on the edge of the table now also on a caudally directed movement of the trunk the left, ipsilateral hindleg is vigorously extended. 4. On a forward directed movement the supporting tonus decreases again as soon as the foreleg is directed backwards. Thus standing on one foreleg the position of this leg exerts an influence on the supporting tonus of both hindlegs.

postural tonus of the *left* foreleg not under static strain (Fig. 173, see also Fig. 168), and also in the dorsal position the left foreleg always went into an extended posture on backward movement on one or both hindlegs.

Fig. 172. Decerebellate dog Piccolino. 1. The animal stands only on the hind-legs, forelegs are flexed.   2. On moving the trunk forwards, the forelegs extend.

Fig. 173. Decerebellate, *right-sided*, decorticate dog Vici in dorsal position. 1. Right hindleg passively extended and held forwards. Left foreleg flexed, left hindleg extended.   2. On moving the right hindleg backwards, the left foreleg shows a half extended posture, the left hindleg is flexed. 3. On strongly caudally directed posture of the right hindleg, there is maximum extension posture of the left foreleg, while the left hindleg was again also extended.

Grahe (109) observed similar reaction movements in rabbits in the supine position. On extension of the hindlegs backwards, the forelegs were directed forwards and on the ventral flexion of the hindlegs, the forelegs went slightly backwards. Grahe observed that these movements call to mind jumping. The reactions observed in dogs probably also participate in jumping and galloping movements. Under certain circumstances they will also serve the maintenance of balance, for instance, when the supporting surface for the hindlegs is suddenly lowered or when the animal is pushed forwards or backwards at the time when it has raised one or several legs on running. These reactions can also gain a small though not negligible importance for the adaptation of leg posture to the position of the surface. When a dog stands on a surface which is suddenly tilted downwards at the tail end, the animal will be pulled backwards by the force of gravity, the forelegs are pulled

Fig. 174. Decerebellate dog Piccolino. 1. Tail end of board is moved downwards; by this the forelegs retract at the shoulders, the hindlegs are maximally extended. 2. On upward movement the forelegs prop forwards, the hindlegs are transferred forwards and flexed. 3. The same reactions occur also on raising the tail end with closed eyes. As shown by these illustrations, in the adaptation of the posture to the obliquely inclined supporting surface, the retraction of the forelegs is accompanied by extension of the hindlegs, the forward bracing of the forelegs by a flexion of the hindlegs contrary to the reciprocal influence we have discussed.

forward passively at the shoulders and by so doing produce an extension of the hindlegs in a caudal direction. In a not too rapid downward movement of the supporting base one will see, however, that the forelegs are then not pulled forwards at the shoulders but retract simultaneously with the hip joints so that in the adjusted position all four legs are directed backwards. Contrary to the reactions discussed, in the adaptation to the position of the supporting surface, the backward directed forelegs are accompanied by an extension posture of the hindlegs and the forward directed forelegs by a flexed one (Fig. 174).

Thus, in the adaptation of leg posture to the position of the supporting surface factors must be present which counteract the reciprocal influence of fore- and hindlegs.

The lateral movements of the forelegs, too, produce alterations in supporting tonus of the hindlegs. When a dog is set down with the forelegs on a supporting surface and the trunk is moved to the right, the raised right hindleg under static stress is vigorously and suddenly extended and abducted (Fig. 175, No. 2). The same reaction occurs when the surface, with the forelegs in relation to the trunk are shifted to the left.

In order to produce these reactions in the right hindleg it is not necessary for the animal to stand on the supporting surface with both

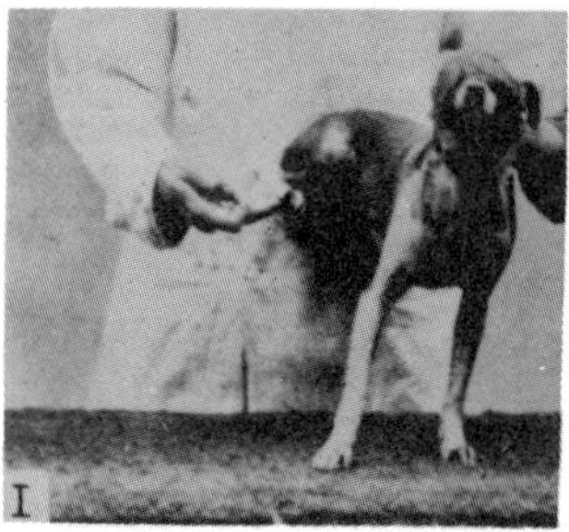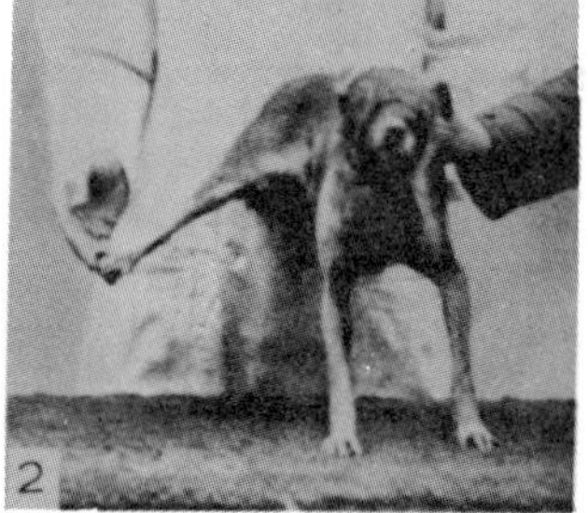

Fig. 175. Labyrinthectomized, *left-sided*, decorticate dog Jenny. Animal with the forelegs on a mat, posterior part of the body passively raised, right hindleg put under static stress by one hand. 1. Anterior part of the body of the animal passively drawn somewhat to the left so that the forelegs are directed towards the right. Right hindleg can easily be flexed, almost no supporting tonus. 2. On passive movement of the anterior part of the body towards the right, so that the forelegs are directed towards the left, the right hindleg is suddenly vigorously extended and abducted.

forelegs. In standing on the left foreleg (Fig. 176, No. 1), the right hindleg is flexed when the supporting leg is in adduction posture and shows no, or scarcely any, supporting tonus. Conversely, when the left foreleg is in abduction posture, it is extended, abducted and shows a strong supporting tonus. In standing on the right foreleg the pattern is reversed (Fig. 176, No. 3 and 4).

The alterations in posture and supporting tonus produced in this way continue as long as the position of the animal is unchanged. Thus, the hindleg shows an increase in supporting tonus on abduction of the

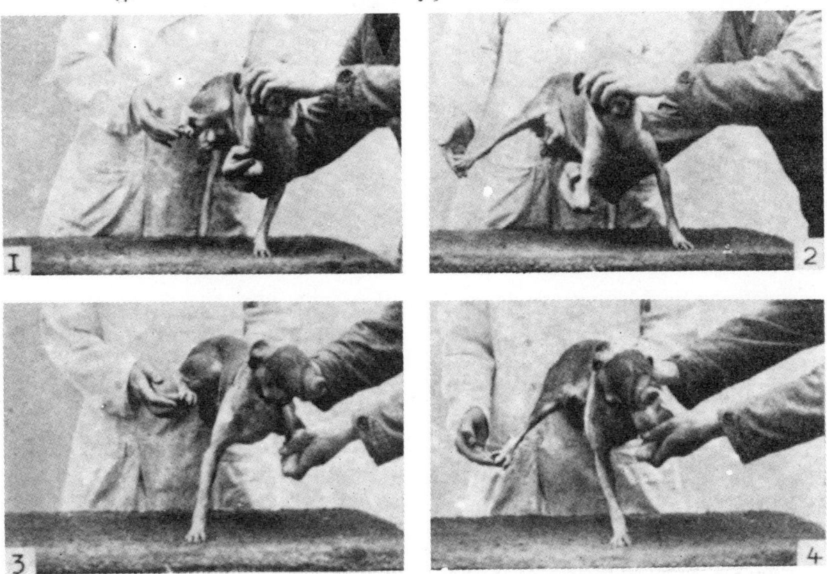

Fig. 176. Dog Jenny, bilaterally labyrinthectomized, left-sided decorticate, standing with only one foreleg on a supporting surface. Left hindleg passively flexed. 1 and 2. The animal stands with the left foreleg on the supporting surface. 1. On movement of the anterior part of the body to the left, flexion of the right hindleg and decrease in supporting tonus as soon as the left foreleg goes into adduction posture. 2. On movement of the anterior part of the body towards the right strong extension with abduction of the right hindleg as soon as the left foreleg reaches a posture of abduction. (Observe also the extension of the elbow joint of the right foreleg.) 3 and 4. Animal with the right foreleg on the supporting surface. 3. Now also decrease in supporting tonus in the right hindleg on movement of the anterior part of the body towards the left: by this the supporting leg is abducted. 4. On movement of the anterior part of the body towards the right, which causes an adduction of the supporting leg, the right hindleg is extended and abducted with strong supporting tonus.

346

opposite foreleg, on adduction of the ipsilateral foreleg and, as we
have seen above, also on abduction of the opposite hindleg. Thus,
when in an animal standing on all four legs and one hindleg is raised, these
three factors, on a sideways movement of the trunk, will have a similarly directed effect and intensify one another. Consequently, one gets
the impression that the supporting tonus of one hindleg increases on
stretch of the abductors of the opposite foreleg and on stretch of the
adductors of the ipsilateral foreleg and on stretch of the abductors
of the contralateral legs. Whether the stretch of these muscles really
and solely produces these reactions is left undecided.

These reactions, too, can play a role in the recovery and maintenance
of balance, for instance, when the animal is pushed to the right while
the right legs are raised or when the supporting surface is lowered on
the left side and also in the sudden dropping of the legs from one side
from the support, etc.

Another effect can be observed when, for instance, a dog is set down
on a supporting surface with the right foreleg and the trunk is moved
to the right so that the supporting leg goes into adduction; the right
hindleg, in this case, at first shows an extension with increase of supporting tonus (Fig. 177, No. 2), which, however, disappears again when
the trunk is moved further to the right so that the supporting leg is
passively adducted very much (Fig. 177, No. 2). Thus, one gets the
impression that a very strong stretch of the abductors has just the
opposite effect to a moderately strong stretch. On a very strong
abduction of the foreleg sometimes, though usually less distinct, a
decrease in supporting tonus of the opposite hindleg can be observed;
thus, here too is an effect contrary to that in moderate abduction.

On alterations in position of the hindlegs, similar reactions appear
in the forelegs; thus the supporting tonus of each foreleg increases on
adduction of the ipsilateral hindleg and an abduction of the contralateral one, while the opposite alterations of posture produce a decrease
in supporting tonus.

In these reactions produced by adduction and abduction of the
supporting leg neither optical nor labyrinthine stimulations play any
role; they still occur distinctly in labyrinthectomized dogs with blindfolded eyes.

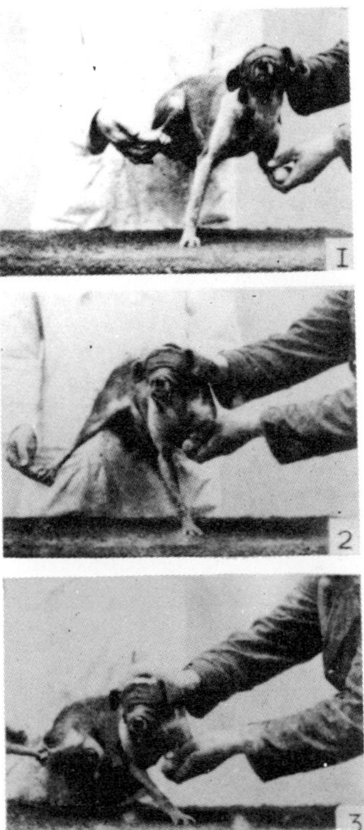

Fig. 177. Labyrinthectomized, left-sided, decorticate dog Jenny with blindfold. The right foreleg stands on a supporting surface. Both left legs passively flexed, the right hindleg put under static stress with one hand. 1. Right foreleg in abduction, right hindleg has no supporting tonus. 2. On moving the trunk to the right the right foreleg is passively abducted, right hindleg is extended with a strong supporting tonus. 3. On a further movement of the trunk to the right there is a decrease in supporting tonus of the right hindleg. (In intact dogs the disappearance of the supporting tonus is always accompanied by an outwards transfer ⌈hŏp⌉ of the right hindleg.)

In investigations on intact dogs, the supporting leg has to be fixed to the supporting surface, otherwise it will be lifted and transferred laterally on the sideways movements of the trunk. Even with the supporting leg fixed, investigations are very difficult since the fixed leg attempts to liberate itself and is alternately flexed and extended. These attempts to free the passively fixed supporting leg are also shown by decorticate and decerebellate dogs. In these animals, however, the supporting leg usually does not need to be fixed for the hop of the supporting leg is delayed (see the following chapter). Therefore, decorticate and decerebellate dogs are particularly suited for these investigations. In decorticate animals, however, only the reactions of the hindlegs to alterations in position of the forelegs appear clearly; the reactions of the forelegs to alterations in position of the supporting hindleg are less pronounced, probably due to the reduced force of supporting tonus in the hindlegs. After the unilateral extirpation of cerebrum, the legs contralateral to extirpation show very lively reactions (Figs. 175, 176, 177). In the legs on the side of extirpation, they are less strong when the opposite leg is the supporting leg. In unilaterally decerebellate animals, a distinct difference between reactions of the right and the left legs cannot be established.

The intensity of the reactions depends on several factors: on the distribution of tonus in the muscles of the supporting leg, on the mode of reaction of the supporting leg to passive abduction and adduction, on the ability to respond and on the initial tonus of the leg taking no weight, on the reaction of the muscles of the leg taking no weight to static stress, on the presence or absence of preparation for standing, etc. This explains the fact that after unilateral extirpations, for instance after the unilateral extirpation of the cerebrum, the reactions can be different every time when each of the four legs is tested when unsupporting.

The reactions are absent after decerebration, also in the first weeks after total or unilateral extirpation of cerebellum and in the first weeks after birth.

When a dog is set down on a supporting surface which is tilted, for instance, on the right side (Fig. 207, No. 3), the left legs are extended and abducted, the right ones flexed and adducted. Thus, the abduction of the left legs does not produce an extension of the right ones; in the adaptation of leg position, therefore, factors must take effect which counteract the influence of the abduction posture on the supporting tonus of the opposite leg.

Thus, in summing up, the following occurs :

1. On a forwards movement of one or both forelegs: increase in supporting tonus in both hindlegs.

2. On a backwards movement of one or both forelegs : decreasein supporting tonus in both hindlegs.

3. On a forward movement of the hindlegs : decrease of supporting tonus in the forelegs.

4. On a backward movement of the hindlegs : increase of supporting tonus in the forelegs.

5. On a moderate adduction of one leg : decrease of supporting tonus in the opposite legs, increase in supporting tonus of the ipsilateral leg, while a strong adduction produces contrary reactions.

6. On moderate abduction of one leg : increase of supporting tonus in the opposite legs, decrease of supporting tonus of the ipsilateral leg, while a strong abduction produces contrary reactions.

Thus, a moderate adduction and a strong abduction produce similar reactions which are contrary to those on strong adduction and moderate abduction.

XIII. SUPPORTING TONUS OF ONE LEG AFFECTED BY THE POSTURE OF THE SAME LEG, LEG-BRACING REACTIONS, SLACK-LEG REACTIONS AND PROPRIOCEPTIVE CORRECTION MOVEMENTS

A. THE LEG-BRACING REACTION.

When a dog is standing on one leg, the strength of its supporting tonus is dependent on its posture. When, for instance, the aniaml stands on a strongly abducted foreleg (adductors strongly stretched),the leg shows only a weak supporting tonus and the elbow joint gives way completely on the slightest pressure on the shoulder. When, by a passive movement of the trunk, the posture of abduction is brought into an intermediate posture in which the adductors are relaxed and the abductors stretched, the leg will resist this movement by bracing or propping sideways (Fig. 178, No. 2) and the supporting tonus shows a distinct increase (Fig. 178, No. 3).

Also, in a strongly forward directed movement (protractors of shoulder joint stretched, retractors of shoulder joint relaxed) the supporting tonus is only slight (Fig. 179, No. 1). On a forward movement of the trunk in which the leg is again brought into an intermediate

350

position, a considerable increase of supporting tonus and an extension take place; in addition, the leg is braced forwards (Fig. 179, No. 2).

Similarly, with a foreleg directed backwards, a caudal movement of the trunk causes an increase in supporting tonus and a backward bracing of this leg. The hindlegs under static stress show similar reactions when they are brought *passively into the intermediate posture* from a posture of abduction or from a forward or backward directed position.

Leg-bracing reactions are purely reflex reactions and as such are present not only in intact and decerebellate dogs, but also in decorticate ones. However, they are totally absent in decerebrate and spinal animals. No details are known on the factors producing these reactions. In the leg-bracing reactions forwards and backwards, extension or relaxation of the flexors and extensors of the shoulder and hip joints, as well as extension of the muscles of distal joints (the elbow, wrist, knee, tarsal, metacarpophalangeal and metatarsophalangeal joints),

Fig. 178. Decerebellate dog Piccolino with blindfold, the posterior part of the body is supported by one hand, the right foreleg on a balance. 1. right foreleg in a strong abduction position. The animal lies on the supporting hand almost entirely and supports itself only a little on the supporting foreleg. Although the counterpressure of the supporting surface is only 1 kg, the foreleg is flexed at the elbow joint. The supporting tonus is feeble. 2. On passive movement of the trunk to the right, the right foreleg is propped outwards so that the balance is almost knocked over. At the same time, the elbow joint is extended. 3. The trunk is passively moved so far to the right that the right foreleg now stands in an intermediate position. In this movement the animal supports itself more on this leg so that, finally, it exerts a pressure of more than 3 kg on the balance. In spite of the increase of the counterpressure by 2 kg , the leg is not yet totally extended. Strong supporting tonus is now present.

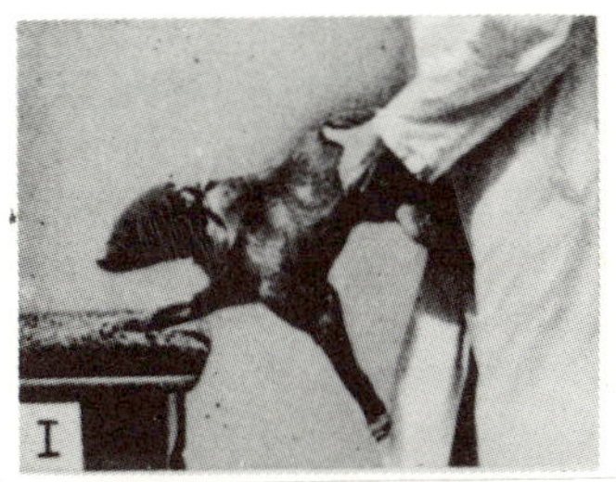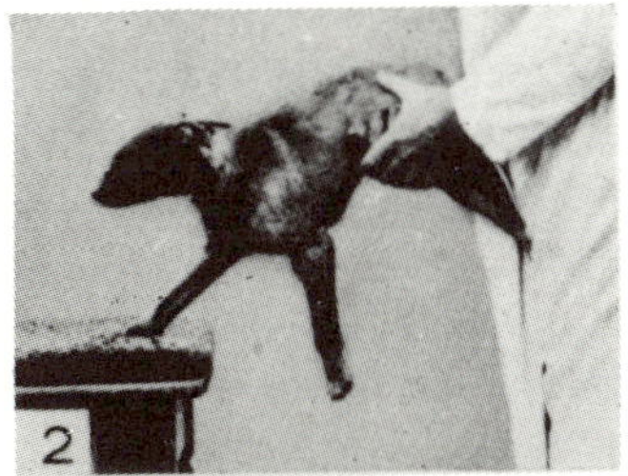

Fig. 179.   Decerebellate, left-sided decorticate dog Daümling. 1. Animal stands with forward directed right foreleg on a supporting surface.   Feeble supporting tonus of the leg. Elbow joint bent under the weight of anterior part of body. 2. On movement of the trunk forwards, the leg braces forwards and is extended.

probably play a certain role.   The laterally directed leg-bracing reactions are possibly produced by an extension of abductors and a relaxation of adductors of the legs under static stress, but thorough investigations by Sherrington's recording methods are still lacking.

The leg-bracing reactions can be distinctly observed when the dog stands on all four legs. On an attempt to displace such an animal, for instance to the left, the left legs are braced sideways, i.e., towards the left, and there appears a concavity of the spinal column towards the left (Fig. 180).

By exerting pressure on the right side of the thorax, the left vertebral muscles are stretched. One could now imagine that due to this stretch-

Fig. 180.   Normal dog. 1. The animal is pushed to the left by a pressure on the right side of the chest; left legs are braced outwards, spinal column is curved concave to the left. 2. With a strong pressure on the right side of the chest so that the concavity is passively counteracted (or even a convexity to the right appears) the left legs are still braced outwards. 3. On a pressure to the right, the *right* legs are braced outwards.

ing, a myotatic reflex of these muscles would be produced which would cause the concavity of the spinal column as well as secondarily the leg-bracing reaction. It became evident, however, that leg bracing reactions do not appear as a result of the curvature of the spinal column. When by a stronger pressure the concavity is counteracted or even turned into a convexity, distinct leg-bracing reactions still occur. The concavity of the spinal column is secondary to the leg-bracing reactions, although, of course, the participation of a myotatic reflex of the vertebral muscles cannot be excluded. As we have seen (P. 291)a passive abduction posture of the legs on one side produces a concavity of the spinal column to the same side; an active abduction probably causes a similar alteration.

The lateral leg-bracing reactions of the left legs occur when the animals are pushed to the left (Fig. 180, No. 1 and 2, and Fig. 181, No. 1) as well as when they are pulled to the left (Fig. 181, No. 3). Thus a pulling to the left causes leg-bracing reactions of the left legs towards the left, pulling to the right, leg-bracing reactions of the right legs to the right, while in pulling forwards a leg-bracing reaction forwards, in pulling backwards a leg-bracing reaction backwards can be

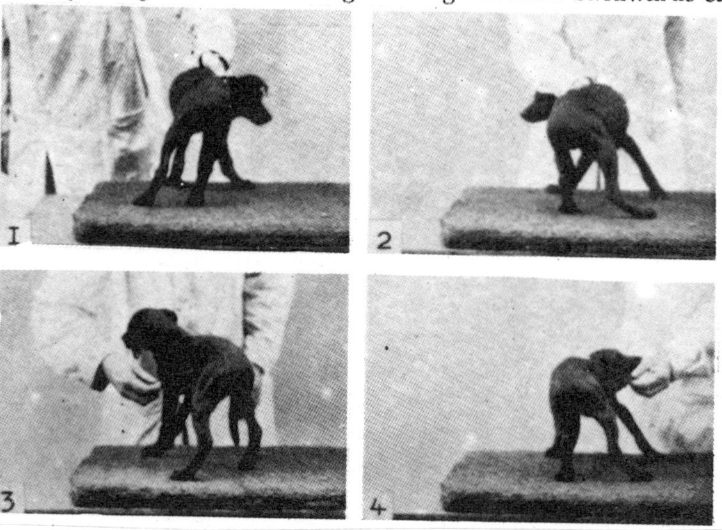

Fig. 181.   The same dog as in Fig. 182. 1. The animal is seized by the right side of the chest and pulled towards the left, the left legs are braced sideways. 2. The animal is pulled towards the left by a fold of skin of the left side of the chest; now, too, the left legs are braced laterally. 3 and 4. On pulling to the right, the right legs are also braced laterally.

observed in intact, decerebellate and decorticate dogs. For instance, when we pull a decorticate (thalamus) dog backwards by a strap around its chest, the four legs are braced backwards (Fig. 182, No. 1), while on pulling forward they are propped forwards (Fig. 182, No. 2); they react in a similar way when they are pulled forwards or backwards by the collar (Fig. 182, No. 3 and 4). Thus, the direction of the pulling and

Fig. 182.   1 and 2. Decorticate (thalamus) dog Fuchs. 1. The animal is pulled backwards by a sling attached to its chest. By bracing the four legs backwards, the animal resists the caudal traction. 2. Animal is pulled forwards by the sling. It now braces the four legs forwards, especially the forelegs and in so doing pulls itself actively caudally. 3 and 4. Thalamus dog Miesel. 3. Animal is passively pulled caudally by the collar. The four legs are braced backwards and propel the animal actively forwards.   The animal is passively pulled forwards by the collar, where upon the legs are braced forwards and the animal actively exerts a caudal traction. (With the caudal as well as with the forward traction passive alterations in position are caused in the shoulder and hip joints as well as in the elbow, wrist, knee, tarsal, metacarpophalangeal and metatarsophalangeal joints. The muscles of these joints are therefore also passively stretched or relaxed. The forward directed traction causes a passive stretch of the protractors of the upper arm and thigh, of the extensors of elbow and tarsal joints, the flexors of wrist and knee joints and of the volar and plantar flexors of the toes.)

not the point of application plays the principal role; the stronger the pulling, the more vigorous the bracing of the legs.

On a forward pull, the forelegs especially are braced forwards. Here, the reactions of the hindlegs differ according to the position of neck and spinal column. When the animal is pulled by the collar forwards and *upwards*, the hindlegs are flexed; when forwards and *downwards*, they are extended and braced forwards. In traction forwards and upwards, head and neck are raised and the back is hollow; in a traction forwards and downwards head and neck are ventrally flexed and the back is curved convexly.

The kind of supporting surface (slipping), too, plays a role. In a displacement to the left, the left legs brace vigorously outwards; the right ones sometimes also brace to the left, but are usually flexed and become more or less flaccid. The reactions of the right legs, in this case, depend on the speed and force of traction, on the kind of supporting surface, and on the initial posture of the legs.

In intact dogs, the leg-bracing reactions do not occur constantly on pulling forwards, backwards or sideways, i.e., they are dependent, firstly, on influences of the cerebrum and conditioned reflexes, secondly, on the initial posture of the legs and on the point of application of traction. When, for instance, a dog is pulled forwards by the collar by its master, no leg-bracing reactions will usually occur, but they will almost always appear when a stranger tries to pull the animal forwards. They also usually occur when the animal is pulled, for instance, towards a lake or towards a fire or when the attempt is made to pull it away from its feeding bowl, but are absent when the animal is drawn towards it. Thus, the leg-bracing reactions can be inhibited by optical or other impulses emanating from the cerebrum and conditioned by other reactions.

In decerebellate dogs[1] in the stage of permanent impairment leg-bracing reactions usually appear regularly, although they are also influenced by the initial position of the legs, the point of application of the traction and by cerebral reactions. One gets the impression here that the cerebral reactions have a less inhibiting effect. Leg-bracing

---

(1) In the period immediately following extirpation of the cerebellum, the leg-bracing reactions are absent.

reactions are conspicuously strong and abundant in decerebellate dogs, particularly on pulling backwards or sideways (Fig. 183), probably due to their broad-based standing on exaggeratedly extended legs and the extraordinary tension of the muscles of the extremities in standing posture.

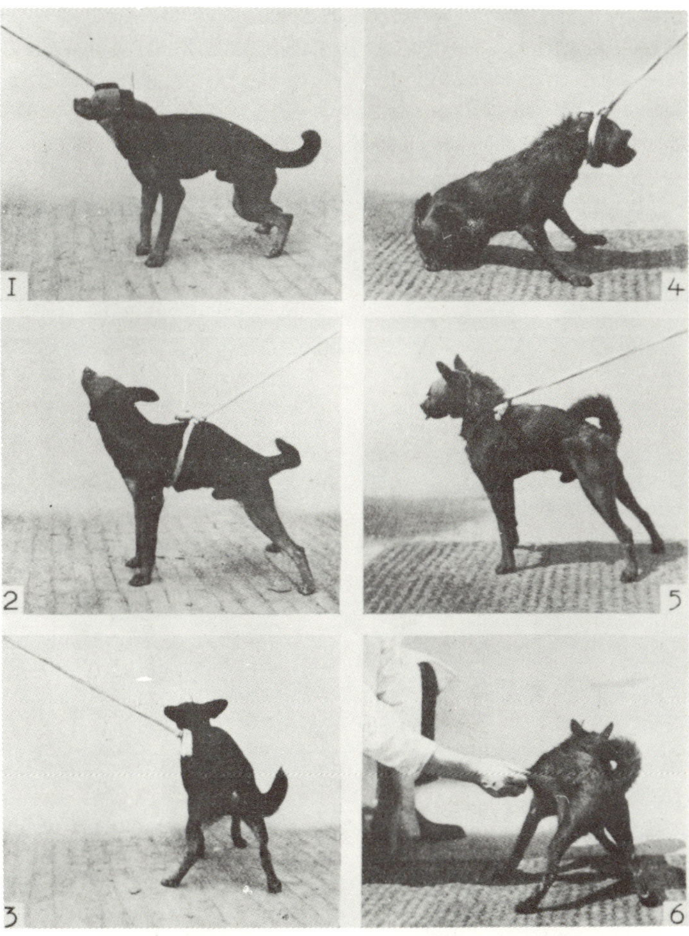

Fig. 183. Leg-bracing reactions in blindfolded decerebellate dogs. 1-3. Decerebellate dog Moor pulled forwards, backwards and to the left by a strap fixed around its thorax. On forward traction the animal distinctly braces its forelegs forwards, on backward pulling the hindlegs are braced backwards, on pulling to the left the left legs are vigorously propped to the left and outwards. 4 and 5. Decerebellate dog Erik pulled forwards and backwards by the collar. Here, too, distinct and vigorous leg-bracing reactions forwards and backwards. 6. In pulling by a fold of the skin to the left, the left legs are vigorously braced laterally.

Leg-bracing reactions elicited by traction backwards in decerebellate dogs are usually stronger and more constant than leg-bracing
reactions to forward movements, which is probably due to the fact
that decerebellate animals, standing freely, keep their four legs directed
backwards (see Fig. 44). Only in certain initial postures of the legs and
in gradual traction do leg-bracing reactions to forward displacement
appear strong and constant (Fig. 183, No. 1 and 4).

In unilaterally decerebellate animals, the leg-bracing reactions are present

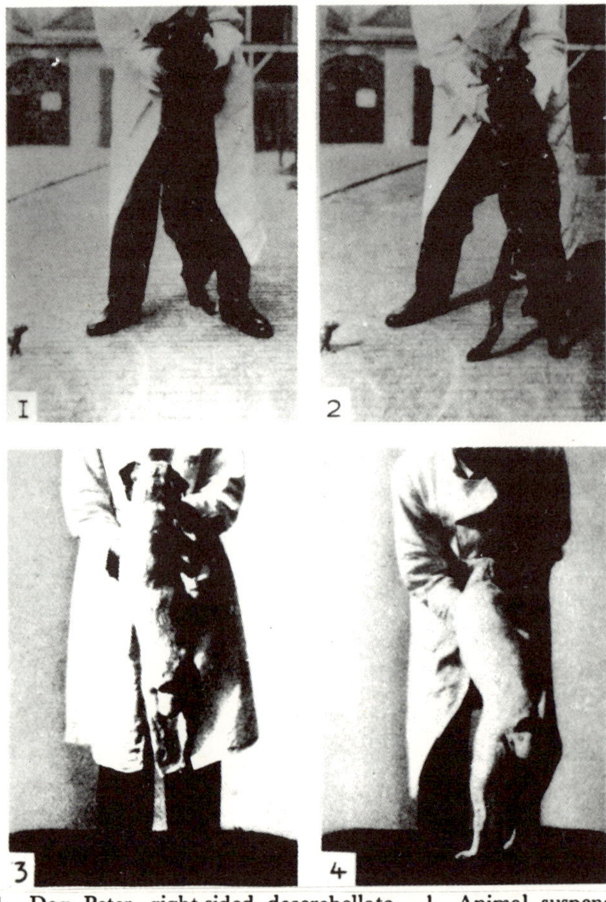

Fig. 184. Dog Peter, right-sided decerebellate. 1. Animal suspended head
upwards, kept in the air by the forearms. Dorsal side of the pelvis
rotated towards the left. 2. Set down on the hindlegs, the animal
transfers the *left* hindleg outwards and backwards. By this, the
dorsal side of the pelvis is turned towards the right; moreover, a
concavity of the lumbar spinal column towards the left appears.
3 and 4. Dog Fox, also right-sided decerebellate. Corresponding
reactions under similar conditions.

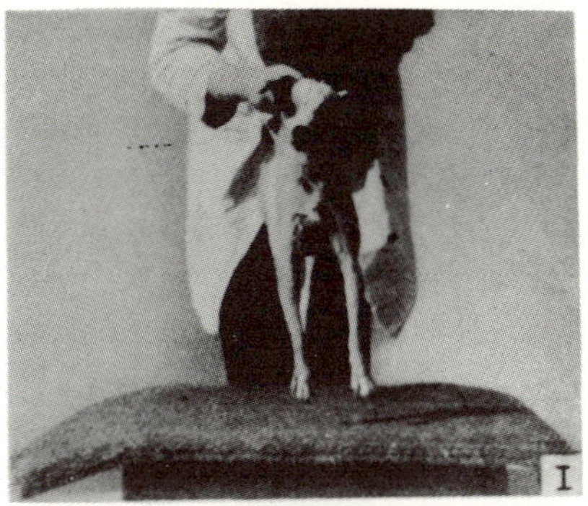

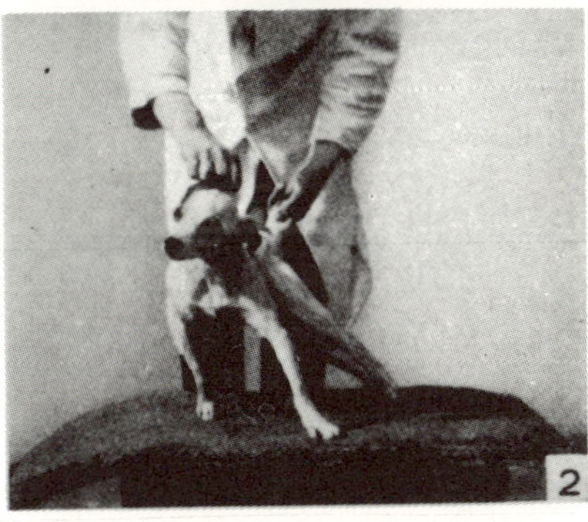

Fig. 185.   Dog Fox, 1 month after extirpation of the right half of the cerebellum. 1. Animal is suspended in the air by the tail and the skin of the
neck. 2. On moving downwards, the left foreleg is braced vigorously
forwards and outwards, the left hindleg outwards and backwards
as soon as the soles touch the supporting surface. By the reactions of
the left legs, the trunk, when held in the air by tail and skin of the
neck, is pressed towards the right and the spinal column shows a concavity towards the left.

on both sides;[1] however, they appear stronger and more evident on
the side with the intact half of the cerebellum than on the side of
extirpation.   The contralateral legs show an increased tendency to
abduction and lateral bracing already on touching a supporting surface
(Figs. 184, 185).

The difference becomes particularly evident when the animals,
standing only on their hindlegs, are alternatively pulled to the right
and to the left. When the animals are lowered from a position suspended
head upwards, the leg contralateral to the extirpation braces laterally

---

(1) Directly after unilateral extirpation, leg-bracing reactions are sometimes absent
on both sides; they are always absent in the first four weeks on the side of extirpation.

as soon as the animal supports itself on this leg (Fig. 184). When teh animal is pulled towards the side of the intact half of the cerebellum, the leg braces even more vigorously and a passive movement encounters strong resistance. On further traction leg-bracing reactions and resistance finally cease and the leg makes a limping step sideways, to brace outwards again instantly. On drawing towards the other side, the resistance is less strong and on more vigorous traction the hindlegs are alternatively transferred so that the posterior part of the body hops sideways. On traction towards the side of extirpation, the animal steps laterally. In unilaterally decerebellate animals the fore- and hindlegs contralateral to the extirpation show a pronounced tendency to abduction and bracing laterally (Fig. 185) even early after extirpation when the leg-bracing reactions are still absent in the legs on the side of extirpation.

The leg-bracing reaction is strongest in the contralateral hindleg. The increased tendency to abduction of the contralateral hindleg can also be distinctly observed when the animals stand on the supporting surface with this leg only. When, for instance, the dog Fox whose *right* half of the cerebellum was removed a month earlier, is held in the air by the head and the right hindleg and set down on a supporting surface on the left hindleg (Fig. 186, No. 1) this leg will remain in a vertical posture as long as the dorsal pelvis is directed to the left due to the lifting of the right thigh. When, however, the pelvis is brought into a symmetrical position to the rest of the trunk due to a lowering of the right thigh, the hindleg makes one or more steps sideways and braces outwards (Fig. 186). Due to the lateral bracing of the hindl_g, the pelvis is pushed more and more towards the right so that the posterior part of the body finally falls into a right lateral position (Fig. 187).

Later, when the animals are able to stand and to run, there is also an increased tendency to a bracing laterally of the contralateral legs which, among others, is clearly perceptible in running. When, for instance, a right-sided decerebellate dog comes running on call, one can clearly see that the longitudinal axis of the trunk does not remain in the direction of movement. Every time the left leg is set down and simultaneously the right one is raised the left hindleg displaces the pelvis towards the right; consequently, the right hindleg on extension is set down on a spot situated more towards the right and the posterior part of the body deviates more and more in this direction. The same is true

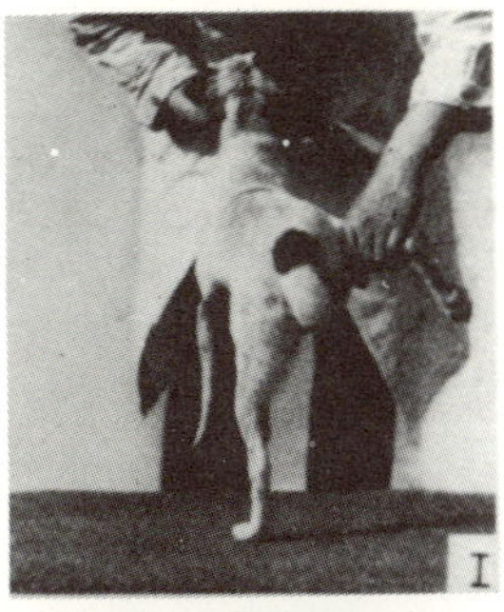

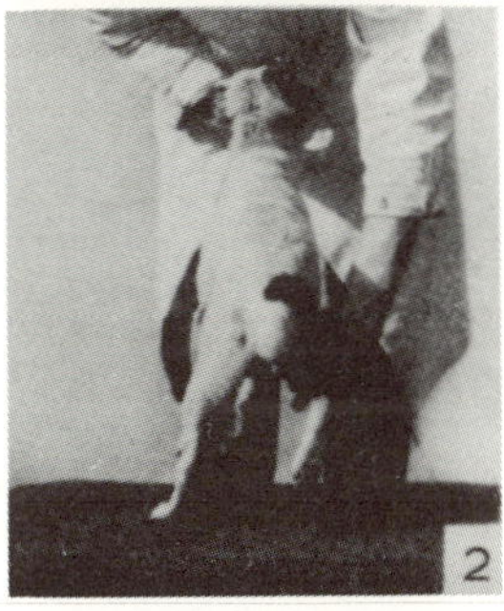

Fig. 186. Dog Fox, 1 month after extirpation of right side of the cerebellum, set down on a supporting base with the left leg only. The right half of the pelvis is strongly raised by a hand which simultaneously clasps the right thigh, so that the dorsal side of the pelvis is directed towards the left. As long as the pelvis retains this posture, the left hindleg stands in vertical position. 2. Right half of the pelvis is passively moved downwards so that the pelvis is placed symmetrical to the trunk. The left hindleg makes a step outwards to the right and braces towards this side.

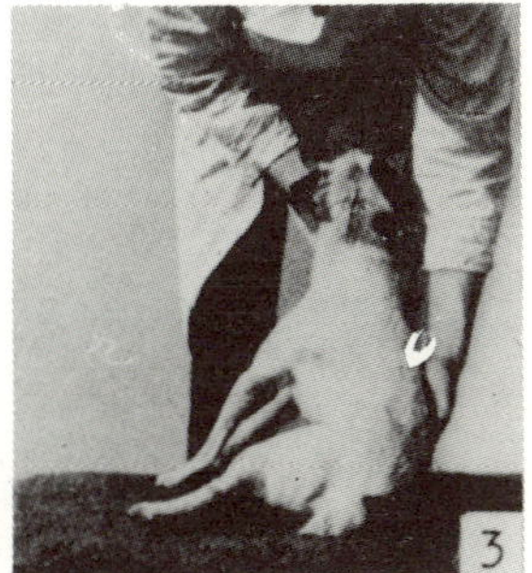

Fig. 187. Dog Fox, 1 month after extirpation of the right half of the cerebellum. 1. As soon as the animal is set down on the hindlegs, the left hindleg is braced outwards. By this the pelvis is moved towards the right and pressed against the hand which supports the posterior part of the body from the right. 2 and 3. On withdrawing the supporting hand, the pelvis moves more and more to the right till the posterior part of the body falls to the right.

for the anterior part of the body but not in such a pronounced way, so
that the right half of the body, although the animal keeps its head in
the original direction, is finally turned towards the caller (Fig. 188)

Fig. 188. Dog Fox, right-sided decerebellate. 1. Animal is held in such a way
that the longitudinal axis of the trunk is directed towards the camera.
2. The animal is called by a person standing behind the photographer
and runs towards him. In so doing, the trunk is more and more
deviated to the right due to the strong lateral bracing of the left legs.
By this the longitudinal axis of the trunk makes an increasing angle
with the direction of the movement so that finally the posterior part of
the body approaches with lateral steps towards the right.

Fig. 189. Deviating to the right while walking (with closed eyes) of a patient
with right-sided cerebellar disease. 1 and 2. The patient supports himself on the left leg. 3. The patient supports himself on the right leg.
On comparing the angle between the outer side of the trunk and the
outer side of the legs, one can see that the right leg, when supporting,
is kept more abducted (Fig. 3) than the left when the latter is the
supporting leg. From K. Goldstein: Das Kleinhirn . *Handbuch
der Normalen und Pathologischen Physiologie.* Chapter
X, p. 271. Verlag Julius Springer, Berlin, 1927.

and the posterior part of the body approaches with steps that are partly directed laterally to the right.

These observations are remarkably demonstrated in motion pictures of a patient with right-sided cerebellar disease (Fig. 189), published by Goldstein.

As we have seen, *in decerebellate dogs*, the direction of the leg-bracing reaction depends on the direction of traction and not on the plane of application (thus, for instance, pulling forwards produces a forward bracing of the legs, pulling backwards a backward bracing, etc.). This, however, is only true within certain limits because, by pulling the tail, the ear or the muzzle, the reactions can behave quite differently. When a decorticate dog is grasped by the tail, the animal runs forwards and tries to liberate itself by a rearward bracing of the legs (Fig. 190, No. 3 and 4). When the animal is pulled sideways by the tail, the legs brace outwards and backwards. Conversely, it was never possible to produce a forward bracing of the legs by a forward pulling of the tail. When grasped by the ears the animals always make attempts to escape by moving the head ventrally and bracing the legs forwards (Fig. 190,

Fig. 190. Decorticate dog Fuchs. 1. Seizing of the muzzle causes a raising of the head and a forward bracing of the four legs. 2. On being seized by the ear, even without pulling, the animal braces its legs forwards and moves its head ventrally. 3 and 4. When seized by the tail, the animal tries to liberate itself by a rearward bracing of the legs and by running forwards.

362

No. 2). Similarly, grasping of the muzzle always causes a forward propping of the legs and an attempt by the animal to run backwards. In the attempt to draw the animal backwards by the ears or to displace it backwards by the muzzle, leg-bracing reactions could never be produced in decorticate animals.

In decorticate dogs, leg-bracing reactions forwards, backwards and laterally are present in fore- and hindlegs. In the hindlegs, whose strength of supporting tonus is reduced by the extirpation, the force of leg-bracing reactions is usually distinctly diminished, particularly in a later bracing.

After the unilateral extirpation of the cerebrum, the legs contralateral to extirpation sometimes show particularly lively and ample leg-bracing reactions on lateral traction, even if the strength of supporting tonus of these legs is distinctly reduced; they are sharply and widely abducted (Fig. 191). On drawing the animals towards the side of extirpation, one sometimes encounters a greater resistance than when they are

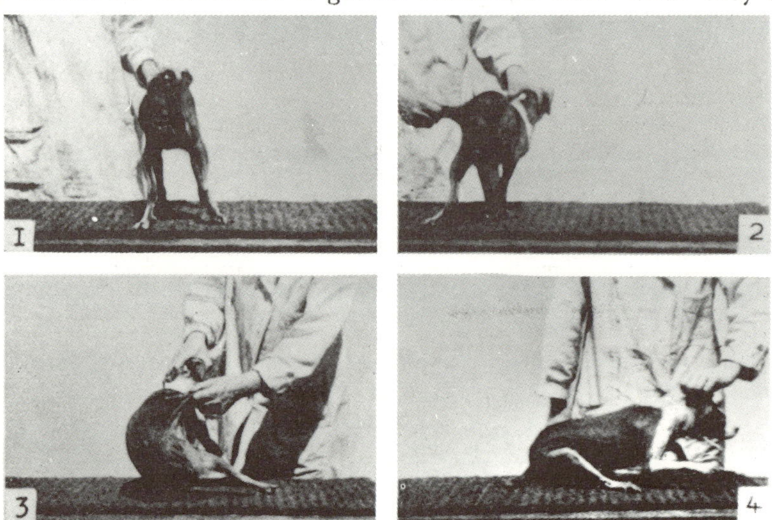

Fig. 191. Leg-bracing reactions in the dog Jenny, whose two labyrinths and the *left* half of the cerebrum were removed. 1. The animal seen from behind with the head held symmetrically to the trunk. 2. When the animal is pulled to the left by a fold of the skin, the left legs show distinct leg-bracing reactions laterally. 3. A pulling to the right, on the contrary, causes exaggerated strong leg-bracing reactions of the *right* legs which are abducted particularly strongly and widely, while the left ones give way so that the animal falls down on its left side. 4. On a forward traction, the right legs are also braced much further forwards than the left ones.

drawn towards the other side. The legs contralateral to extirpation slip in the course of the sudden lateral bracing, while in a lateral bracing of the ipsilateral legs the soles retain their position on the supporting surface and abduction only produces an alteration in posture of the trunk. Slipping produces a particularly strong abduction posture of the laterally braced contralateral legs; it is readily understandable that this posture greatly impedes traction to the side. The reason, however, why the lateral bracing reactions of the hindlegs usually appear so slight in decorticate dogs and, on the other hand, are sometimes particularly lively and vigorous in the contralateral hindlegs of unilaterally decorticate animals is not yet settled.

THE IMPORTANCE OF THE LEG-BRACING REACTIONS FOR THE ADAPTATION OF LEG POSTURE TO THE POSITION OF THE SUPPORTING SURFACE

In intact, decerebellate, decorticate and labyrinthectomized animals, leg-bracing reactions play a large role in the adaptation of leg posture to the position of the supporting surface, even when the eyes are blindfolded. For instance, when a dog stands on all four legs and the tailend of the supporting surface is lifted, the force of gravity will draw the animal forwards and corresponding leg-bracing reactions forwards will occur (Fig. 192, No. 1 and 2).

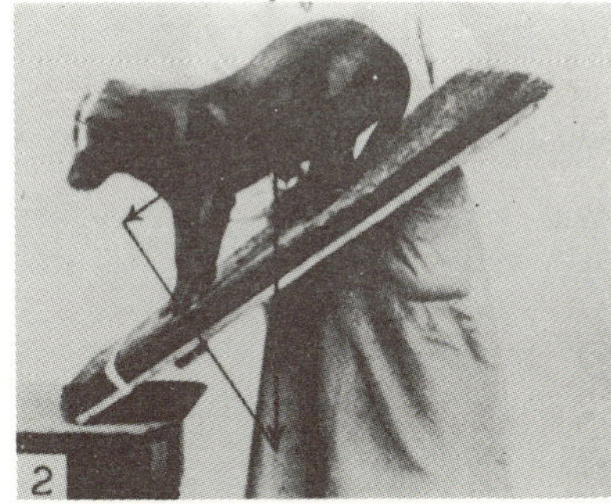

Fig. 192. Intact dog, blindfolded, on a board which is raised by the tailend. 1. As soon as the tailend of the board is higher than the headend, the force of gravity will pull the animal forwards; simultaneously the four legs are propped forwards. 2. On raising the tailend further, the forwards pulling force increases and, similarly, the legs are braced forwards even more (compare posture of the limbs of the above illustrations with those of Fig. 182, No. 3 and 4).

The leg-bracing reactions will fail to appear when, on lifting the tailend, the animal is pulled backwards by a strap around its chest and by this traction the force of gravity is compensated. When the traction backwards is stronger than that of the force of gravity, leg-bracing reactions backwards will appear even on the board lifted by the tailend (Fig. 193, No. 1 and 2).

On lowering the tailend or lifting the headend of a supporting board, the force of gravity will draw the animal backwards, producing a back-

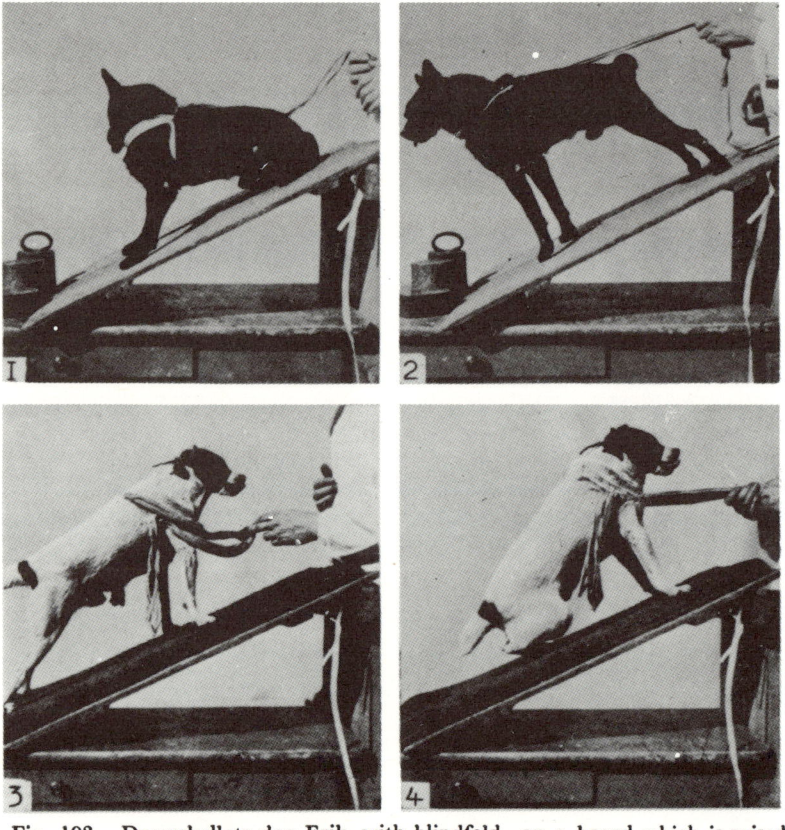

Fig. 193. Decerebellate dog Erik, with blindfold, on a board which is raised at the tailend.   Typical adjusting reactions: the animal's forelegs are propped forwards while the hindlegs go into flexion.   2. With unchanged inclination of the board, the animal is pulled backwards by a *strap* around its thorax. In this case, the legs are propped backwards instead of forwards and the animal stands as if on a board raised from the headend. 3. Unilaterally decerebellate dog Fox, with blindfold, on a board raised by the headend.   The animal props all four legs backwards. 4. On a forward pulling of the animal, the legs are braced forwards and the animal stands as if on a board raised by the tail-end.

ward bracing of the legs. In the same way, upon lowering the supporting surface on the right side, or lifting it on the left side, the force of gravity will cause a traction to the right and by this a bracing of the right legs towards this side (Fig. 194).

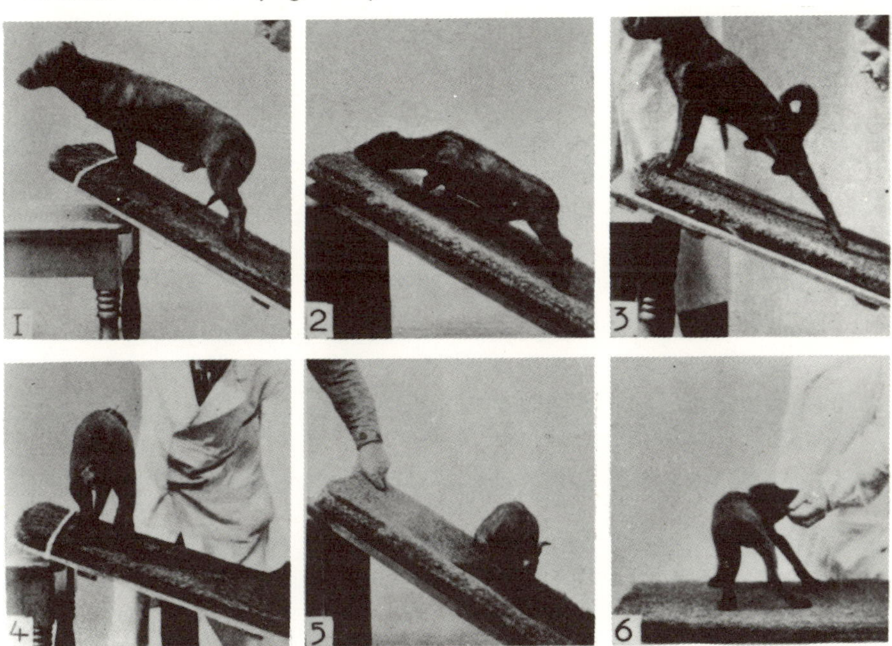

Fig. 194.   1-3. Reactions on lowering the tailend of the supporting base. 1. Normal dog with blindfold. 2. Labyrinthectomized dog with blindfold. 3. Decerebellate dog. All three animals are propping their legs backwards like a dog pulled backwards (compare with Fig. 182, No. 1 and 2). 4 and 5. Reactions of an intact (4) and a labyrinthectomized dog (5) both with blindfold, on a board lowered from the right side. 6. Intact dog in standing posture, pulled to the right by a fold in the skin. The animals in Figs. 4, 5 and 6 show a similar limb posture.

Just as unilaterally decorticate dogs usually show very strong laterally directed leg-bracing reactions in the contralateral legs, in the ipsilateral, however, only weak ones appear. These animals show strong leg-bracing reactions only when the supporting surface is lifted *on the side of extirpation*, but on lifting the other side, the reactions fail to appear almost entirely (Fig. 195).

In intact, decorticate and decerebellate as well as in labyrinthectomized dogs, leg-bracing reactions also participate in the adaptation of leg posture in standing on a supporting surface which can be shifted in a horizontal direction. When, for instance, the base is suddenly shifted forwards, the legs, too, are drawn forwards and the trunk lags behind a little. The movement of the supporting surface thus causes passive alterations in position of the hip and shoulder joints similar to

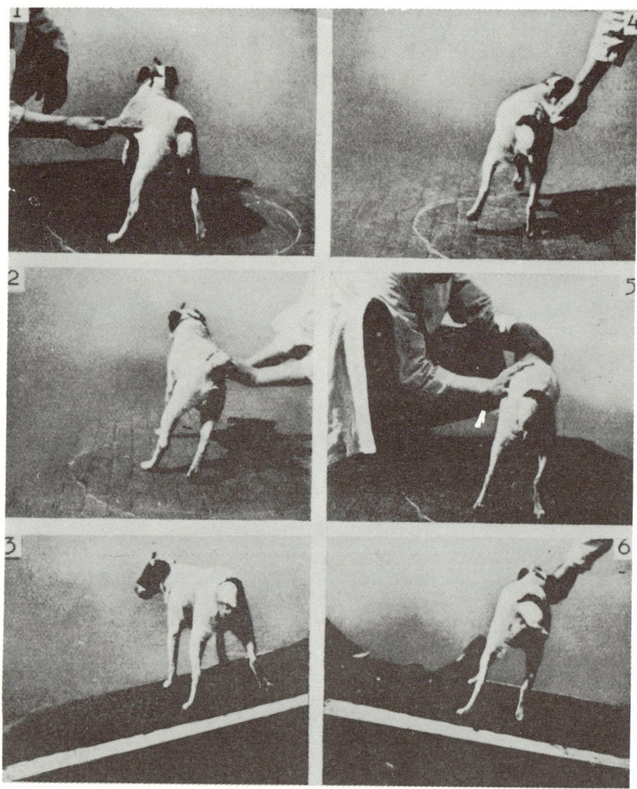

Fig. 195.   Dog Fox, 15 months after extirpation of the *right* half of the cerebellum. 1-3. On pulling (1) as well as on pushing (2) to the left, and also on raising the supporting surface on the right side (3) the left legs are strongly braced outwards. 4-6. The right legs show neither on pulling (4), nor on pushing (5) to the right side, nor on lifting the supporting surface on the left side (6) any distinct lateral leg-bracing reactions. On raising the supporting surface on the side of the intact half of the cerebellum (6) there is not sufficient adaptation of leg posture, so that the animal, when not supported, would fall to the right side. (With open eyes the adaptation is not always as bad as shown in Fig. 6.)

those caused by a passive backward traction of the trunk and is similarly followed by a backward bracing of the legs. Conversely, a shifting of the supporting surface backwards causes a forward bracing of the legs, while shifting of the surface to the right brings about lateral bracing of the left legs.

On cessation of the movement, contrary reactions occur. On cessation of the forward movement, for instance, the trunk moves forwards due to inertia, the legs go passively backwards to the trunk (i.e., the same alteration of position as to a forward traction), causing a reflex bracing forward of the legs. Finally, the legs return into an intermediate posture.

Standing on a turntable also causes leg-bracing reactions to appear. For instance, when the animal is set down on a radius of the turntable with the head towards the center and the disk is rotated in a clockwise direction, the legs of the animal will be drawn passively to the left causing a sideways bracing of the right legs. In addition, the centrifugal force will pull the animal backwards so that the legs are braced backwards. When the animal stands with the head towards the outer rim of the disk, the left legs, in the same direction of rotation, will brace laterally and all four legs will brace forwards. When the animal stands with the right side towards the outer rim of the disk, the centrifugal force will produce an outwards bracing of the right legs and the horizontal shifting of the supporting surface a forwards propping of all four legs. These reactions appropriate for the maintenance of balance can be distinctly observed when labyrinthectomized dogs are placed on a turntable. In intact, decorticate and decerebellate dogs, they are less distinct, since in these animals the labyrinthine reactions on rotation may complicate the picture and mask the leg-bracing reactions.

Leg-bracing reactions occur only when the supporting surface is not moved too quickly or the turntable rotated too rapidly, otherwise leg-slackening reactions are produced.

B. THE LEG-SLACKENING REACTIONS

When a leg is passively moved more and more from its intermediate position the leg-slackening reaction occurs, i.e., the supporting tonus abates, and the leg is lifted and makes a step.

Thus, the passive moving of a leg under static stress towards the intermediate position produces leg-bracing reactions and away from the intermediate position leg-slackening reactions.

For instance, when the foreleg of a dog with passive fixation of the sole of the foot on the surface is brought from an intermediate position into an adduction posture by a lateral movement of the trunk, the leg gives way under the weight of the anterior part of the body (Fig. 196). Thus, the supporting tonus of the foreleg as well as of the hindleg is less in a strong adduction position than in intermediate posture.

The collapse of a leg under static stress on strong adduction of this leg represents a component of a complex reaction which occurs when the sole of the foot is not attached to the supporting surface. In this

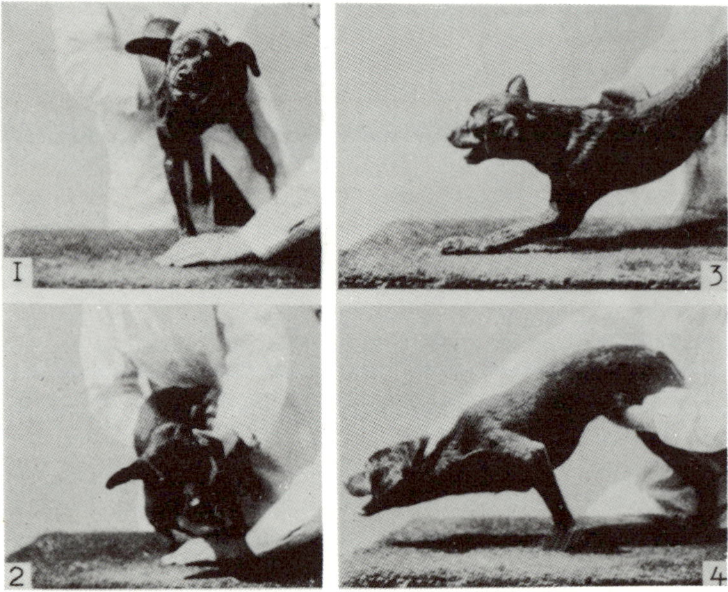

Fig. 196. Alterations in supporting tonus of one leg on passive movement of this leg inwards, forwards and backwards. 1. Decerebellate dog Moor set down on a supporting surface on the right foreleg. The leg, almost in an intermediate position, shows a strong supporting tonus and does not give way to a pressure on the shoulder. 2. On passive movement of the trunk to the right by which the leg is more and more adducted, with decrease in supporting tonus, the leg gives way under the weight of the anterior part of the body. 3. The supporting tonus also decreases when the trunk is moved backwards so that the leg is strongly moved forwards at the shoulder. 4. Similarly the elbow joint gives way more and more on a movement of the trunk forwards and finally the supporting tonus completely disappears.

case, one will see that the leg, simultaneously with the lessening of the supporting tonus, is lifted from the surface, abducted, extended and set down again laterally. Thus, on adduction from an intermediate position in which the adductors are strongly stretched, the abductors relaxed, an abatement of supporting tonus and a transfer outwards of the leg take place (outwards directed hopping reaction; Fig. 197).

These reactions appear when the leg under static stress is led into an adduction posture by passive displacement with one hand (Fig. 197, C) or by passive movement of the trunk (Fig. 197, B) or by movement of the supporting surface (Fig. 197, D).

When the trunk of a dog standing on one leg is progressively moved forwards instead of sideways, so that the leg is directed more to the rear in the shoulder or hip joint and the protractors (forwards movers) of the shoulder joint, or the flexors (forwards movers) of the hip joint are increasingly stretched, corresponding reactionsoccur; first the supporting tonus decreases, then the leg is lifted and transferred forwards (Fig. 198).

Also, when the supporting leg, instead of being moved medially or to the rear, is moved laterally or forwards, there occurs first a decrease in supporting tonus in the foreleg as well as in the hindleg, which is followed by a slackening and hop, in this case inwards and backwards.

Thus, the supporting tonus of a leg appears strongest in the intermediate posture of the leg. Each strong alteration of position in shoulder or hip joint in the sense of a removal from the intermediate position, be it outwards, inwards, forwards or backwards, causes a distinct decrease in supporting tonus. In a slight adduction from an abduction posture an increase in supporting tonus is evident (leg-bracing reaction), in a wide adduction movement beyond the intermediate position, a decrease in supporting tonus and a lifting occurs (leg-slackening reaction); In the first movement, the abductors are extended only slightly, in the second, passively extended with vigor. Here, too, it seems that the strong extension produces opposite reactions to those produced by moderate extension.

We deduce from the observations just discussed that the strength of supporting tonus depends strongly on the position of the legs in relation to the trunk, and in the comparison of strength of supporting tonus one has to take great care that both legs occupy the same position in relation to the trunk, or false interferences will be unavoidable. When,

Fig. 197. Laterally directed leg-slackening reactions of the foreleg in the decerebellate dogs Piccolino and Erik. A. Decerebellate dog Piccolino. Animal stands only on the right foreleg which is held to the supporting surface with one hand. On passive movement of the trunk towards the right the supporting tonus decreases and the leg gives way (A 5-8). B. When the leg is not fixed, the passive movement of the trunk causes a raising abducting of the leg (B 5), followed by an extension and setting down of the leg on a spot situated more to the right (B 6-7). C. In the decerebellate dog Erik which stands with bot hforelegs on the supporting surface, the right foreleg is more and more adducted by the hand. As soon as the leg reached a stronger adduction posture (C 2), the leg is raised (C 3), strongly abducted (C 4), and set down more to the right (C 5-6). D. Decerebellate dog Piccolino, with only the left hindleg on a turntable. The anterior part of the body and the right hindleg are pas-

Fig.197. (cont.) sively raised. By the rotation of the table, the left hindleg is led into an increasing posture of adduction (D 1-4). Now, too, the leg is raised (D 5) as soon as it is strongly adducted, then vigorously abducted (D 6), extended and set down laterally (D 7).

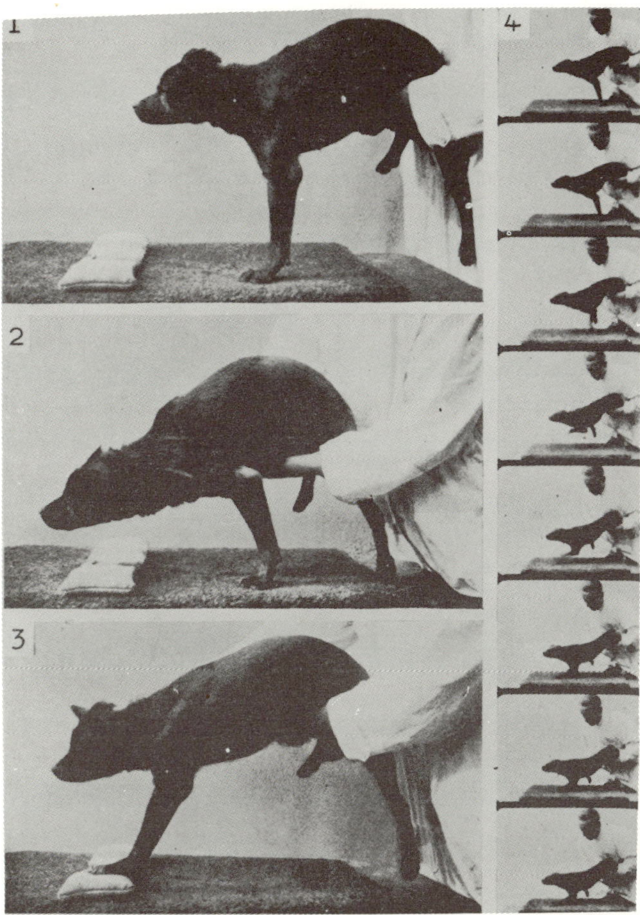

Fig. 198. Leg-slackening reaction forwards of the foreleg in the decerebellate dog Moor. 1. Animal with only the right foreleg on a supporting surface. The leg is almost in an intermediate position and shows a strong supporting tonus. 2. On a passive movement of the trunk forwards (the leg is passively moved backwards in the shoulder), the supporting tonus decreases, the elbow joint is flexed. 3. On further movement forwards, the leg is suddenly raised and set down again on the surface in a more forward direction. 4. Moving pictures of the leg-slackening reaction of the same dog with blindfold.

for instance, the head of a standing dog is rotated, the thorax will rotate too, by which the position of the forelegs in relation to the trunk will become asymmetrical, though the position of the legs on the supporting surface, however, remains unchanged. The alteration in position towards the trunk causes an alteration of supporting tonus, thus simulating an influence of tonic neck reflexes on the supporting tonus of the legs (Fig. 199).

In intact dogs, even with eyes blindfolded, lateral and forward leg-slackening reactions in the forelegs and lateral and backwards in the hindlegs are always very lively; the other leg-slackening reactions, too, usually appear promptly but they are more easily inhibited by experimental conditions (cerebral influences).

In decorticate dogs, too, leg-slackening reactions are present. They are thus reflexly produced. In these animals the onset is delayed, that is, they appear only with a strong alteration in the position of the leg and are mostly hypermetric, i.e., the leg is raised a little too high and

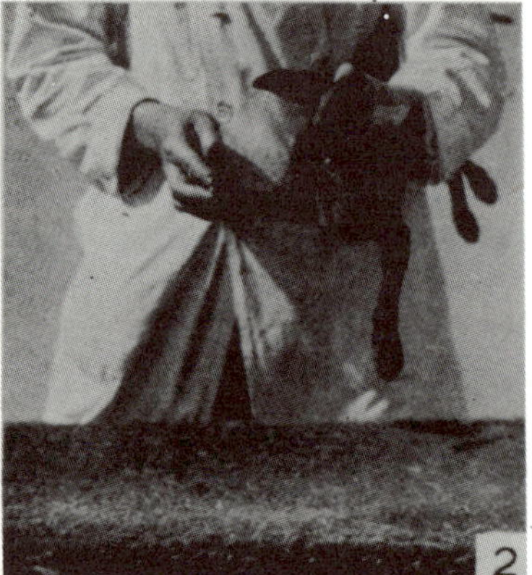

Fig. 199. Decerebellate dog Piccolino: the head is rotated to the left by 90°, the right foreleg (mandibular leg), put under static stress with one hand, is held motionless somewhat abducted. 2. Head passively rotated to a right lateral position. The fixed right foreleg, which is now the cranial leg, shows a distinct decrease in supporting tonus. Thus one has the impression that the supporting tonus has decreased under the influence of the tonic neck reflexes. But as shown in the illustration, the rotation of the head produced a rotation of the thorax, by which the foreleg became more strongly abducted. The decrease in supporting tonus was mainly caused by the alteration in position of the leg in relation to the thorax.

the leg-slackening step is too large. The delay is most evident in the leg-slackening reactions backwards and inwards of the forelegs and forwards and inwards of the hindlegs. Some time after extirpation of the cerebrum the delay is only slight, but in my animals it never disappeared completely (Fig. 218, also see Chapter XVII).

After unilateral extirpation of the cerebrum only the leg-slackening reactions of the contralateral legs is delayed; in addition, in these reactions these legs are lifted higher and perform larger limping steps than the ipsilateral ones. When a unilaterally decorticate dog is made to run from one edge of the table to the other with hopping steps, one will see that the leg contralateral to extirpation will need a smaller number of steps (usually about half as many) than the homolateral one (Fig. 200). With a rapid passive movement of the trunk, the delay is less. The leg-slackening reactions of the contralateral legs in this case take place already on smaller alterations in position than on a slow forwards movement of the trunk, but the delay can also be clearly observed when the animals are set down alternatively on the left and the right legs.

After the total extirpation of the cerebellum, leg-slackening reactions are at first entirely absent, most certainly in rigid animals in the first 2-3 weeks. They always return later but constantly show some deviations. First of all they are retarded and require strong alterations in position. Secondly, the leg is raised extremely high and is moved abnormally strongly, forwards and outwards (Fig. 197). Thirdly, it is always set down conspicuously strongly and with an audible impact of the sole on the surface. Since the hops are usually abnormally large, a decerebellate dog needs a smaller number of steps to run a certain distance than an intact dog of the same size. This difference becomes particularly evident when the animals stand on a turntable with one leg only. With an increasing speed of rotation, decerebellate as well as intact animals perform the hopping steps more rapidly, but the number of steps, on rotating the disk by 360°, is less almost by half in the decerebellate animal than in the intact one, with smaller as well as with higher speeds of rotation.

After the unilateral extirpation of the cerebellum, the leg-slackening reactions and hopping are temporarily absent on both sides. They reappear on the side of the intact half of the cerebellum after several days and on

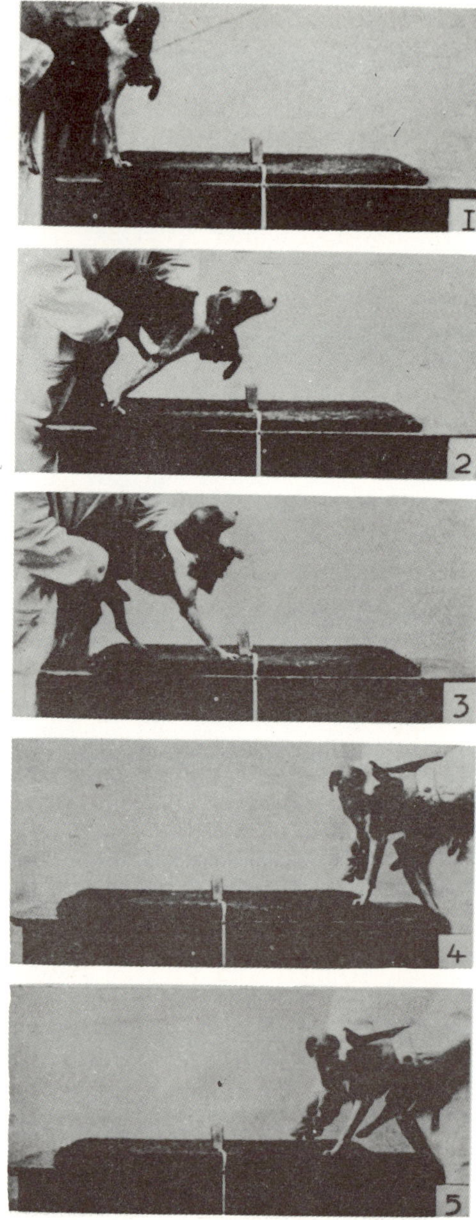

Fig. 200. Left-sided, decorticate, bilaterally labyrinthectomized dog Jenny. 1. Animal with the *right* foreleg on a mat. Leg almost in an intermediate position. 2. Trunk of the animal slowly moved forwards. Although the leg is strongly directed backwards, no leg-slackening reaction occurs. 3. With a further movement, the leg-slackening reaction occurs, the animal makes a large hop forwards, almost as large as half the length of the mat. 4. Animal on the mat with the *left* foreleg. 5. On a slight, passive forward movement of the trunk, the left foreleg makes a much smaller hop forwards.

Fig. 201. A. Decerebellate dog Piccolino with the left foreleg on a turntable rotating anti-clockwise and thus moving the leg caudally (A 1 and 2). Only with a strongly backward directed position of the leg (A 2), leg-slackening reaction forwards occurs in which the leg is raised abnormally high (A 3), is extended abnormally forwards (A 4) and is set down abnormally far forwards (A 5). On rotating the disk by 180° (A 1-12) the animal makes only two hopping steps. B. Decerebellate, right-sided decorticate, dog Vici, with the right foreleg on the turntable. The leg-slackening reactions show the same deviations. Sometimes the paw is raised to the height of the ear, so that it is not visible from the left side of the animal (A 9).

the contralateral side usually only after 2-6 weeks. Strangely, the leg-slackening reactions of the ipsilateral leg are only present when the animal is set down on this leg, but are absent when it is set down on all four legs. This is probably brought about by the fact that the supporting tonus in the ipsilateral leg decreases as soon as the feet of the opposite legs touch the surface and the animal supports itself on them.

376

The leg-slackening reactions of the ipsilateral legs are absent in the animal standing on all four legs, even when the legs of the other side show again lively supporting reactions and an increased tendency to abduction and lateral bracing. In this case, when the animal is set down on its four legs, the contralateral legs will push the trunk towards the side of extirpation and the ipsilateral legs are led more and more into a posture of adduction. Since the posture of adduction does not produce any leg-slackening reactions outwards, the animal will fall towards the side of extirpation (Fig. 202, No. 2). At a later stage the falling down is prevented by these reactions.

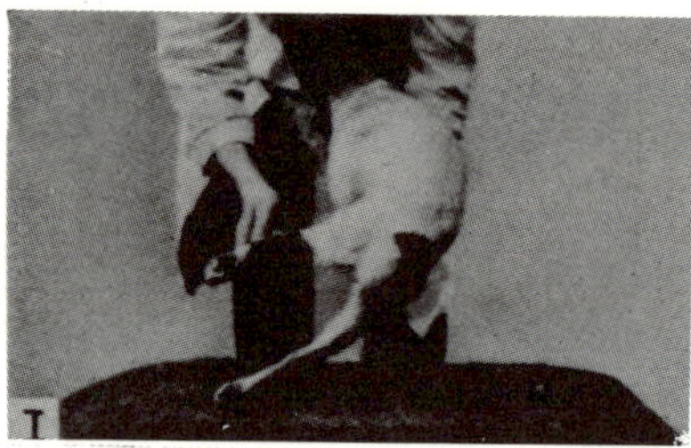

Fig. 202.   Dog Fox, 1 month after extirpation of the right half of the cerebellum. 1. Animal standing on a supporting surface with the right hindleg only, which is brought into strong adduction by a passive movement of the trunk. Although the posterior part of the body rests on the strongly adducted leg, no leg-slackening reaction occurs. 2. Animal standing with both forelegs on a supporting surface. On contact of the sole with the supporting surface a strong supporting tonus instantly appears and the leg is braced outwards. By this the thorax is pressed to the right and the right foreleg goes passively more and more into adduction. Since the adduction posture does not produce any leg-slackening reactions the animal, without being supported, would fall on its right side (see also Fig. 187).

Some time after the unilateral extirpation of the cerebellum the ipsilateral legs again show leg-slackening reactions during standing on these legs alone as well as in standing on all fours. Now the animal is able to stand and run freely and securely and does not become deviated laterally towards the side of extirpation. However, the leg-slackening reactions of the ipsilateral legs are always retarded, i.e., they appear only on wide alteration in position, the leg is lifted too high, is moved more forwards and outwards and is set down with a clearly audible impact. Through this the hopping steps in the ipsilateral legs are larger than those of the contralateral ones (Fig. 203) which is shown particularly distinctly in slow passive alteration in position of the legs in relation to the trunk.

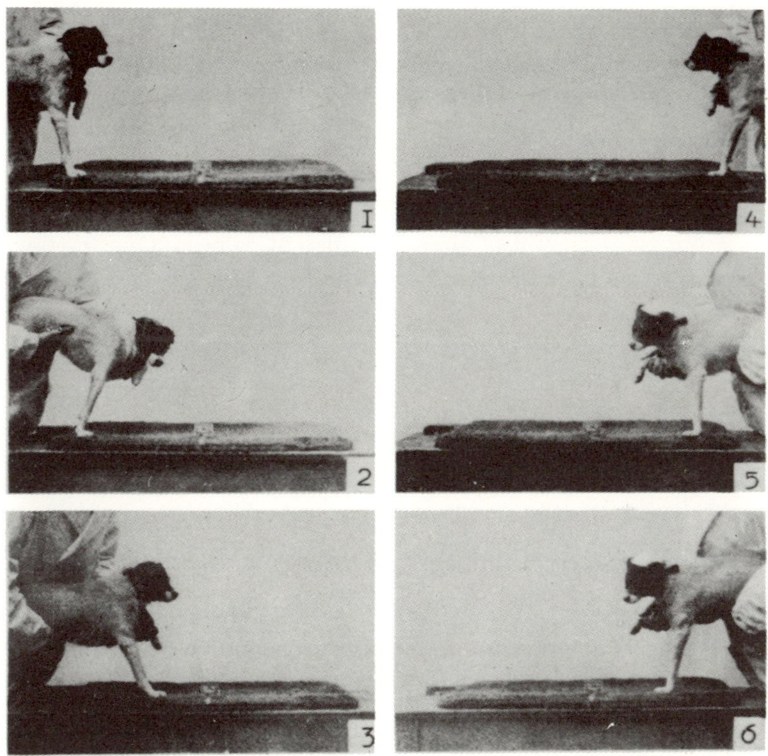

Fig. 203. Dog Fox, 6 months after extirpation of the *right* half of the cerebellum. 1. The animal stands with the *right* foreleg which is directed forwards on a mat. 2. The trunk is slowly moved forwards. Although the leg remains standing strongly directed backwards, still no leg-slackening reaction forwards occurs. 3. On further movement, the leg is suddenly lifted high, moved forwards and set down again. The size of the step is more than a quarter the length of the mat. 4. The animal is set down with the *left* foreleg directed somewhat forwards on the mat. 5. On a slight forward movement of the trunk, the leg makes a hop forwards. 6. On a further forward movement, this hop is followed by a second one. The size of both these hops is inferior to the size of one hop of the right foreleg.

Owing to the different behavior of leg-bracing and leg-slackening reactions in the ipsilateral and contralateral legs, unilaterally decerebellate animals hop in a different manner when they are pulled alternately to the right and to the left. When, for instance, a right-sided decerebellate dog is pulled to the left, there appears at first strong leg-bracing reactions of the left legs which are vigorously braced sideways. These reactions are intensified by further pulling, then the *right* legs are *passively* drawn from the ground and go more or less into flexion due to the decrease in supporting tonus (Fig. 195, No. 1). When the traction continues till the left legs are passively adducted, these will make a short hop outwards to the left and then instantly brace outwards again. Sometimes the leg-slackening reactions already occur when the legs are still in a posture of abduction and have not yet reached the intermediate position. (Thus one gets the impression that a very slight protraction of abductors is already sufficient to produce a leg-slackening reaction when the muscles are strongly stressed by the leg-bracing reaction.) Simultaneously with the lifting of the left legs, the right ones are extended (crossed extensor reflex) and set down on the supporting surface. In the movement towards the left, which encounters a strong resistance, the reactions proceed in the following way:

1. Strong leg-bracing reaction of the left legs with passive lifting of the right legs from the supporting surface.

2. After a long interval lifting of the left legs with simultaneous extension and setting down of the right legs.

3. Instant setting down of the left legs on a spot situated only slightly to the left, then outwards bracing of these legs.

Thus, in traction towards the side of the intact half of the cerebellum, the sideways running proceeds in an abnormal rhythm and in an atypical succession of reactions. The animal runs with short, audible hops of the left legs towards the intact left side while the right legs join in the flexing and extension movements. Thus, the animal runs like an obstinate intact dog when it is pulled sideways. Totally decerebellate dogs show the same rhythm and the same succession of reactions when they are pulled sideways to the right or left, they also will run sideways in hops which are abnormally large.

On the other hand, when the right-sided decerebellate dog is pulled to the right, only slight leg-bracing reactions of the right legs occur, which soon subside so that the legs almost immediately go into an increasing posture of adduction (Fig. 195, No. 4). Simultaneously with the passive adduction of the right legs, the left ones are *actively* raised and on a further adduction, set down medial to the right legs (influence of displacement reaction). Thus, in pulling to the right, towards the side of extirpation, the left legs make the first steps. The right legs are thereupon also raised and set down again more laterally to the right. The raising occurs only on a strong adduction posture; the leg is lifted abnormally high, moved too far laterally and set down on a spot too far to the right (retarded and exaggerated leg-slackening reaction).

Thus, on pulling to the right, towards the side of extirpation, the reactions proceed in the following succession:

1. Weak leg-bracing reactions of the right legs.

2. Raising of the left legs.

3. Extension of the left legs and setting down medial to the right ones.

4. Raising of the right legs.

5. Setting down of the right legs further to the right.

Thus, rhythm and succession of reactions behave as in an intact dog which is pulled sideways; but the steps are excessively large. Moreover, the right legs are set down loudly, an action which is performed almost inaudibly in intact animals.

On traction towards the side of the preserved half of the cerebellum, the legs are moved in an abnormal rhythm and abnormal succession, and the movement encounters a strong resistance; the animal hops with short steps sideways, in the course of which the preceding legs brace vigorously outwards with every step. When pulled towards the side of extirpation, the unilaterally decerebellate dog *runs* sideways with particularly large steps. In doing so, first the legs of the intact side are raised and, after a short interval, softly set down and then, after an equal interval, the preceding legs are raised and set down loudly. Moreover, on moving towards the side of extirpation, one encounters less resistance than on moving to the other side. Thus, impairments appear on traction to both sides, but they are different from each other. Of course, these defects in movement are not of equal strength in all unilaterally decerebellate dogs; in shy animals the reactions are more or less modified. These differences can usually be best observed when the animals are set down on their hindlegs only and are then alternatively pulled to the right and to the left.

As we have seen, when totally decerebellate dogs are pulled sideways, they show the same rhythm, the same succession of reactions and as strong a resistance as unilaterally decerebellate dogs which are pulled towards the intact side. Moreover, the steps are as large as those of a unilaterally decerebellate dog when pulled towards the side of extirpation.

When one observes a unilaterally decerebellate dog while standing and running, one will see that the contralateral legs are strongly braced outwards so that the ipsilateral legs are passively brought into a posture of adduction which often causes a flexion and a hopping reaction outwards. This can be seen very distinctly when the animals approach running and suddenly stand still. In this case, they will not immediately stand quietly but will make a step laterally with the ipsilateral legs. Similar reactions can be observed when the animals are set down on a supporting surface from a ventral position in the air (Fig. 204).

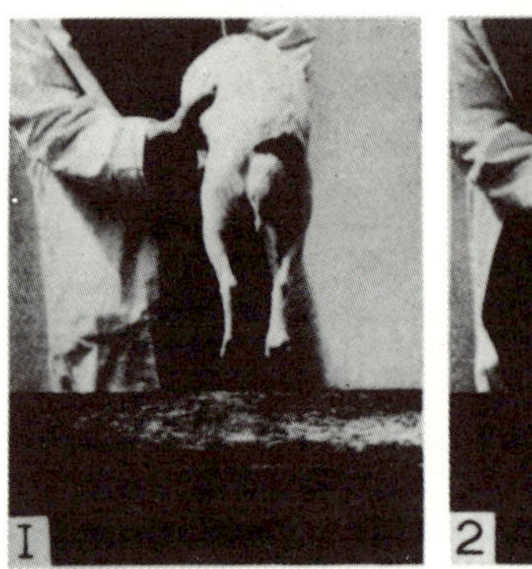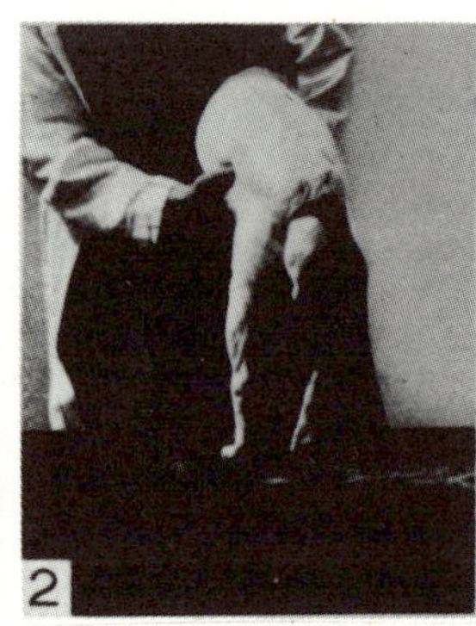

Fig. 204. Dog Fox, 15 months after extirpation of the right half of the cerebellum. 1. Animal in ventral position in the air. 2. Animal is moved downwards till the hindlegs touch the supporting surface. 3. As soon as the animal supports itself on the hindlegs, the left hindleg braces outwards and the right hindleg, led passively into adduction, makes a hop to the right. (See also Figs. 184 185 and 186.)

After the unilateral extirpation of the cerebellum, the leg-slackening reactions soon appear in the contralateral legs, but only after a considerable time in the ipsilateral ones. At this stage, the animals will instantly fall on the side of extirpation when they are set down on their legs (Fig. 187). At a later stage when the leg-slackening reactions have already returned in the ipsilateral legs and the animals start to stand and run freely, the laterally directed hopping reactions of the ipsilateral legs occur with such a delay that the animals continually stumble and fall towards the side of extirpation. Only when the hopping reactions appear so promptly as to prevent falling are the animals able to stand and run freely without losing their balance. Since the altered hopping reactions still occur at the stage of permanent impairment, the posterior part of the body will stumble and sometimes fall towards the side of extirpation, for instance, when the animals, while playing with others, run around wildly, jump and turn around quickly.

The leg-slackening reactions *inwards* appear in the ipsilateral legs usually with an even greater delay than those directed outwards. When an intact dog is set down with both forelegs on a supporting base and the right leg is shifted outwards, the thorax moves to the right so that both legs go into a posture of abduction and the **abduction** of both sides are strained. On a further shifting, the leg is suddenly raised and set down more medially. When a right-sided decerebellate dog is set down on a supporting surface immediately after the reappearance of the leg-slackening reactions and then is displaced so that its right leg shifts more and more outwards, and only the left leg will make a hop inwards. In the stage of permanent impairment, too, when the delay of the hopping reactions in the ipsilateral legs is no longer very great, this phenomenon still sometimes occurs (Fig. 205), but in this case also the displaced leg often shows a hop inwards, just as in intact, decorticate and decerebellate dogs.

The hopping reactions forwards also show distinct delays in the ipsilateral fore- and hindlegs. In the ipsilateral foreleg, the hopping reaction backwards takes place even later than the reaction forwards; conversely, in the ipsilateral hindleg, the hopping forwards occurs after a longer delay than one backwards. *In the decerebellate, right-sided decorticate dog Vici*, hopping reactions were also evident on both sides. In the right legs the reactions behaved like those of exclusively decerebellate dogs(Fig.201,B), and they appeared retarded and were performed hypermetrically. In the left legs, they could not be established with certainty. In the foreleg which was kept extremely caudally extended and on standing constantly slid backwards, true hopping reactions were never observed, but only a distinct decrease in supporting tonus on the passive alteration in posture of the limb away from intermediate position. The left hindleg, however, showed distinct hopping reactions inwards and outwards, forwards and backwards, even when the animal was set down with this leg on a turntable. With increasing speed of rotation, they occurred more rapidly but always with a much greater delay than the reactions in the right legs. The delay is so considerable that the reaction was not able to fulfil its purpose, maintenance and recovery of balance. Besides, as in the right legs, they exhibited strong hypermetria.

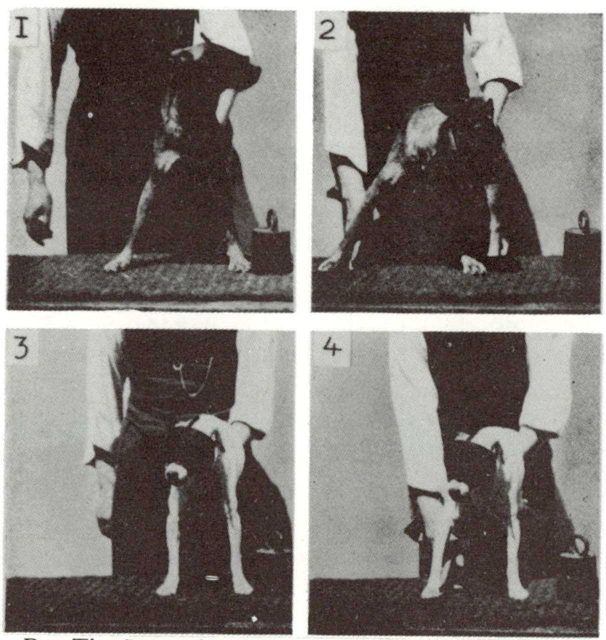

Fig. 205. 1. Dog Tip, 7 months after extirpation of the right half of the cerebellum, with the forelegs on a supporting surface. 2. On lateral displacement of the right foreleg, the left foreleg was raised from the surface and set down again more to the right and inwards. 3. Dog Fox, 15 months after extirpation of the right half of the cerebellum, with the forelegs on a supporting surface. 4. Lateral displacement of the right foreleg produces a hop inwards of the left foreleg.

In the decorticate and decerebellate dog Robbie, too, the hopping reactions returned after a certain time, at least partly. When, for instance, the animal was set down on its right hindleg, it would make a hop sideways on a movement to the right and a backwards step on a movement to the rear. Unfortunately, the animal died of distemper $5\frac{1}{2}$ weeks after the last extirpation (extirpation of the cerebellum) so that it could not be established whether the other hopping reactions, too, could be produced without cerebrum and cerebellum.

After the bilateral extirpation of the labyrinths, all hopping reactions were present on both sides; they were absent, however, *in decerebrate dogs* (only acute experiments, Fig. 206), *in spinal dogs* and *also after severance of the posterior roots belonging to the legs.* They were also absent in seven *new-born dogs* within the first weeks of their birth, but they were almost all present again in the second week (the hopping reactions inwards are usually the last to appear).

Fig. 206.   Decerebrate dog B (see Fig. 6)   1. The animal is set down on a supporting base with the left foreleg.   Both hindlegs and the right foreleg are held high with one hand. 2. Trunk is slowly moved forwards passively till the paw lies on the dorsum.   The alteration in posture does not produce any leg-slackening reactions.

THE IMPORTANCE OF HOPPING REACTIONS FOR THE MAINTENANCE OF BALANCE AND FOR THE ADAPTATION OF LEG POSTURE TO THE SUPPORTING SURFACE

Like the leg-bracing reactions, the leg-slackening reactions, too, are typical balance reactions. According to circumstances, falling is prevented by different reactions. When, for instance, a dog is pulled to the right by a fold of its skin, the right legs will brace outwards. In so doing, the center of gravity of the body is shifted to the left, preventing the animal from falling to the right. The right legs press the trunk towards the left and thus bring the left legs into a posture of adduction. Now, when the animal is suddenly released or is given a push to the left, the trunk shifts even more to the left, the adduction posture of the left legs increases and produces hopping of the left legs outwards, preventing falling to the left. When the animal is pulled more and more vigorously to the right so that finally the laterally braced legs are passively adducted more and more, the leg-bracing reaction slackens and hopping of the right legs to the right takes place preventing falling towards the right side. According to whether a weak or a strong, a gradual or a sudden traction is exerted and according to the initial posture of the legs, sometimes leg-bracing reactions and sometimes leg-slackening reactions towards the right are produced. When, by pulling to the right, the legs are already in a strong adduction position, only the leg-slackening reactions will appear and the balance is preserved. When the animal is pulled forwards gently and gradually, the legs are braced forwards and thus prevent the animal from falling head forwards. On a gradually increasing traction the legs are braced more and more vigorously forwards, till finally the strongly forwards directed posture of legs to trunk will produce hopping reactions backward. This is particularly the case when the strongly forward braced legs slip forwards. In this case, the animal starts to run backwards and after each backward step braces the legs again forwards. When the traction again becomes so strong that the forward braced legs are moved passively backwards, the leg-bracing reactions cease and, instead, hops fowards occur, but even now the animal braces its legs forwards with every step.

Thus, depending on the force of the forward traction either leg-bracing reactions forwards, or hopping reactions backwards, or hopping reactions forwards will occur. The different effects of the force of traction can best be observed in decorticate dogs in which the reactions are not impaired by conditioned reflexes, and also in decerebellate dogs in which the cerebral influences usually have a less disturbing effect.

In decorticate dogs, it can be observed particularly clearly that in addition to the initial posture of the legs, and the direction and force of traction, the *plane* of application, too, exerts a great influence. When the animal is seized by the muzzle, the legs brace forwards more and more strongly even when the head is kept motionless, so that finally the paws slip forwards and hopping reactions backwards are produced. After each step the paw is braced forwards anew, slips again and is transferred backwards. This is the case to a stronger degree when the animal's muzzle is pinched. When the hand holding the muzzle relaxes a little, the animal runs backwards, bracing the legs forwards with every step. Hopping reactions forwards can only be produced by a sudden seizure of the muzzle and an instant, sharp pull forwards. Hopping reactions backwards and leg-bracing reactions forwards produced by the forwards pulling of the muzzle are instantly stopped by a pinching of the tail: in this case, the animal runs forwards and presses against the hand clasping the muzzle.

Conversely, by pulling the tail almost no hopping reactions backwards can be produced (forwards pulling by the tail caused a running forwards; grasping of the tail and backwards pulling by the tail causes leg-bracing reactions backwards and when, in so doing, the legs slip to the rear, hopping reactions forwards).

Leg-bracing and leg-slackening reactions appear also when the animals stood on a supporting surface that was shifted in a horizontal plane, so as to preserve the maintenance of balance. When, for instance, the surface was shifted to the left, the legs went passively to the left, i.e., the right legs went passively into adduction, the left in ot abduction and the line of the center of gravity was transferred to the right in relation to the supporting surface. By the passive adduction, leg-bracing reactions were produced in the right legs which braced outwards to prevent a falling to the right. On a very sudden shifting of the supporting surface the right legs were passively adducted so much and the left ones abducted so strongly that hopping reactions were produced, those of the right legs outwards, those of the left legs inwardly and falling towards the right side was prevented by hops towards the right. The leg-slackening reactions also occurred when at the start of the displacement the right legs were already in a posture of adduction, the left in a posture of abduction, for instance when the supporting surface was at first moved to the right and then suddenly to the left. When the supporting surface was at first shifted slowly and then more and more rapidly to the left, the right legs braced more and more outwards, causing the left legs to go passively into a stronger adduction. The trunk was pressed more to the left so that it finally leaned over and threatened to fall to the left side. The strong adduction of the left legs and abduction of the right ones, however, simultaneously produced hopping reactions to the left which prevented falling. A shifting of the supporting base to the left could, according to circumstances, also cause leg-bracing reactions to the right, leg-slackening reactions to the right or hopping to the left; just as a forwards displacement of the base could, according to circumstances, cause leg-bracing reactions backwards,[1] or hopping reactions forwards,[2] or hopping reactions backwards.[3]

Standing on a turntable, too, leg-bracing and leg-slackening reactions participate in the maintenance of balance. This is shown most clearly by labyrinthectomized dogs where the reactions are not complicated or impaired by the labyrinthine rotating reactions. When a labyrinthectomized dog stands with the right side along the outer rim of the turntable which is rotated slowly clockwise, leg-bracing reactions forwards will be produced since the disk moves the legs backwards on the trunk, which prevents falling head-forwards caused by the passive alteration in position of the legs. With a gradual increase in rotation speed the legs brace forwards more and more and, when the legs finally slip forwards due to the strongly forwards directed posture, hopping reactions backwards will take place, sometimes only in the forelegs while the posterior part of the body sits down. When the turntable is rotated with a greater speed so that the legs are passively and strongly moved backwards all at once, hopping reactions forwards occur by which a plunge head forward is made impossible. When the disk is rotated so rapidly that after each hopping step the paw is pulled again backwards, hopping reactions are again produced and the animal runs forwards against the direction of disk rotation. In addition, the animal (right side parallel to the outer rim) is pulled to the right by the centrifugal force which causes the right legs to brace outwards and, with a gradual increase in force, hopping reactions of the four legs towards the left. On the other hand, a strong increase in force produces leg-slackening reactions to the right. Due to these latter reactions, the animal runs sideways towards the rim and tumbles down from the disk.

When the animal stands on the radius of a turntable rotating in a clockwise direction, with the head directed towards the center, it will be pulled caudally by the centrifugal force, through which the legs brace backwards and, with an increasing speed of rotation, perform hopping reactions forwards so that the animal runs towards the

---

(1) On a slow forwards shifting.

(2) On a forwards shifting with increased speed.

(3) On a sudden shifting with great speed.

center. At the same time the right legs brace outwards, since they are pulled to the left by the disk, while with a gradually increasing speed of rotation hopping reactions of the four legs to the left are produced and the animal, on running towards the center, deviates to the left. With a high speed of rotation, the centrifugal force will cause hopping reactions backwards, the strong shifting of legs to the left produced by the disk will cause hopping reactions to the right, and, in this case, the animal, while performing hopping reactions to the right and backwards, will be thrown off the disk.

On the other hand, when the animal stands with the head towards the outer rim, the centrifugal force produces leg-bracing reactions forwards, and the posterior part of the body sometimes sits down. With increasing speed of rotation hopping reactions backwards occur so that the animal runs backwards towards the center, while with a high speed of rotation hopping reactions forwards and outwards and, due to the linear shifting, to the left occur and the animal tumbles off the disk.

Due to the leg-bracing and hopping reactions, the labyrinthectomized animal is capable of maintaining its standing posture on the turntable even with a considerable speed of rotation.

The hopping reactions participate also in the maintenance of balance in standing on a slanting supporting surface. When the supporting surface on which the dog is standing is slowly lowered from the right side, the force of gravity will draw the trunk towards the right and the right legs are passively moved to the left in relation to the trunk. This causes the production of leg-bracing reactions of these legs outwards (Fig. 207, No. 1).

On further lowering, the leg-bracing reactions of the right legs press the trunk more and more towards the left, and the left legs are brought passively more and more into adduction, which causes a decrease in supporting tonus and a giving way of the left legs (Fig. 207, No. 1). When the base is lowered even more, hopping reactions of the four legs towards the left are produced, caused by the posture of strong abduction of the right and strong adduction of the left legs, especially when the paws begin to slip; the animal runs up the inclination of the board to the left (Fig. 207, No. 2) and in doing so the right legs are braced outwards with every step. The laterally directed hops of the hindlegs are nearly always larger than those of the forelegs so that

the posterior part of the body precedes in the sideways movement
(Fig. 207, No. 2 and 4). With a poor initial supporting tonus and
inhibited hopping reactions (in intact, decerebellate and labyrinthecto-

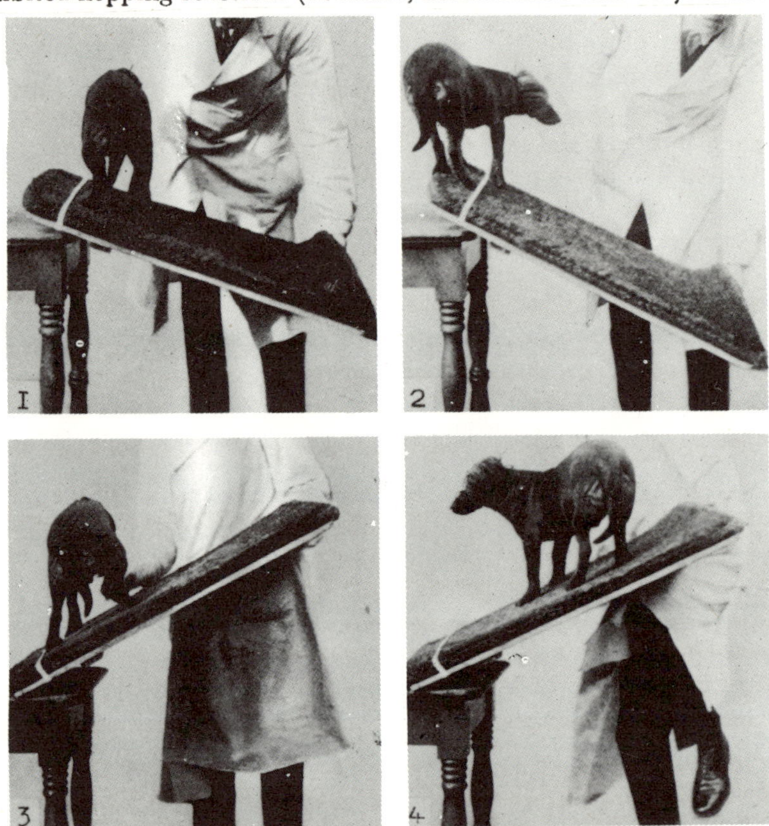

Fig. 207. 1. Normal dog with blindfold in standing posture on a board which
is lowered from the right side. The right legs are braced laterally,
press the trunk towards the left and lead the paws passively into a
posture of adduction. Due to the alteration in position, decrease in
supporting tonus and giving way of the left legs. 2. On further lowering, the abduction of the right legs and adduction of the left ones
gradually increase and finally cause hopping reactions to the left so
that the animal runs increasingly more up the inclination of the
board. (A decorticate animal does not stop running sideways at the
edge of the board like an intact one.) 3. Board is raised on the right
side of the animal. The force of gravity pulls the animal to the left.
Leg-bracing reactions of the left legs outwards to the left. 4. On
further raising, hopping reactions towards the right, the animal runs
again up the inclination of the board, i.e., to the right. (A slight
lowering followed the raising of the board which brought the running
upwards to a standstill.)

mized dogs with blindfolded eyes this is often the case due to cerebral inhibiting influences) the left legs totally collapse prior to the appearance of hopping reactions and the animal, in this case, goes gradually into a lateral position to the left (Fig. 106).

When the right legs stand in a posture of strong adduction at the start of the downwards movement of the supporting surface or when the board is suddenly lowered so quickly that the force of gravity draws the animal vigorously all at once to the right, hopping reactions to the right will take place which prevent falling towards this side. On a very rapid lowering of the surface towards the right (to which intact dogs are quite able to adapt themselves due to the effect of labyrinthine reflexes), labyrinthectomized dogs will fall into a lateral position to the right. Bracing reflexes occur in time, but hopping reactions occur too late to prevent falling.

On a slow lowering of the tailend of the supporting board, the animal is drawn backwards by force of gravity; in agreement with this, leg-bracing reactions backwards at first (Fig. 208, No. 1) then hopping reactions forwards (Fig. 208, No. 2) will occur so that the animal runs up the inclination of the board. On a sudden, rapid lowering of the base the trunk is strongly thrown backwards all of a sudden and hopping reactions backwards take place.

Only these reactions can prevent a backward tumbling in labyrinthectomized dogs. In intact, decorticate and decerebellate dogs, however,

Fig. 208. Normal dog with blindfold, on a board that is slowly lowered by the tailend. 1. With a slight lowering, all four legs are braced backwards. 2. With a further lowering, the legs are directed even more backwards (compare the position of the right upper limb in 1 and 2) and produce a hopping reaction of the initially more backward directed (see No. 1) left foreleg.

labyrinthine reactions (extension of the hindlegs backwards, movement towards the rear and slight flexion of the forelegs) appear, produced by the passive rotation round the bitemporal axis. The same is the case on a rapid lifting of the head of the board end. Here, too, it appears particularly evident that at a certain speed of movement of the board the hopping reactions occur too late in relation to the labyrinthine reflexes for the maintenance of balance. In this case, labyrinthectomized dogs turn a somersault backwards, while intact and decerebellate animals, at an equal speed, are able to maintain their standing posture and balance due to the influence of labyrinthine reflexes.

On a slow lowering of the headend or a slow lifting of the tailend, the forward traction produces at first leg-bracing reactions forwards (Fig. 209, No. 1) and then hopping reactions backwards (Fig. 209, No. 2) so that the animal again runs up the inclination of the board, but this time to the rear.

Fig. 209. Normal dog with blindfold, on a board which is slowly raised by the tailend. 1. On raising the tailend, forward bracing of the two hindlegs and of the left foreleg, and hopping reaction backwards of the right foreleg, simultaneously the two hindlegs are slightly flexed. 2. On further raising both forelegs make steps backwards, while the hindlegs are totally flexed so that the posterior part of the body goes into a sitting posture.

When a dog is pulled forwards by a sling or by the force of gravity (see, among others, Fig. 182, No. 4) the forelegs, in particular, are actively braced forwards and press the trunk caudally, while the hindlegs are moved forwards at the hip joint by the bracing forwards of the forelegs, at least in most cases, but sometimes only passively. Due to this passive alteration in position the supporting tonus of the hindlegs decreases, they flex and the posterior part of the body goes into a sitting posture (Fig. 209, No. 2). Sometimes active leg-bracing reactions forwards and backwards directed hopping reactions of the hindlegs also occur when supporting tonus and hopping reactions are not inhibited by the experimental conditions (blindfolding, etc.). The position of the head in relation to the trunk also plays a role in this case (Fig. 182, No. 2 and 4).

The participation of leg-bracing and hopping reactions in the adaptation of leg posture to the position of the supporting surface can be established in intact, decorticate, decerebellate (Figs. 97, 98, 112, 174, 194 and 249) and labyrinthectomized (Figs. 99, 106, 107 and 194) dogs, but not in decerebrate (Fig. 9) or spinal dogs.

It is remarkable that adaptive reactions, among others leg-bracing, still occur in a labyrinthectomized dog which lies with flexed limbs on an obliquely tilted base (Fig. 210). How the reactions are brought about in this case is still obscure. Such adaptation does not take place any more when the labyrinthectomized animal is blindfolded and lies on a narrow board with its forelegs dangling down over the edge.

Fig. 210.   1. Labyrinthectomized dog Moritz with blindfold, in ventral position on a board held horizontally. 2. On raising the right side of the supporting base, leg-bracing reactions of the left legs which are propped outwards, pressing the trunk into a right lateral position. 3. On lowering the supporting surface from the right side similar reactions. By the outward bracing of the right legs, the trunk is kept in the left lateral position, contrary to the influence of body righting reflexes on the body, as long as the oblique position of the supporting base continues to be present.

392

As already mentioned, in the decorticate and decerebellate dog Robbie, part of the hopping reactions, particularly of the hindlegs, was present again four weeks after extirpation of the cerebellum so that the animal, when set down on a horizontal or obliquely tilted supporting base, showed to some degree corresponding adaptive reactions. When the animal was lowered from being suspended head upwards, towards a supporting surface, the hindlegs due to the weight of the trunk were passively displaced forwards more and more in the hip joint (and, in addition, at the same time the tarsal joints were more flexed). These alterations in posture produced hopping reactions backwards, first in one and then in the other hindleg, so that the posterior part of the body, according to the inclination of the base, would run up (Fig. 211, No. 1 and 2) or down (Fig. 211, No. 3 and 4) the inclined board. When one hindleg was also raised with the hand and, thus, the animal stood only with one leg on the surface, a hop to the rear was performed.

In summing up, the observations reported in the last chapters demonstrate that, in the adaptation of leg posture to alterations in position of the supporting surface, the following factors can participate, according to circumstances:

1. Labyrinthine reactions due to rotation round the bitemporal axis or longitudinal axis.

2. Labyrinthine progressive reactions (lifting reaction, readiness to jump).

3. Labyrinthine righting reflexes with associated neck righting reflexes.

4. The influence of the position of proximal joints on the position of the distal joints of the extremities (Fig. 55).

5. The influence of curvature of the back on the supporting tonus (Fig. 146).

6. Hopping reactions.

7. Influences produced by alterations in position of one leg on the supporting tonus of the other three legs.

8. Leg-bracing reactions.

9. Leg-slackening and hopping reactions.

Fig. 211. Decorticate and decerebellate dog Robbie, 4½ weeks after the extirpation of the cerebellum. 1. Animal suspended head upwards, above an inclined supporting surface with the back towards the raised end of the surface. 2. The animal is lowered downwards towards the supporting surface. By the weight of the posterior part of the body, the hindlegs are passively moved forwards in the hip joints, causing hopping reactions backwards. The posterior part of the body runs backwards up the inclined board. 3. Animal suspended head upwards, above the inclined surface, but this time with the back towards the lowered end. 4. On setting down the hindlegs, the weight of the trunk causes a passive forwards movement of the hindlegs in the hip joint producing hopping reactions backwards. Thus the posterior part of the body runs backwards down the inclined board.

On a rapid, sudden oblique tilting of the supporting surface, the labyrinthine rotating and progressive reactions play the principal role. Labyrinthectomized dogs, therefore, in distinction from intact ones, show no distinct adapting reactions and roll towards the lowered side, tumble head forwards or turn a backward somersault.

However, on a slow tilting of the supporting surface, labyrinthectomized animals promptly adapt themselves, probably mainly due to leg-bracing and hopping reactions which are here fundamental for the maintenance of balance. In the same way leg-bracing and hopping reactions play the principal role in the maintenance of balance when the animal is pulled or pushed sideways, forwards or backwards, or when it is passively pulled by one or several legs (Fig. 248), also in standing on a turntable and in standing on a supporting surface which is shifted in a horizontal plane.

In man, too, leg-bracing or leg-slackening and hopping reactions can be observed on pushing or pulling forwards, backwards or sideways which, just as in animals, can appear less promptly under pathological conditions (Fig. 213).

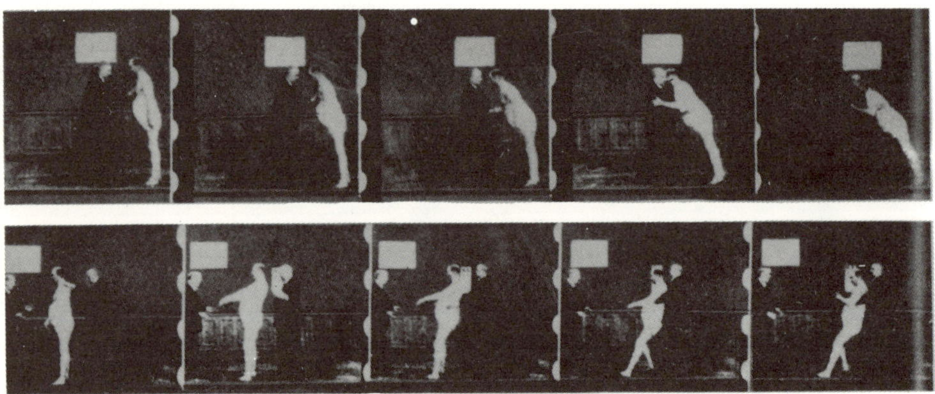

Fig. 212.   Patient with multiple sclerosis. A. The patient is asked to lean against the hands of the physician which are placed on the patient's chest. On withdrawing the hands he will fall forwards, because the hopping reactions are absent, which are shown by intact persons under the same circumstances and which prevent the falling forwards. B. Hopping reactions backwards, however, occur when the patient is pushed or pulled backwards, with a considerable delay, and the hop is too small to prevent the falling backwards. (In Fig. B 5, the line of the center of gravity falls behind the supporting surface of the legs.) Moving pictures from A. Thomas : *Pathologie du cervelet, Nouveau Traite de Medecine*. G. H. Roger, F. Widal, P. I. Teissier, Vol. 19, Pathologie de cerveau et du cervelet, p. 755. Masson et Cie, Editeurs, Paris, 1925.

C. PROPRIOCEPTIVE CORRECTING MOVEMENTS

We have already mentioned the remarkable fact that decorticate dogs, in spite of the absence of preparation for standing on optical stimulations and stimulations from the body surface, show neither on standing nor on running any abnormal postures of the limbs. Contrary to animals in which the posterior roots belonging to the legs have been severed, and to spinal animals, which almost always drag their feet over the ground on the dorsal side of the toes, decorticate animals set down their paws with the soles correctly on the supporting surface. Only in the first stage of shock manifestation, when the muscle tonus is still considerably reduced, do they sometimes place the foot on its dorsum on standing and running. Further we have also seen(P. 98) that a passively imposed dorsum-down position is corrected as soon as the decorticate animal moves the leg forwards at the shoulder or hip joint as a result of any stimulation. In this case, the correction is brought about by the influence of the position of the proximal joints on the posture of the distal joints (Fig. 56).

Further investigations on decorticate animals showed that any passive alteration in position that places the leg into an abnormal posture nearly always brings about stimulation to produce movements, produced by lesser alterations in position the longer the time since extirpation. When, for instance, the leg is passively moved backwards, forwards, inwards or outwards in relation to the trunk, the hopping reactions we have discussed above occur in decorticate animals. On a passive movement of the foreleg backwards (or of the trunk forwards) the limb is raised and moved forwards at the shoulder and, in doing so, the toes move dorsally as soon as a good muscle tonus is again present (Fig. 56) so that the pads of the feet are placed on the supporting surface when the paw is set down. We have already expressed the supposition that these reactions are produced by the passive caudally directed alterations in position of the shoulder in which case the forward movers of the upper arm, among them m. biceps, are stretched. The hopping reaction occurs not only when the animal places on the surface with the sole of the foot, but also with the dorsum of the foot (Fig. 213) and, at the same time, the trunk is passively moved forwards, which is a proof that the alteration of posture in the proximal joints indeed plays in important role.

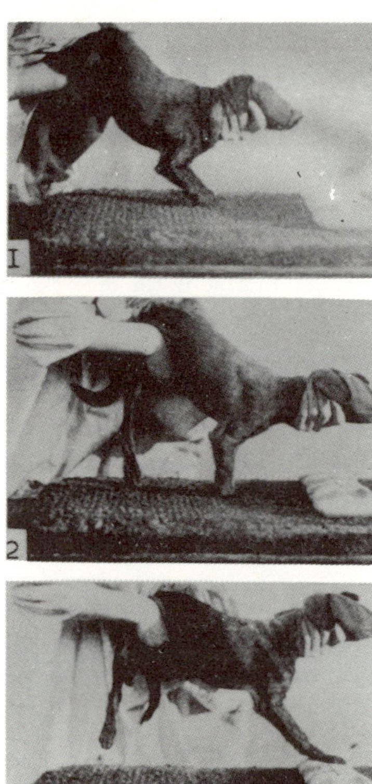

Fig. 213. 1. Dog Sepp, left-sided decorticate with the dorsum of the right forepaw on a supporting surface. 2 and 3. On a passive movement of the
trunk forwards there are hopping reactions forwards in the course
of which the paw is set down correctly with the sole on the supporting surface. (This reaction is also shown by totally decorticate
animals.)

When the forearm above the wrist joint of a decorticate dog is also
held against the edge of a table (Fig. 214, No. 1) and the trunk is passively moved forwards so that only the position of the shoulder and
elbow, but not of the distal joints, is passively altered, the hopping
reaction forwards still occurs; in this case, too, the foot is at first drawn
up in a flexed position, then moved forwards and set down with the
sole on the table (Fig. 214, No. 3 and 4).* These reactions also take
place on slow movements of the trunk and also when, with a fixed trunk,
the forearm is moved caudally with one finger. Thus, they must be

* Cf. the "proprioceptive placing reaction" of Bard (Ed.).

produced either by the passive caudal movement of the upper arm at the shoulder or by the passive stretching of the elbow joint (in both cases m. biceps is stretched) or by both factors together.

When with the dorsum of the paw down the trunk is moved instead of forwards, for instance, outwards or the paw is passively moved inwards at the shoulder, the hopping reaction takes place outwards and in this case, too, on stretching the elbow joint the toes are moved dorsally, so that the paw is set down with the sole on the surface.

Hopping reactions can also be produced by alterations in position of the distal joints. When, for instance, the dorsum of the foot of a decorticate dog is held against the edge of a table (Fig. 215) and the trunk is moved forwards while movements of upper arm and forearm are passively restrained, the paw, when the wrist joint is more or less passively flexed, is also raised, extended forwards and set down with the sole on the surface.

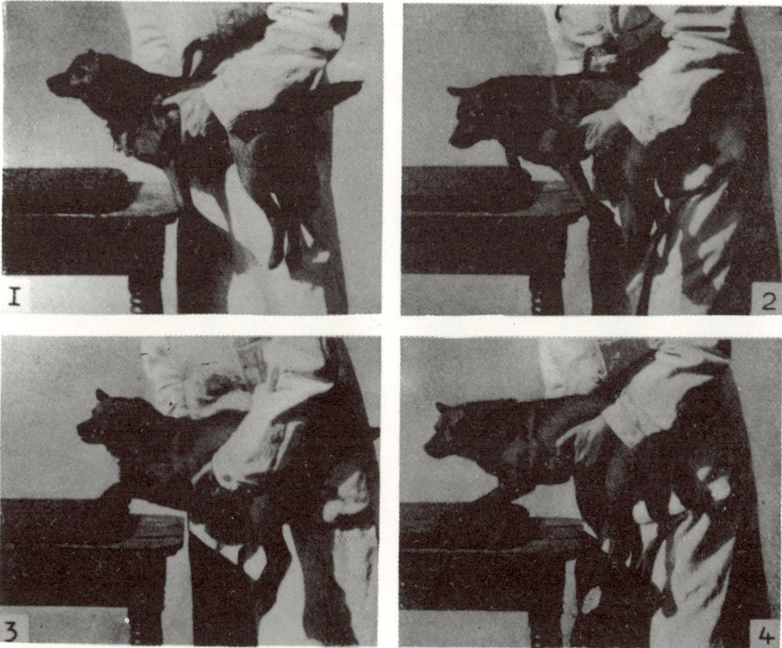

Fig. 214. Decorticate dog Fuchs. 1. The left foreleg above the wrist joint is held against the edge of a table. Due to the absence of preparation for standing, the foot is not set down on the table. 2. The trunk is passively moved slowly forwards so that the upper arm is pulled backwards and the elbow joint is maximally stretched. 3. When the upper arm is strongly directed backwards, the paw is suddenly drawn up and moved forwards. 4. Then again extended and set down with the sole on the table.

When the trunk of the decorticate animal is held motionless in the air and the wrist joint is passively flexed by a pressure with one finger on the volar surface of the metacarpus, the paw is also withdrawn, moved forwards above the finger and again extended. The animal makes in the air a step over the finger. In intact animals these reactions are difficult to examine, since these animals put the paw on the table already on touching the edge, due to the preparation for standing (placing). Similarly, in an intact animal held in the air, touching of the dorsum of the foot causes a withdrawal of the paw before the finger has brought about a flexion of the wrist joint (Munk's touch reflex, preparation for standing ?). Due to the hopping reaction which

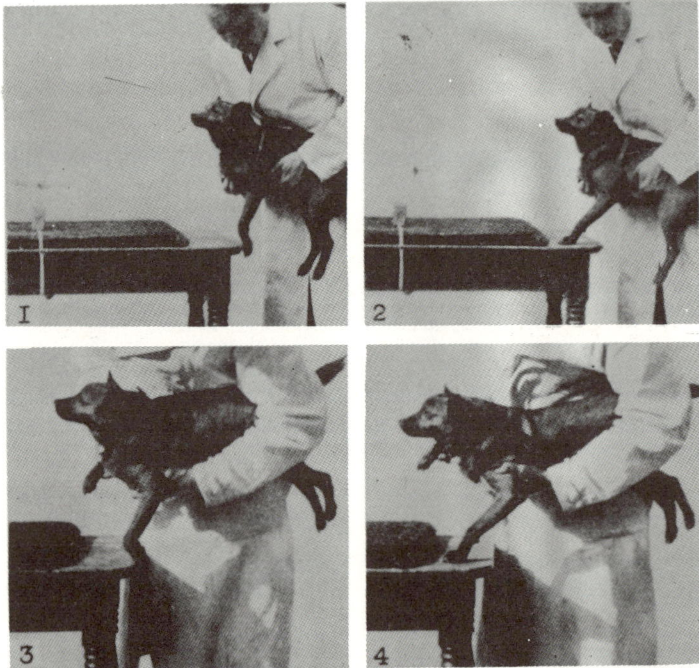

Fig. 215. Decorticate dog Fuchs. 1. The dorsum of the left forepaw is held against the edge of a table. Due to the absence of preparation for standing (placing), the paw is not set down on the table. 2. On moving the trunk forwards producing a flexion of the wrist joint, the paw is suddenly raised, moved forwards, extended and set down with the sole on the table. 3 and 4. Even when any caudal movement of the upper caudal and forearm is prevented, as soon as the movement of the trunk causes a flexion of the wrist joint of more than 45°, an active flexion of the leg at the elbow joint will take place, followed by an extension and setting down of the paw on the surface in a normal position.

is produced by the flexed posture of the wrist joint (by which the *extensors* of the wrist joint are stretched), it is no longer possible, a certain time after extirpation of the cerebrum, to set down decorticate animals on a surface with the back of the foot down because the posture is instantly corrected.

Finger and toe position can also produce hopping reactions. When a decorticate dog is set down with the toes of one foreleg on the edge of a table, this leg will show a strong supporting tonus (positive supporting reaction). When the toes are further depressed till overstretching of the fingers takes place (Fig. 216, No. 2), the leg is raised, moved forwards and set down with the sole on the table (Fig. 216, No. 3). Thus the hopping reaction here occurs when the finger flexors, which at the same time are *flexors* of the wrist joint, are strongly stretched.

The effect of the overstretched toe flexors also appears when the animal is set down with one foreleg on a swinging board above the axis of rotation (Fig. 217) and the headend of the board is raised while the trunk is fixed. In this case the board presses the fingers more and more dorsally till the animal makes a hop forwards (Fig. 217, No. 2). While standing on all four legs on a supporting surface which is slowly raised from in front the legs are braced more and more backwards, and, in so doing, the toes are moved dorsally in the metacarpo- and the metatarsophalangeal joints (Fig. 208). On a stronger raising of the board, hopping reactions forwards appear, probably partly due to alteration in position of fingers and toes.

With increased loading of the shoulders, too, the toes are increasingly dorsiflexed, the wrist joint more and more overstretched (i.e., the flexors of phalangeal and wrist joints are increasingly stretched) and correspondingly, as we have seen in Chapter V, the paw is raised, transferred forwards and laid on the ground in a flexed position. (In a strongly forward directed posture, the foreleg easily gives way, Fig. 164, No. 1.) Moderate stretch of the flexors of the wrist and phalangeal joints produces supporting tonus (positive proprioceptive supporting reactions), while a strong stretching apparently causes an opposite reaction (raising of the paw).

The hindlegs of decorticate dogs, too, show hopping reactions on alterations of position in the proximal hip joint and in the distal tarsal

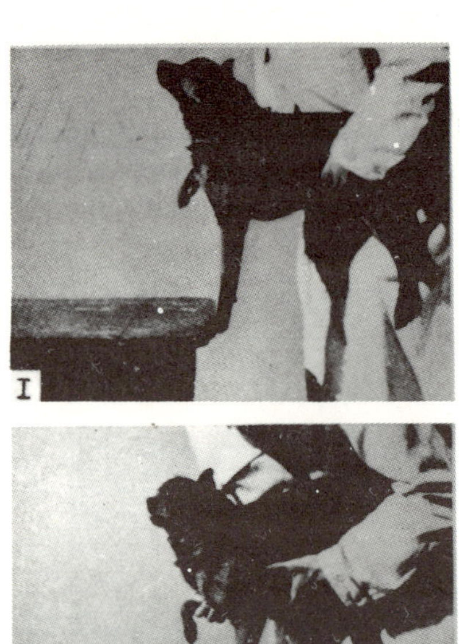

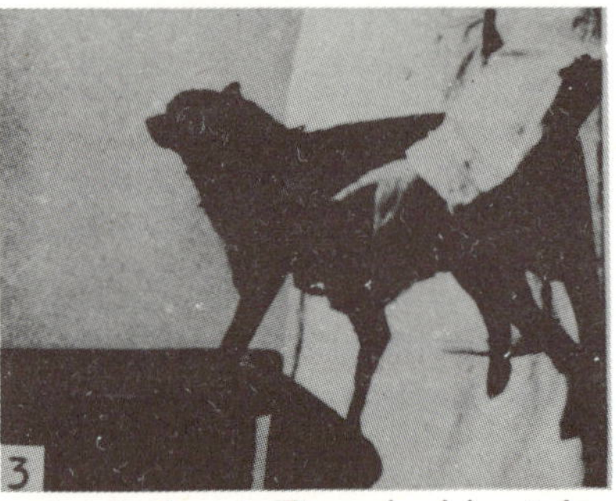

Fig. 216. Decorticate dog Fuchs. 1. The animal is set down with toes of the left forepaw on the edge of a table. By the weight of the anterior part of the body the toes are moved dorsally to the position customary for standing. The leg is extended and shows a strong supporting tonus. 2. The weight of the anterior part of the body has moved the toes dorsally to the point of over-stretching. 3. On an even stronger dorsal movement of the toes, the paw suddenly is lifted from the supporting surface, moved forwards and is set down again with the sole on the table a little more forward.

and toe joints; the paw is always set down in a normal posture, i.e., with the sole on the surface.

When, for instance, a dog is set down with one foreleg on a surface and the trunk is moved forwards, not only the upper limb is moved backwards at the shoulder, but also an overstretching of wrist joint and metacarpophalangeal joints takes place (see Figs. 200 and 203). Several factors appear which, each by itself, can produce hopping reactions. In spite of the cooperation of the various factors, the hopping reaction appears retarded, i.e., it occurs only after a wide alteration in posture (Fig. 218). There is less delay with a rapid alteration in position of the trunk then with a slow one; the delay is especially great when the animal becomes "sleepy," smaller when the animal is previously excited by pinching or pulling the tail.

In the earliest period after extirpation, the reactions appear greatly delayed, later the delay gradually decreases but can still be present some time after extirpation. At the same time, the hopping step becomes smaller, but still remains abnormally large.

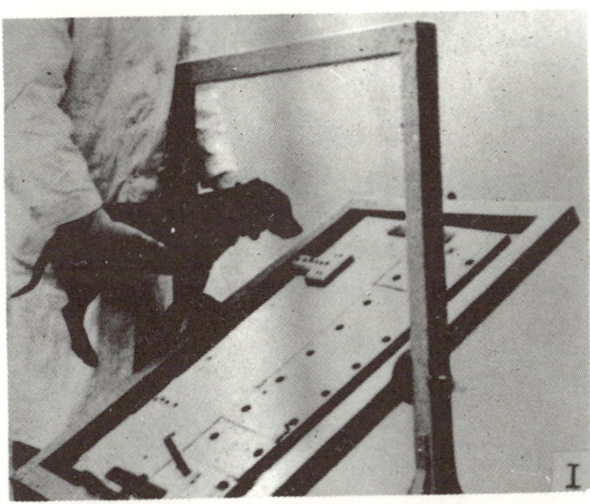

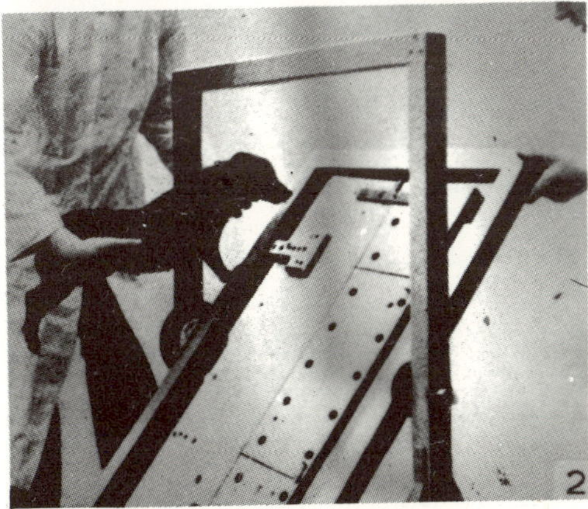

Fig. 217. 1. Decerebellate dog Piccolino with blindfold; one foreleg set down on a swinging board just above the axis of rotation. 2. On raising the head end of the board the fingers are displaced more and more dorsally, the wrist joint becomes overstretched, finally causing a hop forwards. (Decorticate dogs, too, show the same reaction.)

It is not yet known why this delay occurs; whether by the absence of preparation for standing or by the absence of a sensation of discomfort on overstretching, whether by a different distribution of muscle tonus at the start of the alteration of posture or by a change in behavior of reactions in the muscles on stretching. It should be remembered that in decerebellate dogs, too, where preparation for standing is present, the hopping reactions to the different alterations in position also are delayed in appearance.

On the basis of astereognosis often observed in persons with cerebral lesions on the one hand, and the prompt reaction of decorticate animals to touch (only the Munk's touch reflexes are absent) and pain stimulations on the other hand, the impairment of the correction of abnormal paw position has until now been ascribed to a defect in sensations which emanate from the deeper parts of the body, i.e., to an impairment of the proprioceptive sensibility (see, among others, Dusser de Barenne [61]). As we have seen, a part of the impairments of correction of limb position, for instance those which appear when the animals are set down on a lattice, are based on a failure of correction movements on stimulations from the body surface and reactions to exteroceptive stimulations (absence of preparation for standing, placing). The reactions produced by proprioceptive stimulations due to abnormal postures, however, are present in decorticate animals but are delayed. Thus, the impairments of proprioceptive reactions cause only a delay but not a nonappearance of correction.

The proprioceptive correction movements are probably caused by stretching of the muscles. We know very little about whether normal stimu -lations from the deep parts of the joints, the articular surface and articular capsules also participate in these correction movements and whether and how these stimulations take effect in decorticate animals.

In the reactions just discussed one repeatedly obtains the impression that a strong stretch of muscles brings about reactions contrary to those on moderate stretch. When the trunk is moved forwards with forward directed forelegs so that, among others, the protractors of the upper arm and the elbow extensors are moderately stretched, bracing reactions occur (extension with increase of supporting tonus). With a further forward movement of the trunk and consequently a stronger stretch of the muscles, hopping reactions appear (flexion and decrease of supporting tonus). On prevention of the hopping reactions by fixation of the sole of the foot to the supporting surface alternating flexion and extension movements appear as if the animal is attempting to liberate its paw.

On the dorsal movement of the toes in which the toe flexors are stretched, extension and supporting tonus appear in the paw (positive proprioceptive supporting reactions). A stronger dorsal movement, however, produces flexion and decrease in supporting tonus, while a fixation of the toes in overstretched position again causes alternating flexor and extensor movements.[1] Strange to say, on a stronger volar-

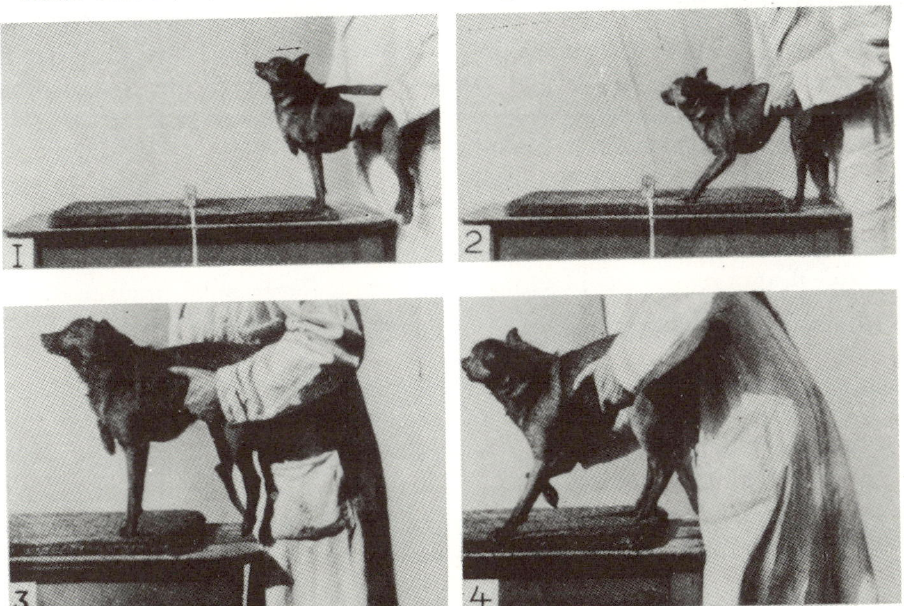

Fig. 218. Decorticate dog Fuchs. Hopping reactions forwards of the left fore-leg, 3 months after extirpation of the cerebrum (1 and 2) and 4 months after extirpation of the cerebrum (3 and 4).

(1) Investigations carried out jointly with Dr. Hoogerwerf, but not yet published, showed that in decerebrate cats in a lateral position, on dorsal movement of the toes the action currents of biceps and triceps are very much increased in force, especially those of triceps (components of supporting reactions?). On a further dorsal movement, however, the action currents diminish, first those of the biceps and then also of the triceps (diminishing of extensor rigidity or of supporting tonus). When the fingers are moved even more dorsally, the triceps currents fail to appear while, on the contrary, very strong action currents appear on the biceps (flexion of the paw, component of the leg-slackening reaction or ipsilateral flexor reflex), which, on fixation of the toes in strong dorsally directed posture, disappear and reappear periodically (alternating flexor and extensor movements).

404

flexion of the toes which is accompanied by stretching of finger and wrist joint extensors, no opposite reactions appear; thus, on a slight flexion, a flexing of the paw takes place. A fixation of the fingers in a strongly flexed posture, too, causes alternating stretch and flexion movements, especially when a pressure is exerted. This manifestation also appears in decorticate animals. On the passive volarflexion of the toes of intact human beings, the leg is only drawn up when the flexion causes a sensation of pain. In decorticate animals, the sensation of pain as such cannot be the producing factor of course but it is still possible that the strong volarflexion of the toes creates nociceptive stimuli which produce the flexed posture and flexion and extension movements, all the more because decorticate animals sometimes react with vocalization, like whimpering, especially when the fingers are pressed into a strongly flexed posture. If the drawing in of the leg into a flexed position on the strong passive flexion of toes were really exclusively due to pain stimulation, flexion would be produced by quite different stimulations than the lifting of the paw in the hopping reaction. The active flexion of the negative supporting reaction is to be compared to Sherrington's ipsilateral flexor reflex on pinching the toes. The same is true for the "phénomène des raccourcisseurs" observed in man by Pierre Marie and Foix.

The question whether perhaps the contrary reactions are produced by pain stimuli can probably be answered in the negative since, on passive flexion of the wrist joint or caudal movement of the forelimb at the shoulder, hopping reactions occur on alterations in position which can hardly cause any pain stimulations.

Finally, the observations made by Liddell and Sherrington on decerebrate animals should also be mentioned, namely that the reflex tension of the stretch muscles detached from their insertions at first becomes stronger on extension (myotatic reflexes) but with a very strong extension decrease, and even an alternating contraction and relaxation can appear in the muscle with a continuous strong extension.

The different proprioceptive correction movements just described are present in intact, decorticate, decerebellate and labyrinthectomized dogs. They are, however, absent in decerebrate and spinal dogs and after severance of the posterior roots belonging to the feet.

In man it seems that for the production of hopping reactions forwards and backwards, alterations of position in the ankle joints are of greater importance than those in the hip joints. When the lower leg of a standing person is pulled backwards, at once, after the passive alteration in posture of the ankle joints, the dorsiflexors of foot and toes (m. tibialis ant., ms. ext. digit., m. ext. digit., m. ext. hallucis longus) go into contraction so that the tendons of the muscles become visible on the dorsum of the foot; the toes and the tips of the toes are lifted and only the heels remain on the ground. At the same time, m. quadriceps contracts which pulls the patellae upwards and fixes them. On a stronger traction, this tension is followed by a hopping reaction backwards. However, when the person undergoing the experiment is pulled forwards by the lower leg, a tension of the calf muscles with a raising of the heels occurs and on stronger pulling, leg-slackening and hopping forward reactions take place. Thus hopping reactions occur on slight alterations in position of the ankle joints, while alterations in position of the thigh to the trunk must be very large to produce hopping. (With hopping reactions outwards and inwards, one has the impression that alterations of position in the hip joint play the principal role.)

Under pathological conditions in man, hopping reactions produced by passive alteration of position in the ankle joints can be impaired (see Fig. 212, A: in spite of lifting the heels no hopping reaction forwards appear on falling forwards). Foix and Thévenard (80) made very careful investigations on the behavior or reactions on pushing forwards or backwards ("phénomène de la poussée") in various cerebral diseases in man, but we cannot enter here into particulars of their very important findings.

XIV. SUPPORTING TONUS AFFECTED BY STIMULATION ARISING FROM THE BODY SURFACE AND BY ACOUSTIC AND OPTICAL STIMULATIONS

In Chapter V, we discussed in detail the great importance of a part of the body surface, i.e., of the soles of the feet, for the production of supporting tonus (magnet reaction). Furthermore, we have already mentioned that intact and decerebellate dogs in ventral position show, on static stress, a strong supporting tonus which in a dorsal position on a supporting surface (when the animals do not make any defensive movements) is present only very feebly and may even be sometimes totally absent in the hindlegs. This observation is worth emphasizing again for the reason that, according to Magnus and De Kleyn, the labyrinths in dogs are in their maximum position for tonic labyrinthine reflexes when the head is kept in a dorsal position with the muzzle $\pm$ 45° above the horizontal. Due to these reflexes, decerebrate dogs in a dorsal position with the muzzle above the horizontal show an extensor rigidity (Figs. 7 and 8) which is distinctly stronger than in a ventral position or in a standing posture when the muzzle is held 45° below the horizontal. Thus, for instance, the investigation of stretch tonus of a decerebrate dog (Dog A, Fig. 6, body weight 5.2 kg) showed that the forelegs in a dorsal position withstand a pressure of 10 kg on the soles while in a standing posture give way already to a counterpressure of the supporting surface of 8 kg.

Dog A.

Before decerebration:

Supporting tonus of the forelegs in standing
posture   +15 kg — 17 kg

After decerebration:

Extensor tonus of forelegs in standing posture   + 6 kg — 8 kg

Extensor tonus of forelegs in dorsal position   +10 kg — 11 kg

Thus decerebrate dogs show to some extent (especially those with strong tonic labyrinthine reflexes) supporting reactions opposite to intact and decerebellate animals.

In decerebellate dogs in the stage of permanent impairment, in a dorsal position on a supporting surface, there is indeed a decrease in supporting tonus which, however, is less remarkable than that of intact and decorticate dogs (see Chapter VI) and never leads to a total disappearance of supporting tonus.

Immediately after extirpation of the cerebellum, the transfer from a ventral to a dorsal position causes, in rigid animals, an increase in extensor rigidity, just as in decerebrate animals, while in dogs with flaccid extremities after extirpation a distinct supporting and extensor tonus is absent in ventral and dorsal positions.

These findings can be explained by the fact that, in intact and decorticate dogs in a dorsal position, certain stimulations emanate from the surface of the back which cause a decrease in supporting tonus. When the animals are lifted from the dorsal position, supported with one hand by the neck and pelvis, an increase in supporting tonus on static stress instantly occurs as soon as the contact of the back with the supporting surface is eliminated, while a decrease in supporting tonus again appears when the animal is put down. In decorticate animals, even the seizing of a fold of the skin of the back, in a standing posture, causes a strong decrease in supporting tonus so that the limbs give way under the weight of the trunk (Fig. 219).

On seizing a fold of skin above the pelvis only the supporting tonus of the hindlegs disappears, above the shoulders only that of the forelegs, while on seizing a large fold of skin in the center of the back all four legs collapse.

We want once again to mention here the observation made on the right-sided decerebellate dog Fox one month after extirpation. The animal, when transferred from a standing posture into a dorsal position on a supporting surface, showed no decrease in supporting tonus in the right hindleg; in the left, however, a very strong one (of $5\frac{1}{2}$ kg, see Table 27).

Table 27. The dog Fox ($6\frac{1}{4}$ kg) 1 month after extirpation of the right half of the cerebellum.

	Strength of supporting tonus in	
	Right hindleg kg	Left hindleg kg
In standing posture on one leg	$+ 6\frac{1}{2}$ ($-7\frac{1}{2}$)	$+ 8$ (-9)
In dorsal position	$+ 7$ ($-7\frac{1}{2}$)	$+ 2\frac{1}{2}$ (-3)

Thus, in a dorsal position the right hindleg showed a stronger supporting tonus than the left one, just contrary to a standing posture. At a later stage, however, a decrease in support tonus also appeared in the ipsilateral right hindleg of the animal in a dorsal position which was inferior to that in the contralateral left hindleg (see Chapter VI).

Thus, the hindleg on the side of the preserved half of the cerebellum which showed a strong decrease in supporting tonus behaved like that of the intact or decorticate animal. The hindleg of the side of extirpation, which showed only a slight decrease, behaved like that of the totally decerebellate animal(immediately after the extirpation an increase even occurred in this leg which thus behaved similarly to that of a decerebrate animal).

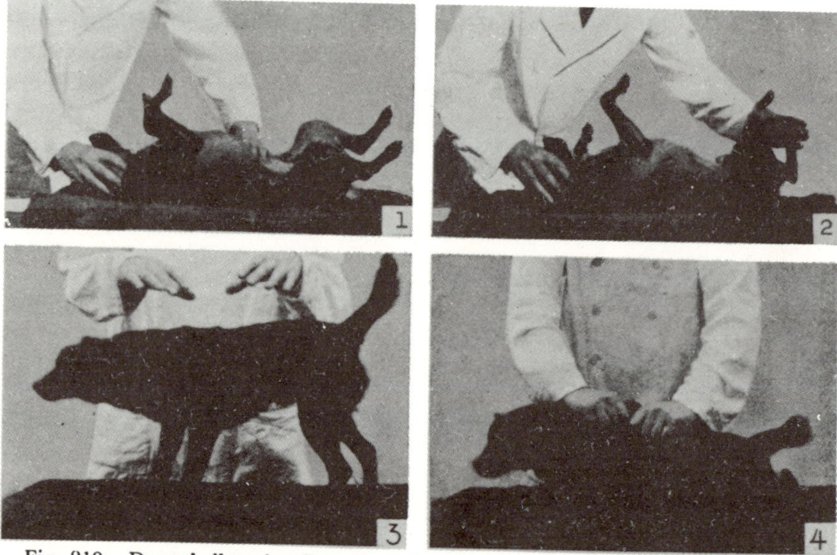

Fig. 219.   Decerebellate dog Fox.   1. Animal in a dorsal position on a supporting surface with the muzzle 45° above the horizontal, legs in flexed position.   2. Hindlegs under static stress not extended, but even flexed. No supporting tonus. 3. Animal in standing posture. Distinct supporting tonus in the legs.   4. Instant flexion of the limbs on seizing a skin fold even when no pressure is exerted on the back.

Thus, in a dorsal position on a supporting surface, stimulations emanate from the surface of the back which reduce the supporting tonus. These inhibiting influences are most evident in intact and decorticate dogs and less strongly in decerebellate animals. In unilaterally decerebellate animals, they are not very strong in the legs on the side of extirpation and, conversely, they are usually very strong in the opposite legs, while they are totally absent in decerbrate animals.

In normal human beings in a dorsal position, the legs can be flexed rather easily by a pressure on the soles of the feet, when they are not making definite movements. Conversely, in a dorsal position, they can hardly be overcome by pressure on the shoulders and even offer a much stronger resistance than spastic legs. It is thus possible for dock workers to carry loads up to 100 kg and more on their shoulders more or less automatically, without their legs giving way. As in animals, intact persons in a dorsal position also show no distinct positive supporting reactions. Conversely, a supporting tonus appears as soon as the contact of the back and the seat with the supporting surface ceases and the soles of the feet stand on the ground.

In patients suffering from paralysis agitans, as was emphasized recently by Froment, the increased tension in the muscles of the extremities decreases when the patients lie with their entire back on a supporting surface, while on standing the rigidity is extremely strong (even on leaning the back against the wall of a room the contraction in the arm muscles often distinctly decreases according to Froment).

In agreement with observations made on animals, Foerster, Stenvers and myself were able to show that patients with diseases of the cerebellum, in contrast to normal persons lying in a supine position, sometimes show distinct positive supporting reactions.

In hemiplegics, the healthy leg often shows a stronger supporting tonus on standing. In the dorsal position, however, the hemiplegic leg offers a stronger resistance to pressure on the sole of the foot. Thus, transfer to a dorsal position does not bring about a distinct decrease in supporting tonus in the hemiplegic leg. Also, in cases of typical decerebrate rigidity in man, a dorsal position does not bring about a decrease in rigidity. It seems that in man, too, the dorsal position exerts an inhibiting influence on supporting reactions and this influence, just as in animals, appears impaired under pathological conditions (cerebellar diseases, hemiplegia, decerebrate rigidity).

Also, in lateral positions on a supporting surface, a decrease in supporting and stretch tonus normally takes place, probably as a component of Magnus' body righting reflexes on the body. In a lateral position in the air, the animal usually keeps its legs more or less extended; conversely, in a lateral position on a supporting surface, the legs are at first totally drawn in to the trunk, even when the head is held in a lateral position. Body righting reflexes acting on the body then cause the trunk to rotate into a ventral position and the body is finally raised into a standing posture due to supporting reactions (Fig. 229). This succession of reactions in a lateral position on a supporting surface can be observed in intact as well as in decorticate, decerebellate (in the stage of permanent impairment; Fig. 229) and labyrinthectomized dogs.

When a tired dog lies lazily in a lateral position in front of a fireplace, it usually keeps its four legs extended; the mechanism of the body righting reflexes on the body and the mechanism for decreasing the stretch tonus (flexion) are then put out action by the sleeping condition. Similarly, a decrease in extensor tonus and righting of the trunk fail to appear when decerebrate animals are transferred from a lateral position in the air to a lateral position on a supporting surface; sometimes, a transient increase in extensor rigidity then appears. This is particularly evident when the animal is dropped on a supporting surface in a lateral position. Also shaking the animal to and fro in a lateral position, as with any other strong stimulation, usually increases rigidity temporarily.

In decerebellate animals which became rigid following extirpation, the lateral position on a supporting surface also does not cause any body righting reflexes on the body in the early stage after extirpation and the legs remain in a pronounced extended posture. In these animals, the righting reflexes generally return simultaneously with the disappearance of the extensor rigidity. In those animals which show a rigidity for a long period after extirpation, the body righting reflexes on the body also fail to appear for a longer time; they conversely appear earlier when there had only been a temporary and slight rigidity.

Similar conditions are found in unilaterally decerebellate animals. In the first 1 or 2 days the animals show an extensor rigidity in all four legs in both lateral positions, the body righting reflexes on the body are also absent in the right as well as in the left lateral position. In right-sided decerebellate animals, the body righting reflexes on the body soon reappear in lateral position to the left; similarly, the four legs are totally flexed (found in several animals already on the day of the operation); in lateral position to the right, however, they remain extended. In a lateral position to the right, the animals soon (sometimes from the start) draw in their left legs in a flexed position while the ipsilateral legs remain in extended posture.[1] In this case, the inhibiting influence manifests itself in lying on the side of extirpation only on the legs of the intact side. Two to four weeks after extirpation, the animals, like intact ones, show a flexed posture of all four legs in both lateral positions, as well as body righting reflexes on the body.

In the dog Däumling, whose cerebellum was first removed, followed 52 days later by the right half of the cerebrum, the following findings worth mentioning were brought to light 4 days after the last extirpation. In the lateral position to the right on a supporting surface, the animal kept its right legs rigidly extended, the left ones flexed; it showed the same leg position when it was held in the air by a fold of skin of the right side. In a left lateral position on a supporting surface or held in the air by a fold of skin of the left side of the body, on the other hand, the left legs showed an extended posture, the right ones were flexed. When the animal, from a right lateral position, was lifted by a skin fold of the left side, the extended right legs went into flexion, the flexed legs were extended. When the animal in a ventral position in the air was held supported from the ventral side of the body, it kept all four legs totally extended. Seizing a large fold of skin of one side produced a decrease in stretch tonus and a flexion of the opposite legs. Thus, in this animal, stimulations from one side of the body surface only caused inhibition of extensor tonus and flexion in the opposite leg. The body righting reflexes on the body were absent in both lateral positions. Later, the body righting reflexes on the body reappeared and all four legs were flexed in both lateral positions.

Thus, stimulations emanating from the body surface can:

1. reduce the extensor or supporting tonus;

2. produce body righting reflexes on the body.

---

(1) The peculiar behavior of the leg position after unilateral lesion of the cerebellum, i.e., flexed posture of all four legs in lying on the intact side, flexed posture of the contralateral legs coupled with extended posture of the ipsilateral ones when lying on the side of extirpation, was observed and reported already by André Thomas (308: Figs. 36 and 37).

412

The decrease in supporting tonus probably represents a component of the body righting reflexes on the body.

This influence reducing the extensor tonus appears in intact, decorticate, decerebellate and labyrinthectomized dogs but is absent in decerebrate animals and in decerebellate animals in the first period after the extirpation. In unilaterally decerebellate dogs, it fails to appear at first on stimulations from the right as well as from the left side of the body. Later, however, it shows itself only on stimulations from the side of the preserved half of the cerebellum and finally also on stimulations from the side of extirpation, and sometimes at first only in the contralateral legs while later on also in the ipsilateral ones.

Other parts of the body surface, too, are able to produce alterations of tonus in the legs under static stress. Thus, standing decorticate dogs show distinct alterations of tonus in the extremities on seizing the tail, ears, or muzzle (Fig. 190).

Acoustic stimulations can produce alterations of tonus in two ways. With certain noises decorticate dogs lying quietly in a lateral position sometimes show a pricking up of the ears, a raising of the head from the supporting surface and an extension of the legs. Thus the alterations of tonus in the muscles of the extremities can be brought about by subcortical reactions. Similar alterations of tonus in the muscles of the extremities are shown by intact dogs when they sleep on the ground in a lateral position and are startled by noises. Moreover, intact dogs also show other reactions on acoustic impulses. For instance, when one suddenly hits the table loudly with the hand, one can observe simultaneously with the appearance of startle movements a flexion of the limbs of a standing dog. Since decorticate dogs in this case do not show any decrease in supporting tonus, it can be assumed with great probability that the decrease in supporting tonus in intact dogs is brought about by way of the cerebrum. This assumption is supported by the "conditioned" occurrence of the decrease, which fails to appear after a frequent repetition of the noise.

Shy dogs start back when they are snarled at, in which case the supporting tonus is reduced; this is particularly the case when these animals had been snarled at before and had then been given a thrashing.

When, however, the animal is given a piece of meat for several consecutive days immediately after being snarled at, the animal will no longer start back and the legs will not flex. Thus the decrease of supporting tonus is caused by a conditioned effect which produces the inhibition of supporting reactions.

In man, too, reflex muscle tensions can be inhibited by acoustic stimulations. It is well known, for instance, that, in terror caused by an explosion, the legs may give way and that the tension of the abdominal wall produced by palpation can be abolished by energetic persuasion on the part of the examiner.

Optical stimulations can also produce conditioned alterations of supporting tonus in dogs with an intact cerebrum. This is very clearly shown in my decerebellate dogs which, in the third week after extirpation, were let loose on a floor tiled with lingeous granite. Here, the animals still crawled around after some time with flexed limbs as if the legs were atonic and the animals were no longer able to run around on extended legs. When one day, however, a heap of straw was placed in a corner of the room, the animals crawled with flexed limbs towards this heap, but shortly before reaching it suddenly extended their legs and with extended legs made a few quick steps to throw themselves on the straw. Outside on a lawn, they were able to run with extended legs for a time, over a rather long distance. In this case, the legs were even kept in an exaggerated extended position, while the animals when brought back into the room again crawled around with flexed limbs.

Thus, optical impulses can influence the supporting tonus. In the above-mentioned decerebellate dogs, the impulses produced by the hard floor tiled with lingeous granite were the cause of a reduction of supporting tonus. As with acoustic stimulations, a sudden appearance of optical stimulations can also produce withdrawal reactions and collapse of the legs in animals with intact cercbrums.

From the above-mentioned investigations, we concluded that not only intact but also decerebellate dogs are subject to the influence of "conditioned" effects of the cerebrum on the supporting tonus. It seems, however, that the inhibiting influence on supporting tonus, magnet and supporting reactions is essentially inferior to that in intact animals. At any rate, these reactions were more constant. It could not be established whether this is perhaps caused by the fact that the subcortical magnet and supporting reactions are abnormally strong in decerebellate animals or by the fact that the "conditioned reflexes" were altered by the influence of the repeated falling and collisions (particularly in the period when the animals started to stand and run again). We will learn more about this subject when we succeed in keeping alive a few combined decerebellate-decorticate dogs for longer periods.

XV. GENERAL INFORMATION ON BALANCE REACTIONS AND THEIR SPECIAL BEHAVIOR IN DECEREBELLATE ANIMALS

There exists a large number of theories on the functions of the cerebellum which, unfortunately, remain nothing more than hypotheses. Through Luciani's brilliant experiments (1882 and later), we know indeed which functions are preserved in the decerebellate higher mammal and, on the other hand, which abnormal phenomena are shown by these animals. Luciani (181) was the first to succeed in keeping alive higher mammals for a long time after total extirpation of the cerebellum and observed these animals so closely that later investigators like Munk (213, 214), Lewandowsky (167, 168), André Thomas (300), Dusser de Barenne (61) and others, who repeated his experiments, could hardly add anything new to his findings. Unfortunately, neither Luciani nor any other investigator succeeded in finding a satisfactory explanation for the mechanism of the observed defects, much less to deduce from these impairments the function of the cerebellum.

A great number of authors (Magendie, Serres, Ferrier, Babinski, Munk, André Thomas, Ingvar and others) adhered to the theory that the cerebellum is an organ which regulates the balance of the body in the various functions like standing, running, and jumping.

Magendie's conception (187) is based on observations after partial lesion of the cerebellum. He repeated Pourfour du Petit's experiments and severed the cerebellar peduncles in rabbits. According to the severance of the right or the left peduncle, he observed the occurrence of rotating movements to the right or to the left round the longitudinal axis. He saw similar reactions after an asymmetrical sagittal bisection of the cerebellum, as well as after a vertical lateral severance of the transverse fibres of the pons. Moreover, after a lesion of the cerebellum there was often an irresistible inclination to run backwards and to turn a backwards somersault, while, on the other hand, transverse lesions of the brainstem through the corpora striata produced a strong inclination to walk forwards.

Based on these investigations Magendie assumed that, in the brain four moving forces which should regulate the balance of the body, two should regulate the balance in the frontal plane and two in the sagittal plane. He placed the sideway and forward directed moving forces in the cerebellum (in addition to the participation of pons and medulla) and those directed backwards in the corpus striatum.

Serres (271) agreed with this theory based on his own observations in patients and experimental animals. In patients with cerebellar haemorrhage he found a pronounced inclination to fall backwards. He also described for the first time a case of compulsive rotation round the longitudinal axis. Most important in his reported observation was that, in experimental animals, the backward somersaults, the backward running and the backward stretching of the neck following a cerebellar lesion ceased instantly after bisection of the neck muscle, so that the animals could very well run forwards with their head lowered between the forelegs.

Ferrier (66), too, arrived at a similar conception based on experiments with electrical stimulation and extirpation of certain parts of the cerebellum. According to him, cerebellar injuries caused impairments in standing and in locomotion. The muscles are not paralysed and the animals are still able to perform coordinate movements which, however, are no longer in harmony with the position of the body in space. In his experiments with electrical stimulations, he observed movements of the eyes as well as of the extremities which he considered as compensatory movements for the avoidance of impairments of balance in opposite directions.

According to him, there are various centers of balance in the cerebellum of which each particular one is necessary for the restoration of the balance of the body when this threatens to be lost in a certain direction. Each half of the cerebellum is only capable of exerting an influence on the same half of the body so that a section in the midline of the cerebellum from front to back can bring about no perceptible impairment of balance. Conversely, asymmetrical lesions, depending on their location, can produce different impairments of balance, thus with a lesion of the anterior part of the middle lobe, the vermis, animals show an inclination to fall head forwards, with lesions of the posterior part of the vermis an inclination to fall backwards and a compulsive dorsal posture of the head, while a lesion of one of the lateral lobes causes a falling sideways. Ferrier assumes that the animals are later able, by conscious efforts, to compensate for the loss of this mechanism and, by this, are able to maintain their balance little by little, even if less safely than before. Ferrier's observations are based on reactions which he saw with electrical stimulations of the cerebellar cortex and immediately after cerebellar lesions. How prudent one has to be in the estimation of findings in experiments with electrical stimulations is shown by a review of Ferrier's own work; stimulation of pyramid, for instance, produced different reactions in different species of animals.

Moreover, Clarke and Horsely showed that the phenomena after electrical stimulation of the cerebellar cortex are due to currents spreading to the nerves of the brainstem and the nuclei and they assert that the cerebellar cortex cannot be stimulated at all electrically. Irrespective of whether this assertion is correct or not, Horsley and Clarke's experiments, as well as those of later investigators, show that the method of investigation by electrical stimulation can by no means be regarded as incontestable and precise for the physiology of the cerebellum.

Ferrier's findings obtained after partial lesion and extirpation are not convincing, as Luciani already emphasized rightly, apart from the fact that it is always risky to jump to conclusions from phenomena appearing immediately after an operation. To support his theory, he mentioned his observations on two experimental animals whose cerebellums he had injured, a monkey in which he had injured the caudal end of the vermis by means of a cautery and another monkey in which the left lateral lobe was partially destroyed by the cautery. The first animal showed a strong inclination to turn backward somersaults, while the other was inclined to fall to the right and backwards. An inclination to fall head forwards could not be observed. Later, Ferrier himself felt that he had gone too far in his deductions, because, in a later work in cooperation with Turner (67), he was already more prudent and recognized the fact that his experiments could not solve the problem of cerebellar function.

Ramon y Cajal (38) adopted a similar opinion based on anatomical findings. According to him, in all probability the cerebellum regulates the compensatory movements of trunk, head, and eyes; the cerebellar hemispheres the compensatory movements of these parts of the body round the anteroposterior axis, and the vermis the flexion and extension movements, i.e., the movements round the transverse axis.

Recently, Ingvar (124), too, agreed with the theories of balance, that is, based on his own anatomical investigations and in consideration of observations made by Rynberk (246 - 251), Rothmann (255 - 261), Thomas and Durput (304) and Barany (10 - 12) in their attempts to solve the problem of localization of cerebellar function in the cortex cerebelli. According to Ingvar, the arrangement of functions in the cerebellum does not correspond to the body region it influences, as asserted by Bolk (26), but to the different directions of movement. The center of gravity of the body can be drawn into every direction, therefore the cerebellum must possess centers for all possible angular displacements. Lobus ant. and post. cerebelli (for nomenclature see Ingvar's illustrations, Figs. 220 and 221) form, to a certain degree, a ring-shaped structure. This basal annular structure relates to the static and kinetic displacement of the center of gravity of the mass of the trunk. In the lobus anterior, all those muscular contractions (synergies) are influenced by reflexes which maintain the balance of the body forwards and counteract forward drawing forces, in other words, which pull the body towards the rear; while in lobus post. medianus

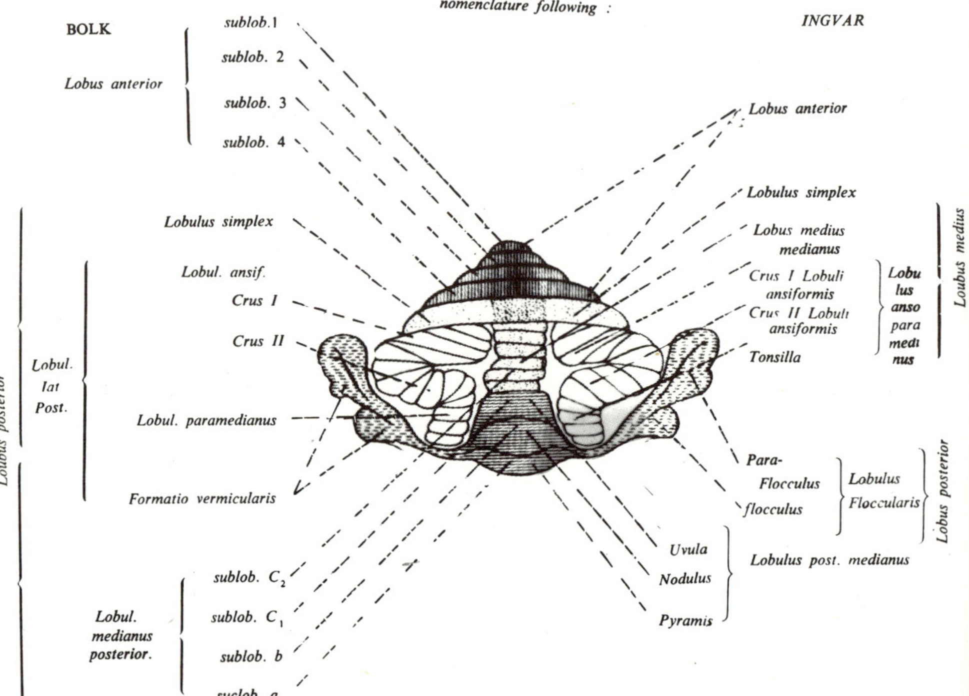

Fig. 220. Diagram of the cerebellum of mammals by Sven Ingvar.

(pyramis, uvula, nodulus) there are centers which innervate the muscular synergies for the balance of the body backwards. Adjusted to the trunk are the extremities set up on it which present, within certain limits, free mechanical systems with their own centers of gravity and therefore require an independent representation in the cerebellar cortex.

According to Ingvar, the experimental physiology has shown us that individual regions for the extremities are present in the lobus ansoparamedianus and that crus I lobuli ansiformis comprises the centers for the equilateral posterior ones. Crus I lies nearest to lobus ant., crus II to lobus post. According to Ingvar, here, too, the principle of localization is arranged according to the principle of movement "since the anterior extremities play a much greater role than the posterior ones in the balance forwards. For the balance backwards, however, the posterior extremities" play a greater role. He places the centers for adduction into the most lateral section of lobus ansiformis at the point where crus I merges with crus II and the centers for abduction of the two equilateral extremities into the medial section of the crura. In his opinion, the cerebellum is "the organ which, for the purpose of maintenance of balance of the body, has to regulate by reflexes and to compensate and resolve the interplay of forces which are woven into total body performance. These forces, gravity and inertia, are of a static and kinetic nature. Every active muscular activity alters the static and dynamic conditions and every alteration of these conditions produces stimulations in the labyrinths and in the body which, without attaining consciousness, flow into the cerebellum, the principal ganglion for the regulation of balance." Ingvar considers these stimulations specific for the cerebellum and emphasizes that his concept corresponds to that for Lotmar (178) which shows that the sole impairment of sensibility to be found in cerebellar lesion or disease is an impairment of estimation of weight by the extremities. By this observation which is confirmed by Maas (186), Goldstein (98), Reichmann (99) and others (which is, however, contradicted by Gordon Holmes 121, André Thomas, Dusser de Barenne and others), Ingvar considers proven that the elementary components of innervation with which the cerebellum performs its reflex activity and which reach the cerebellum by way of the nervous vestibularis and of Gower's and Flechsigs tract, are of a static and kinetic nature. The influences of the forces of gravity and inertia of the body, which threaten the balance of the body in the different actions and positions, are regulated by way of the cerebellum by unconscious stim-

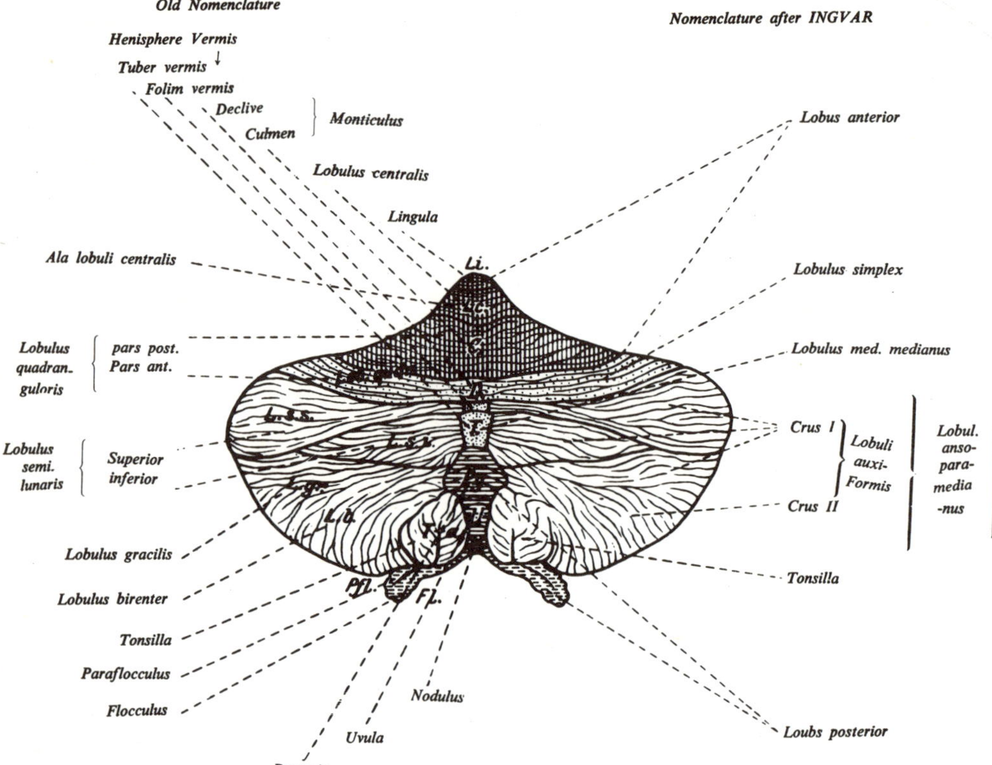

Fig. 221. Diagram of the cerebellum in man by Ingvar.

Lobulus quadrangularis anterior	= Lob. quatrilaterus anterior	= Lob. lunatus ant.	= Anterior crescentic lobule
,, ,, posterior	= ,, ,, posterior	= ,, ,, post.	= Posterior ,, ,,
Lobus bircnter	= Lobus digastricus	= Lobus cuneiformis	
,, gracilis	= Lobe grelc.		

ulations produced by the forces of gravity and inertia themselves and which are comprised by Ingvar under a sense of inertia. "The cerebellum is an organ in the service of an unconscious inertial sense which, by reflexes, has to counteract and to resist the gravitation and inertia of the body masses within appropriate limits to the purpose of maintaining the balance of the mechanical system of the body." Ingvar endeavored to prove his theory by animal experiments and, for this purpose, injured the lobus anterior in five rabbits (no microscopic control). One animal showed no impairments at all, the second was killed instantly, the third remained motionless for 8 days and never started to run again. The two other animals remained alive for 5 and 4 days; one showed, in running, a sideways slipping of the forelimbs so that the head came to lie on the ground, while the second, on the third day, made jumping movements which led to falling head forwards. Somersaulting forwards was never observed; once he even saw a temporary compulsive falling backwards in a rabbit with a destroyed lobus ant. In some other rabbits, he destroyed the lobus posterior and, after that, saw several times (but not always) a falling backwards, but only on the day of the operation. Only one animal showed a distinct falling backwards in the two days it stayed alive; the post mortem examination revealed an extensive subdural bleeding which reached over the corpora quadrigemina, the convexity of the cerebral hemisphere, the ventral side of the pons, the posterior part of the fossa rhomboidea and along the dorsal plane of the spinal cord to the thoracic cord. Ingvar made no experiments relative to the role of the cerebellum for the maintenance of balance to both sides. One can hardly call these findings convincing. Ingvar also attempted to support his conception with comparative-anatomical results. With the great versatility and variability of nature, it is usually equally easy to find species of animals whose conditions of brain and body can serve to support a theory as it is to find those whose conditions are in direct opposition to the theory.

Man is the only mammal who stands and walks on two legs and is even able to stand with stilts on his feet and his hands in the pockets of his trousers without falling backwards. Yet, in man, the lobus posterior medianus (pyramis, ovula, nodulus) which, according to Ingvar, regulates the balance of the body backwards is very poorly developed. Among mammals, the goat possesses a very poorly developed cerebellum; nevertheless, it must have a very highly developed balance mechanism when one considers how this animal climbs, stands upright on the hindlegs and reaches for high-hanging leaves. "In the turtle, the cerebellum has a more highly developed shape than that of snakes and frogs";[1] nevertheless, the turtle is not able to turn over its carapace when lying on its back and must wait for death in this posture. When a turtle is dropped from the air in a dorsal position, it will touch the ground with its back and not turn round in the air. It seems to me that these examples speak against the theory of balance with the same justification as, according to Ingvar, the strong development of cerebellum in the "massive" elephant and crocodile speaks for the support of this theory.

(1) Quoted from C. U. Ariens Kappers: Die vergleichende anatomie des Nervensystems der Wirbeltiere und Menschen (Comparative anatomy of the nervous system of vertebrates and man) 2, p. 682, Haarlem: De Erven F. Bohn, 1921.

The just discussed theories on the functions of the cerebellum all contain, on the whole, similar or equal modes of viewing the mechanism of balance which is regulated by various forces in the body, forces which are capable of drawing the body forwards, backwards and to both sides, which compensate each other and which carry out their functions by way of the cerebellum (only partly according to Magendie). Moreover, these theories are most exclusively based on phenomena which appeared immediately after the cerebellar lesion and from which one drew conclusions as to the function of the cerebellum. In my opinion, this lacks any justification, since today we know for certain that, after a lesion of the central nervous system, phenomena may appear which are not produced directly by the absence of functions of the injured or extirpated centers but by shock, diaschisis, alterations in the flow of blood, or haemorrhagic pressure on the intact parts of the central nervous system.

All these theories are based actually on only two observations made after lesions of the cerebellum, i.e., the falling backwards and the rolling sideways; a falling head forwards has never been observed after lesions of the cerebellum, so that the assumption that the balance forwards is also regulated by the cerebellum is totally unfounded. Just as unfounded are the reasons which support the theory that the centers for the regulation of balance backwards are to be found in the caudal parts of the vermis, lobulus medianus post. (uvula, nodolus and pyramis). The fact that sometimes after lesion or extirpation of these parts a backward falling can be observed, is in itself no proof because, firstly, the occurrence is not constant; secondly, it appears only temporarily and sometimes only in the first hours after the operation and, thirdly, it can also be observed occasionally after other lesions of the cerebellum and also after extirpation of the entire median part, the entire "old" vermis and after total extirpation. Ingvar himself observed a backward falling even after a lesion of the median part of lobus anterior, while Thiele (299) saw it occur regularly after the exclusive removal of the tentorium cerebelli in cats. Assumedly, backward falling is produced by the absence of forward drawing forces which prevent the rearward shifting of the center of gravity of the body. Until now there is no proof that the forward balance mechanisms are really absent. In this respect, Serre's observations are of importance, namely, that after bisection of the neck muscles the animals are able to run forwards quite well, that the dorsiflexion of the neck and the extension of the anterior extremities relaxed and that the trunk shows no inclincation to fall backwards. Simonelli (282, 283), by a simple passive ventral flexion of the head, totally eliminates the falling backwards; thus, according to these observations, the backward falling must be caused only by an enormous tension of the dorsiflexors of the neck which move the head more and more dorsally. In this way, there appears tonic cervical and labyrinthine reflexes on the muscles of the extremities and the spinal column which finally cause the backward somersaulting of the animals.

The theories of regulation of balance in the sagittal plane are thus based solely on a phenomenon appearing *inconstantly* and for *a short time* immediately after the operation *in ths state of shock*, which is produced in the first place by an increased contraction of the levators of the neck and which appears after lesions of the various parts of the cerebellum as well as in the neighborhood of the cerebellum.

The supposition that the cerebellum also regulates the maintenance of balance in the frontal plane is based on the sideways rolling of the animals after the unilateral extirpation or lesion of the cerebellum. After lesion of the cerebellar cortex rolling movements rarely occur, most frequently (but not always) they appear after the unilateral bisection of the cerebellar peduncles with or without following unilateral extirpation of the cerebellum. In this case, the line of intersection lies directly dorsal to the vestibular nuclei, Deiter's and Bechterew's nuclei, and associated lesion of these nuclei is hardly avoidable. However, without such lesion, either the shock or the lesion of the blood vessels in the cerebellar peduncles and the following alterations of the circulation of blood in the vestibularis region would also be deterimental to the function of these nuclei. Therefore, it should not be surprising that the direction of the rolling movements and of the usually simultaneously occurring strabismus and nystagmus is the same as in the same manifestations due to extirpation of the ipsilateral labyrinth. After extirpations of the labyrinth, a part of the manifestations (strabismus and a turning of the head downwards, when the animal is suspended with blindfolded eyes) remain permanent, while after bisection of the cerebellar peduncle the appearance is of a short duration. Rothmann succeeded in severing a brachium conjunctivum alone in a dog without causing any forced movements in the 27 days the dog stayed alive. Similarly, another dog in which the entire cortex of the left cerebellar hemisphere was removed without injuring the cerebellar nuclei (microscopically controlled) showed no trace of any forced movements.

424

In summing up, the above-mentioned observations justify the following conclusions:

1. After lesion of the cerebellar cortex, a falling sideways occurs only exceptionally and is of short duration, so that the assumption of the presence in the cerebellar cortex of centers for the maintenance of balance in the frontal plane is not justified.

2. The bisection of the cerebellar peduncles always produces a falling sideways and often[1] also rolling movements. However, these symptoms, too, occur only temporarily and it is not at all proved whether these manifestations are produced on account of the bisection of the efferent and afferent cerebellar tracts or by shock in the vestibular and other nuclei situated in the brainstem, or by both factors together.

Munk and André Thomas, too, consider the cerebellum as a balance regulating organ and based their conception on the *permanent* impairments observed in dogs and monkeys after partial or total extirpation of the cerebellum.

Munk (214) points to the fact that decerebellate dogs perform some movements entirely normally. The decerebellate dog pricks up its ears normally, wags its tail normally, scratches itself normally with fore- or hindpaws, licks itself clean and snaps fleas without any perceptible impairments. The decerebellate monkey reaches in a normal way for a piece of carrot lying near it, puts it into its mouth with its hand and, after having eaten it, licks the remains sticking on its hand. It scratches and licks itself, even catches flies on its body, climbs, all in a normal way. Munk further confirmed Luciani's statement that decerebellate dogs can swim very well and points to the fact that, when held in the air, they make well-coordinated running movements and when supported on one side can run along a wall in an almost normal way. Marked defects were observed only in free standing and running, but even then the paws are set down and moved normally. Based on these observations Munk says: "Thus a great number of isolated arbitrary movements are again carried out normally, not exceptionally once in

---

(1) Of five animals (four dogs and one cat) in which, after bisection of the peduncle of the cerebellum, I removed the right half of the cerebellum, three showed rolling movements to the right, while, in the other two (the cat Josephine and the dog Peter, see Fig. 296), these movements were totally absent.

a while but often, and one sees them taking place again and again without a perceptible trace of asthenia, atonia, astasia or ataxia. Besides, the normal movements extend to all parts of the body, to the vertebral column and to the extremities where one would have least expected to find them. Therefore, there can be no question of a general impairment of motility and sensibility; and it has to be decided that the finer details of maintenance of balance are the function of the cerebellum. The cerebellum is the central organ for unconscious, coordinate, cooperative movements of the vertebral column and extremities in general, and for the finer maintenance of balance of animals in particular." Munk's observation that decerebellate animals are capable of carrying out various movements correctly, confirmed, among others, by Luciani, is naturally of great importance. Moreover, his assertion appears to me to be absolutely correct in that one is not justified to regard general impairments of motility as caused by the loss of the cerebellum, but I cannot agree with him in his opinion, to conclude from the absence of general impairments of motility and sensibility that an impairment of balance must absolutely be present. In my opinion, he would have been somewhat justified to arrive at this conclusion only after establishing the complete absence of certain balance reactions. Just as little proof follows from the normal setting down of limbs and paws in free standing and running that the impairments of balance have to be shifted to the veretebral column and the proximal extremities, as asserted by Munk, because he never could observe a constant absence of certain movements of the spinal column and extremities participating in the maintenance of balance. Munk says: "Adequate observation material for the vertebral column is lacking and will be difficult to provide." In his work, however, not only this observation material is lacking, but also no observations can be found on the absence of balance reactions in the extremities.

André Thomas (300), based on numerous animal experiments and observations on cerebellar patients, particularly with cerebellar atrophies, also comes to the conclusion that the cerebellum represents a center for balance reflexes and explains the falling down into a lateral position after the unilateral extirpation of the cerebellum as the abolition of a reaction complex which normally prevents the transfer of the center of gravity towards the side of extirpation. When a dog raises its foreleg, a balance reaction occurs, according to Thomas, which consists of a rotation of the trunk and neck with an inclination of the head towards the other side and an adduction of the raised leg. Similar reactions occur on raising a hindleg or when the supporting surface of the animal is moved up and down from one side. Thomas asserts that the rotation of the trunk and neck is brought about by a contraction of the vertebral muscles on the side of the raised leg and that these contractions are normally present bilaterally and compensate each other at rest. After the unilateral extirpation of the cerebellum, however, only the muscles and the adductors of the extremities of the contralateral side would be stressed,[1] while the compensatory rotation of trunk and neck towards the intact side and the adduction of the ipsilateral legs would not occur, due to the absence of contraction of the muscles of the side of extirpation and because of this the animals would fall and roll towards the side of extirpation.[2]

Thus, according to André Thomas, the abnormal abduction posture of the legs on the side of extirpation promotes a falling down towards this side, while he considers the strong abduction posture which is usually shown by these animals also in the opposite legs as an action of compensation for the prevention of falling down towards the side of extirpation.

After a total extirpation of the cerebellum these balance reactions are absent on both sides. André Thomas places the central mechanisms for the maintenance of balance and of the statics of the trunk into the vermis into which terminate the great majority of spinocerebellar fibres and the connection with the vestibular nuclei. Up to this point, his theory essentially agrees with Munk's: For the maintenance of balance, the cerebellum regulates the contractions of the upper part of the body and the adductors of shoulder and hip joints. Diverging from Munk, however, Thomas assumes that the cerebellum also regulates the balance reactions of the other parts of the body, among others, of the distal extremities and that, caused by the cessation of these reactions after the extirpation of the cerebellum, decerebellate dogs, on altering their position, are not able to regain instantly a total stability.

---

(1) As we have seen, in the unilaterally decerebellate animal, just the opposite occurs: the vertebral muscles on the side of extirpation and the abductors of extremities on the contralateral side are particularly strongly contracted (see also Chapter XVIII).

(2) To prevent falling towards the right side, in my opinion, the right legs should brace outwards, thus the abductors should contract.

André Thomas further asserts that the cerebellum regulates the balance reactions under the influence of peripheral stimulations and central impressions and does not assume, like Munk and Lewandowsky, that the deep sensibility is impaired by the extirpation of the cerebellum. He does not consider the cerebellum an organ of sensibility, but it receives various stimulations to which it reacts specially with balance reactions; he considers it "a reflex center for balance." The ability to maintain balance is, however, not totally eliminated but returns again. André Thomas is of the opinion that the lost balance reactions are partly compensated by the cerebrum. He thinks, however, that due to the absence of cerebellar balance reactions, the regulation of balance remains constantly deficient and that the restoration of balance takes place slowly and is often insufficient.

If the various permanent impairments in decerebellate animals were really due, in the first place, to the absence of cerebellar balance reactions one should expect that after a total extirpation of the cerebellum at least one balance reaction would be permanently and constantly absent, even after the incomplete restoration of balance by cerebral influences. In the protocols of decerebellate animals observed by Thomas, however, we do not find records of the failure of a single specific balance reaction and just as little on reactions of the neck and spinal column. Thomas further points out the importance of studying cases on cerebellar atrophy for elucidating the physiology of the cerebellum. Together with Dejerine, he described, for the first time, the clinical picture and the microscopical anatomical alterations of olivopontocerebellar atrophy and we also have from him the first description of a case of lamellar atrophy of Purkinje's cells. According to Thomas, these patients show severe impairment of balance and are only able to remain in an upright position with an effort, but in their case histories we find no mention of a constant and permanent absence of a specific balance reaction and of a compensatory movement for the preservation of balance, while in the illustrations added to the description distinct compensatory movements, particularly of the arms, can be distinguished (Fig. 222).

André Thomas himself noticed that these patients, in spite of their unsteady running, were relatively well able to keep their upright posture when, while running, they were unexpectedly pushed or pulled. Placed on a turntable, they also showed no distinct differences from normal persons; moreover, the asynergies described by Babinski are not constantly

428

present. André Thomas is even of the opinion that the asynergies observed by Babinski which, under certain circumstances, as in a backwards bending of the upper part of the body, produce a loss of balance, are not due to a failure of cerebellar functions and emphasized that Babinski's patient with his typical asynergies does not present a clear case of cerebellar disease.

Like Munk, André Thomas also places an inhibiting and a motor- and tonus-increasing function in the cerebellum. Cessation of the first produces a dysmetria of movements, while due to the tonus-increasing function all postures and movements deviate from the normal. In decerebellate animals, not only the balance reactions are ditsri buted by the absence of these two functions but, according to Thomas, there also appears an impairment of "tonus synergy" of those muscles which go into effect in these balance reactions. Thomas formulates his conception on the functions of cerebellum as follows : "The cerebellum assures the measure and continuity of movement, the stability and the reactions of balance by a special tonic action" ("le cervelet assure la mesure et la continuité du movement, la stabilité et les reactions d'équilibration par une action tonique speciale"). From this definition, it is not clear whether the removal of the cerebellum produces a loss of unspecified balance reactions or whether the balance reactions do not cease but only occur with an abnormal distribution. When he assumes the latter, the centers of balance reactions must lie outside the cerebellum and, in this case, the cerebellum is not an organ of balance but only an organ regulating the muscle tonus, the distribution of muscle tonus and the muscle tonus synergy.

Fig. 222.   Case of lamellar atrophy of the Purkinje's cells.   See the strong compensatory movements of the arms. From: André Thomas, La Fonction Cerebelleuse.   Paris: G. Doin, Editeur, 1911.

In his later studies, André Thomas brings the tonus regulating functions increasingly more into the foreground. In 1897, the conclusion of his book quotes: "The cerebellum is a reflex center of balance"; in 1911, he gives the just mentioned definition, while in 1914, based on observations on war-wounded, carried out together with Durupt, he gives even more emphasis to the impairment of cooperation of agonists and antagonists (anisosthenia). In his last publication (1925), he also emphasizes the special importance of "anisosthenia," i.e., hyposthenia (reduced resistance) of muscles acting in a certain direction paired with hypersthenia (increased resistance) of contrary acting muscles.

The conception that the cerebellum is a balance-regulating organ has been contested by numerous authors, among others by Luciani. Luciani and other investigators advanced a counter-argument that decerebellate animals are able at any moment to avoid falling down by distinct "compensatory actions" and are very well capable of maintaining their balance even in water.

The theories of balance are accepted in particular by many clinicians and it is indeed specious to a certain extent to assume an impairment of balance regulation when one sees a patient with "cerebellar ataxia" staggering forwards and backwards, to the left and to the right while running. On the other hand, one will always have to doubt this theory when one looks on and sees how the patient, each time he loses his balance, is able to avoid falling and again and again restore his equilibrium from the most varied postures.

In my opinion, the question whether the cerebellum is an organ of balance can only be answered when it is established once and for all what is meant by an organ of balance or a center of balance.

An organ or center of balance is an organ or center by way of which balance reactions are brought about. A balance reaction is a reaction (complex of movements) by which the balance is maintained or restored, i.e., by which the line of the center of gravity of the body in a standing posture or in a ventral position is kept within the supporting surface or is brought back to it.[1]

---

[1] How urgently this definition was needed became evident at the meeting of German physiologists in Dusseldorf (1920) where as a principal subject "balance and balance impairments were" discussed. It could be perceived clearly from the lectures on various reactions and impairments as, among others, compensatory movements of the eyes, impairments of eye movements, nystagmus, etc., that no clear understanding had yet been reached as to the conception of balance reactions and impairments.

From this definition it follows that a removal or total destruction of an organ of balance must cause the cessation of one or several balance reactions, if there is in the central nervous system no other organ with a similar function. An impairment of balance in the form of a retarded, abnormally strong or weak balance reaction cannot be caused by the absence or inactivity of a center of balance.

When an animal after extirpation of a part of the central nervous system still shows a certain balance reaction, be it in a normal or an abnormal way, the center of this reaction must still be present and situated somewhere in the intact parts of the brain.

The impairment of balance of an animal can be restored:

1. by displacing of the leg;

2. by an alteration in position of the head and neck, by a curvature and rotation of the trunk;

3. by extension and flexion and further by abduction and adduction of one or several legs without displacing them.

Each of these reactions represents a movement. One could also think that the maintenance of balance could be brought about by an isometric contraction of muscles. When, for instance, in an animal standing on its four legs the line of the center of gravity falls within the triangle formed by the points of support of the two forelegs and the left hindleg, the balance will not be impaired when the animal raises its right hindleg. In doing so, however, the three other legs become more burdened. When, in this case, the right foreleg has no sufficient supporting tonus, the leg will give way, the animal will lose its balance and will fall to the right side. By a straining of the muscles which fix the joints of the forelegs, i.e., by an increase of supporting tonus, falling down can be avoided and the balance is preserved without the animal making a single movement. Such contractions (see Chapter V) which cause an adaptation of supporting tonus to the load and which also occur in decerebellate animals are not counted among balance reactions.

The question about which role is normally played by the cerebellum in the maintenace and restoration of balance can, of course, not be solved by investigations in decerebellate animals on present or absent balance reactions because these observations give information only on the function of the intact part of the brain and of the central nervous system without the cerebellum. When a decerebellate animal shows a certain reaction, this only means that the center of this reaction lies in the intact parts of the brain, but it does not mean that the cerebellar functions are not normally taking part, because the possibility cannot be excluded that this reaction in the intact animal can be brought about in two ways, one of them proceeding via the cerebellum and the other bypassing it. Neither can the absence of a reaction after extirpation be taken as a proof that the center is normally situated within the cerebellum or that the tract leading to and from it must proceed via the cerebellum since such a serious operation can be the cause for even the centers situated outside the cerebellum to lose temporarily or permanently their ability to function.

Considering these different interpretations, in my opinion, one is only justified, to a certain extent, to assume a center of balance in the cerebellum when a constant, permanent cessation of one or more balance reactions is established. Moreover, by establishing a constant absence of balance reactions in a great number of decerebellate animals, not only the dependence of this reaction upon the cerebellum is made probable, but also the possibility then arises to prove, by partial extirpations and lesions, dependence of these reactions on the cerebellum and to investigate in which part of the cerebellum the center of this reaction is situated.

A. THE BEHAVIOR OF RIGHTING REFLEXES IN DECERE-BELLATE ANIMALS

A decerebellate animal, decorticate as well as midbrain animal, is able by reflexes to occupy its "basic or normal posture" from any position. In this righting function, as shown by Magnus, five groups of reflexes participate, the stimulations of which emanate from the labyrinths, the deeper parts of the neck, the body surface and (this only in intact animals) from the eyes. By way of these righting reflexes, the body is led back into a normal posture from positions with impaired balance. In this case, they represent balance reactions. However, righting reflexes are not solely balance reactions, they do more, they also bring back the body from a lateral into a normal position, i.e., from one balance position into another; in the air, too, the animal even without any supporting surface can from any position return to its normal posture by way of its righting reflexes.

432

Since the righting reflexes, under certain circumstances, represent balance reactions, their presence after extirpation of the cerebellum was investigated. Magnus and De Kleyn (146) have already investigated this question in decerebellate cats, but only in acute experiments (the animals were kept alive at most for several days). Magnus and Dusser de Barenne further examined two decerebellate dogs which were still alive 5½ and 9 months after extirpation, as to their righting reflexes. The results of the investigation will be discussed later together with our findings.

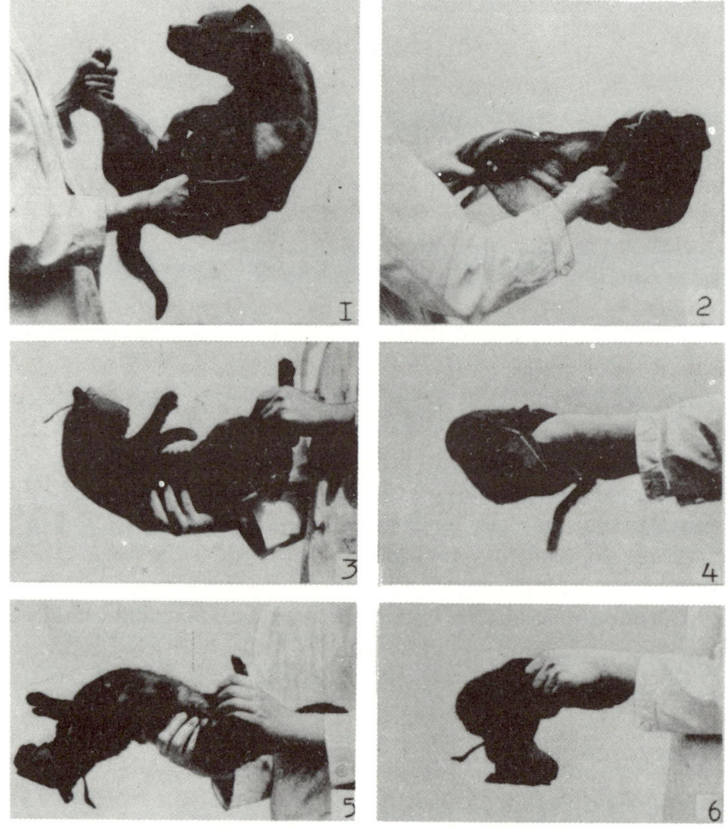

Fig 223. 1 and 2. Decerebellate dog Moor, with blindfold, in a dorsal position (1) and in a left-side lateral position (2), held in the air. In both positions head in normal posture. 3 and 4. Decerebellate cat Pierrette, with blindfold, in dorsal position and in right-side lateral position in the air, head also in normal posture. 5 and 6. Labyrinthectomized cat Peggy (3 weeks after bilateral labyrinthectomy), with blindfold, held in the air, in dorsal as well as in lateral position lets its head dangle downwards, without bringing it into normal posture.

1. THE LABYRINTHINE RIGHTING REFLEXES

The labyrinthine righting reflexes cause the maintenance and restoration of the normal posture of the head. *They are present without exception in all decerebellate animals* and make it possible for all animals held in the air with blindfolded eyes in left-side as well as in right-side lateral positions, in a dorsal position and suspended head upwards or downwards, to turn back their head promptly into the normal posture (Fig. 223).

The labyrinthine righting reflexes can usually be soon observed again after the ablation, but no exact time can be indicated since the animals should be left in peace for the first days after the operation. In several animals, among them the dogs Pim and Susanne, raising of the head from a lateral position in the air could be observed within the first 48 hours after extirpation; conversely, from a dorsal position, the head was raised much later because in this case a stronger effort is necessary. Furthermore, most of the animals (not all of them!) show in the first days after operation an increased tension of the levators of the neck which is strongest in the dorsal position due to the effect of the tonic cervical reflexes and counteracts the vental flexion of the neck. Several animals brought the head into a normal posture by a rotation of 90-180°; other decerebellate dogs were able to raise their heads by a ventral flexion already within the first weeks (dog Susanne on the first, dog Pim on the second, dog Däumling on the fourth and dog Tommy on the fifth day after extirpation).

These observations correspond to those of Magnus and De Kleyn who observed in a decerebellate cat (preparation III) labyrinthine righting reflexes in lateral position on the day of extirpation and in a dorsal position on the following day; similarly, Dusser de Barenne established labyrinthine righting reflexes in his decerebellate dogs.

In unilaterally decerebellate animals, too, labyrinthine righting reflexes are present in all positions in the air, as well as in three animals (cat Esperance, dogs Vici and Däumling) where together with the cerebellum the right half of the cerebral hemisphere was also removed (Fig. 224), while in the decorticate and decerebellate dog Robbie the labyrinthine righting reflexes were apparent only from the two lateral positions. (In the dorsal position, the animal, during the 6 weeks it stayed alive, showed no labyrinthine righting reflexes but only a very strong forced attitude of retraction of the head.)

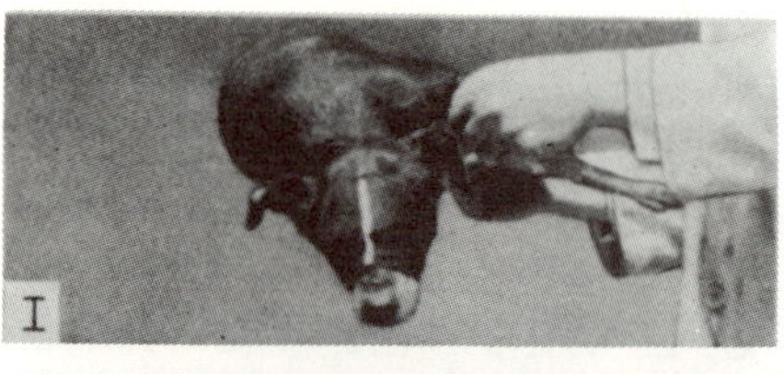
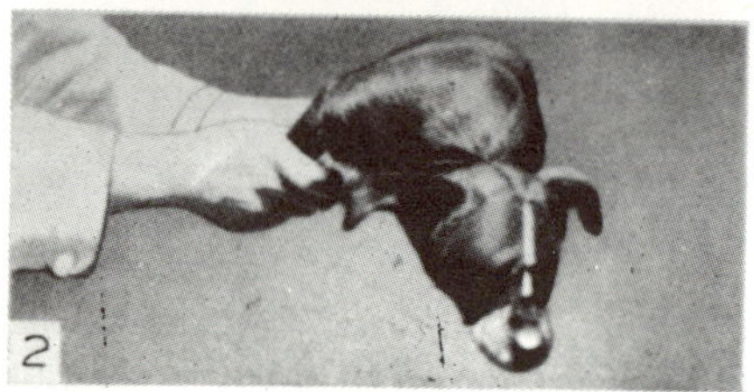
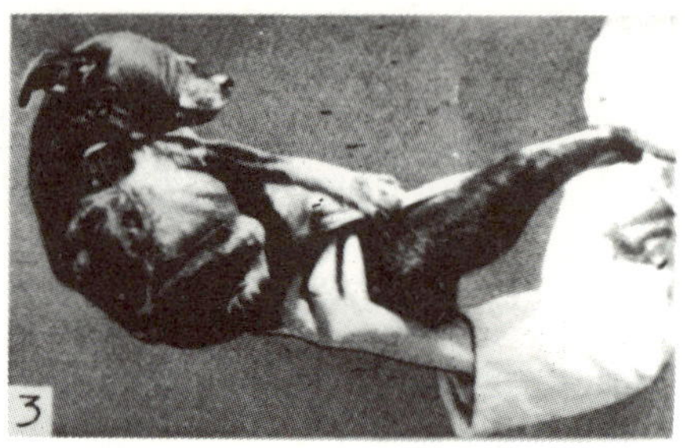

Fig. 224. Labyrinthine righting reflexes in the dog Vici after extirpation of the cerebellum and the right half of the cerebrum. 1. In right-side lateral position in the air. 2. In left-side lateral position in the air. 3. In dorsal position in the air.

The observations in the dog Robbie prove that after extirpation of the cerebellum, the labyrinthine righting reflexes can appear in a lateral position without a compensatory effect of the cerebral cortex.

Decerebellate animals sometimes also show in a standing posture, even in a lying position when they actively raise the head, a typical shaking, trembling and wobbling of the head which usually does not appear in animals held in the air, not even when the animals in a dorsal position hold their heads in a normal posture and the ventral flexors of the neck and spinal column are strongly contracted due to the labyrinthine righting reflexes (Fig. 223, No. 1 and 3, Fig. 141).

On the role of the labyrinthine righting reflexes in the adaptation of leg posture to the position of the supporting surface, see Chapter VII. In the following, we will see that they also participate in the production of a special balance reaction, the pulling-up reaction.

2. THE NECK RIGHTING REFLEXES

On turning of the head, stimulations are produced in the deeper parts of the neck which cause compensatory rotation of the cervical, thoracic and lumbar parts of the spinal column. In intact and decorticate animals, the trunk and pelvis tend towards a normal posture as soon as the head has been turned. When the animal falls sideways, the labyrinthine righting reflexes bring the head back to a normal posture and the neck righting reflexes then restore the balance of the trunk.[1]

Rotation of the cervical, thoracic and lumbar parts of the spinal column, due to active or passive rotation of the head, occurs with varying force in intact and decorticate animals, depending on the initial position. When the head is rotated into a normal posture from a lateral position of the animal the compensatory rotation occurs promptly; trunk and pelvis instantly go into a ventral position, since the effect of the neck righting reflexes in this case is facilitated by the body righting reflexes acting on the body. Conversely, when, in a standing posture or ventral position, the head is rotated to a lateral position, the trunk will not usually go into a lateral position, since in this case the effect of the neck righting reflexes is prevented by the body righting reflexes on the body, the leg-bracing reactions, etc. On rotation of the head from a dorsal into a lateral position, the trunk follows into the lateral position since in this case the cervical righting reflexes are not subject to any interfering influences from other reactions. Conversely, when the head is now rotated back into a dorsal position (for instance from a left-side lateral into a dorsal position), the trunk will only follow incompletely, i.e., only with the anterior part, since in this case the body righting reflexes on the body attempt, in a sense contrary to the cervical righting reflexes, to bring back the trunk into the normal posture. On a further rotation of the head into the other lateral position on the right side, however, the cervical righting reflexes are aided by the force of gravity and the body righting reflexes on the body and the trunk suddenly results in a turn to a dorsal and then to a lateral position on the right. Thus, the behavior of the neck righting reflexes can best be examined when the head is passively rotated from a dorsal into a lateral position, since in this case the neck reflexes can exert their influence without the interference of other reactions.

Neck righting reflexes are present in all decerebellate animals and can usually be demonstrated again on the day of operation. In the first period after extirpation of the cerebellum, when the body righting reflexes on the body are still absent, the influence of the cervical righting reflexes is abnormally great, especially in animals which show no

---

(1) Moreover, the labyrinthine reactions also participate due to a rotation round the longitudinal axis.

436

extensor rigidity of extremities.[1] In these animals, on rotation of the
head from a normal posture into a lateral position, trunk and pelvis go
over instantly into a lateral position; when the head is rotated from
one lateral position to the other via a normal posture, the trunk will
promptly follow and roll from one side to the other by way of the ventral
position (Fig. 225).[1]

Not only by rotation but also by a turning of the head the position
of the trunk can be almost automatically regulated at this stage. When,
for instance, the animal lies on the left side, then on turning to the left
of the upright held head, the posterior part of the body at first goes
into a ventral and then into a lateral position on the right (Fig. 226),
while on turning back the head towards the right, the posterior part
of the body rolls back into a lateral position on the left side by way of
the ventral position.

Fig. 225.   1. Decerebellate dog, 3 weeks after extirpation of the cerebellum, in
lateral position on the left side.   2. On rotation of the head
into normal posture the anterior part of the body goes into ventral
position, the pelvis rotates 45° to the right.   3. On rotation of the
head into a right-sided lateral position the whole trunk goes into this
lateral position.

In the stage of permanent impairment, the neck righting reflexes
are particularly strongly evident on rotation of the head of an animal
lying on its back. However, in this stage the influence of cervical righting reflexes in a standing posture or in a ventral position is distinctly
reduced due to the reappearance of leg-bracing reactions and of body
righting reflexes acting on the body. In this case the decerebellate
animals, just as intact animals, cannot be brought into a lateral position
from this posture by a passive rotation of the head. The posterior part

---

(1) In rigid animals the associated rotation of the trunk is prevented by the
extended posture of the rigid legs.

of the body, too, no longer rolls from one lateral position to the other
by way of the ventral position on a turning of the head.

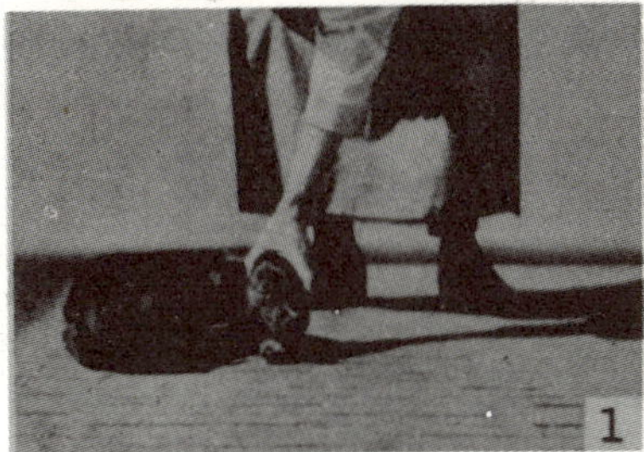

Fig. 226. Decerebellate dog, 3 weeks after extirpation of the cerebellum. 1.
The animal lies with the posterior part of the body in left-sided lateral
position, head in ventral position turned towards the right. 2. On
passive turning of the head towards the left, the posterior part of the
body goes first into a ventral and then into a right-sided lateral
position.

When in a decerebellate dog, in a standing position or lying on its
belly, the head is turned with the vertex of cranium to the right, the
right legs will be braced and the dorsum of the pelvis rotated to the
left, i.e., contrary to the influence of the cervical righting reflexes (Fig.
101). The same is the case in intact dogs in a standing posture or in a
ventral position. The normal synergy on rotation of the head towards
the right in a standing posture thus consists of a rotation of the dorsal
side of the pelvis to the left. Conversely, when in decerebellate animals
in the stage of absence of leg-bracing reactions and body righting
reflexes acting on the body, the head is rotated to the right in a standing
posture or ventral position of the animals, the trunk and pelvis under
the exclusive effect of neck righting reflexes will also rotate to the right
and the animal will fall towards the right side. Thus, an asynergy or
better a dyssynergy is present which is produced by the fact that the
normal synergies of the neck reflexes are not counteracted by the
leg-bracing reactions and the body righting reflexes on the body,
as is normally the case.

As already mentioned earlier, in decerebellate dogs in the stage of
permanent impairment, a passive rotation of the head, in a standing
posture, produces such a strong rotation and such a considerable alteration in position of the pelvis, that the animals in most cases can only
maintain balance in the trunk by shifting the legs (hypersynergy, Fig.
245).

Fig. 227. Decerebellate dog Piccolino, with blindfold, in dorsal position. Anterior part of the body fixed by the forelegs as well as possible. 1. On rotating the pelvis to a lateral position on the left side, the head, too, goes into this position. 2. Rotation of the pelvis into a right-sided lateral position causes a right-side lateral position of the head.

The neck righting reflexes are a chain reaction, i.e., spreading from the rostral to the caudal parts of the vertebral column. Conversely, an opposite succession can be produced by a passive rotation of the pelvis. When the pelvis of a dog lying in a lateral position to the right is rotated at first to a dorsal and then to a lateral position to the left, the trunk and head will rotate as well and also go into a lateral position to the left. The head will show the reaction even when the anterior part of the body is fixed by the forelegs (Fig. 227). These reactions appear more distinctly when the eyes of the experimental animal are blindfolded. A rotation of the head can be observed more constantly in decerebellate animals than in intact ones.

Observations by Peiper and Isbert (Fig. 228) show that infants, too, can be moved from one lateral position via the dorsal position to the other lateral position by a rotation of the pelvis (Fig. 228, No. 1-3). Fixation of the thorax with one hand still leaves rotation of the head in response to that of pelvis, but, contrary to our observations in dogs, in the opposite direction the lateral position of the pelvis to the right is accompanied by a lateral position of the head to the left (Fig. 228,

No. 4 and 5). In adult, intact persons no distinct influence in position of pelvis on the position of the head can be observed. Under pathological conditions, however, as shown in a case by Stenvers (293), a passive rotation of the pelvis from the dorsal to a lateral position could instantly produce a brusque rotation of the head towards the same side. Even when the shoulders were fixed, a rotating of the pelvis to a lateral position to the right caused a rotation of the head into a lateral position to the right, corresponding to observations in animals.

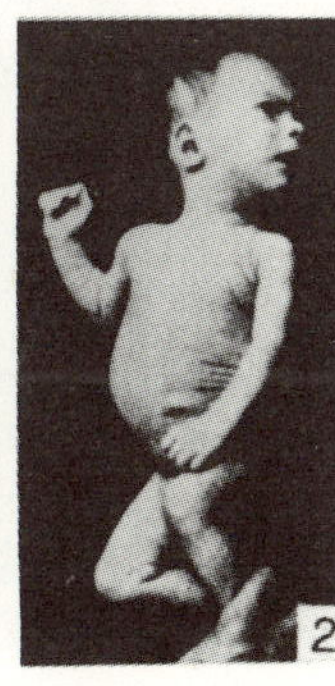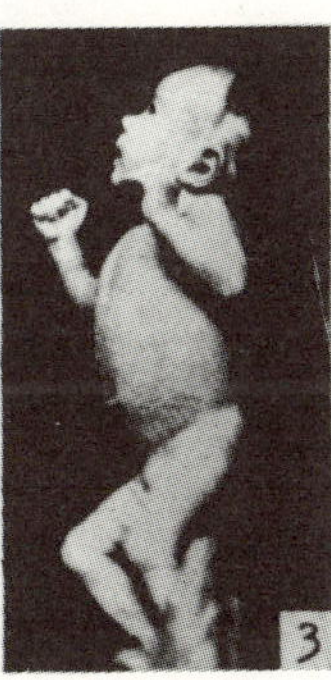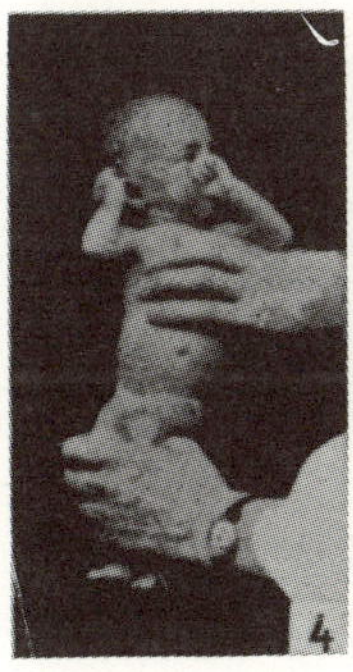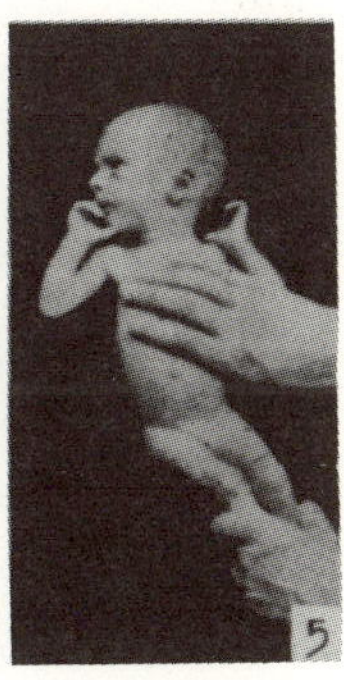

Fig. 228. Influence of the position of pelvis on the position of head and thorax in infants. 1 to 3 with free thorax, 4 and 5 with fixed thorax. 1. Initial position. 2. Pelvis passively rotated into dorsal position, trunk in dorsal position. 3. Pelvis rotated into right-sided lateral position, head and trunk in the same lateral position. 4 and 5. With fixed thorax the rotation of the pelvis was accompanied by opposite rotations of the head. From A Peiper and H. Isbert: (On the body posture of the infant) Uber die Korperstellung des Sauglings. Jb. Kinderheilk. Vol. 115, 158, 159 (1927).

Finally, we would like to mention the reports made by Goldstein and Riese, Zingerle, Hoff and Schilder, according to which cerebellar patients show particularly strong neck righting reflexes on passive rotation of the head. (In several of these patients, a rotation of the trunk and head appeared on a passive rotation outwards of a leg.)

3. THE BODY RIGHTING REFLEXES ACTING ON THE BODY

When a decorticate animal is placed in a lateral position on a supporting surface and the head is held fast in a lateral position, the animal will instantly attempt to raise its body contrary to the effect of the neck reflexes and to the force of gravity. The necessary stimulations are provided by an asymmetrical stimulation of superficial sensation by the pressure of the supporting surface. These stimulations, too, can participate in the restoration of balance when, for instance, on stumbling, a part of the trunk falls into a lateral position.

All decerebellate animals—dog, cats and one monkey—are able to right their trunk into a normal posture even when the head is fixed in a lateral position; without exception they all show body righting reflexes acting on the body.

These findings are in accord with Dusser de Barenne's observations, but, on the other hand, the body righting reflexes on the body were absent in the decerebellate cats of Magnus and De Kleyn (only in acute experiments).

The body righting reflexes on the body usually reappear later than the labyrinth and neck reflexes after cerebellar operation. The delay differs in the different animals. They were observed in the cat Karolus after 14 days, in the cat Pierrette after 20, in the cat Friedal after 31 days, while they appeared in the dog Pim after 3, dog Tommy after 4, dog Däumling after 5 days, in the dogs Moor and Piccolino after 1 month, in dog Erik after 2 and in Caesar only after 3 months.

The duration of the latent period depends on various factors and primarily on the extirpation itself. When the operation was performed quickly, without profuse bleeding and great manipulations, the convalescence is shorter and the various reflexes can be demonstrated sooner. Intercurrent diseases during convalescence, too, like sepsis of sutures, diarrhoea caused by milk feeding, etc., delay recovery. The first dogs to be operated all developed conjunctivitis, an after-effect of anesthetic ether. Later, this could be prevented by a previous application of boric-acid vaseline to the eyes, which caused a shortening of the convalescence and a much earlier appearance of reflexes. In animals which after the extirpation showed a forced dorsiflexion of the neck and an extensor rigidity of the forelegs, the body righting reflexes on the body remained absent for a longer time than in animals which did not show this extensor posture, since the extended limbs themselves prevent the raising of the animal from a ventral position. To this should be added the fact that the increased tension of extensor muscles occurred almost exclusively in animals on which the operation was performed less carefully. The age of the animals, too, plays a certain role. The dogs Pim, Tommy and Däumling, which showed body righting reflexes acting on the body already 5 days after the extirpation, were young dogs which, as is known, are less a subject to the effects of shock than older dogs. The race and character of the animals also exerts a certain influence.

Dogs with long legs start standing up and running later than short-legged one; in shy animals, the different reflexes are sometimes produced only with difficulty, as can also be also observed in intact dogs. Some animals, when placed on their side, remain in this posture "shamming dead," even when they are shaken to and fro, some let themselves fall on their side in a contorted heap when one attempts to touch them. In obstinate animals, the body righting reflexes acting on the body are less inhibited by cerebral influences. When one considers that these reflexes have to bring the trunk from a lateral position with a laterally fixed head into a normal position, i.e., contrary to the effect of neck reflexes and to the force of gravity, and this in animals with an abnormal distribution of muscle tonus, then it is not astonishing that the animals are not able to do it for the first time after such a serious operation as an extirpation of the cerebellum. It is, on the contrary, amazing that the other righting reflexes reappear so soon.

Following a unilateral extirpation of the cerebellum, the body righting reflexes acting on the body are absent in both lateral positions. Soon, however, this is only shown on lying on the side of extirpation and, finally, they are lively in both lateral positions in all unilaterally decerebellate animals (in the dog Peter already present on the tenth day). Also, in the three animals (cat Esperance, dogs Vici and Däumling) in which *in addition to the cerebellum the right half of cerebrum* was also extirpated, the body righting reflexes on the body appeared in both lateral positions, though not equally strong and rapid. From a lateral position on the right side with a fixed head, the animals instantly brought the trunk into a ventral position and sometimes remained in this posture, but sometimes the trunk rotated further until it reached the lateral position on the left. From the left lateral position with a fixed head, the animals usually succeeded only temporarily in bringing the trunk into a ventral position. Generally, the trunk soon rolled back and sometimes, in doing so, fell on the right side. Thus, the animals were not able to keep the trunk in a normal posture when the head was fixed in a lateral position to the left. One has to consider, however, that in this case the left legs had an abnormal distribution of supporting tonus, showing no distinct leg-bracing reactions and carried out all movements particularly abruptly, hypermetrically and awkwardly.

In the decorticate and decerebellate dog Robbie, the body righting reflexes on the body did not return within the 5 weeks the dog remained alive after extirpation of the cerebellum.

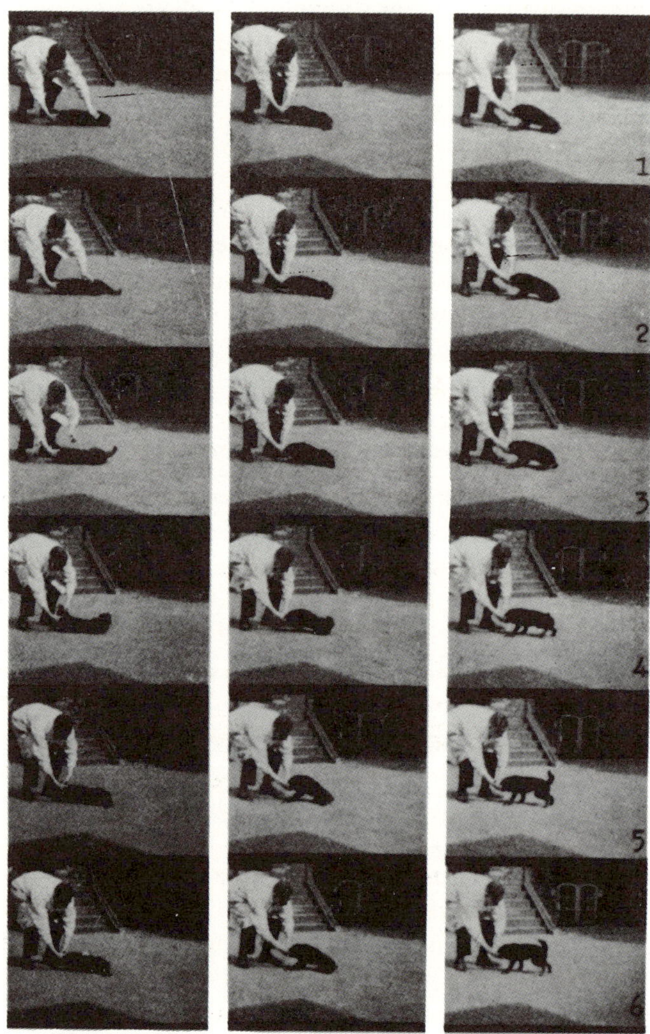

Fig. 229. Cinematographic pictures of the body righting reflexes on the body in the decerebellate dog Erik, 5 months after extirpation of the cerebellum. A 1 and 2. The animal is placed into a lateral position on the right side. A 3 and 4. The trunk is released, but the head remains held in lateral position with both hands. A 5 to B 3. In spite of this position of the head, the trunk rotates into a chest ventral position (body righting reflexes on the body). B 4 to C 2. First the anterior part of the body goes into a standing posture (supporting reactions of the forelegs) and, C 3 to C 6, soon also the posterior part of the body stands up (supporting reactions of the hindlegs).

4. THE BODY RIGHTING REFLEXES ACTING ON THE HEAD AND OPTICAL RIGHTING REFLEXES

A symmetrical stimulations from the body surface not only bring the trunk but also the head into a normal position. The normal position of the head in intact animals is secured by three mechanisms: by the labyrinthine righting reflexes, the body righting reflexes on the head and by optical righting reflexes.

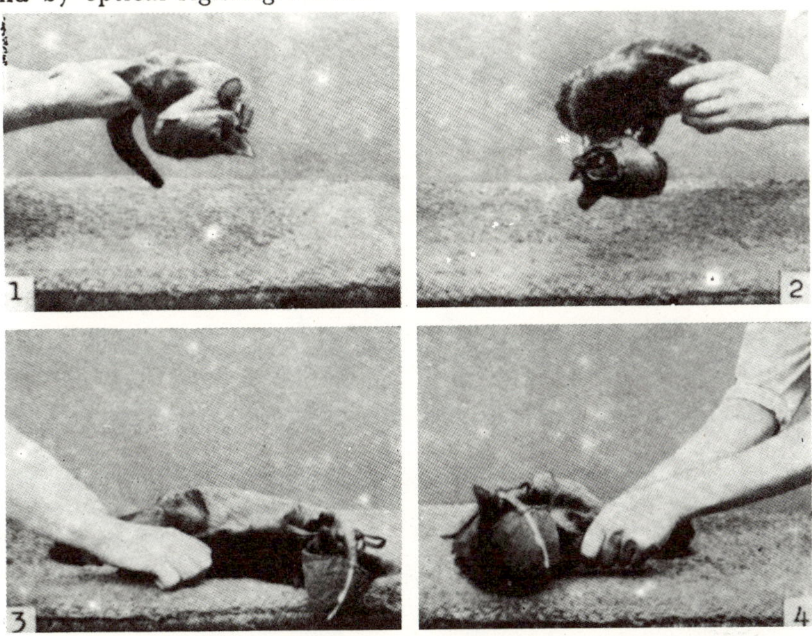

Fig. 230. Body righting reflexes on the body in the labyrinthectomized and decerebellate cat Peggy with blindfold. 1 and 2. Held in the air in lateral position the animal lets its head hang down in either lateral positions (1) or with the cranial vertex downwards, i.e., in dorsal position. Due to the absence of labyrinthine righting reflexes not a trace of righting function of the head. 3 and 4. On contant of a lateral surface of the trunk with a supporting surface the head is instantly turned into a normal position (body righting reflexes on the body). Pictures taken on the 2nd day after extirpation of the cerebellum.

The presence of body righting reflexes on the head can be established with certainty only in labyrinthectomized animals with blindfolded eyes. De Kleyn extirpated both labyrinths in a cat (Peggy) after which the animal did not show a single labyrinthine reflex. Three weeks later, after cessation of the acute disturbance, the cerebellum was extirpated and already two days after the body righting reflexes acting on the head were again present. With blindfolded eyes the animal was totally disorientated in space and, when held in the air, let the head hang down; on touching a supporting surface, however, the head was instantly raised up straight (Fig. 230). When the blindfold was removed from the animal in the air, the head also instantly returned to a normal position; thus, optical righting reflexes were also present (Fig. 231). Autopsy showed a total absence of the cerebellum in this cat Peggy (Fig. 294).

Dusser de Barenne removed the cerebellum of a dog (while preserving a part of lob. ant.) and De Kleyn extirpated both labyrinths. In this animal the body righting reflexes on the head were only sluggish at first; after 12 days, however, they were totally present. When held in the air with blindfolded eyes, its head fell down, too, but was raised on touching a supporting surface as well as on removal of the blindfold.

It is concluded from these observations that the body righting reflexes on the head and the optical righting reflexes can still appear after a total removal of the cerebellum.

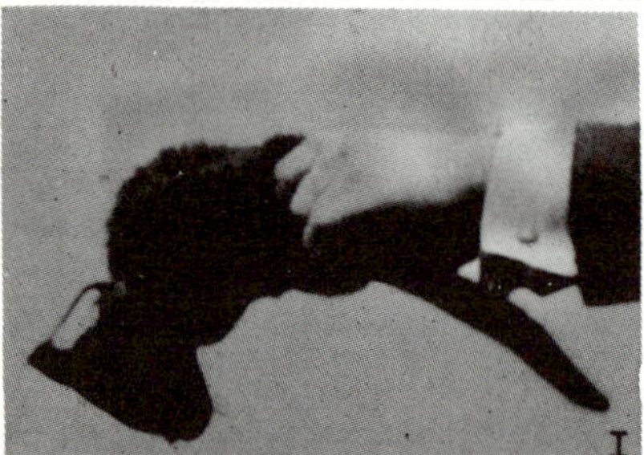

Fig. 231. Optical adjusting reflexes in the labyrinthectomized and decerebellate cat Peggy. 1. Animal with blindfold in right-sided lateral position in the air, the head hangs down with the cranial vertex downwards. 2. After removing the blindfold, the head is lifted up into a normal posture.

When an intact animal falls into a lateral posture, four mechanisms will lead back the head into a normal posture and thus participate in the restoration of balance:

1. the labyrinthine righting reflexes,

2. the body righting reflexes acting on the head,

3. the optical righting reflexes,

4. the labyrinthine rotating reflexes of the head (for instance, when the animal falls down to the right, it performs a rotation towards the right round the longitudinal axis, thus stimulating a rotation of the head towards the left).

When decerebellate animals are again able to bring the head and the anterior part of the body easily into a normal posture and labyrinthine righting reflexes and the labyrinthine rotating reactions, too, are present, the animals, however, in moving forwards, tend to fall forwards. The head will usually in this case make an audible impact with the ground. Whether the impact of the head is caused by the rigidity of the neck or by incomplete body righting reflexes on the head, or by the retarded appearance of other reactions or perhaps by the rapidity of falling is not yet known.

B. THE BEHAVIOR OF LABYRINTHINE REFLEXES IN DECEREBELLATE ANIMALS

The following are identified as balance reactions:

1. the labyrinthine righting reflexes,

2. turning of the head produced by rotation round the dorso-ventral axis,

3. rotation of the head produced by rotation round the longitudinal axis,

4. reactions of the extremities produced by rotation round the longitudinal axis (see Chapter IX),

5. reactions of the extremities produced by rotation round the bi-temporal axis (see Chapter IX),

6. flexing of the legs caused by the lifting reactions.

As seen in Fig. 232, a flexion of the legs can also restore the balance.

The labyrinthine reflexes serve to maintain the balance when the animals stand on a supporting surface which is rapidly moved up and down from one side. They also contribute to the preservation of objective orientation in space and also to the fact that an intact or decorticate animal, in dropping freely in the air, i.e., from a position without a supporting surface, again occupies a normal position and reaches the

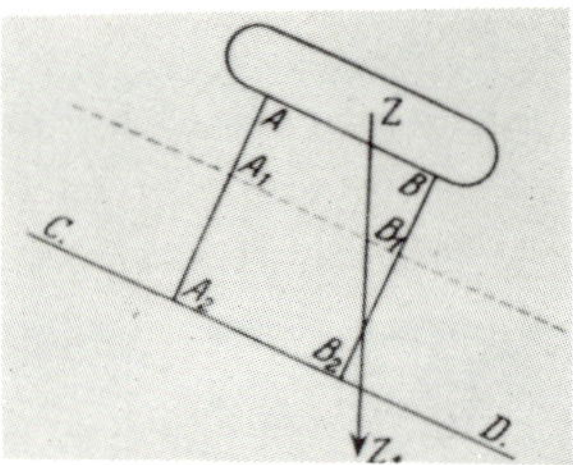

Fig. 232. Animal (Z) with extended legs (A A$_2$ and B B$_2$) on an obliquely
inclined supporting base C D. Line of center of gravity Z Z$_1$ falls
outside the supporting surface. By flexing the legs, for instance due to
lifting reactions, by means of which they get a length of A A$_1$ or
B B$_1$ the balance is restored.

ground in an upright posture with a labile balance. The labyrinthine
reflexes do the same in the case of jumping, i.e., while passing over
from a situation of labile balance via a position without supporting surface into a new situation with a labile balance. Therefore, it is clear that
the labyrinths play an important role in the theories of cerebellar function in preserving equilibrium, for example that of Ingvar.

The fact that labyrinthine reflexes under all circumstances do not
promote the maintenance of balance can be seen when a labyrinthectomized and an intact dog are placed on a turntable the speed
of which can continuously or suddenly be increased and reduced or
the direction of rotation suddenly changed. In this case, the labyrinthectomized animal will be more capable of maintaining its balance
than the intact one.

As a result of our investigation of decerebellate animals we can
state that *after removal of the cerebellum all labyrinthine reflexes can be present,
namely*:

1. eye rotating reactions with ocular after-reactions, head nystagmus
   and head after-nystagmus,

2. head rotating reactions with head after-reactions, head nystagmus
   and head after-nystagmus,

3. compensatory deviation of the eyeballs,

4. caloric deviation of the eyeballs with caloric nystagmus,

5. lifting reactions,

6. preparation for jump,

7. tonic labyrinthine reflexes (see Chapter IX),

8. labyrinthine righting reflexes,

9. reactions of the extremities on rotation round the longitudinal axis (see Chapter IX),

10. reactions of the extremities on rotation round the bitemporal axis (see Chapter IX).

These findings agree with those by Magnus, De Kleyn and Dusser de Barenne who also observed the occurrence of all labyrinthine reflexes in decerebellate animals. The ocular reactions (1 and 3) as well as the head reactions (2) are nearly always present again on the first day after extirpation, sometimes tonic labyrinthine reflexes can also be observed on the first day. The preparation for jump and the reactions of the extremities produced by rotation as well as the labyrinthine righting reflexes from a dorsal position, however, usually remain absent for a considerable time.

Lovenberg already stated in 1873 that the reactions of the semicircular canals in producing head movements on stimulation of the labyrinths in pigeons remain after extirpation of the cerebellum. Spamer (1880) after a nearly total extirpation of the cerebellum in pigeons saw the occurrence of typical reactions on local galvanic stimulation of the labyrinths. Hogyes found that in rabbits the eye-rotating reactions and eye-rotating nystagmus were preserved after incomplete extirpation of the cerebellum. Wilson and Pike were able to produce deviations of the eyeballs and nystagmus by irrigating the auditory canal with hot water after extirpation of the cerebellum in dogs. Barany, Reich and Rothmann observed labyrinthine head-rotating reactions forwards and backwards after incomplete extirpation of the cerebellum.

These observations show that Bechterew's assertion, "After the elimination of the cerebellum all forces which otherwise are effective on the labyrinthine arcs now proved ineffective, i.e., they do not produce the usual results," is surely not correct. With this falls one of the most important supports of his cerebellar theory and, from what has been said, one cannot do otherwise but definitively abandon the concept that the cerebellum is the central apparatus for the labyrinths, as Magnus and De Kleyn have already done. The labyrinthine reflexes must possess reflex arcs whose supply and discharge tracts do not proceed by way of the cerebellum.

In the *decorticate and decerebellate dog* Robbie, the various labyrinthine reflexes were not systematically investigated; still the labyrinthine righting reflexes from a lateral position, eye-rotating reactions, eye-rotating nystagmus and tonic labyrinthine reflexes were certainly present, which proves that these reflexes must possess reflex arcs which bypass the cerebrum as well as the cerebellum.

Of course, with the rejection of the above cerebellar theory we do not say that the cerebellum does not participate at all in these reflexes. The question whether the labyrinthine reflexes in decerebellate dogs appear in a totally normal manner or whether they indeed show any kind of deviation in the sense of a retarded, inhibited or abnormally lively appearance is extremely difficult to answer, since failure to establish a deviation never proves, particularly in this case, that the reflexes really proceed normally. On the other hand, an abnormality would not be a proof of an alteration in the central mechanism of the labyrinthine reflexes, since it could be based on a change in the distribution of muscle tonus. The following observations, too, which in my opinion can in no way be utilized for the question of relationship between labyrinth and cerebellum, distinctly show that it is impossible to ascertain reliable differences in the findings of investigations in intact and decerebellate animals.

The ocular reactions to rotation, with after-reactions, nystagmus and after-nystagmus always appear in exaggerated form in decerebellate animals and after a rotation round a rather narrow angle. But also in intact animals, these reactions can show varying intensities and duration. Together with De Kleyn, we determined the number of after-nystagmic beats after a rotation of 360° in decerebellate cats and several intact animals. Thus, an intact cat after 10 rotations in 15 seconds showed 4 after-nystagmic beats, another 6 and another 22, while with the same rotation there appeared in the decerebellate cat Pierrette 17 and in the decerebellate cat Nikker 18 beats. In the decerebellate cat Karolus, which continuously tried to free itself during rotation, after-nystagmic beats appeared irregularly; thus, after 10 rotations in a clockwise direction, it showed 4 horizontal beats to the left, then 10 to the right and, after 10 rotations in an anti-clockwise direction, first 4 beats to the right, then 16 beats to the left. Constant and distinct differences between intact and decerebellate cats, thus, cannot be established.

In the cat Josephine, too, 2½ years after extirpation of the right half of the cerebellum, the number of after-nystagmic beats has been determined in a clockwise as well as in an anti-clockwise direction.

Cat   Josephine

10 anti-clockwise rotations in 16 sec.	6 after-nystagmic beats
10 anti-clockwise rotations in 16 sec.	8 after-nystagmic beats
10 anti-clockwise rotations in 16 sec.	8 after-nystagmic beats
10 clockwise rotations in 16 sec.	15 after-nystagmic beats
10 clockwise rotations in 15 sec.	13 after-nystagmic beats
10 clockwise rotations in 18 sec.	18 after-nystagmic beats

After anti-clockwise rotations, only a few but regular after-nystagmic beats appear and after clockwise rotations several but irregular beats take place in groups.

According to Bauer and Leidler (17) the rotary nystagmus is increased after extirpation of the vermis.

The head rotating reactions with associated lateral curvature of the vertebral column are very strong in decerebellate animals. One has the impression that turning of the head occurs on lower vertebral disks than in intact animals and that the following lateral curvature of the spinal column is abnormally strong. After cessation of the rotation, the reactions (clockwise movements), too, are overactive in decerebellate animals, particularly in decerebellate cats. It is uncertain whether this phenomenon is due to a disinhibition of labyrinthine effects or to an abnormal distribution of muscle tonus. At any rate, the particularly strong lateral curvature of the spinal column is surely caused by the latter, since it also appears on the passive turning of the head.

The compensatory deviations of the eyeballs as well as the labyrinthine reflexes are present as in the normal animal.

The preparation for jump is always lively in decerebellate animals. When an animal with blindfolded eyes is suspended in the air head downwards, a slight downward movement is sufficient to produce a strong extension of the forelegs with a spreading of the fingers, and similarly, suspended head upwards, a slight downward movement is accompanied by a distinct extension of the hindlimbs with a spreading of the toes (Fig. 233, B). In the forelegs, the extension is usually excessively strong, as can be clearly seen in cinematographic pictures of jumping downwards or falling decerebellate animals (Fig. 233). The excessive extension can naturally be due to an inhibited effect of labyrinths as well as to increased excitability of extensors.

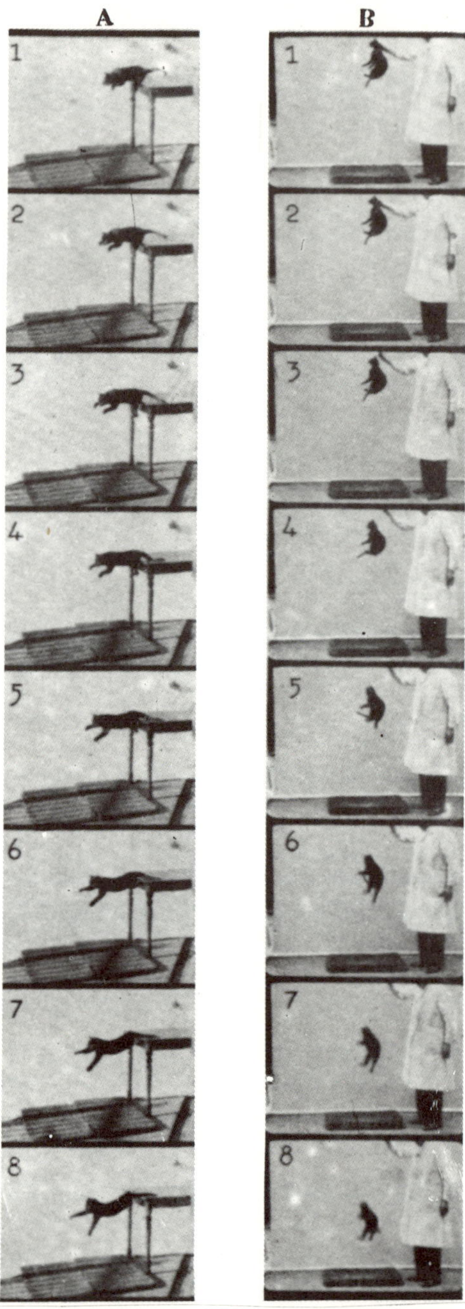

Fig. 233. A. Decerebellate tom cat Carolus, jumping down from a table. Slow-motion photo. Note the strong extension of the forelegs due to readiness to jump (7 and 8). B. Decerebellate cat Pierrette falls from a position suspended head upwards. Cinematographic picture taken at normal speed. Preparation for jump causes a strong extension of the hindlegs (6 to 8).

The *lifting reaction* is generally difficult to investigate in decerebellate animals, since they usually lie down flat on the chest and belly with each movement of the supporting surface, especially when their eyes are blindfolded. Sometimes, a distinct lifting reaction can be observed and on cessation of the upwards movement often an excessive exaggerated extension takes place. This was particularly the case in some rabbits with cerebellar lesion which, on cessation of the upwards movement for a short time, stood as if on stilts.

The flexion at the end of the downward suspended vertical movement does not proceed with the same elasticity as in intact animals and sometimes is very abrupt so that one gets the impression that they fall in ventral position; on the other hand, it can also be only slight, as in some dogs, due to the strongly extended limbs. In these animals, vertical movement thus produces only a distinct increase of the extended posture (at the start of the downwards movement and at the end of the upwards movement).

When the animals held by the skin of neck and back are moved up and down in the air in a ventral position, all four legs, not under static stress, show a lively extension on a downwards movement and on an upwards movement a drawing in to a flexed position.

The investigations on the behavior of *tonic labyrinthine reflexes* are particularly remarkable, since they prove the great importance of distribution of muscle tonus for the appearance or the absence of labyrinthine reflexes. In animals which after extirpation show a rigidity, distinct tonic labyrinthine reflexes in the rigid extremities can nearly always be observed, whereas they cannot be detected in animals with flaccid limbs. These tonic labyrinthine reflexes usually appear stronger when the rigidity has somewhat subsided; in this case, the legs not under static stress show a vigorous extensor posture in the dorsal position of the head, the muzzle $\pm 45°$ above the horizontal. In a ventral position of the head, the muzzle $\pm 45°$ below the horizontal, however, a flexed posture is shown. A certain time after extirpation, in the so-called stage of permanent impairment, distinct tonic labyrinthine reflexes in the limbs not under static stress appear as feeble in decerebellate animals as in intact and decorticate ones. On changes of distribution of muscle tonus by static stress, however, the reflexes return so that an alteration of head position in space is again accompanied by a distinct change in supporting tonus. As we have seen in Chapter IX, these changes appear more constant and distinct in decerebellate animals than in intact ones, possibly because in this case the experimental conditions have a less inhibiting effect.

When standing on a hinged board which is rapidly raised and lowered from one side, the legs under static stress of intact and decerebellate animals show lively *labyrinthine rotating reactions*. On the other hand, a rotation in the dorsal or lateral position of the animals produces distinct alterations in posture and supporting tonus of the four legs only in decerebellate animals (see Chapter IX). Thus, the labyrinthine reactions on the extremities by rotation are less inhibited by experimental conditions in decerebellate animals than in intact ones.

The labyrinthine reflexes for the maintenance of balance appear in decerebellate animals for an *appropriate* reaction. Labyrinthectomized dogs in water are disoriented and drown (Ewald, Thomas); decerebellate dogs, however, are able to save themselves by swimming (Luciani, Thomas and others).

Magnus has already pointed out that the body righting reflexes on the body produced by asymmetrical stimulation of the body surface cannot take effect in swimming and thus swimming in decerebellate animals can be considered as proof of the excellent functioning of labyrinthine reflexes.

Decerebellate pigeons are able to fly well (Lange) while labyrinthectomized ones cannot do so at all.

When one drops a blindfolded labyrinthectomized animal from the air in a dorsal position, it will fall down like a dead mass and hit the ground with the dorsum of the head and with the back. A decerebellate animal, however, will turn around in the air while falling and will arrive on the ground on all four legs in a ventral position (Fig. 234).

The cinematographic picture shows how first the head is turned into normal posture (labyrinthine righting reflexes) followed by a rotation of the trunk, first from the anterior part and followed by the posterior part (neck reflexes). On falling further, a limb position produced by the preparation for jump can be observed. Finally, the animal reaches the ground on all four legs by touching the floor either with all four limbs simultaneously or with the fore- or hindlegs first.

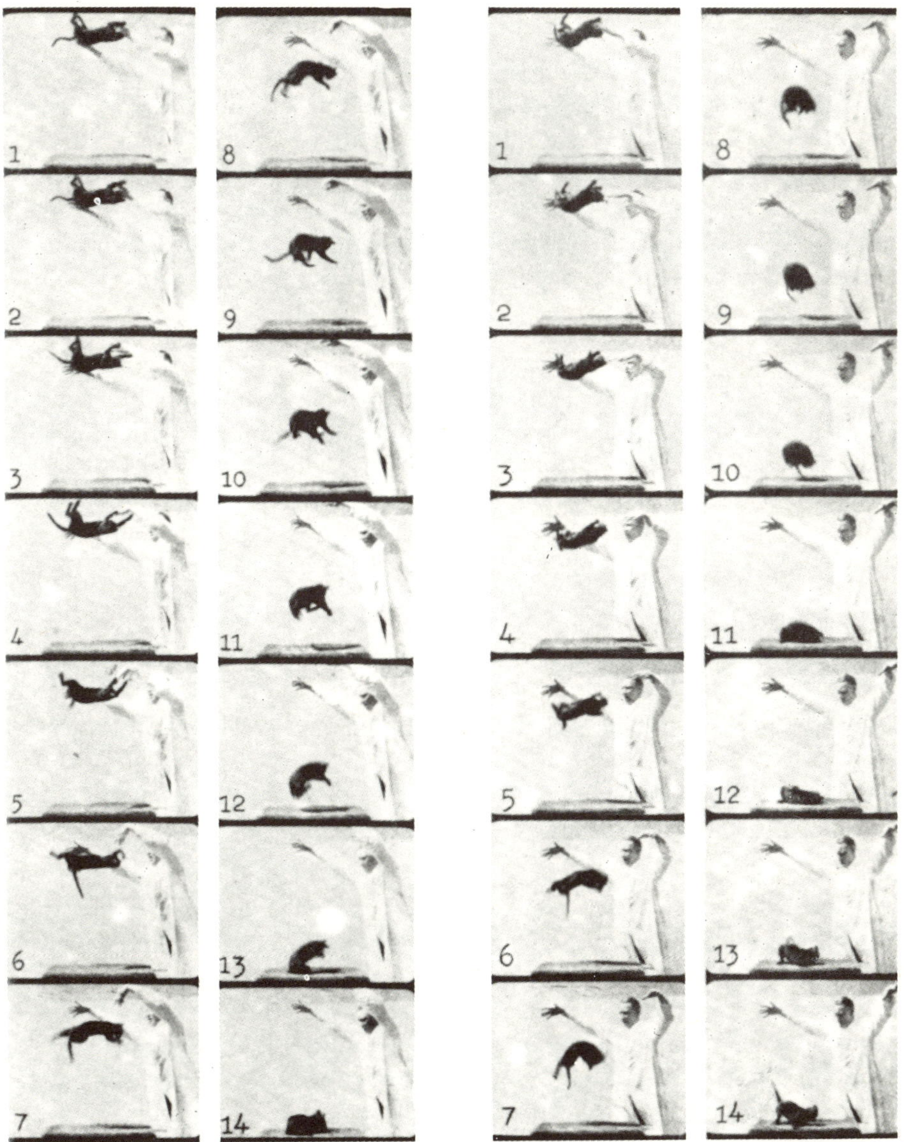

Fig. 234. A. Decerebellate cat Pierrette in a dorsal position in the air (1) is dropped (2). Owing to the labyrinthine righting reflexes the head rotates into a normal position (4-5); a rotation first of the anterior (6), then of the posterior (7-8) part of the body follows (cervical righting reflexes), so that the animal continues its fall in a ventral position. In this case the legs are extended due to the preparation for jump (9-10) till it finally arrives on the ground in a ventral position. B. The same cat with blindfolded eyes, dropped from a dorsal position. This time, too, the head rotates first (6) then the anterior and posterior parts of the body (7) into a ventral position, so that it arrives on the ground in this position.

Decerebellate animals in a free fall turn around in the air promptly, even with blindfolded eyes, so that they reach the ground in a ventral position and unlike labyrinthectomized animals who fall on their backs. Decerebellate animals suspended head upwards (Fig. 235), with or without blindfold, will likewise fall on their legs while labyrinthectomized animals in a dorsal postion will hit the ground with a hard impact on their back.

When the eyes of the animals are occluded with a blindfold, the occurrence of the various reactions during the fall is impaired in some degree, not only in decerebellate but also in intact animals. Blindfolding the eyes has also a strongly inhibiting effect on other reactions. Moreover, cats nearly always try to pull down the blindfold while falling (Fig. 234 B, 8-9) and rolling their body to and fro to pull their head out of the blindfold.

Also, on falling from a position suspended head downwards and in jumping down, for instance from a table, decerebellate animals clearly differ from labyrinthectomized ones.   Decerebellate cats on jumping down (from a height of 1 3/4 m) land correctly on their paws (Fig. 236) with the center of gravity line within the supporting surface. The decerebellate dog Piccolino also jumped down sometimes from a chair and arrived correctly upright on its paws; the other dogs refused to jump.

With closed eyes, too, decerebellate cats will land on their paws on jumping down as well as on falling from being suspended head downwards, contrary to labyrinthectomized animals which generally learn to jump down again from a chair or table with open eyes. With closed eyes, however, on falling from being suspended head downwards or jumping down, they will always tumble head forwards and hit the ground hard with their heads.

When a labyrinthectomized animal is placed on a swinging board and the board is quickly raised from one end, the animal, on the rapid raising of the headend, will turn a somersault backwards and, on the the rapid raising of the tailend, will turn a somersault forwards causing the head and shoulders to reach the ground with an audible impact. On the rapid raising of the right side, however, the animal is instantly flung into a lateral position to the right and will roll over its back. Conversely, the decerebellate animal is able to maintain its balance in a standing posture as well as in a ventral position.

Fig. 235

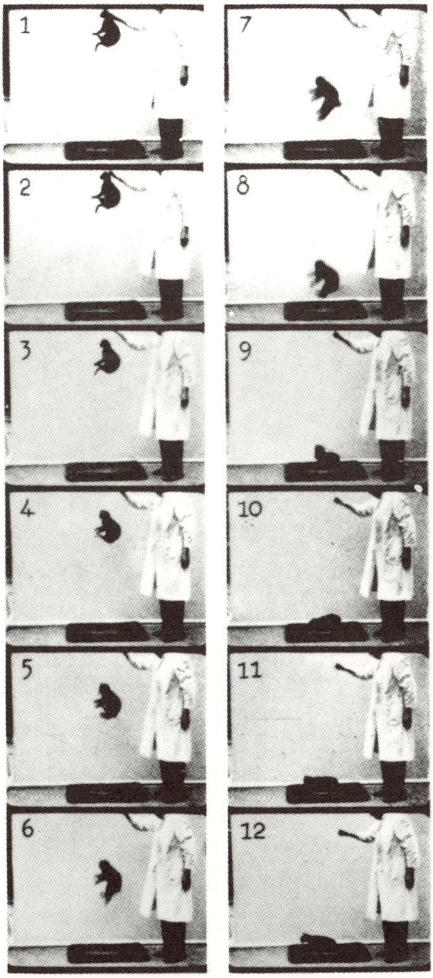

Fig. 236

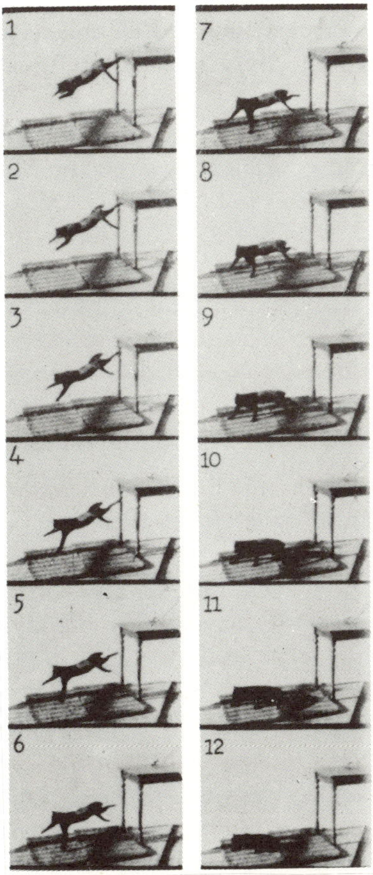

Fig. 235. Decerebellate cat Pierrette falling from a suspended position head
upwards (1). The animal reaches the ground on its limbs.

Fig. 236. Decerebellate tom cat, slow-motion picture, while jumping down
from a table. The animal jumps down in an oblique direction (1),
strongly extends its forelegs (2-4, readiness to jump), keeps its head
raised from the trunk (labyrinthine righting reflexes and lifting reaction) and finally lands correctly on its paws (9). Then the head is
ventrally flexed and the legs are flexed (lifting reaction on cessation
of lifting movement).

In decerebellate as well as in intact animals the balance is naturally impaired by a particularly abrupt and rapid lifting. On a rapid lifting on the right side, decerebellate animals as well as intact ones go into a lateral position to the right.

According to Bogumil Lange's observations, pigeons are not only able to fly very well after destruction of the cerebellum, but are also capable of checking their flight and alighting on a perch. These animals also walk along a rod, like tightrope-walkers, stand on one leg and very cleverly keep their balance when one holds them on a finger and tries to throw them off by sudden movements. When the labyrinthine reflexes were not able to function in an appropriate manner, the animals were not capable of performing these complicated movements and postures. Thus, labyrinthine reflexes are not only present in decerebellate animals, but they also occur with a particular liveliness and efficiency, which does not mean, of course, that their appearance is totally normal and that decerebellate animals do not show any impairments at all in falling, jumping down or standing on a rocking board.

While intact cats jump down from a table almost vertically and land close by it, decerebellate cats jump obliquely. They also jump recklessly and with too large a leap so that they land far away from the table, sometimes collide with objects and usually hit the ground with the posterior part of the body with an audible impact. Whether this impact is perhaps caused by the lack in extensor tonus of the hindlegs on the dorsal movement of the head, or by a too abrupt flexion of the hindlegs at the end of the downwards movement, or by a retarded appearance of supporting reactions is difficult to decide. In this connection, we should mention the observations made on the right-sided decerebellate cat Josephine which, when jumping from a table, would flex the right hindleg more strongly than the left one so that it would sometimes fall to the right on the posterior part of the body, but the animal could usually avoid falling down. This cat jumped twice out of a window and the decerebellate monkey Corrie once, from a height of 7 m onto a stone floor; unfortunately it could not be observed in what position the animals arrived there. They immediately ran away and showed no ill effects after the leap. On falling from a dorsal position, too, the unilaterally decerebellate cat Josephine turned round in the air and landed correctly on its feet.

When escaping, for instance, decerebellate animals sometimes leap horizontally and reach the ground quite well, but jumping from one object to another could never be observed as from a table to a chair, from one cage to another, neither in decerebellate cats nor in the monkey Corrie. Rarely can they leap over an object onto another, for instance, from the ground onto a chair, a performance that labyrinthectomized animals are able to learn again.

Another difference between bilaterally labyrinthectomized and totally decerebellate animals consists in the fact that labyrinthectomized animals are able to learn again quite well to climb stairs while decerebellate animals do that only very imperfectly. They further behave completely differently on a turntable; on standing they show a different distribution of muscle tonus; labyrinthectomized animals never show the "unrestrained" movements of decerebellates.

The still very widespread theory that the cerebellum is the central apparatus for labyrinthine function is chiefly based on the alleged conformity of phenomena after the unilateral lesion of the cerebellum, particularly the unilateral section of the cerebellar peduncle, with those of the unilateral extirpation of labyrinth. After the hemiextirpation of the cerebellum, the animals sometimes show in the first days rolling movements, deviation of the eyeballs and eye nystagmus. Pourfour du Petit already described in 1766 the compulsive movements made after a cerebellar lesion. Magendie (1825) described the rolling movements and deviation of the eyeballs after the unilateral section of the peduncles of the cerebellum, especially when these were severed close to their attachment to the brainstem. Flourens was the first to point out the similarity of phenomena with those after lesion of the semicircular canals and interpreted the symptoms after lesion of the semicircular canals as cerebellar manifestations. For the same reasons, Luciani and several other authors also assumed that the labyrinths exert their influence by way of the cerebellum.

Following the unilateral extirpation of the cerebellum, several of my animals also showed *rolling movements* round the longitudinal axis,[1]

---

(1) The direction of the rolling movement after the unilateral extirpation of the cerebellum has long since been a disputed question. Some authors write "from the healthy side to the lesioned," others "from the side of lesion side towards the healthy one." These statements as well as indications on rolling movements from right to left are inappropriate, since each rolling animal will roll from the left side to the right and from the right side back to the left, i.e., in both directions of rotation, change from the healthy side to the diseased as well as from the diseased to the healthy side. According to observations by the author, the initial position, the position of rest of the animal

458

with deviation of the eyeballs and nystagmus just as after the extirpation of the ipsilateral labyrinth. *Still, not all the symptoms after unilateral extirpation of the cerebellum were analogous to those after the unilateral extirpation of labyrinth.* Firstly, the distribution of muscle tonus is different after the two operations. When a unilaterally decerebellate animal is placed on its back, the head fixed symmetrically to the trunk, it will keep the legs ipsilateral to extirpation extended, the contralateral ones more or less flexed, at the same time the ipsilateral paws will offer more resistance to passive flexion. After the unilateral extirpation of the labyrinth, however, the animals with the same experimental arrangement (i.e., in a dorsal position with the head fixed symmetrically to the trunk) exhibit flexion of all four legs and, according to statements made by several authors, show a poor stretch tonus in the ipsilateral legs.

A second difference appears on examination of the labyrinthine righting reflexes. When a right-sided labyrinthectomized animal is held in the air with blindfolded eyes, in a lateral position to the right, it will not hold its head in a normal posture even a long time after extirpation. The head remains in a lateral position owing to the impairment of the labyrinthine righting reflexes. When a few days

(1) (Cont.) is of importance. Among three right-sided decerebellate animals, two were lying with the trunk to the right side, the third, however, to the left side. The first two animals could not endure the lateral position to the left; when placed on the left side, they would change immediately into a ventral and then a lateral position to the right, with the head turned with the vertex of the cranium down. The third animal could not endure the lateral position to the right; when placed on the right side, it turned the head entirely downwards with the cranial vertex in a dorsal position. The animal then made kicking movements, the head rotated further, the trunk followed and rotated via the back to the left side. The animal then remained with the trunk in a lateral position to the left, the head with the lower jaw flat on the ground, the longitudinal axis of muzzle and neck almost vertical on the longitudinal axis of the trunk. A fourth animal which has also the right half of the cerebellum extirpated could not, strange to say, in the first two days endure the right position. In the following days it could not endure the left lateral position. Although in all animals the right half of the cerebellum was extirpated, the initial starting position of the rolling movement was not the same in the various animals. Nevertheless, the succession was always the same: ventral position—lateral position to the right—dorsal positi—onlateral position to the left, etc. The direction of rotating movements, thus, was the same in all of them, in spite of the difference in initial position (on the cause of the various initial positions see Chapter XVIII) and was similar to that observed after extirpation of the right labyrinth.

after unilateral extirpation of the cerebellum, animals with closed eyes are held in the air, they will hold their head erect in a normal posture in both lateral positions.

When suspended head downwards, with closed eyes, the animals will also show widely differing behavior after the two operations. The unilaterally labyrinthectomized animal keeps the vertex of its head constantly turned towards the side of the operation (basic reaction), while the unilaterally decerebellate animal will hold its head totally or nearly symmetrically related to the trunk already a few weeks after the extirpation. After a unilateral labyrinthectomy, symptoms appear which are absent after the unilateral extirpation of the cerebellum. Conversely, the latter is accompanied by impairments which cannot be observed after a labyrinthectomy, as for instance, the hypermetria of movement of the ipsilateral legs. Not only following bilateral labyrinthectomy and a total extirpation of the cerebellum but also after a unilateral labyrinthectomy and a unilateral extirpation of the cerebellum, the clinical pictures are also completely different.

The unilateral extirpation of the cerebellum is sometimes, *but not constantly*, followed by labyrinthine symptoms. After two unilateral extirpations of the cerebellum (dog Peter and cat Josephine), we did not observe rolling movement or deviations of the eyeballs, which indeed are nearly always absent after extirpations.

Risien Russell (264) and Munk (214) already stated that in the monkey (Macacus rhesus) and dog strabismus and deviation of the eyeballs are absent after a total extirpation of the cerebellum and only appear in experiments which are complicated by haemorrhage, secondary injuries or inflammations. Dusser de Barenne repeatedly performed unilateral as well as total extirpations of the cerebellum in dogs and cats without having observed even a trace of deviation of the eyeballs or of eye-nystagmus. Explanation for the labyrinthine symptoms which sometimes appear after hemiextirpation cannot for the moment be made; perhaps the severance of the fibres leading from the cerebellum to the vestibular nuclei and the cessation of their function may sometimes upset vestibular balance and sometimes not. Alternatively, the temporary inactivity of the vestibular nuclei may be caused by a change in blood pressure, by the pressure of blood clots, desiccation and other detrimental influences during the operation, or by small lesions, slight haemorrhages in and around the vestibular nuclei. Probably several of these factors play a role. At any rate, it is certain that the labyrinthine symptoms are only absent when the hemiextirpation was performed with great care and without heavy bleeding or trauma.

If the cerebellum were the central apparatus for labyrinthine reflexes, extirpation of the labyrinths would not produce any more labyrinthine symptoms in decerebellate animals. Magnus and De Kleyn, however, have shown that unilateral labyrinthectomy can produce in decerebellate animals the same symptoms as in intact ones: rotating and turning of the head, rolling, deviation of the eyeballs, nystagmus and decrease in tonus of the muscles of the extremities on the side of extirpation. In the same way, they could establish very clearly Bechterew's compensation after the extirpation of the second labyrinth. This last finding was confirmed recently by Spiegel. Conversely, the extirpation of the cerebellum in labyrinthectomized animals brings about the same impairments as in intact ones. In the cat Peggy in which De Kleyn had removed both labyrinths, I have extirpated the cerebellum three weeks later, when the animal recovered so much that it ran around fairly normally and leapt from its cage (i.e., from a height of 1 m) safely to the ground. After the extirpation, the animal showed a much stronger "cerebellar ataxia" than the exclusively decerebellate cats; an ataxia that could not be reduced by blindfolding the eyes. In an attempt to crawl, to stand up or to eat, the animal always hit the ground with its head with such an impact that its muzzle became sore. During the first month, the animal was so helpless that food had to be put in its mouth. Later, the animal succeeded in eating only when it was supported from the side, but even then only with exaggerated movements of the head and repeated mistakes. Drinking was not possible at all without help and so was running. On crawling it always rolled from one side to the other and each time the head hit the ground wih an impact.   t

Bogumil Lange made corresponding observations with pigeons. He, too, saw that the extirpation of labyrinths in decerebellate pigeons, as well as the extirpation of the cerebellum in labyrinthectomized ones, did not only produce the consequent phenomena typical for that extirpation, but also that the symptoms appearing after extirpation of cerebellum were stronger in labyrinthectomized pigeons and showed less inclination to subside than they did after the extirpation of the cerebellum in intact animals.

All these observations speak against the assumption that the cerebellum is the central apparatus for labyrinthine reflexes.

In summing up, the investigations on the behavior of labyrinthine reflexes in totally or unilaterally decerebellate animals showed the following facts:

1. That in totally decerebellate animals, all labyrinthine reflexes, including those representing balance reactions, are present and occur briskly and appropriately.

2. That the clinical symptoms after unilateral and total extirpation of the cerebellum differ sharply in many points from those after unilateral or bilateral labyrinthectomy.

3. That after unilateral or total extirpation of the cerebellum, the labyrinthine symptoms such as deviation of the eyeballs, nystagmus of the eyes and rolling movements may be totally absent.

4. That a unilateral and bilateral extirpation of the labyrinth can produce labyrinthine symptoms also in decerebellate animals and that, on the other hand, extirpation of the cerebellum in labyrinthectomized animals causes typical, even very severe cerebellar impairments.

C. THE PULLING-UP AND PROPPING-UP REACTIONS OF THE FORELEGS

If, when running along a gutter, a cat slips with its hindlegs so that the posterior part of the body drops down, it will hold fast with its claws, pull up the posterior part of the body with its forelegs till the center of gravity of the trunk lies above the supporting surface of the forelegs and then set down the hindlegs close by the forelegs in the gutter. Thus, the animal restores the balance through a complex series of reactions. Decerebellate cats, too, with and without blindfold, are able to restore the balance in the same way when, for instance, they hang with the forepaws from the edge of a table (Fig. 237).

Decorticate cats, too, pull up the trunk under the same circumstances but they do not place the hindpaws on the supporting surface, due to the absence of preparation for standing (placing). Thus, the pulling up of the trunk is brought about purely as a reflex. The question is now raised on how this pulling-up reaction is produced. When a decorticate dog is lifted up by the forelegs, the trunk is also pulled up

462

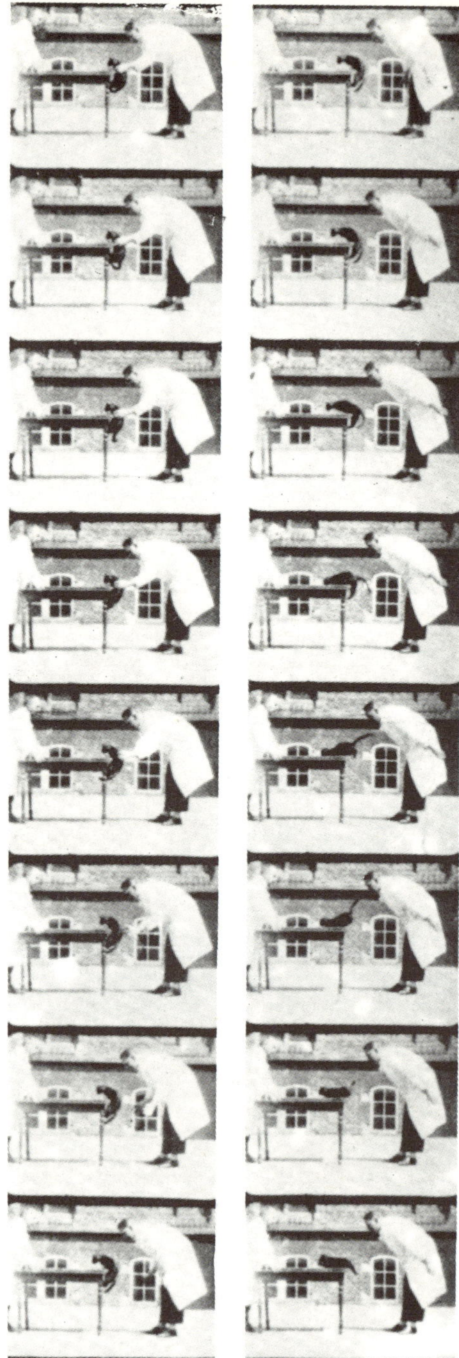

Fig. 237. Decerebellate cat Pierrette. The animal hangs with its forelegs from the edge of a table (1), when released, pulled (2) with its forelegs the anterior part of the body on the table (4-11) and only then places at first the right (12) and then also the left hindleg (13 and 14) onto the table.

by a flexion of the forelegs in the shoulder and elbow joints (Fig. 238, No. 2). In the passive lifting up of the animal from the ground, the retractors of the upper arm and the flexors of the elbow joints are stretched more and more (Fig. 238, No. 1), so that the pulling-up reaction is possibly brought about by myotatic reflexes (stretch reflexes) of these muscles.

Fig. 238. Thalamus (decorticate) dog Fuchs. 1. The animal is pulled up by the forelegs; in so doing, shoulder and elbow joints are passively extended; the retractors of upper arms and flexors of elbow joints are more and more stretched. 2. As soon as the hindlegs leave the ground, the trunk is pulled up by a flexion of the forelegs in the shoulder and elbow joints.

The validity of this hypothesis is confirmed by the following observations. When one grasps the forelegs and during the lifting-up maneuver holds them directed backwards, in doing so the shoulder and elbow joints are further flexed passively, the protractors of the upper arms and the extensors of the elbows are extended further (Fig. 239, No. 4). The animal now moves the trunk upwards as soon as the hindlegs have left the ground; this movement now takes place not by flexion but, in agreement with the stretch of the extensor muscles, by an extension of the shoulder and elbow joints (propping-up reactions, Fig. 239 No. 5). Propping-up reactions of the forelegs are shown by intact, decerebellate and decorticate dogs.

It is, however, remarkable that the active contraction of shoulder and elbow muscles does not appear immediately on the passive extension or flexion of these joints, but only when the contact of the hindlegs with

Fig. 239. Pulling-up (1-3) and propping-up reactions (4-6) in the decerebellate dog Piccolino. 1. Animal pulled upwards by the forelegs. Shoulder and elbow joints passively maximally extended. 2. As soon as contact of hindlegs with supporting surface ceases, trunk is pulled up by flexion of shoulder and elbow joints (pulling-up reaction). 3. On touching the soles of the hindpaws with two fingers, the flexion of these joints relaxes, the trunk sinks downwards till flexors of the shoulder and elbow joints are again maximally stretched. 4. The animal is standing on its hindlegs: the forelegs are seized with two hands by the forearms and then moved passively backwards and upwards. By the upwards movement shoulder and elbow joints are more passively flexed, the protractors of the upper arms and the elbow extensors are more extended. 5. As soon as the contact of the hindlegs with the supporting surface is interrupted, the trunk is pulled upwards, not by flexing, however, but by extending the shoulder and elbow joints (propping-up reaction). 5. On touching the soles of the hindlegs with two fingers, the extension of these joints relaxes and the trunk sinks down. These reactions appear with (4-6) as well as without (1-3) blindfold. See also Figs. 35 and 36.

the supporting surface is lost. Thus, it appears as if the static stress on the hindlegs would inhibit the extensor reflexes of the muscles of the forelegs. Accordingly, the tension of the flexors of shoulder and elbow joints (Fig. 239, No. 3) and the extensors (Fig. 239, No. 6) relaxes when the soles of the hindlimbs of the animals suspended in the air are put under static stress or touched with two fingers.

It results from further information that other factors, too, participate in these reactions. When an animal is lifted from the ground by the forelegs, the posterior part of the body is moved ventrad by the force of gravity (Fig. 238); when this movement is counteracted, the pulling up of the trunk fails to take place (Fig. 240).

Pulling-up reactions appear instantly when the posterior part of the body is released or slowly moved ventrad. In so doing, the head is passively rotated backwards, thus producing labyrinthine righting reflexes and labyrinthine reactions due to a rotation round the bitemporal axis which move the head ventrally in relation to the trunk (Figs. 238 and 240). This alteration in position of the head towards the trunk produces, as is known, tonic neck reflexes. Moreover, the head is brought back into the minimum position of tonic labyrinthine reflexes.

Fig. 240. Thalamus dog Fuchs. 1 and 2. The animal is pulled up by the forelegs till the hindlegs leave the ground. In so doing, the ventral movement of the trunk due to the force of gravity is prevented with two hands. Under these circumstances there is no flexion of shoulder and elbow joints and the trunk is not actively pulled up. 3. In return, instant flexion of joints of forelegs and pulling up of the trunk when the released posterior part of the body moves ventrally. In so doing, there is not only a flexion of the joints of the forelegs but the head and neck are ventrally flexed to the trunk (see also Fig. 35).

Since, corresponding to the effect of tonic neck reflexes, the forelegs are drawn into a flexed position simultaneously with the ventral flexion of the head, it is probable that these reflexes, too, participate in the pulling-up reaction. This supposition is confirmed by the following observations. Firstly, the flexion of shoulder and elbow joints instantly relaxes when the head is passively moved dorsad (Fig. 241); secondly,

Fig. 241. After occurrence of pulling-up reactions the head is passively moved dorsally to the trunk, followed by a relaxation of flexion of shoulder and elbow joints and a sinking down of the trunk. 2. This is the case, too, when by backward moving of the head the posterior part of the body is moved ventrally.

the pulling-up reaction fails to appear in labyrinthectomized dogs with blindfolded eyes since they do not raise their heads in a ventral flexion (Fig. 242, No. 2). On the other hand, the flexion of shoulder and elbow joints also takes place instantly in labyrinthectomized dogs as soon as the head is passively flexed ventrad (Fig. 242, No. 3) and also when, after removing the blindfold, the head is actively raised due to optical righting reflexes. Thus, the position of the head to the trunk surely plays a role in the appearance of pulling-up reactions. In this case, alterations in position of the joints of the forelegs agree with the alterations in tonus (decrease in extensor tonus and increase in flexor tonus) on a ventral flexion of the head, due to tonic neck reflexes observed in the decerebrate animal by Magnus and De Kleyn. The propping-up reactions (extension of shoulder and elbow joints) in intact as well as in decorticate dogs, on passive backwards movement of the

Fig. 242. Labyrinthectomized dog Jenny with blindfold.   1. The animal is pulled up by the forelegs.   2. As soon as the hindlegs leave the ground, the posterior part of the body moves ventrad, due to the force of gravity. Through absence of the labyrinthine righting reflexes the head remains directed backwards and the pulling-up reactions fail to take place.   3. On passive ventral flexion of head to trunk, a flexion of shoulder and elbow joints occurs due to tonic neck reflexes, thus pulling the trunk upwards.

head also relaxes, while it also fails to appear in blindfolded labyrinthectomized dogs. Thus, in the propping-up reactions, too, tonic neck reflexes play a role; only in this case the ventrally directed alteration in position of the head to the trunk does not cause an increase in flexor tonus but an increase in extensor tonus. We thus have here an example of Sherrington's reflex inversion of Magnus' pattern of neck reflex probably caused by the different initial tension. In the pulling-up reaction, the tonic neck reflexes, due to a ventral flexion of the head, cause a contraction of the passively stretched flexors; in the propping-up reaction of the passively stretched extensors of shoulder and elbow joints.

As we have seen, the pulling-up and propping-up reactions abate on touching or on static stress on the soles of the hindpaws. Sometimes, a touching of the tip of the tail or of the rump with a supporting surface has the same effect. This is only the case in animals with intact cerebrum; the abating of reactions, under these circumstances, is due to a "conditioned" cerebral inhibition. When, however, a decorticate animal is moved downwards till at first the soles and then the heels of the hindlegs touch the supporting base and then the hindlegs are passively moved forwards by the supporting surface, backward leg-slackening reactions of the hindlegs appear (Fig. 211), the animal stands up on the hindlegs and runs backwards and, in doing so, the head moves dorsally in relation to the trunk, causing an abatement of pulling-up and propping-up reactions.

To sum up our observations:

When the posterior part of the body falls down in an intact cat, there occurs a complicated balance reaction in which grasping reflexes, pulling-up and propping-up reactions of the forelegs and readiness to stand of the hindlegs all participate. The pulling-up and propping-up reactions are brought about by the cooperation of the following factors:

1. Cessation of an inhibiting influence by elimination of contact between the soles of the hindfeet and the supporting surface.

2. Activation of flexors and extensors of shoulder and elbow joints due to passive stretch.

3. Ventral flexion of head and neck due to labyrinthine reflexes produced by the passive backward movement of the animal by the force of gravity (eventually also due to optical righting reflexes).

4. Contraction of the activated muscles due to the alteration in position of head to trunk.

This complicated balance reaction appears in normal and in decerebellate cats; in decorticate cats, however, its occurrence is incomplete due to the absence of readiness to stand. When hanging by the forelimbs, decorticate cats indeed pull or prop themselves up but the hindlegs are not placed on the supporting surface beside the forelegs. In dogs, the grasping reflex is absent. Pulling-up and propping-up reactions are present in intact, decerebellate and decorticate dogs; however, they are absent in the decerebrate and new-born; in labyrinthectomized animals, they only occur when the eyes are open. In new-born children, only the grasping reflex (according to Peiper and Isbert: tonic grasp reflex*) is present. Pulling-up reactions and laby-

---

* The tonic grasp reflex appears at 4 weeks of age; the response discussed here is now called the "traction response." See T. E. Twitchell, The automatic grasping responses of infants, Neuropsychologia, 3 : 247-259, 1965 (ed.).

rinthine righting reflexes are absent, as is very clearly visible in the illustrations published by Peiper and Isbert (Fig. 243).

The pulling-up reaction probably also plays a role in climbing. When a decorticate (thalamus) cat hangs with the forepaws on the edge

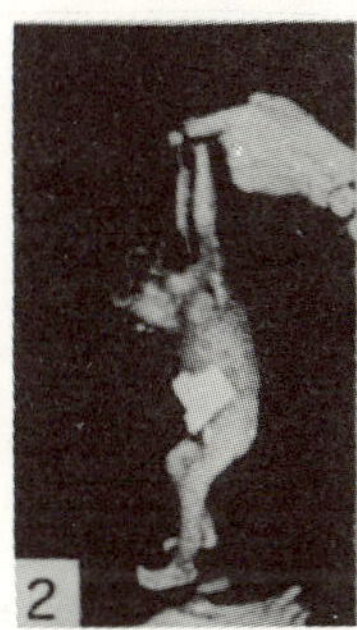

Fig. 243. Note the absence of pulling-up reactions and of labyrinthine righting reflexes. "Tonischer Hand reflex bei Neugebornen" (tonic hand reflexes in new-born infants). From A. Peiper and H. Isbert: Uber die Korperstellung des Saunglings. Jb. Kinderheilk. Vol. 11, 160 (1927).

of a table, one will see that, simultaneously with the pulling-up movements, it will usually make alternating movements with the hind-legs which can be interpreted as attempts to climb. In this case, the flexed forelegs are usually alternatively raised and again apply the claws to the table. Decerebellate monkeys are also able to climb up the poles and screens of their cages (as in our decerebellate monkey Corrie) even if not as elegantly and quickly as intact animals and carry out absolutely adequate and appropriate climbing movements. Decerebellate cats can also climb (Figs. 237 and 244).

Fig. 244. Decerebellate cat Pierrette. The animal clings to the white coat of the researcher, instantly climbs up on it and sits down on his shoulder.

D. THE MAINTENANCE OF BALANCE BY MEANS OF HOPPING REACTIONS, LEG-BRACING REACTIONS AND LEG-SLACKENING REACTIONS IN DECEREBELLATE ANIMALS

When an animal, which has raised one or two legs from the ground, is pushed to the right, hopping reactions will occur which maintain or restore the balance. When, however, an animal standing on all four legs is pulled or pushed to the right, balance is maintained by means of lateral bracing (leg-bracing reactions) or of lateral displacement of the leg (leg-slackening reaction outwards), i.e., by an *abduction movement*. On the other hand, when a paw slides sideways, the animal will avoid falling by a rapid lifting, adduction and displacement inwards of the sliding paw (leg-slackening reaction inwards). Thus, according to circumstances, balance can be restored either by abduction or by adduction movements.[1]

With regards the behavior of hopping, leg-bracing and leg-slackening reactions after total extirpation of the cerebellum, we observed that:

1. hopping, leg-bracing and leg-slackening reactions are absent in the first period, but are later present again;

2. the hopping reactions appear lively and brusque and are accompanied by ample flexor and extensor movements (a retarded or premature appearance could not be established with certainty);

---

(1) According to Thomas and Ingvar, a failing to the right side, in addition to rotation of the vertebral column and inclination of the head towards the other, left side, is prevented by raising and adducting the right legs. After the unilateral extirpation of the cerebellum, according to their opinion, the adducting reactions of the extremities ipsilateral to the extirpation are absent, which promotes the falling towards the side of extirpation. According to Ingvar, the centers for adduction are situated in the most lateral section of the equilateral lobulus ansiformis, the centers for abduction in the median part of crura lobuli ansiformis, and he asserts that lesions of the lateral part of crura, therefore, cause a falling towards the side of the lesion and destruction of the median part, falling to the opposite side. However, as we have seen, falling towards the side of extirpation is nearly always prevented by a bracing outwards, or transferring outwards, i.e., by abduction of the legs ipsilateral to extirpation and only in rare cases, as for instance in slipping, by adduction. The absence of centers of abduction of the ipsilateral legs would thus promote falling towards the side of the operation, in opposition to Ingvar's assertion. Besides, the statement is also wrong that, after unilateral extirpation of the cerebellum, the adducting reactions in the ipsilateral legs are absent.

3. the leg-bracing forwards, backwards and outwards appear constantly lively and strong (here, too, no retardation or prematurity of appearance could be established with certainty);

4. leg-slackening reactions also take place abruptly; in this case, the legs are raised abnormally high and make abnormally large steps but the reactions are usually retarded.

The question is how far these reactions are appropriately carried out for the maintenance of balance and how much they contribute to the maintenance of balance in decerebellate animals.

It is nearly impossible to make a decerebellate animal fall down by a violent rotating, dragging or turning of the head, which is all the more remarkable because in a standing decerebellate animal every rotation of the head is accompanied by many alterations in position of the pelvis (Fig. 245) that threaten the stability of the posterior part of the body.

Fig. 245. Decerebellate dog Piccolino. 1. Head in lateral position to the left; and, by the outward bracing of the left legs, falling to the left side is prevented. 2. Head suddenly rotated from the left to a right lateral position, causing such a strong rotation of the spinal column and pelvis that the animal can avoid a fall only by displacing the hindleg to the right followed by an outward bracing of this leg.

Even when the anterior part of the body is lifted from the ground by the head so that the animal stands only on its hindlegs and the trunk is then pulled to and fro, one rarely ever succeeds in throwing down the rear part of the body, since falling is prevented by prompt alterations in posture of the hindlegs. Similarly, any attempt to induce the animal to perform a backward somersault by a sudden backward pulling of the head fails, because the rear part of the body preserves its standing posture either by a rapid lateral stepping or by quickly running backwards. Even when one flings the animal in the air by the head and then attempts to place the animal on the ground in a lateral position by rotating and pulling the head downwards (Fig. 246) does one rarely succeed.

Not only on horizontal, motionless supporting surfaces but even on inclined planes and on a turntable, decerebellate animals are able to preserve their balance by bracing and displacing their legs when the head is passively pulled or rotated (Fig. 247).

Similarly, decerebellate animals rarely lose their balance when their tail is pulled to the right or to the left, forwards or backwards or by a rapid rotation of the tail round the dorso-ventral axis. When the posterior part of the body is lifted by the tail and pulled to and fro, the anterior part of the body will also remain in a standing posture by a prompt displacement of the forelegs. It is impossible to drag the animal upwards or forwards by a sudden pull on the tail or make it turn a somersault forwards or to throw it down by pushing or by pulling the skin forwards, backwards or sideways.

When, in a decerebellate dog, a limb is slowly displaced forwards, backwards, outwards or inwards, the limb is raised at first and is then transferred more to the front, to the rear, inwards or outwards. Although these reactions, particularly displacement inwards[1] are retarded, the animal does not lose its balance. Even when, in the course or running, suddenly one or two legs are pulled away, the animal usually corrects this in time, so that tumbling down is prevented (Fig. 248). In standing on a supporting surface which is shifted forwards, backwards and sideways, decerebellate animals also promptly show leg-bracing and leg-slackening reactions which serve to maintain balance. Due to these reactions also, they do not lose their balance on

---

(1) Also, in intact animals, the leg-slackening reactions inwards sometimes occur late; in man, too, the correction by adduction is more difficult than that by abduction.

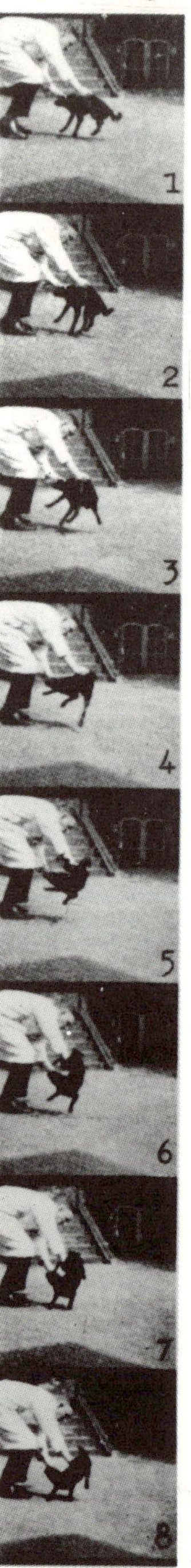

Fig. 246. Decerebellate dog Erik. Head first rotated to a lateral position to the right (1), the animal is then flung in the air by the head and skin of the back (2-5) and an attempt is made to place it on the ground in a lateral position to the right. As soon as the right foreleg touches the ground (7) it braces outwards and the trunk is rotated into a normal position in spite of the fact that the head is still fixed in a lateral position to the right (7-8),

Fig. 247. Decerebellate cat Pierrette and dog Piccolino on the turntable. During the rotation of the disk, an attempt is made to bring the trunk of the animal into a lateral position by a passive rotation of the head, which, however, does not succeed even when the anterior part of the body is lifted from the disk by the head (B 9-13).

a slow inclination of the supporting surface (Fig. 249) and do not fall down even on an inclination of the supporting surface by 45°, in spite of the fact that they are very restless in standing due to a continual defective control in movement (Fig. 249).

When a decerebellate animal is set down on a turntable with one foreleg, the displacement of the leg, even with a considerable speed of rotation, will be so rapid that the head and the anterior part of the body do not fall.

Fig. 248. Decerebellate dog Moor. While running, sometimes one and sometimes both legs are suddenly pulled away from under the animal; the animal does not fall. A5. During the raising of the right foreleg, the left one is suddenly pulled backwards. B2-4. Both right legs pulled strongly to the right. C4-5. Left hindleg pulled backwards. C3 and 4. While the right foreleg is lifted, both hindlegs are pulled backwards.

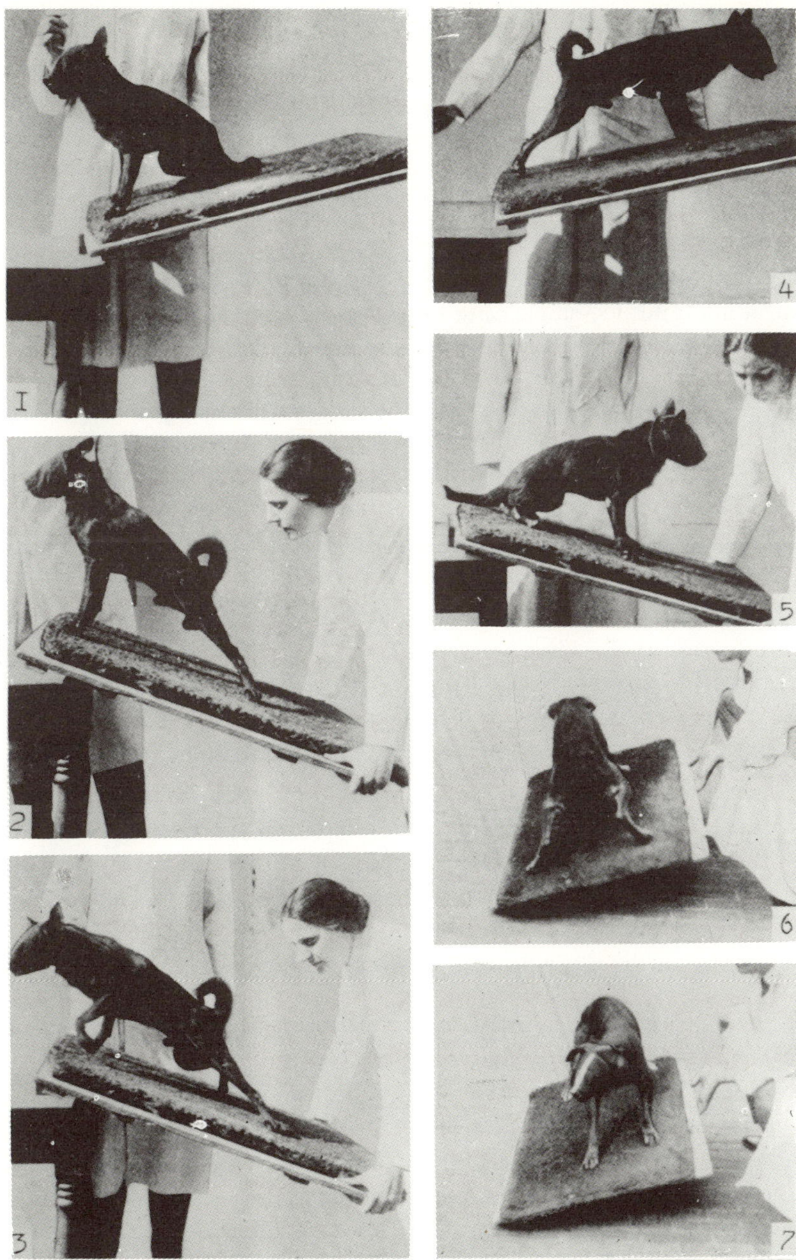

Fig. 249. 1 to 5. Decerebellate dog Erik with blindfold, on a board which is lifted and lowered from the tailend.   1. On lifting the tailend, the forelegs promptly brace forwards while the hindlegs are flexed. 2. On lowering the tailend, all four legs promptly brace backwards. 3. On a further lowering, leg-slackening reactions occur and the animal runs forwards.   4 and 5. On lifting and lowering of the headend similar reactions. 6 and 7. Decerebellate dog Moor with blindfold, on a board that is lifted from the right and the left side. 6. On lifting

Fig.249. (Cont.) the board on the right side, the left legs promptly brace outwards, while the right ones are flexed. 7. On raising the board on the left side leg-bracing reaction on the right legs, the left give way. Notice also the rotation of trunk and pelvis in No. 6.

In summing up, we can say that the hopping reactions as well as leg-bracing and leg-slackening reactions in the decerebellate animal are carried out in the most appropriate manner for the maintenance of balance.

With regard to these observations, André Thomas' findings are noteworthy in that patients with cerebellar atrophy, in spite of a high degree of cerebellar ataxia, are able to offer good resistance to pulling and pushing while running so that usually neither falling nor propulsion, retropulsion or lateropulsion occurs.

E. THE MAINTENANCE OF BALANCE OF DECEREBELLATE ANIMALS ON THE TURNTABLE

Decerebellate cats and dogs are also able to maintain their balance on a turntable, not only on a sudden change in speed of rotation but also on a sudden change of direction of rotation, with open eyes as well as blindfolded, by means of propping and transferring of the legs (Fig. 250). With closed eyes, however, the animals usually do not remain upright but sit down or go into a ventral posture.

Thus, decerebellate animals are able to maintain their balance in spite of the fact that rotation causes a strong turning of the head, followed secondarily by an especially vigorous and abrupt curvature of the spinal column and an alteration in position and supporting tonus in the extremities. By these reactions the leg-bracing and leg-slackening reactions, produced by inertia and the force of gravity, are counteracted under certain circumstances and the maintenance of balance is rendered more difficult.

When, for instance, an animal stands on the radius of a disk, with the head directed towards the center on rotation of the disk in a clockwise direction, inertia will draw the animal to the right and produce a *bracing outwards, i.e., an abduction of the right legs*. On the other hand,

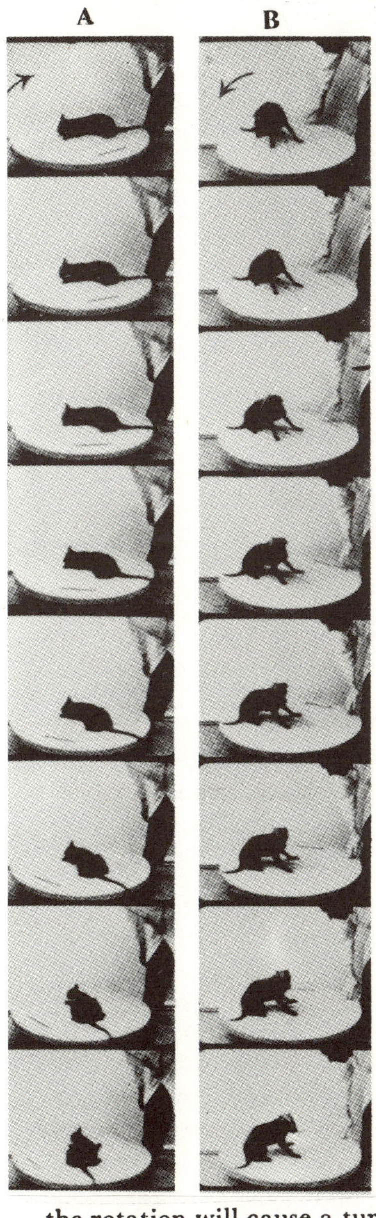

Fig. 250. A. Decerebellate cat Pierrette with blind-fold on a disk rotating in a clockwise direction. The animal does not fall down. The rotation causes a turning of the head towards the left followed by a strong concave curvature to the left of the spinal column. B. Decerebellate dog Piccolino, also with blindfold, on a disk rotated anti-clockwise. It also does not fall but is able to retain its sitting position. Now the rotation causes a turning of the head towards the right with a concave curvature of the thoracic spinal column to the right.

the rotation will cause a turning of the head towards the left, followed by a concave curvature of the spinal column to the left, an abduction of the left and an *abduction of the right legs*, and this adduction will counteract the outwards bracing necessary for the maintenance of balance.

This explains why the regulation of balance in decerebellate and intact animals in a standing posture on the turntable is not the same as in labyrinthectomized animals. In too rapid a change of speed of rotation or direction of rotation, decerebellate animals lose their balance

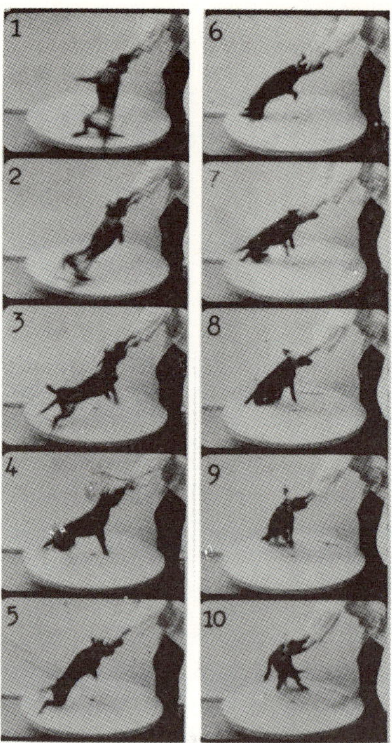

Fig. 251. Decerebellate dog Piccolino on a turntable rotating in a clockwise
direction. The animal is seized by the head with both hands in an
attempt to pull the animal backwards against the direction of rotation (1). (On Fig. 1 one sees the ventral side of the animal.) The
animal, however, is able to bring back the posterior part of the body
into a normal position by rotation of the trunk to the head (2 and
3) and to preserve this position also on rotation of the head to the
right (5) and by turning the head to the left (10).

more easily than labyrinthectomized ones. Still, one is amazed by
the efficiency of the maintenance of balance in decerebellate animals
in spite of the uncontrolled wobbling movements and strong labyrinthine rotating reactions on the turntable. As already mentioned, it is
nearly impossible to bring decerebellate animals, even on the turntable;
into a lateral position by rotating and by tugging the head (Fig. 251;
see also Fig. 247), the tail or the skin of the back.

Decerebellate animals are also capable of restoring their normal
position during rotation. When they are placed in a lateral position,
they will rise instantly even with blindfolded eyes (Fig. 252, A) and
set up their trunk straight, even when the head is fixed in a lateral position. When decerebellate cats are dropped on the turntable from a
lateral or dorsal position in the air, they will arrive on the table on
their feet and instantly maintain their erect posture (Fig. 252, B).

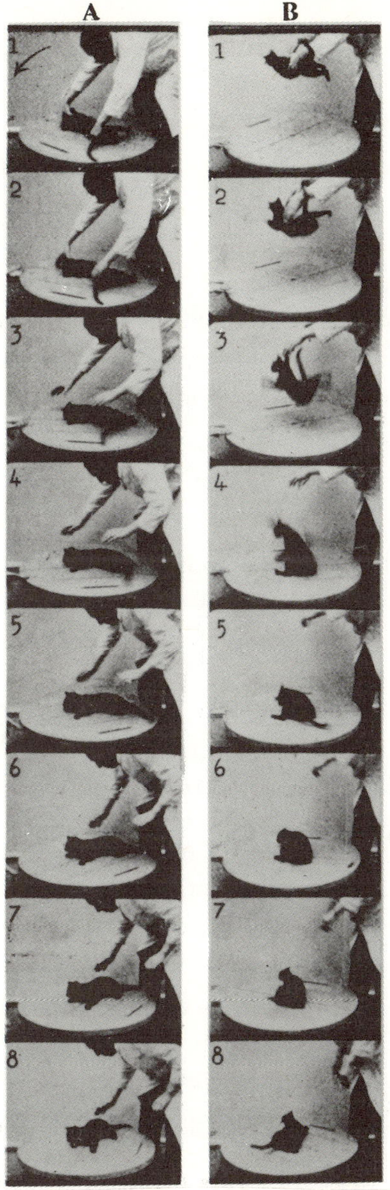

Fig. 252. A. Decerebellate cat Pierrette with blindfold, placed on the rotating disk in a lateral position (A1). Animal instantly raises first the head (A2), then the trunk (A4). B. The same cat, also with blindfold, held in the air in a lateral position (B1), then released above the disk (B2). The animal turns in the air (B3) so that it lands on the disk with its paws (B4 and B5). On reaching the surface, it is instantly able to maintain its balance.

F. THE REACTIONS OF SPINAL COLUMN AND PELVIS FOR THE MAINTENANCE OF BALANCE IN DECEREBELLATE ANIMALS

Munk and André Thomas assert that in decerebellate animals the compensating actions of spinal column and pelvis for the maintenance and restoration of balance are impaired.

However, we made these observations :

1. Decerebellate animals,when brought into a lateral position, instantly turn into a ventral position even when the head is retained in a lateral position (Fig. 229, body righting reflexes on the body).

2. In the attempt to bring the animals into a lateral position, for instance to the right, by a rotation of the head towards the right, the pelvis performs a distinct, ample rotation to the left, by means of which falling down is prevented (Fig. 245).

3. In standing on an inclined plane which is raised from the right side, decerebellate as well as intact animals rotate head, trunk and pelvis towards the raised side, i.e., towards the right and at the same time show a concavity of the spinal column towards the left (Fig. 249, No. 6 and 7). These reactions occur with closed as well as with open eyes, on a slow as well as on a rapid raising or lowering of the supporting surface, in standing posture as well as in a ventral position of the animals.

4. Decerebellate animals also show distinct rotations of the pelvis on pulling sideways, for instance, by a fold of the skin (Fig. 183, No. 6) and on a sideways shift of the supporting surface.

When in a standing decerebellate dog the toes of one hindpaw are seized, the animal will instantly draw up its paw and perform a distinct rotation of the pelvis (Fig. 253, No. 1). On the spontaneous raising of the hindleg on urinating or in running, this rotation of the pelvis also appears in decerebellate animals and generally in an exaggerated manner (Fig. 253, No. 2 and 3).

It is worthy to note that André Thomas, in agreement with these findings in animals, could observe in patients with cerebellar injuries distinct rotations of the pelvis on crawling on hands and knees and

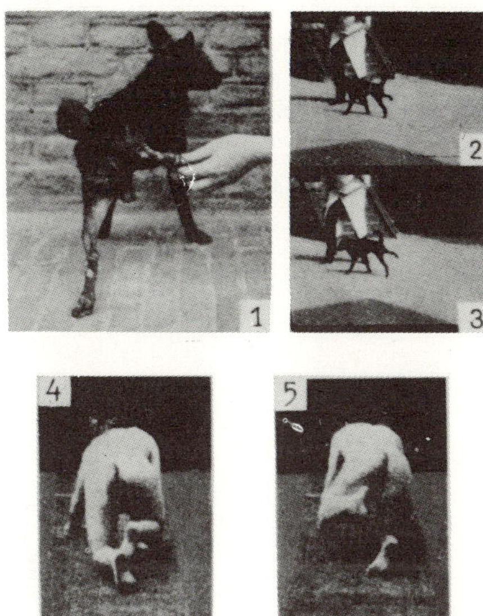

Fig. 253.   Decerebellate dog Erik.   1. On seizing the toes of one hindleg, the animal draws this leg up and supports itself on the other three.   In doing so, the pelvis strongly rotates towards the other side.   2 and 3. Also, in running on raising the right hindleg, distinct rotation of the pelvis to the left.   4 and 5. Patients with severe lesion of the left half of the cerebellum from a gunshot injury, while crawling on all fours. On raising of the left as well as of the right leg the pelvis is turned to the other side, on raising the left leg more abnormally.   From André Thomas: Etude sur les blessures du cervelet. Paris: Vigot Freres, 1918.

that the rotation is abnormally strong on raising the leg on the injured side (Fig. 253, No. 4 and 5).   In decerebellate cats not only the strong movements of the pelvis but also the excessive compensatory movements of the tail while running attract our attention (Fig. 254).

In summing up, our investigations show, *that neither in decerebellate cats nor in decerebellate dogs could any constant and lasting absence even of any balance reaction, righting reflex or labyrinthine reflex be noticed.*

From this one can conclude that there is no reason for the supposition that the cerebellum is to be regarded as the central apparatus for the regulation of balance, for the righting functions or for the labyrinthine reflexes.   The balance reactions, the righting and labyrinthine reflexes must have centers situated outside the cerebellum, as well as tracts leading to and from it.

Fig. 254. Decerebellate cat Pierrette while running. Notice the excessive compensatory movements of the tail.

The absence of almost all balance reactions in the first days after extirpation of the cerebellum is not necessarily based on a loss of cerebellar function but can be conceived as a shock effect on the preserved centers, particularly because they are sometimes again already present within the first weeks after extirpation.

Similarly to André Thomas, many other authors assert that the restoration of the ability to maintain balance is brought about in decerebellate animals by the compensatory activity of the cerebrum. Some possible validity of this assumption cannot be excluded, though observations on the decerebellate-decorticate dog Robbie, which already 5 weeks after extirpation of the cerebellum showed again labyrinthine righting reflexes from a lateral position and leg-slackening reactions of the hindlegs outwards and backwards, prove that these reactions can be produced without the compensating influence of the cerebrum.

Similarly, alterations in the distribution of muscle tone which serve for the maintenance of balance and prevent falling have been observed in decerebellate animals. Thus, alterations in muscle tone appear very clearly on raising a leg, on the passive rotation and turning of the head, on the inclination or passive shifting of the supporting surface and on the passive sideways pulling, rocking and to and fro displacement of the animals. There is little foundation to André Thomas' supposition that the centers for a special synergy of tonus for the regulation of balance are situated in the cerebellum.

By this, of course, we do not want to say that the cerebellum does not normally exert any influence on the balance reactions, righting and labyrinthine reflexes or that in decerebellate animals the balance reactions proceed normally or that these animals possess a stable equilibrium equal to intact ones. As we have seen, balance reactions appear particularly abrupt and hypermetric; moreover, leg-slackening reactions are retarded.

Whether these deviations are caused by the absence of influence which in intact animals is exerted by the cerebellum on the central mechanism of these reactions or whether they are brought about by an abnormal distribution of tonus is as yet quite unclear.

Decerebellate animals fall more frequently and lose their balance more easily than intact ones. The cause of this cannot be attributed to the absence of balance reactions because, when we observe decerebellate animals running, we see that, owing to "cerebellar ataxia," i.e., due to abrupt, unrestrained movements, they get into extraordinary postures and, therefore, look as though they will fall down more easily. But nearly always they succeed in restoring their balance by appropriately performed balance reactions in which head, neck, thoracic spine, limbs, pelvis and tail participate.

Similar observations are found often in the literature with persons with healed cerebellar abscesses, injuries or atrophies of the cerebellum (Oppenheim, v. Monakow, André Thomas, Gordon Holmes and many others). Ingvar's assertion, "The patient without a cerebellum obeys the same mechanical laws as the puppet," is certainly incorrect.

XVI. GENERAL INFORMATION ON MUSCLE TONE, PARTICULARLY ON ITS BEHAVIOR IN DECEREBELLATE ANIMALS[1]

The idea that the cerebellum extensively influences muscular force was expressed nearly simultaneously for the first time by clinicians and physiologists. In 1861 Leven and Olliver (164), on the strength of observations in 76 cases of cerebellar diseases, supposed that the impairments of movement in cerebellar patients were based on muscular weakness. Luys, in 1864, came to the same conclusion based on material from 100 cases. He believed that the principal symptoms of cerebellar diseases were due to a general condition of weakness and a progressive decrease in muscular force. "Not a paralysis but an asthenia." In the same year, Poincare (227) gave a similar description of cerebellar symptoms and compared the debility and muscular weakness of these patients with that of progressive paralysis. Among experimentators Dalton (1861) was the first to point out the remaining general weakness after the partial extirpation of the cerebellum even when the strong impairments of motility have already disappeared. Weir-Michell (325) continued Dalton's investigations, made partial extirpations of the cerebellum in a great number of pigeons, injected mercury into the cerebellum, painted the surface of the cerebellum with ferric chloride, etc. Supported by these observations, he arrived at the conception that the cerebellum was an organ of strengthening a source of energy for the spinal centers in regard to voluntary and involuntary movements. This theory, however, did not gain numerous supporters until after Luciani's fine investigations in dogs and monkeys. According to Luciani, the loss of function of the cerebellum makes itself felt in three different types of impairment, i.e., in neuro-muscular asthenic, atonic and astatic

---

(1) For detailed descriptions on the present state (1927) of the various tonus problems, see the excellent monographs by S. Cobb (46), J. F. Fulton (87) and E. A. Spiegel (288), further Vol. 8 of the Handbuch der normalen und pathologischen Physiologie (Handbook of the normal and pathological physiology), Berlin, 1926.

manifestations; consequently, in his opinion, one is allowed to assume that the influence normally exerted by the cerebellum on the central nervous system is to be found in "sthenic, tonic and static neuro-muscular activity," i.e., in a complicated activity by which:

1. *the potential energy* which is at the disposal of the neuro-muscular apparatus is increased (sthenic activity);

2. *a degree of tension* during the pauses in function develops (tonic activity);

3. *the rhythm of elementary impulses* in the course of functioning is accelerated and the normal transition and regular constancy of action is brought about (static activity).

Moreover, Luciani ascribed a *trophic function* to the cerebellum. Since every functional stimulus acting on living elements is necessarily coupled with an alteration in their nutritional status, it is clear that the complicated activity carried out by the cerebellum through its efferent pathways is intimately connected with direct or indirect trophic activity. The *direct* trophic activity is sufficiently proved by the distinct degeneration and sclerotic conditions in the brain which appear as a result of cerebellar ablation; the *indirect* by the progressive degeneration in muscles and skin, by the various general and local dystrophic effects, by the slower growth and reduced resistance against external harmful interventions through which the animal easily sickens and lives for a comparatively shorter time.

According to Luciani, the cerebellum is only an auxiliary or intensifying system and does not possess a specific field of action, that is exclusively its own and which is not influenced by centers of the cerebrospinal system. The intensifying effect is derived from the entire cerebellum, whose different sections have the same function as the entirety, only each half of the cerebellum supplies or influences predominantly the muscular system of the ipsilateral side.

Munk, Probst, von Bechterew, André Thomas and others, too, based on experiments, favor the opinion that a cessation of the cerebellar function will cause atonia and asthenia of muscles; they, however, see in the intensifying effect not a specific function of the cerebellum and do not believe that the various phenomena after extirpation of the cerebellum can be entirely explained by the cessation of this effect.

According to Luciani, all these manifestations are based *exclusively* on the cessation "of a supporting or intensifying influence which raises the degree of tension of the neuro-muscular apparatus during the functional pause *or during rest* (tonic effect); increases the energy employed in the various voluntary, automatic and reflex activities (sthenic effect); accelerates the rhythm of individual impulses by which the activities cooperate and smoothing their normal fusion and regular continuity (static effect).

Luciani distinguishes tonus and force of a muscle. "Muscle tonus is the *active* tension of a muscle *at rest*; muscular force is the energy employed by the muscle in the various voluntary, automatic or reflex activities."

This distinction at first seems rational, although an *active* tension *at rest* is somewhat of a contradiction in terms. However, what is a muscle at rest? This question is all the more justified when one attempts to understand what Luciani means by the existence of an atonia and asthenia. He concludes the presence of atonia of the ipsilateral muscles after unilateral extirpation of cerebellum from the following four manifestations:

1. "When a bitch, deprived of the right half of the cerebellum, was lifted in the air and seized by the flanks, one could see that the muscles of the right posterior extremity were more relaxed than those of the opposite side."

2. "When one lifted the soles of the feet of the animal with the flat of the hand, one could perceive that the leg of the healthy side offered a stronger resistance to passive movement than the leg on the side of the operation."

3. "When the bitch was investigated while eating, to which purpose it kept itself upright with widely spread legs to enlarge the supporting surface and while its attention was concentrated on its food, one observed that the extremities of the lesion side gradually and slowly gave way."

4. "From the frequent falling of the animal towards the operated side in the first period after the unilateral extirpation, since falling depends on a muscular relaxation which the animal is not yet able to modify by suitable actions of compensation."

"For this complex of manifestations," says Luciani, "I employed the generally adopted term of atonia." Thus, according to Luciani, the muscles of the extremities are at rest when the paws hang down (1), further when they are moved passively (2) and in standing posture (3 and 4) of the animals.

On the following three manifestations of a unilaterally decerebellate animal Luciani speaks of asthenia.

1. "When food is held at a certain height above the head, the animal raises itself to an upright position; however, it falls instantly due to the giving way of the posterior extremities."

2. "When one induces an animal to drag a load attached to its tail, it will fall frequently towards the damaged side."

3. "When a clip is attached to the ear of the intact side, the animal is able to remove it by appropriate movements with the ipsilateral anterior extremity, when the clip is attached to the other ear, the animal does not even try to use the leg of this side but restricts itself to shaking its head violently and, in doing so, will lose its balance and fall towards the operated side."

Thus, when the eating animal falls down while standing on its four legs, it is due to an atonia of muscles and, when it falls while standing on two legs, it is due to asthenia. When the animal is pulled into a lateral position by its own bodyweight, it is atonia and, when pulled by the weight attached to its tail, the cause is asthenia. I think I will be understood better if I do not adopt the distinction between muscular atonia and muscular asthenia in the sense of Luciani.

The definition of muscle tonus as the state of tension of a muscle at rest is today entirely customary in physiology and neurology (see Wertheim-Salomonson [329], Dusser de Barenne [60B], Spiegel [288] and many others). However, when is a muscle at rest and how can the state of tension be investigated without rousing it from its position of rest? When a muscle causes the movement of a body part, it is certainly not at rest. Is it indeed in a state of rest in passive movement which, as is well known, produces reflex tensions? When the tension of the biceps is increased voluntarily without flexing the elbow, is in this case the muscle at rest? Do isometric tensions, caused by reflexes abolish rest? In ventral and dorsal postures, the muscle tension of the extremities is differently influenced by reflexes from the labyrinths and the body surface; in which position of the trunk and in which posture of the head are the muscles at rest, and in which are they not? Are the muscles at rest while sleeping, although in this case they still respond to some stimulations? What is the state of rest in muscles of horses and other animals which sleep in a standing posture ? Alterations in temperature, acoustic and optical impulses, stimulations from the labyrinths, from the body surface, from the deeper parts of the neck, from other muscles, from the interior organs can produce alterations in tension. Which of these stimulations must be absent to produce the muscular state of rest? Even in deepest narcosis, it is never possible to eliminate *all* reflex influences; dogs still make alternating flexing and extending movements with their legs in deep narcosis (Graham Brown). It seems, therefore, more than doubtful whether it is at all possible for a muscle to be at rest as long as it is alive. Fibrillar tremors and undulatingly contractions can be observed even some time after death; mechanical, electrical, thermal and chemical stimulations as well as alternations in humidity of the environment cause movements of the muscle fibres even on excised muscles.

Practically, the muscle tonus is determined by measuring of muscular, resistance in passive alteration in position of parts of the body to each other.[1]

In the following we understand by muscle tonus the tension with which a muscle resists to the passive alteration in position of a part of the body.[2] *By muscular force we understand the maximum force with which the muscles can perform a movement; it is measured by determining the force which is able to prevent this movement.*

Other authors, too, give a similar definition for muscle tonus, however, with the difference that with muscle tonus they only mean the involuntary resistance or involuntary tension. In my opinion, this limitation is incorrect since there is no objectively discernible difference between voluntary and involuntary contraction. Apart from the fact that it is very doubtful whether one can really speak of a volition and voluntary action and muscular contraction in animals (for example, conditioned reflexes). In human beings, certain muscular contractions are termed voluntary without reason and sometimes quite erroneously. The muscles of the abdominal wall, for instance, can become so strongly reflexly contracted on palpation in highly excitable persons that an examination of the organs of the abdomen is made impossible. On

---

(1) On the hardness of a muscle and method of determination, see Riesser (245).

(2) The tonus (resistance) is, in this case, the total muscular elasticity and a tension which is dependent on the central nervous system.

strong persuasion the tension often relaxes. The clinician says in this case that, at the beginning, examination was rendered difficult by voluntary contraction, although the patient had desired this examination and had no intention at all to prevent it by straining the abdominal muscles and probably, on the contrary, had tried to achieve a relaxation of the tension.[1]

Whether by the use of strong persuasion or in the attempt to eliminate arbitrary action by "diverting the attention," inhibiting or promoting cerebral influences are put in or out of action is still entirely unknown. By "diverting the attention," some reflexes, among them the patellar reflexes, can be more easily produced, while others, for instance the contraction of the muscles of the abdominal wall on palpation and the muscular contractions of the extremities on passive movement of the latter, are reduced in force. Add to this the fact that "diverting the attention" effects are themselves variable which seems to exert an influence in many cases on the process of the reflexes. The changes in muscular tension due to affects (for instance, fear or nervousness during the examination) are nearly always interpreted as voluntary, although we know from the pseudo-affective reflexes of decorticate animals (Sherrington) that they are not necessarily brought about by way of the cerebrum. In decorticate dogs, too, which are made frantic by pinching, reflexes and reflex muscular contractions are then increased. Moreover, we know by Pavlov's experiments that conditioned stimulations when producing cerebral reactions, at the same time, inhibit subcortical reactions.

Finally, each voluntary action is caused by preceding stimulations and the transition from a voluntary to an involuntary reaction is so gradual that, in my opinion and with regard to muscular tensions, voluntary and involuntary cannot be sharply distinguished. To this should be added that each voluntary muscular tension is always accompanied by reflex alterations in tension of other muscles.

In man, one generally speaks of a voluntary contraction when the intention to contract has been announced beforehand or when the contraction has been carried out to order. How often, however, do we observe in children or mentally diseased patients that they announce an action without carrying it out or even carry out an opposite action.

---

(1) Decorticate dogs, too, show in a dorsal position a strong tension of the abdominal muscles on palpation.

Did they, in this case, intend this action or the announced one? On command, they sometimes do not react at all, sometimes according to the ordered action and sometimes in a totally different way. In the latter case, is the reaction voluntary? Besides, reactions on command (unconditioned) can represent acoustic reflexes,and decorticate animals indeed still show reactions on acoustic stimulations. Moreover, let me remind you of the conditional reflexes on acoustic stimulations and of the reactions of animals trained to command.

Hypertonia is an increased resistance and hypotonia a reduced resistance of muscular tension against passive alterations in position of parts of the body to each other. Thus, hypertonia and hypotonia are opposite conditions. Hypertonia is a positive manifestation, i.e., its appearance proves that a central reflex mechanism must be present, while hypotonia can be caused by a total absence of these mechanisms. The point they have in common is that they represent deficiency phenomena.[1]   The transition from hypo- to hypertonia does not take place by way of normal muscle tonus, though normal muscle tonus, due to the cessation of central mechanisms, is sometimes transformed into hypotonia and sometimes into hypertonia. The succession of the different stages of tonus is atonia ⟶ hypotonia or hypertonia ⟶ normal muscle tonus. Hypo- and hypertonia thus are originally very closely related phenomena and sometimes even merge into one another, a hypotonia into a hypertonia or vice versa, for instance, after an apoplexy. In this case, there is no intermediate stage with normal muscle tonus; the latter will appear only after complete cure.

Hypertonia is an increased resistance of muscular tension to passive alteration of position, but this increased resistance is only relative. The force of resistance against passive flexion of the rigid foreleg of a decerebrate animal is not stronger than the strength of supporting tonus of this leg before decerebration(See P. 174)or of the foreleg of a decorticate dog of the same size. The resistance against passive flexion of the hypertonic, noncontracted leg of a hemiplegic patient is not stronger than the resistance of the healthy opposite leg in a standing posture. *Hypertonia is the continuance of a strong muscular tension in circumstances, among others the influence of reflexes, which would normally cancel it.*

With an inactivation of all central mechanisms, as for instance section of the motor nerves, there is a true hypotonia of muscles.

---

(1) Hypertonia is at the same time a deficiency symptom and a positive manifestation.

In man, the exclusive preservation of spinal mechanisms, as after the transverse lesion of the spinal cord, can account for the appearance of a hypertonia, in addition to hypotonia, particularly in the adductors, the quadriceps and the calf muscles or, in other words, an increased tonic resistance against passive extension (Riddoch 244, Lhermitte 171). Similar manifestations can be observed in spinal animals.

Since hypotonia as well as hypertonia can be present after elimination of all stimulations from centers situated oral from the spinal cord, one has to allow the conclusion that a total destruction of the cerebrum, cerebellum or of the red nuclei or of other centers of the brainstem excludes neither hypertonia nor hypotonia.

In a previous work I concluded that the occurrence of decerebrate rigidity is based in the first place on a functional loss of the red nuclei. This, however, does not mean that a destruction of these nuclei excludes a hypotonia due to a simultaneous inefficiency of centers situated more caudally. Whether the cerebellum has a tonus promoting or a tonus reducing influence excludes neither the occurrence of a hypotonia nor of a hypertonia, as shown by numerous cases of cerebellar diseases described in the literature, which partly show hypotonia and partly hypertonia of the muscles. In the case of cerebellar tumors and abscesses, in gunshot wounds and other injuries of the cerebellum (particularly in those with secondary infections), hypotonia can generally be observed.* However, several cases of bleeding, abscesses, tumors and softening of the cerebellum have been described in which a distinct hypertonia was present. Thus, among others, Guillain, Alajouanine and Marquis described a patient with an almond-sized bleeding in the right cerebellar hemisphere, who showed tonic spasms and rigidity of the legs. André Thomas, too, did not see any hypotonia in his cases of cerebellar atrophy. Ley (165), Guillain, Mathieu and Bertrand (113) described cases of cerebellar atrophy in which the patients showed a typical Parkinson's syndrome with maximum rigidity and plasticity.[1]

---

* The classical account of acute hypotonia and asthenia in man in the English language is that of Gordon Holmes in 1922 (ref. 121) (ed.).

(1) The patient described by Ley was also examined by Pierre Marie. Based on this examination, Pierre Marie reached the following conclusions : "If the symptoms given as characteristic for a lenticular syndrome are correct, this patient shows a typical lenticular syndrome, or we must renounce diagnosing this disturbance." However, in the microscopic examination of the brain, no deviation in corpus striatum of substancia nigra were found. Also, in the case described by Guillain, Mathieu and Bertrand, any kind of alteration was found lacking in the corpus striatum and in the substancia nigra.

Add to these the numerous cases of cerebellar abscesses mentioned in the literature in which, after the operation, almost no impairments, not even in muscle tonus, could be observed.

Experimental investigations into the question of participation of various parts of the central nervous system in the maintenance of normal distribution of muscle tonus are extremely difficult and should be utilized only with great care, since it can often be hardly decided whether the distribution of tonus is normal or whether there are transitions to hypo- or hypertonia. Moreover, a hypo- or hypertonia appearing after a partial extirpation of the central nervous system is no proof at all that the centers of a normal distribution of muscle tonus are situated in the extirpated part, since both hypertonia and hypotonia are deficiency symptoms which can also be caused by shock or diaschisis of The remaining centers. Besides, an abnormal distribution of muscle tonus does not necessarily have to be produced centrally, as shown by Sherrington's observations on a dog with a severed spinal cord. The dog constantly kept one limb in a flexed and the other in an extended position due to a decubitus injury on the paw of the flexed limb. In the same way, thalamus dogs while running sometimes keep one limb, which had been chafed by walking, drawn up in a flexed position.

Feverish diseases, too, among others cystitis and decubitus injuries in various parts of the body, are able to exert a certain influence on the muscle tonus. Thus, they sometimes prevent the reappearance of muscle tonus and of the various spinal reflexes after transverse section of the spinal cord or they cause the disappearance of already returned tonus and reflexes. In animals which were rigid after extirpation of the cerebellum, the rigidity does not disappear as quickly when a suppuration of wire suture, purulent conjunctivitis or rhinitis, etc., occur, while these complications delay the reappearance of muscle tonus in animals which had become flaccid after extirpation. In the stage of permanent impairment, infections can also strongly reduce the supporting tonus in decerebellate animals. Every clinician knows that also persons during or after the occurrence of fever are often hardly able to keep themselves upright on their feet, just as the appearance of abnormal muscle tensions in arthritis, anal fissures, intestinal diseases as, for instance, gastric ulcer, gall-, renal- or bladder- stones, or peritonitis. After the scanning of the literature on the causes of alterations of muscle tonus, particularly of extensor and flexor contractures,

I have the impression that too little attention was paid to the influence of eventually occurring decubitus wounds, or cystitis on the alteration of muscle tonus, and the causes were looked for too exclusively in the central nervous system. Although Pierre Marie and Foix (199) have shown in paraplegics and in cases of patients with transverse myelitis that exteroceptive stimulations in man as well as in animal, depending on the point of application,[1] can cause sometimes a tonic extension and sometimes a tonic flexion. The erroneous conception by Bastian and others that a transverse section of the spinal cord always and absolutely causes atonic paralysis was probably based on observations in cases with inflammatory processes.

Finally, the investigation on the central mechanism of distribution of muscle tonus is rendered more difficult by the fact that, for the establishment of an eventual hyper- or hypotonia, intact animals have to be used for comparison, whose distribution of muscle tonus always varies strongly due to cerebral influences. When investigating the muscle tonus of the extremities in a number of intact dogs, one will see that a number of animals, as soon as they are touched, lie down "shamming dead," start back and draw in their paws in a flexed position (in cats this happens almost always). In this case, the legs only show a strong flexor tonus, while the extensor tonus is absent. Others submit quietly to a turning into a dorsal position and offer only a very slight resistance against passive flexion and extension. Others again are obstinate, make defensive movements and show at times an excessively strong extensor tonus and at times a particuarly strong flexor tonus. It is also striking how much the distribution of tonus can be altered by blindfolding the eyes.

Human beings can voluntarily either reduce muscular tension to such a degree that passive movement hardly encounters any resistance, or increase it to such an extent that a passive movement has to overcome a resistance which is not less strong than the maximum resistance of a hypertonic patient. Cerebral influences, of course, are absent in spinal, decerebrate and decorticate animals.[2] In most of the diseases in man which are accompanied byhyper- and hypotonia, their effect

----

(1) According to Foerster (71), it depends, in the first place, on the force of stimulus whether a flexion or an extension takes place.

(2) On cerebral influences on tonus and contractions of muscles of the extremities after section of posterior roots, see in addition to Mott and Sherrington (276), Munk (214) (experiments on monkeys) and particularly Foerster (71).

is also strongly reduced. Conversely, all preceding statements, including those concerning cerebral influences, equally apply to the distribution of muscle tonus in totally and partially decerebellate animals. Therefore, it is not astonishing that the opinions on the influence of cerebellum on the muscle tonus differ widely. Luciani's conception that the cerebellum has a strengthening influence on the muscle tonus, and which is shared by Munk, Lewandowsky and others, was disputed for the first time by Laborde. Laborde saw that cocks after extirpation of the cerebellum still were able to lift weights attached to their legs and that on seizing them, the legs were drawn up with great force. Ferrier and Turner were also unable to find any atony or astheny in decerebellate monkeys. The monkeys grasped energetically with their extremities, climbed excellently and their muscle tonus was rather increased and not reduced. Bickel, too, did not acknowledge atonia and asthenia as a result of extirpation of the cerebellum, nor did Dusser de Barenne, based on his observations that the symptoms of his decerebellate dogs stood in direct contrast with muscular atonia and asthenia. According to André Thomas, a partial destruction of the cerebellum causes "anisosthenia of the antagonists," i.e., the muscles effective is a certain direction show hypostheny (reduced resistance) on passive alteration in position, the muscles effective in the opposite direction show hyperstheny (increased resistance).

By muscle tonus we understand the resistance caused by muscular tension against passive alteration in position of the body parts to each other. This resistance is normally brought about by:

1. the specific elasticity of the muscle substance, i.e., by the tension which is still detectable in the denervated muscle;

2. reflex tensions, i.e., contraction produced by subcortical stimulations;

3. the tension produced by cerebral influences (conditioned and unconditioned cortical reflexes).

Flexor tonus is the muscular resistance against passive extension movements and extensor tonus a muscular resistance against a passive flexion movement. In a passive extension movement, the flexors are stretched, in a passive flexion movement the extensors, and according to the reaction of the muscle to the stretch, the tonus, i.e., the muscular resistance against this passive movement, will be very different. Thus, an investigation of the resistance against passive movement is, in the first place, an investigation into the behavior of the various stretch reflexes.[1]

The stretch reflexes can be divided into:

1. *Stretch reflexes first order, these are the reflex contractions of the stretched muscles themselves* (the myotatic reflexes according to Liddell and Sherrington, the specific reflexes according to Hoffmann).

The stretch reflexes first order in animals can best be investigated by Sherrington's method (muscles detached from their insertions). Investigations of that kind (Liddell and Sherrington) showed that *in the decereberate animal* particularly strong and, by maintaining the extended state, long lasting extension reflexes first order appear in the knee extensors. Phasic as well as postural contractions are particularly distinct here.

Liddell and Sherrington observed in decerebrate cats in the quadriceps, among others, increases in tension up to $3\frac{1}{2}$ kg on a stretch of 8 mm, $2\frac{1}{2}$ kg on a stretch of 4 mm and even more than 2 kg on a stretch of less than 1 mm, while even on a stretch of $\frac{1}{2}$ mm a distinct reflex contraction appeared. On the rectus femoris, length 116 mm, distinct reflex contractions were observed on a stretch of less than 1 mm.

Thus, the stretch reflexes of knee extensors in decerebrate cats appear strongly, are long lasting and already at a slight stretch and after a short latency (20 ms). They are all the stronger, the greater the stretch. Whether the contractions are stronger in the various increases in length than they are in intact animals has not yet been investigated.

As already mentioned, the legs of decerebrate dogs cannot carry a load on their shoulders or pelvis heavier than that carried by intact dogs without giving way. Liddell and Sherrington investigated the stretch reflexes in the dorsal position in which the legs of decerebrate animals showed more resistance to pressure on the soles than those of intact animals. Contrary to knee extensors, knee flexors in decerebrate animals show no first order stretch reflexes.

After severance of the posterior roots belonging to the legs there are no first order stretch reflexes, neither in the flexors nor in the extensors.

---

(1) On the role of stretch reflexes in flaccid and spastic paralysis, see publication by O. Foerster (70, 71): Schlaffe und spastische Lahmung, Vol. 10 of the Handbuch der normalen und pathologischen Physiologie (Handbook of normal and pathological physiology), p. 893. Berlin, Julius Springer, 1927.

498

After transverse section of the spinal cord, Liddell and Sherrington observed a reduction in the increase of postural tension of the knee extensors; on continuance of the stretch condition the reflex contractions did not last. In dogs whose spinal cord was severed 2-3 months previously, at the level of D 12, Denny-Brown and Liddell (52) found, on the other hand, that the gastrocnemius (length 12-14 mm) with an extension of physiologic size (4 mm and less) showed not only an increase in "phasic" but also a "tonic," "static" and "postural" reflex tension, which lasted as long as the muscle remained in a stretched condition; in this case, a reflex increase in tension by 1.25 kg and a total tension of 2.1 kg could be recorded. In these dogs, an extension of quadriceps gave a weaker response (too low a lesion). Tibialis ant. never caused a reflex increase in tonic tension; flexor muscles showed only a phasic increase of an abrupt and rapid stretch ("plucking reflex").

In the midbrain of subthalamus animal (i.e., after transverse section 4 to 5 mm above the corpora quadrigemina anteriora), the quadriceps showed on stretch a strong and lively. occurring reflex contraction with tonic duration (Liddell and Sherrington) which could not be established in the knee flexors.

In the "thalamus" cat, Schoen observed tonic stretch reflexes in the flexor digit. prof. of the foreleg, while the triceps brachii showed only a phasic contraction on stretch. In the hindleg, the stretch of the gastroncnemius muscle caused a tonic contraction (Pritchard).

In man, too, distinct first order stretch reflexes can be observed in numerous muscles. On stretch with maintenance of the stretched condition, the human muscles sometimes show a phasic and sometimes a tonic contraction and sometimes even under pathological conditions clonic contractions. This varying behavior does not always depend on pathological conditions. In a normal human being with an intact central nervous system, muscles can sometimes also react with phasic and sometimes with tonic contractions. Thus, certain muscles react in a standing posture with a tonic reaction and in a dorsal position with a phasic reaction. Although hitherto systematic investigations on the behavior of first order stretch reflexes in man are lacking, I cannot omit reporting some interesting observations in this regard.

First order stretch reflexes of m. quadriceps are examined by passive flexion of the knee joint, by passive distal movement of the patella or by tapping of the patellar tendon.[1] By these methods of investigation in the supine position, under pathological conditions, tonic and clonic contractions of the knee extensors may appear when in normal persons in the same position only phasic contractions are produced. In standing with extended knees with the upper part of the body somewhat bent forwards, the quadriceps is usually totally flaccid. Now, when the knee joint is passively flexed, the quadriceps reacts with a tonic contraction.

First order stretch reflexes of the knee flexors can be seen most distinctly when with the subject in a prone position the feet are at first passively raised up to a right-angled flexion of the knee joint and are then suddenly released. As the legs fall a stretch of the knee flexors occurs. Normally, the lower legs go down at first by $\pm 25°$, till by a sudden contraction of the stretched knee flexors the fall is stopped, in which case the tendons of gracilis and semitendinosus distinctly bulge outwards, and the legs are then gradually lowered towards the supporting surface. With reduced stretch reflexes, however, the lower leg drops down like a dead mass and hits the supporting surface with an impact. When the stretch reflexes are retarded, the contraction of the flexors only occurs when the feet touch the supporting surface and only then the leg springs back elastically with distinctly visible contraction of the flexors. This observation was made repeatedly on the paralysed leg of hemiplegics.

The first order stretch reflexes of the calf muscles are produced by passive dorsiflexion of the foot and by tapping of the Achilles tendon. Under pathological conditions, here also with the patient lying supine, distinct tonic and sometimes clonic conditions will appear. When in standing posture these muscles are stretched at the ankle joints by a forward bending of the body, they will contract tonically in normal and in pathological conditions, causing the heels to be lifted from the ground.

The first order stretch reflexes of the flexors of the ankle joint and the toe extensors appear with passive plantar flexion of the foot. With the subject in the supine position, this movement normally causes only slight phasic contraction which may become tonic or clonic under pathological conditions. On tonic contraction, a further plantar flexion of the foot encounters a strong resistance, the toes are dorsally flexed

---

(1) According to Liddell and Sherrington, and to Hoffmann, tendon reflexes are typical stretch reflexes.

and the tendons of the extensors distinctly bulge out below the skin of the dorsum of the foot. The contractions of ankle joint flexors and toe extensors can best be observed in a standing posture when by a passive backward movement of the body they are stretched at the ankle joints. In this case, the balls of the toes are lifted from the ground, while the toes are dorsally flexed and the tendons of the muscles project from the skin of the dorsum of the foot.

In the muscles of the arm, too, first order stretch reflexes can be observed. Under pathological conditions, tonic contractions of the elbow extensors on passive flexion of the elbow and of the flexors on passive extension are distinctly visible. (André Thomas pointed out for the first time a tonic contraction of the triceps on the passive flexion of the elbow.) Passive and active bending forwards of the upper part of the body in sitting and standing posture normally causes a distinct tonic contraction of the ms. sacrolumbales (by which further ventral flexion of the upper part of the body is prevented; équilibre de suspension de Foix), while a movement of the upper part of the body towards the right is accompanied by a tonic contraction of the left, and towards the left by a tonic contraction of the right ms. sacrolumbales.

Babinski and Jarkowski, under the name "phenomene des antagonistes," described the tonic contraction of m. deltoideus which appears when the arm is passively raised and then suddenly released, a phenomenon which could probably be also a first order stretch reflex.

First order stretch reflexes generally appear particularly lively, unrestrained, tonic and strong in a spastic paralysis *in almost all positions and postures* of the patient; in flaccid(atonic) paralysis, their appearance is weak, retarded or totally absent. Foerster even asserts: "When a paralysis is accompanied by an elimination of stretch resistance, we speak of a flaccid (atonic) paralysis, and when this resistance is abnormally increased, of a spastic one." Although this assertion is in itself correct, it can lead to misunderstandings, since, in the case of a spastic hemiplegia, it is not the first order stretch reflex of *all* muscles which is increased, but only in part; those of the others, for instance of tibialis ant. and of the knee flexors, are also usually reduced.

2. *Stretch reflexes of second order are reflex contractions of synergic muscles caused by the stretch of a certain muscle.*

Liddell and Sherrington observed that in decerebrate cats, on stretch of a knee extensor, only the extended muscle showed a reflex increase in tension. When after the longitudinal cleavage of a muscle only a part was stretched, only this part showed a contraction.

Contrary to that are Schoen and Pritchard's observations on muscle recordings of intact cats under slight ether narcosis and of decorticate thalamus cats. Extension of m. flex. digit. prof. of the foreleg, which is at the same time flexor of fingers and wrist joint, also caused a reflex contraction of flex. carpi ulnaris and of flex. carpi radialis, which are also flexors of the wrist joint (Schoen) and a stretch of gastrocnemius (extensor of the ankle joint) also causes a contraction of soleus (extensor of the ankle joint) (Pritchard). On longitudinal cleavage, too, a stretch of a part of these muscles produced a reflex contraction of the other part. Thus, on extension of one head of triceps, a transmission to the other two occurs which, however, show only a phasic contraction (Schoen). Extension of the biceps causes strong action currents in m. brachialis in the decerebrate cat (Hoogerwerf and Rademaker).

3. *Stretch reflexes of third order are reflex muscular contractions produced by extension of antagonists.*

In the decerebrate cat, the stretch of the knee extensor does not cause a reflex contraction of the knee flexor and stretch of the knee flexor fails to produce contraction of the knee extensor. The stretch of the knee flexors even inhibits the stretch contractions of the knee extensor (Liddell and Sherrington). On the other hand, it was observed that, in the course of supporting reactions, the various flexors and extensors of the extremities contract at the same time in intact, decorticate and decerebellate animals. Pritchard and Schoen, in investigations relating to this, carried out on isolated muscle preparations in intact cats under slight ether narcosis and on thalamus cats, found the following results:

1. Stretch of flexor digit. prof. of the foreleg (finger and wrist flexor) causes tonic reflex contractions:

> in ext. digit. communis;
>
> in ext. digit. lateratis;
>
> in ext. carpi ulnaris;
>
> in ext. carpi radialis.

2. Stretch of elbow flexors causes a tonic contraction of triceps (Schoen).

502

3. Stretch of triceps brachii causes a tonic contraction of biceps (Schoen).[1]

4. Stretch of the muscles of the Achilles tendon causes a tonic contraction:

> in ext. digit longus;
> in tibialis anticus;
> in peroneus (Pritchard).

5. Stretch of quadriceps causes a tonic contraction:
> in semitendinosus;
> in biceps (Pritchard).

Thus, the stretch of a muscle can produce tonic reflex contractions of the antagonistic acting muscle.

In man, the case is probably similar, as is shown by manifestations observed by Foix and Thévenard, compiled under the title "Réflexes de posture" (postural reflexes).[2] When in some normal, but better in extrapyramidal disorders, persons lying in a supine position the foot is moved more and more dorsally and at the same time rolled inwards (so that, among others, the flexors digit., the muscles of the Achilles tendon and the peronei are stretched) the relaxed muscles will contract, the extensor digit. long., the extensor hallucis longus and tibialis ant. and the tendons of these muscles distinctly protrude below the skin of the dorsum of the foot (paradox contractions by Westphal). On the exclusive dorsal flexion of the foot, only ext. digit. longus and ext. hallucis longus contract and, on the exclusive inward rotation of the foot, only tibialis anticus contracts while an exclusive outwards rotation (stretch of m. tibialis anticus, relaxation of peronei) causes a contraction of peronei.

Since in the contractions produced by the dorsal movement at the ankle probably the extension of flexores digitorum is the cause, the following observations by Schoen should be mentioned. Schoen observed in isolated muscles of cats that a passive extension of the lateral and median part of m. flex. digit. profundus of the foreleg produces different reflex contractions; thus, for instance, an extension of the lateral part caused a contraction of flex. carpi radialis, but not of flex. carpi ulnaris,

---

(1) Extension of triceps brachii in the decerebrate cat causes the appearance of strong action currents in m. biceps (Hoogerwerf and Rademaker).

(2) It should be noted that the intensity and length of stretch in these experiments were not controlled. Others have had difficulty in repeating them in normal subjects (ed.).

while an extension of the median part, on the contrary, caused a contraction of the latter, but not of the former.

Also, in the knee and elbow joints in man, muscular contractions on passive movements which probably can be interpreted as third order stretch reflexes can sometimes be observed. A strong flexion of the knee (in which the quadriceps is stretched) causes a strong contraction of the knee flexors and a strong passive extension causes a contraction of the knee extensors, i.e., in conformity with Pritchard's observations in cats (see above). Strong passive flexion of the elbow causes a contraction of the biceps while a strong passive extension, one of the triceps (compare Schoen's corresponding observations in cats).

Whether these contractions in man are really caused by a stretch of the antagonists is not yet proved. Foix and Thévenard interpret them as Sherrington's "shortening reactions." It is well-known that this "shortening reaction" of m. quadriceps appears in decerebrate cats in a dorsal position when this muscle is passively shortened by the passive lifting of the lower leg; on releasing of the lower leg, the shortened state of the muscle continues and the lower leg does not fall down. On releasing the lower leg, however, m. quadriceps is passively stretched; thus, the shortening reaction probably represents a first order stretch reflex (see also Sherrington 278, Fulton 87, Samojloff and Kisseleff 264 B). In the above-mentioned manifestations in man, however, the reflex contractions of the passively shortened muscles can also be observed when the shortened condition is maintained and the foot or lower arm is not released. In this case, they are probably caused by the passive stretch of the antagonists and presumably represent stretch reflexes third order.

The "shortening reactions" of the quadriceps appear even more distinctly in the decerebrate animal when all other muscles of both posterior extremities are cut or severed; however, the tonic contractions of finger and wrist joint extensors, of biceps, etc., observed by Schoen and Pritchard in passive shortening, are absent after severance of the antagonists from their insertions. Thus, these contractions are caused by an extension of the antagonists and, as such, are surely stretch reflexes third order. Whether the "réflexes de posture" observed in man are still present after severance of the antagonists is not known.

Stretch reflexes third order are present in intact and decorticate cats. In decerebrate cats, the passive flexion of the elbow joint causes the appearance of strong action currents in the biceps which fail to appear

504

after extirpation of the extensors; thus, they also show these stretch reflexes.

The "réflexes de posture," according to Foix and Thévenard, are eliminated or reduced in hemiplegia, paraplegia, neuritis, tabes, polio-myelitis, muscular atrophies and muscular dystrophies, cerebellar diseases, Friedreich's disease and also often in syringomyelia, and multiple sclerosis. They appear, however, exaggerated in paralysis agitans, postencephalitic Parkinsonism, Wilson's disease and pseudosclerosis.

Goldflam (96), too, observed in these latter diseases a strong paradoxical contraction of Westphal and pointed to the fact that the contraction of tibialis anticus usually appears distinct only when the passively dorsally moved foot is suddenly released and falls down plantarwards. The contractions of tibialis ant., ext. digitorum and ext. hallucis longus observed by Goldflam thus are probably stretch reflexes first order and are not the phenomenon observed by Westphal and others. In correspondence with this assertion, Delmas Marsalet (50) in the investigation of Westphal's paradoxical contractions found two reflex contractions, one on the passive dorsal movement (probably stretch reflex third order) and another on releasing the foot (probably stretch reflexes first order). Goldflam believed that he must compare Westphal's paradoxical contraction of tibialis anticus with the contraction of knee flexors which occurs when, in the prone position of the experimental subject, the lower leg is passively raised and then released (in our opinion a stretch reflex first order). According to Goldflam, the contraction of tib. anticus on releasing of the foot as well as the contraction of the knee flexor on releasing of the lower leg is increased in paralysis agitans and, conversely, it is reduced in the so-called diseases of the pyramidal tract as well as in tabes and chorea minor.

4. Stretch reflexes fourth order are reflex contractions of muscles which have neither a similarly directed nor an opposite effect caused by the stretch of a certain muscle.

On investigation of supporting reactions in intact, decorticate and decerebellate dogs there occurred, in static stress on the soles of the feet, a reflex contraction of almost all muscles of the extremities as well as of the muscles of scapulae, vertebral column and pelvis, even when the influence of exteroceptive stimulations was eliminated.

In further investigation and experiment on these reactions in isolated muscle preparations, it was found:

1. that a stretch of flex. digit. profundus (Schoen) causes reflex contractions in the following muscles:

flexor digit. prof. } palmaris longus	finger and wrist joint flexors	1 order / 2 order
flexor carp. uln. } flexor carp. rad.	wrist joint flexor	2 order / 2 order
ext. digit. comm. } ext. digit. lat.	finger and wrist joint extensor	3 order / 3 order
ext. carpi uln. } ext. carpi rad.	wrist joint extensor	3 order / 3 order
triceps	elbow extensor	4 order
biceps	elbow flexor	4 order
supra spinatus } calvobrachialis	shoulder extensors	4 order / 4 order
teres major } deltoideus } latissimus dorsi	shoulder flexors	4 order / 4 order / 4 order
pectoralis major	adductor of upper arm	4 order
serratus ant. } levator scapulae } trapecius } rhomboideus	muscles of scapula	4 order / 4 order / 4 order / 4 order

2. that a stretch of biceps (Schoen) can produce reflex contractions in the following muscles:

pectoralis major	4.0	4 order
teres major	4. & 3.[1]	order
latissimus dorsi	4. & 3.	order

---

(1) In addition to elbow flexor, the biceps is also a shoulder extensor, while teres major and latis. dorsi are shoulder flexors, i.e., antagonists of biceps.

3. that a stretch of triceps (Schoen)[3] can produce a reflex contraction among others in the muscles:

pectoralis major — 4 order
supra spinatus — 4 and 3 [1] order

4. that a stretch of the extensors of wrist joint and fingers (Schoen)[3] can cause reflex contractions in the muscles:

pectoralis major — 4 order
latissimus dorsi — 4 order
teres major — 4 order

5. that a stretch of the muscles of the Achilles tendon (Pritchard)[3] can cause a reflex contraction in the muscles:

flexor digitorum longus	2	2 order
extensor digitorum longus	3	
soleus	1	and 2 order
gastrocnemius	1	and 2 order
tibialis anticus	3	and 4 order
peroneus	3 [2]	and 4 order
quadriceps	3 [2]	and 4 order
semitendinosus	2 [2]	and 4 order
semimembranosus	2 [2]	and 4 order
biceps femoris	2	and 4 order
adductor longus		4.0 order
glutaeus maximus		4.0 order

6. that a stretch of quadriceps, according to Pritchard, can cause a reflex contraction of glutaeus maximus, according to Liddell and Sherrington, can also produce a reflex contraction of the quadriceps of the opposite legs (extension reflexes fourth order).

In addition to the proprioceptive positive and negative supporting reactions, probably also the tonic neck reflexes, the leg-bracing reaction, the rocking reactions, the increase in supporting tonus in the hindlegs on passive movement of the forelegs forwards all represent stretch reflexes of fourth order.

---

(1) Also, in the supra spinatus, a shoulder extensor, to some extent an antagonist of triceps which, in addition to elbow extensor, is also shoulder flexor.

(2) The muscles of the Achilles tendon are, in addition to tarsal joint extensors, also knee joint flexors.

(3) It should be noted that these results of Schoen and Pritchard still await confirmation (ed.).

When the posterior roots belonging to a leg are severed, a stretch of a muscle of this extremity no longer produces stretch reflexes of any order. These muscles can indeed still participate in stretch reflexes of fourth order, as they show reflex contractions on stretch of muscles of other parts of the body; thus, we see, among others, in a leg whose posterior roots have been severed, distinct flexing and extending movements on adduction or abduction of the opposite legs (hopping reaction).

In man, too, stretch reflexes of fourth order are most probably present. Presumably, the following have to be interpreted as stretch reflexes of fourth order: firstly, the distribution of muscle tonus in a standing posture, secondly, the negative and positive supporting reactions (see Chapter V) which appear under pathological conditions when the patient lies supine; thirdly, the contraction of the left and right sacrolumbalis when one supports oneself alternatively on the right or the left leg; in the fourth place, the contraction of the calf muscles by which the heels are lifted from the ground by a backwards bending of the upper part of the body.

5. Stretch reflexes of fifth order are contralateral reflexes caused by an increase in extension.

As we have seen in preceding chapters, a strong (yet still physiological) stretch of a muscle often produces effects contrary to those produced by a moderately strong one. Thus, as one example, a moderate stretch of digital flexors causes extension and appearance of supporting tonus in the foreleg; overstretching of the digits, however, leads to disappearance of supporting tonus and flexion, while continuation of this posture causes alternating flexing and extending movements of the forelegs. The reactions in the hindlegs on a moderate and a strong passive dorsal movement of the toes are just the same.

Passive plantarflexion of the digits causes a decrease in supporting tonus and a flexion of the foot, while a strong flexion of the fingers or toes produces at first an extension and then an alternating flexion-extension movement. A slight adduction movement of a limb from a position of abduction is accompanied by an increase in supporting tonus and a stretching (leg-bracing reaction); a strong adduction causes the disappearance of supporting tonus and a raising of the paw (leg-slackening reaction); when this posture is maintained flexion-extension movements appear again. The same manifestations occur when the leg directed forwards is slightly or strongly moved backwards, or from a backward position is slightly or strongly moved forwards. On stressing the leg, the supporting tonus increases and on overburdening the strong stretch, among others, of the finger and toe flexors, causes a disappearance of supporting tonus and a raising of the leg. Moreover, on examining the hopping reaction, it was found that a moderate adduction of one leg from the intermediate position brought about a flexion and decrease of supporting tonus in the opposite leg, whereas a strong adduction brought about an extension and an increase in supporting tonus (Fig. 161).

Whether these opposite reactions are really based on the state of stretch and, therefore, represent true stretch reflexes is not yet proved.

In decerebrate cats with a pronounced decerebrate rigidity, triceps and biceps show strong action potentials; those of the triceps are the stronger. On a passive dorsiflexion of the fingers, the action currents in triceps and biceps grow distinctly stronger (positive supporting reaction); on a further dorsiflexion, however, the action currents in biceps and then in triceps disappear at first (decrease in supporting tonus and extensor rigidity; corresponding to the first component of leg-slackening reactions), while on a still greater stretch very strong action currents appear suddenly in the biceps; the triceps, however, remains inactive or shows only weak action currents (flexing of the leg; opposite reaction; second component of leg-slackening reaction).

6. *Relaxation produced by stretch.*

The observations after muscular stretch discussed so far were restricted to the production of a reflex increase in tension in certain muscles. The stretch of one muscle, however, can also reduce the state of tension in other muscles and thus extensively alter the resistance to passive change in position of the various parts of the body to each other. In the negative supporting reaction, the resistance to passive flexion of the proximal joints on passive flexion of fingers or toes decreases. As Schoen proved on isolated muscle preparations of thalamus cats, this decrease in resistance is not only brought by reflex contractions of flexors (see stretch reflexes fourth order, No.4), but also by relaxation of muscles. Thus, a stretch tension of the finger and wrist joint extensors produces a relaxation of, among others, triceps brachii, serratus ant. and levator scapulae; the two latter also relax on stretch of the triceps. This is proof that muscular relaxation in negative supporting reactions is not only based on the absence of stimulations of the positive supporting reactions, but also that the flexion of distal phalanges is able to inhibit the contractions of other muscles. We also want here to call to mind Liddell and Sherrington's observation that in decerebrate cats the stretch contraction of knee extensors abates on stretch of the knee flexors.

Liddell and Sherrington further saw that the reflex contraction of a muscle on stretch increases with increasing stretch, but only within certain limits. With a very strong stretch the contraction weakens and finally totally relaxes.

7. *Alternating contraction and relaxation due to stretch of muscles. Stretch reflexes of sixth order.*

In the preceding chapters numerous observations were mentioned in which the maintenance of strong passive alterations in position of the joints of extremities produces alternating flexing and extending movements, and it was supposed that these movements appeared to be due to the strongly stretched condition of certain muscles. Liddell and Sherrington reported that extension of quadriceps muscle can sometimes produce an alternating contraction and relaxation of this muscle.[1] Just as in intact animals, the forelegs of decerebrate cats on maintained displacement of the digits in a dorsal direction show alternating flexing and extending movements. An examination with a string galvanometer (in cooperation with Dr. Hoogerwerf) showed that in this case an alternating appearance and disappearance of strong action currents in the biceps could be observed while the triceps remained inactive or showed only weak action currents. Thus, the flexing movements are brought about by intermittent contractions of the biceps, while the extension movements on relaxation of biceps contractions are produced by the elasticity of the inactive triceps which was passively stretched by the flexing movement (sometimes, as shown by the weak action currents, the tonus of triceps also participates). Whether the alternating flexion-extension movement, due to holding the digits in strongly dorsally directed position in the decerebrate animal is caused by the strongly stretched state of the finger flexors is not yet conclusively proved (it is well known that these alternating movements also appear on pinching and on a strong plantar flexion of the digits).

Thus, a progressive dorsal movement of the fingers in the decerebrate animal causes successively:

(1) According to a personal report by Liddell, a particularly strong and rapidly performed extension sometimes produces an alternating contraction and relaxation.

510

1. Increase of action potentials in biceps and triceps, especially of the triceps.

2. Disappearance first of the biceps action potentials and then also of those of the triceps.

3. Appearance of particularly strong action potentials in biceps, while the triceps remains inactive or shows only weak currents.

4. On fixation of the fingers in strongly dorsally directed position, alternating disappearance and reappearance of action currents in the biceps; in this case, the triceps remains inactive or shows only weak action currents.

To this group belong probably also the foot and patellar clonus which are often detectable in man under pathological conditions.

From the preceding observations, it follows that the force of muscle tonus of resistance against passive movements is determined in the first place by the various stretch reflexes. As we have seen in the previous chapters, the appearance and the strength of stretch reflexes are influenced by labyrinthine reflexes (tonic labyrinthine reflexes), by stimulations from the body surface (contact with the sole of the feet, influence of dorsal and lateral position) and by cerebral effects (conditioned inhibitions).

Foerster (71) distinguishes stretch, adaptation and fixation reflexes. The adaptation reflex is the reflex contraction of a muscle on approximation to its point of insertion and thus corresponds to the "réflexe de posture" by Foix and Thévenard. In this contraction, the viscosity of the muscle first plays a role, as shown in the degenerated muscle, and secondly the stretch of the antagonist (stretch reflex third order). Thirdly, if the approximation is less, the stretch reflex first order can come into effect (shortening reaction). Foerster and Wachholder (316) assume that stimulations emanate from a passively shortened muscle as well as from a stretched one, causing reflex contractions of the muscle (adaptation tonus, shortening or adaptation reflex). The presence of an adaptation reflex, in my opinion, is not definitively proved. The fixation reflex causes, according to Foerster, that "the muscle can maintain a state of contraction by which the extremity is retained in the position into which it was brought passively." The fixation reflex would thus be the tonic continuation of the adaptation reflex.

Goldflam, under the name of "stretch contraction of the antagonists," describes the following manifestation: When in a healthy person one resists an intended movement of a part of a limb and suddenly ceases this resistance, the segment concerned will continue the movement in the intended direction for a certain distance till it is checked. Sometimes, the intended movement, after a distinct setback, is interrupted by the antagonists. Thus, in this case, the antagonists are stretched by an intended contraction of the antagonists and react to this with a reflex contraction (stretch reflex first order). As Goldflam observed, and already before him Babinski and Jarowski, this contraction is highly increased in patients with Parkinson's syndrome. In cerebellar patients, however, this is reduced, as was specially emphasized for the first time by Stewart and Gordon Holmes.

Recently, Mayer and Reisch reported that in hemiplegics an increased tonic resistance can occur on the nonparalyzed side on the passive pulling of the extremities. They term these contractions, which probably represent stretch reflexes first order, pull and compression reactions.

It is further interesting to note the observation that in man under pathological conditions, on tapping a tendon, the antagonist reacts with a contraction and not the tapped muscle. This phenomenon, in this case, that the antagonist (stretch reflex third order) shows a contraction and that, on the other hand, the stretch reflex first order fails to appear, has been described by other authors (see, among others, Benedak and Thurzo 19) as a paradoxical reflex.[1]

Thus, in the muscular resistance against passive alteration in position of the various parts of the body to each other, the following factors participate, in addition to the elasticity of the muscle (corresponding to that of a denervated muscle):

1. alterations in tension produced by stretch (stretch reflexes 1st to fifth order and muscular relaxation on stretch);

2. alterations in tension by labyrinthine stimulation;

3. alteration in tension by stimulation from the body surface;

4. alteration in tension by cerebral influences (conditioned reflexes and inhibitions by optic and acoustic impulses).

To this can also be added the periosteal stimulation and

---

(1) It should be noted that this phenomenon is associated with conditions of mixed spastic lesions and denervation, as in cervical disk syndromes, and that the "reversed" tendon reflex is then a response to transmitted mechanical stimulus rather than stretch (ed.).

512

stimulations from the surface of joints (see, among others, Sternberg and Foerster), or of the abdominal organs.

In order to obtain a correct picture of the condition of regulation of muscle tonus after lesion of the central nervous system, the behavior of the various stretch reflexes and the effect of labyrinthine stimulations on the stretch reactions, stimulations from body surface (the behavior of reflexes in the various positions) have to be taken into consideration. Unfortunately, the mechanism of these reflexes is not sufficiently analyzed, the latency, force and duration of contractions on more or less strong stretch in intact, decorticate and decerebellate animals is not yet satisfactorily investigated.

BEHAVIOR OF MUSCLE TONUS AFTER TOTAL AND UNILATERAL EXTIRPATION OF THE CEREBELLUM

After extirpation of the cerebellum, the musclet onus behaves very differently according to the time elapsed since extirpation. The alterations in muscle tonus appearing in the first period after extirpation, which can probably be interpreted as effects of diaschisis or of shock in the centers of the brainstem, do not prove that either hypertonia or hypotonia are a direct result of the absence of cerebellar function. Similarly, alterations of the tonus in the case of abscesses, tumors, strong bleedings, etc., in the cerebellum are probably not caused solely by the absence of cerebellar function but have to be interpreted as results of diaschisis, or shock or associated lesion of the brainstem. Cerebellar patients sometimes show a distribution of muscle tonus which more or less corresponds to that of the animal immediately after extirpation of the cerebellum.

a)· *Animals presenting initial rigidity*

Following extirpation of the cerebellum, some animals are quite rigid on awakening from narcosis, like Sherrington's decerebrate animals as, for instance, the dogs Caesar, Erik, Piccolino and Wolf, while the dog Moor, which suffered from a subcutaneous haematoma, only kept its neck stiff. The first four animals kept the neck rigid and extended dorsally, exactly like decerebrate rigidity. A passive ventral flexion movement of the neck at the atlantooccipital joint encountered a strong resistance. When head and neck were released, they remained for a short time in a flexed position, but soon the neck, on some stimulation, again dorsiflexed rigidly. In the forelegs, the elbow and wrist joints were maximally extended and fixed, the upper arm somewhat retracted at the shoulder joints. The digits were in an intermediate or in a slightly flexed position. On a passive forward and backward movement of the rigid foreleg at the shoulder joint, a distinct elastic resistance could be felt and the limb, when released, sprang back instantly into the initial posture. When one attempted to flex the forearm at the elbow joint, one encountered an almost insurmountable resistance. Passive flexion of the wrist joint succeeded more easily, although here, too, the resistance against maximum flexion was considerable and the hand sprang back immediately into the extended position as soon as it was released. The maximum flexion of the wrist joint was always accompanied by an associated flexion of the elbow, but even with a passive flexion of the fingers, it never led to a disappearance of extensor tonus. However, one usually succeeded, after a passive flexion of the wrist joint, in bringing the elbow and shoulder joint into a totally flexed position, in which case the extensor tonus suddenly ceased, corresponding to the "clasp knife phenomenon" typical of decerebrate rigidity. The released paw usually remained in a flexed position until other stimulation again caused a brusque extension.[1] Thus, the extensor tonus showed a typical plasticity. Sometimes evidence of pain also occurred in these attempts to overcome active resistance.

The digits showed an elastic fixation. The dorsal spine was rigid, and in some animals it was convex dorsally, in others it showed opisthotonus. As after decerebration, the position of the hindlegs in rigid animals varied; in animals with opisthotonus the hindlegs were kept semiflexed and in animals with convexly curved backs the hindlegs were kept totally extended. The semiflexed hindlegs offered only slight resistance against passive flexion; on the other hand, they often showed a distinct flexor tonus, i.e., distinct resistance against passive extension In animals with extended hindlegs, the extensor tonus was highly increased and showed a typical plasticity.

The position of the hindlegs in rigid animals was dependent on the position of the head, neck and thoracic spinal column. Ventral flexion of the head and neck caused an increase of extensor tonus in the hindlegs and in the dorsiflexed position, a decrease in extensor tonus (tonic neck reflexes).

The dorsally convex back is accompanied by a strong extensor

---

(1) Sometimes, the extension took place by jerks synchronous with the respiratory movement; often the limb also instantly sprang back into a stretched posture.

tonus and the dorsally concave one by a flexor tonus. The cause of the different curvatures of the back is unknown. We have indeed seen in Chapter X that a dorsal movement of the head in dogs causes at first a concavity of the back with decrease in supporting tonus and that a further exaggerated backward movement causes a convexity of the back accompanied by a ventral flexion of the pelvis and extension of the hindlegs.

Thus, rigid decerebellate animals show exactly the same distribution of muscle tonus as decerebrate animals. In them, too, the rigidity suddenly increases strongly with any general stimulation like a pinching of the tail, to and fro shaking of the animal and, contrary to decerebrate animals, also on visual stimulation. Simultaneous flexor and crossed extensor reflexes are present and show typical "after-discharge," and also tonic neck and labyrinthine reflexes can be distinctly observed. The influence of tonic neck reflexes is evident in all alterations in position (raising, rotating and turning) of the head in relation to the trunk, the influence of labyrinthine reflexes on transition from prone to supine posture when the head moves with the trunk, in which case a distinct increase in rigidity of the neck as well as in the extremities can occur. The influence of tonic neck and labyrinthine reflexes usually is most clearly evident several days after extirpation of the cerebellum when the rigidity has already somewhat subsided. In the first days, the animals, as soon as they are touched, become maximally rigid and remain in this condition in all positions of the head in space and in relation to the trunk.

During the first five days, they make no attempt to rise and they, too, like decerebrate animals, lack the righting reflexes on the head and on the body. The head, too, does not rise to a normal position, although the animals are able to see. (Sometimes they wag their tail when someone approaches the cage and they follow him with their eyes.) Optical impulses cause neither a righting of the head nor preparation for standing (placing) in the legs (Fig. 14, No. 1.) The legs are able to carry the trunk and do not give way under its weight. When the animal is not supported, it instantly falls down and does not make the slightest attempt to prevent falling. Also, on inclination of the supporting surface, the animals immediately fall down due to the absence of displacement reactions and like decerebrate animals they lack readiness to stand, rocking, leg-slackening and leg-bracing reactions and exteroceptive and proprioceptive correction movements.

A similar rigidity also appeared in the decorticate dog Robbie after extirpation of the cerebellum.

The rigidity generally lasts for 3-8 days. It then gradually decreases simultaneously with the return of labyrinthine and body righting reflexes. Infection of the wound or intercurrent infections like purulent conjunctivitis, rhinitis and diarrhea can contribute to its lasting longer. (Boric ointment on the eyes during narcosis, good nursing and care with milk feeding lessen these complications.)

From these observations it appears that:

1. extirpation of the cerebellum can cause rigidity with increased extensor tonus in intact as well as in decorticate dogs;

2. an absence of cerebellar function excludes neither hypertonia nor plastic tonus nor postural tonus;

3. the muscular rigidity after extirpation of the cerebellum shows all the typical characteristics of decerebrate rigidity, although cerebral effects are not totally abolished, as is proved by reactions to optical, acoustical and olfactory impulses;

4. however, the cerebral effects are not able to counteract the rigidity and, probably owing to the abnormal distribution of muscle tonus, cannot produce optical readiness to stand and optical righting reflexes or other numerous conditioned reflexes.

Rigidity after extirpation of the cerebellum was known to Luciani, Munk, Thomas, Sherrington, Thiele and various others. Magnus and Beritoff also observed that rigidity produced by decerebration did not disappear following additional cerebellar extirpation, but continued for as long as eight hours.[1]

There is a question whether rigidity after extirpation of the cerebellum, like decerebrate rigidity, is based on the absence of function of the red nuclei[2] and whether this absence of function is brought about by shock, diaschisis or inhibition. The simultaneous absence of labyrinthine righting reflexes and body righting reflexes speaks for an absence of function of the red nuclei, the inconstant occurrence of rigidity

---

(1) See footnote on page   (ed.).

(2) Rademaker, on the basis of his earlier work, believed that loss of function of the red nucleus was the essential lesion releasing decerebrate rigidity. This view was later criticized, and most physiologists would substitute midbrain tegmentum for red nucleus here (ed.).

for a shock effect. In favor of the latter also speaks the fact that rigidity occurs only in some animals, particularly when extirpation was associated with severe bleeding or when there were other complications. Cats in which extirpation of the cerebellum was performed without a great loss of blood showed the picture of postoperative rigidity much more rarely. This fact, of course, does not exclude diaschisis as a cause of rigidity, it is even more than probable that the severance of the numerous fibers leading from the cerebellum to the red nuclei exert an influence on the ability to function of the red nuclei. These considerations, of course, are only speculative.

Cobb, Bailey and Holtz (45) and Bremer (29) suppose that the cerebellum normally exerts on the muscles an influence inhibiting the stretch tonus, by way of the red nuclei. They observed that a rigidity which appeared after a transverse section of the midbrain rostral to the red nuclei disappeared as a result of electrical cerebellar stimulation.[1] A rigidity produced by decerebration behind the red nuclei was not influenced by the stimulation. Cobb, Bailey and Holtz believe that the stimulations inhibiting the stretch tonus do not emanate from the cerebellar cortex but from the nucleus dentatus and reach the muscles by way of the brachii conjunctivi and the red nuclei.

According to Bremer, proprioceptive stimulations emanate from the muscles which normally bring about a self-regulation of stretch tonus by way of the cortex of lobus anterior cerebelli, nuclei festigi, fasticio-rubral tracts of branchii conjunctivi, red nuclei and rubro-spinal tracts. The question why the intact cerebellum is only able to abolish the rigidity by electrical stimulation is not discussed further. Bremer (30, 31) observed in intact and thalamus pigeons that on unilateral stimulation of the cortex of lobus anterior cerebelli an increase of stretch tonus in the extremities of the contralateral side and a decrease of stretch tonus in the extremities of the ipsilateral side take place. A unilateral destruction of lobus anterior caused the same symptoms ! Contrary to Cobb, Bailey, Holtz and Bremer, Spiegel and Bernis (288) saw that electrical stimulation of lobus anterior still produced flexor movements in the anterior extremities, even after decerebration

---

(1) Some years after Rademaker completed this monograph, the investigations of Stella (1944) and later Granit Holmgren and Merton (1955) showed that this rigidity is indeed different from that produced by intercollicular decerebration. The whole question is well reviewed by Dow and Moruzzi, The Physiology and Pathology of the Cerebellum, Univ. Minnesota Press, Minneapolis 1958 (ed.) .

behind the red nuclei. These contradictory findings prove most distinctly how far we still are from solving the question as to the effect of stimulations which normally proceed from the cerebellum to the muscles by way of the red nuclei.

b. *Animals presenting initial hypotonia*

As we have seen, the dogs Daumling and Pim, the cats Fridel, Karolus and Pierrette and the monkey Corrie showed not a trace of rigidity but only hypotonia after extirpation of the cerebellum. On awakening from narcosis, the animals lay with ventrally flexed head and flexed, flaccid legs. Passive movements of the head and paws encountered hardly any resistance. On passive stretching and static stress, they already gave way at the slightest pressure. The same happened when the animals were set down on their legs (Figs. 65, 262). Touching the sole of the foot and static stress did not produce an extension of the flexed leg (magnet reaction) and caused no distinct reflex muscle tensions. Generally, the resistance was minimal when the legs were moved through a wide range at the shoulder and hip joints. In passive to and fro movements of the animals in the air forwards, backwards or sideways head and legs sometimes show normal flinging movements due to the reduced fixation (passivité according to André Thomas and Durupt). If suspended in the air by the hindlegs, the animals let the anterior part of the body dangle downwards, and fixation of the spinal column did not occur. In transition from a ventral to a dorsal position, no clear influence of tonic labyrinthine reflexes on the extremities could be observed and in a dorsal position in the air as well as on a supporting surface, the limbs were kept totally flexed. This position of the limbs did not change on alteration in position of head to trunk, and neither was there a distinct influence of tonic neck reflexes on the extremities. As in rigid animals, flaccid animals also failed to show the leg-bracing, leg-slackening, and pulling-up reactions and proprioceptive correction movements. Thus, the flaccid animals show a strong reduction of all stretch reflexes from first to fifth order, while in rigid animals extension reactions of the first order in the extensors were very lively. The other stretch reflex effects, however, showed impairment.

In the hypotonia observed after extirpation of the cerebellum, the muscles are not usually completely atonic. In to and fro shaking of the animals, an extension of the legs with distinct resistance to passive flexion can appear, the patellar reflexes can be present; on clasping and especially on pinching of the paws the animal attempts to liberate itself with considerable force. Additional passive movements forwards of the forelegs at the shoulder joints and of the hindlegs backwards at the hip joints produce an extention and a distinct fixation of the joints of the other distal extremities.

The hypotonia usually improves quickly and on static stress of the passively extended legs a distinct resistance brought about by reflex muscle tension soon appears again. The flaccid animals, strange to say, are usually able within the first 24 hours after operation to bring the head and anterior part of the body into a normal upright position. Already on the first day after extirpation, they generally lie with the anterior and posterior part of the body in a prone position and keep their head straight either supported in the air or with the lower jaw flat on the supporting surface. Sometimes, they even occupy the typically curled-up position of intact dogs (Fig. 262, No. 2).

To account for the temporary hypotonia various possibilities have to be considered. In the first place, the centers that promote stretch reflexes, like the centers for decerebrate rigidity, can be put out of function by the effects of shock, with results like those after decapitation or after transverse section of the spinal cord. Secondly, it is possible that the shock effect on the red nuclei was only slight but was stronger in the centers promoting rigidity. There is a third possibility, i.e., that the cessation of function of the tonus regulating centers is brought about in flaccid animals not by shock but by diaschisis in the sense of reduction of excitability, as is often the case after extirpation of the cerebrum.

That hypotonia resulted from shock of the mesencephalic nuclei is suggested by the fact that the righting reactions very soon appear again and also that hypotonia appears particularly prominent in animals in which the extirpation was carried out without severe haemorrhage Flaccid animals recovered more quickly and were capable of running around without support sooner than rigid animals. Sometimes, flaccid animals showed labyrinthine righting reflexes on the day of the operation and were able, one day later, to bring head and trunk again from a lateral to an upright position and within the first week could raise up the trunk from a lateral position. The body righting reflexes on the body, which come about by way of the red nuclei, are thus able to overcome the antagonistic influences of gravitational force and of neck reflexes within the first week.

In flaccid animals, the muscle tonus gradually increased at the same time that the various reflexes influencing the tonus, the stretch reflexes first—fifth order, supporting reactions, hopping reactions, leg - slackening and bracing reactions, gradually returned.

In rigid animals, on the other hand, the hypertonia of extensors gradually decreased, the influence of tonic labyrinthine reflexes on the extensors became less evident, while the stretch reflexes second-fifth order returned. The previously rigid animals began to show positive and negative supporting reactions, hopping, leg-slackening, leg-bracing, pulling-up and propping-up reactions. Similarly, stimulation of the body surface, which appears on shifting the animal into a lateral or dorsal position, or on touching of the soles of the feet again exerted a distinct influence. Thus, the various muscular reflex reactions reappeared in the previously flaccid as well as in the previously rigid animals. However, they were never quite normal again. Indeed, the deviations of these reactions constantly diminished in the first weeks but never disappeared completely and remained still distinctly present in the stage of permanent impairment, six or more months after extirpation of the cerebellum.

In the stage of permanent impairment, decerebellate animals showed a peculiar distribution of muscle tonus which was common to animals which had been flaccid as well as to those which had been rigid after extirpation. This distribution of tonus did not represent a typical hypertonia, and even less atonia. In this stage, the animals lying in a dorsal position on a supporting surface or suspended in the air keep their limbs totally flexed and offer on passive flexion a very slight resistance and on passive extension a somewhat stronger one. This resistance against passive extension-flexing movements is not strong, but it is also not absent; indeed, it can also be felt in intact and decorticate dogs under similar conditions and sometimes only on passive extension.

Patellar reflexes are always lively. Passive movements of the head and neck whether raising, lowering, turning or rotating encounter a considerable and surely not reduced resistance, particularly with the animal in a standing posture, but also in prone or supine lying position. It is striking that this resistance is always considerable in decerebellate animals, while in intact dogs it is sometimes strong and sometimes weak. An intact dog often offers no resistance at all to a passive movement of the head, but in a decerebellate dog one encounters instantly and regularly a lively resistance (Fig. 68).

When the animals are held in the air in a dorsal position by the pelvis and hindlegs, the anterior part of the body is raised until it is upright by a strong tonic muscular contraction (Fig. 270). Here, too, no trace of atony can be detected.

When the animals are held in a dorsal position in the air, the ventral flexors of head, neck and the upper part of the body are passively stretched by the force of gravity and these stretched muscles show a contraction due to the active raising and ventral flexion of the just mentioned parts of the body. Thus, first order stretch reflexes probably participate in this raising. These contractions, however, only occur in dogs with intact labyrinths and in them the labyrinthine righting reflexes bring the head into normal posture by a ventral flexion. Thus, the raising of the head and neck and of the upper part of the body is probably produced by a combination of stretch reflexes first order, labyrinthine righting reflexes and tonic neck reflexes.

Also, when the animals are held in the air in a ventral position by the thighs, they are able to hold the anterior part of the body actively upright by a contraction of the vertebral muscles (stretch reflexes in first order), an action which would be impossible if there was true muscular atonia (Figs. 133, 134, 270).

When decerebellate animals are lifted from the supporting surface by the forelegs, they will pull themselves up vigorously, thus showing strong pulling-up and propping-up reactions (Fig. 239). As discussed in Chapter XV, these reactions, too, are brought about by a combination of stretch reflexes first order, labyrinthine righting reflexes and tonic neck reflexes.

All these observations point to the fact that stretch reflexes first order are not abnormally weak in decerebellate animals in thesta ge of permanent impairment.

The muscular reflex contractions on stretch of the antagonists (stretch reflexes third order) are difficult to investigate in dogs and cats. They can actually only be properly investigated in isolated muscle preparations, but hitherto such investigations on decerebellate animals in the stage of permanent impairment are lacking. Perhaps some incidental observations on the behavior of these reflexes in decerebellate animals can give information to a certain degree. In these animals, a distinct hardening of the biceps can sometimes be felt on a strong passive flexion of the elbow. Further, a stretch of the distal phalanges of a forelimb dorsiflexion of wrist joint with dorsal movement of the digits, as already mentioned, causes an ample and vigorous contraction of all muscles of this extremity, as well as those of the scapula and spinal column (stretch reflexes first, second, third and fourth order). On a dorsal movement of the toes and dorsiflexion of the tarsal joint of the hindlimb (stretch of toe flexors and of the muscles of the Achilles tendon), there also appears considerable reflex shortening contractions of all muscles of the hindleg, in addition to those of the pelvis and vertebral column (also stretch reflexes first-fourth order). These observations indeed speak very strongly for the recovery of stretch reflexes third order and the rest of the stretch reflexs.

In the reactions of the fore- and hindlegs, on dorsiflexion of the distal phalanges, particularly distinct, ample shortening contractions of elbow extensors, knee extensors and of the vertebral muscles (stretch reflexes fourth order) are evident. The occurrence of leg-bracing reactions also supports the presence of stretch reflexes fourth order (and of stretch reflexes first order) since the extension and strong fixation of the propped legs are probably caused by stretch reflexes fourth order. Passive flexion of the distal phalanges of the extremities (stretch of dorsiflexors of the distal phalanges) causes an instant decrease of supporting tonus (extensor relaxation) in intact as well as in decerebellate animals and, in addition, produces contractions of certain flexors (stretch reflexes fourth order). These contractions sometimes occur so vigorously in deceerbellate animals that, through contractions of the flexors, the limb is drawn into a flexed position against the effect of gravity.

Strong dorsiflexion of the distal phalanges also produces in decerebellate animals at first a flexion (stretch reflexes fifth order) and then, alternating flexion-extension movements, contrary to the effect of a moderate extension. Similar movements on forced flexion of distal phalanges and the presence of leg-slackening reactions also speak in favor of the preservation of stretch reflexes fifth order.

To summarize all the above observations on the regulation and distribution of muscle tonus in decerebellate animals in the various positions in the air, in lateral and dorsal positions on a supporting surface, we conclude:

1. Decerebellate animals in the stage of permanent impairment show no symptoms of hypotonia.
2. The postural tonus and the tonus counteracting the force of gravity are present, strong and unimpaired.

3. The stretch reflexes first to fifth order are present, although their special behavior in the stage of permanent impairment needs further exhaustive investigation by recording from isolated muscles.

4. The maintenace of the stretched condition of certain muscles can produce *tonic* stretch reflexes first—fourth order as well as alternating contractions and relaxations (flexion-extension movements) in decerebellate animals.

From this summary, however, it should not be concluded that the regulation of muscle tonus in decerebellate animals is quite normal; on the contrary, it shows very distinct and considerable deviations.

In the first place, the reflex contractions usually cause exaggerated movements (hypermetria), for example, the effect of exteroceptive stimulations (particularly obvious extension of the hindleg) on touching the sole of the foot as well as contractions caused by proprioceptive stimulations (excessive performance of rocking reactions, leg-slackening and hopping reactions). Secondly, the muscular reactions on proprioceptive stimulations, for instance in leg-slackening reactions, often appear too late. Thirdly, the muscular contractions are subject to extraneous influences to a vastly lesser degree than in intact dogs; for instance, the dorsal position of the animals has less effect on the magnet and supporting reactions and there is less influence of experimental set-up on the tonic labyrinthine reflexes and on labyrinthine rotating reactions of the extremities.

The peculiar restlessness, the so-called "unrestrained movements," characteristic for decerebellate animals will be discussed in detail in Chapter XIX.

After unilateral extirpation of the cerebellum, the animals are often just as rigid as after a total extirpation. The distribution of tonus, however, changes within the first days and then most of the unilaterally decerebellate animals show the following symptoms.

In a dorsal position on a supporting surface with the head symmetrical to the trunk, the muzzle pointing upwards, *they keep the legs ipsilateral to the extirpation totally extended,* the opposite legs more or less flexed, the first showing a considerable extensor tonus, the latter a flexor tonus.

A similar position is shown by the legs when the animals, lying on the side of extirpation, are held fast in this position (the head also). Conversely, when placed on the contralateral side (side of the intact half of the cerebellum), all four legs are drawn into a flexed posture.

As we have seen in Chapter XIV, animals which are rigid after a unilateral extirpation of the cerebellum keep all four legs extended at first in both lateral positions, but very soon the extended posture disappears when lying on the side of the intact half of cerebellum. A few days later, also the extended posture of the legs contralateral to the extirpation disappears while lying on the side of extirpation. In this stage, the body righting reflexes on the body appear clearly when lying on the side of the intact half of cerebellum, while they are still absent in the other lateral position. Finally, the extended posture of the ipsilateral legs disappears also on lying on this side and the body righting reflexes, in this case, are again present in both lateral positions.

Also, when the animals are held in a ventral position in the air supported by the thorax, the ipsilateral legs are extended and the contralateral ones are more or less flexed. The ipsilateral legs, thus, in a dorsal position on a supporting surface, lying on the side of extirpation and in ventral position in the air, show an increased tendency to an extended position and in these conditions, when they are grasped from the side and an attempt to flexion is made, a strong resistance and an increased extensor tonus are found, which are absent in the contralateral legs.

The increased stretch tonus of the ipsilateral legs varies considerably on altering the position of the head in space and in relation to the trunk, and the ipsilateral legs show distinct tonic neck and labyrinthine reflexes which usually cannot be observed in the contralateral legs. The increased extensor tonus diminishes when the opposite limbs are under static stress and remain diminished as long as the static stress continues (Fig. 70).

In this stage, not only the extensors of the ipsilateral legs *but also the ipsilateral vertebral muscles show an increased contraction* which was pointed out by Munk. The increased contraction can be most clearly observed when the animals are held in the air in a ventral position supported by the thorax. In this case, the animal keeps its head strongly turned towards the operated side, especially with blindfolded eyes, *while the thoracic spinal column shows a strong concavity towards the same side.* An attempt to counteract the lateral curvature of the cervical and thoracic spine encounters a strong resistance and the concavity of the thoracic spine disappears only a little when the head is placed symmetrical to the anterior part of the body. Also, in a dorsal position in the air and on a supporting surface, the increased contraction of the vertebral muscles ipsilateral to extirpation can be clearly observed. The concavity of the spinal column on the side of extirpation similarly appears in a suspended posture head downwards, with the eyes blindfolded, but, in this case, a rotation of the spinal column also occurs by which the dorsal side of the shoulder and of the vertex of the cranium is directed towards the side of the operation. When the animal is held suspended by the forelegs head upwards, concavity and rotation of the spinal column are also evident. In this case, the muzzle of the animal is directed towards the right (ipsilaterally), the dorsal side of the pelvis towards the left; and now, too, neither rotation nor concavity of the spinal column is abolished when the head is placed symmetrically to the shoulders (Fig. 255; see also Fig. 184).

Fig. 255. Dog Fox, after extirpation of the right half of the cerebellum, held in the air by the forelegs head upwards. Head is held symmetrically to the shoulders. The vertebral column shows a concavity to the right, by this the pelvis, against the effect of the force of gravity, shows a deviation to the right, towards the side of the operation. Moreover, the spinal column shows a rotation which leads the dorsal side of the pelvis slightly to the left.

These phenomena which are evident immediately following the unilateral extirpation of the cerebellum are thus totally incompatible with Luciani's and other authors' assertion that unilateral extirpation produces a general hypotonia of the ipsilateral muscles.

In the first period after unilateral extirpation of the cerebellum, magnet and supporting reactions in the ipsilateral legs are absent; at any rate, they do not bring about any distinct reflex contractions; the extensor tonus, the resistance to passive flexion, however, distinctly increases on pressure on the soles of the feet (stretch reflexes first order, myotatic reflexes of extensors on rapid stretch). Also, absent in the ipsilateral legs, in the first period, are hopping, leg-bracing, leg-slackening, propping-up and pulling-up reactions as well as proprioceptive correction movements; thus, the various proprioceptive muscle reactions are impaired to a considerable degree.

In the weeks following the increased contraction of the ipsilateral extensors of the extremities and vertebral muscles gradually decreases, and concurrently the just mentioned proprioceptive reactions return. In the stage of permanent impairment, the animals in a supine position keep all four legs flexed and do not show any distinct differences between right and left legs in testing resistance to passive flexion and extension. The patellar reflexes are present and lively on both sides and can be produced right and left with the same amplitude.

In intact dogs, the patellar reflexes can vary greatly in amplitude, since they are influenced by experimental conditions, position of the animals, initial posture of the legs, etc. The various deviations of patellar reflexes shown in man under pathological conditions cannot be observed in dogs. This is surely the reason why scarcely any statements on the behavior of patellar reflexes can be found in the publications of various investigators like Luciani, Munk and others.

On altering the position of the head of animals lying on their back, the legs remain flexed; the tonic labyrinthine and cervical reflexes no longer cause distinct alterations in position of the legs not under static stress. In both lateral positions, too, all four legs are now in a flexed position, while in a standing posture the legs of both sides show an equally strong supporting tonus.

On the detailed behavior of supporting tonus in the different stages, see Chapter VI.

As shown by careful observation, the distribution of tonus in unilaterally decerebellate animals still shows some deviations in the stage of permanent impairment. Thus, in a position suspended head upwards or downwards, as well as in a ventral position in the air, usually, *an increased contraction of the ipsilateral vertebral muscles is still detectable*; similarly, there is often, though not constantly, a somewhat increased extensor tendency of the ipsilateral legs in a dorsal position which becomes particularly evident when the animals are shaken to and fro. In all unilaterally decerebellate animals, the maximum extension of both hindlegs is also seen in the dorsal position of the animals (muzzle pointing vertically upwards) when the soles are touched or put under static stress. While in intact and decorticate dogs in a dorsal position, touching or static stress on the soles is usually not followed by a distinct extension of the hindlegs. This response appears instantly and obviously with a distinct supporting tonus in unilaterally decerebellate dogs in the ipsilateral as well as in the contralateral hindlegs. Usually, but not constantly, the supporting tonus is stronger in the ipsilateral than in the contralateral hindleg. (On the force of supporting tonus of the hindlegs in the dorsal position, also see Chapter VI.) Distinct tonic neck and labyrinthine reflexes, too, can be observed in the legs under static stress on altering the position of the head.

Whether the strong reaction of the contralateral hindleg on touching and static stress on the sole in the dorsal position of the animal is caused by an unintentional lesion of the other half of the cerebellum is not certain.

Moreover, the ipsilateral legs show retarded and exaggerated leg-slackening reactions. Similarly, the correction of abnormal limb postures is executed by these paws with retarded reactions with excessive raising. The leg-bracing reactions, which are especially vigorous on both sides in totally decerebellate dogs, are strikingly vigorous in the contralateral legs in unilaterally decerebellate dogs, while in the ipsilateral legs they are far less strong and usually distinctly weakened.

When one sees how decerebellate dogs play on a lawn in the open air and pursue each other for hours, one cannot believe there is asthenia, particularly since the exaggerated movements of ataxia make running even more difficult. Moreover, decerebellate animals are not only able to carry loads on their backs to the amount of their own bodyweight without their legs giving way, but they are often even capable of running at considerable speed over long distances with a heavy load on their backs (Fig. 256; see also Fig. 265).

The force of movement is also not considerably reduced, just as little as fatigue is increased. The dogs and cats which are alive still 2-4 years after extirpation of the cerebellum are strong, fat and healthy without ever showing loss of hair, skin anomalies or any kind of dystrophic change. *Muscular asthenia or constitutional asthenia is in no way detectable in these animals.*

Fig. 256.   (1) Decerebellate dog Erik, body weight 5.3 kg, runs with a sandbag of 5 kg on its back.   (2) Decerebellate dog Moor with a sandbag of 5 kg on its back, running.

XVII. INVESTIGATIONS ON STANDING, THE REGULATION OF MUSCLE TONUS AND THE MOTILITY OF DECORTICATE ANIMALS (1)

Decorticate dogs can stand and run; on superficial observation one even gets the impression that standing and running take place quite normally, with a seemingly normal distribution of muscle tonus (Fig. 257).   In standing, the posture of the trunk and legs shows no impairment; in running, the legs are transferred in a normal succession and set down on the ground in the natural manner.   Decorticate dogs make no

Fig. 257.   (1) Decorticate dog Miesel, in a standing posture.   (2) Decorticate dog Robbie, in a standing posture.   (3) Decorticate dog Fuchs, in a standing posture.   (4) Decorticate dog Fuchs, running.

---

(1) In the various decorticate animals under observation, the corpus striatum was more or less preserved on one or both sides. How far the fact that the corpus striatum is more or less preserved can influence the condition of the reactions discussed in the previous chapters has not yet been systematically investigated.   For a detailed description of the behavior of decorticate animals, see, among others, Goltz (107), Rothmann (254, 262), Dusser de Barenne (60A), Magnus (196), Zeliony (336), Graham Brown (37), Dresel (56) and others.

false steps, do not fall and show prompt righting reactions. Except for the optical reflexes, all other righting reflexes (labyrinthine, neck and body righting reflexes on the head and on the body) are preserved.

However, on investigating the various reactions which participate in normal standing, many defects can be detected. First of all, the readiness to stand is absent, both on optical stimulation as well as on stimulations from the body surface. The absence of readiness to stand is seen in impairment of correction of abnormal limb positions, for instance, the correction of posture of limbs that have stepped through a lattice or fallen over the edge of a table (Fig. 258).

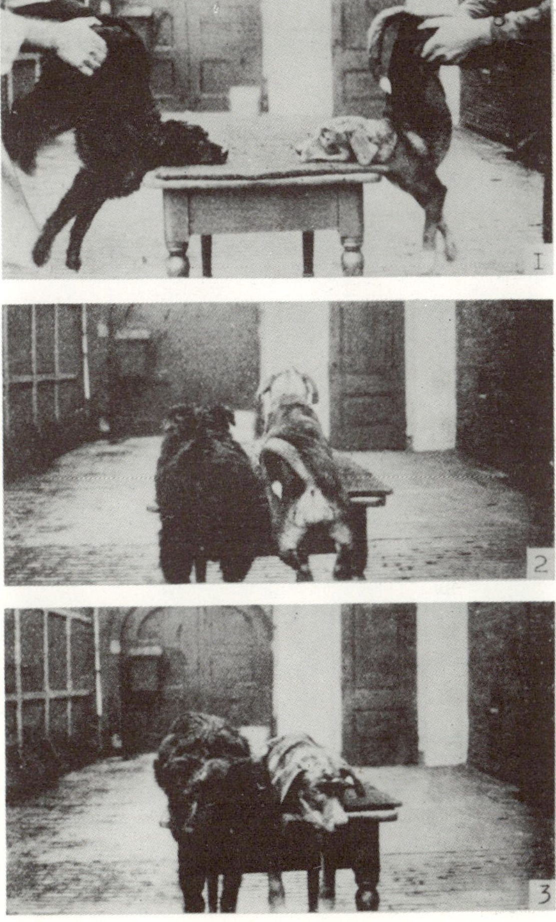

Fig. 258. Absence of preparation for standing (placing) on stimulations from the body surface in the decorticate dogs Miesel and Robbie.

530

Since preparation for standing (placing) in intact animals shows some characteristics of "conditioned" reflexes and, moreover, is impaired in an asymmetrical manner after the unilateral extirpation of the cerebrum, one is permitted to draw the conclusion that it is normally brought about by way of the cerebrum. On the impairments of preparation for standing (placing) after the unilateral and bilateral extirpation of the cerebrum, also see Chapter IV.

In the second place, decorticate dogs show impairments of magnet reactions (exteroceptive supporting reactions). Neither lying in a dorsal position on a supporting surface nor in a ventral position in the air does touching of the soles of the feet produce a distinct extension or fixation in an extended position (Fig. 259, No. 1). On the behavior of magnet reactions after the unilateral and bilateral extirpation of cerebrum, also see Chapter V.

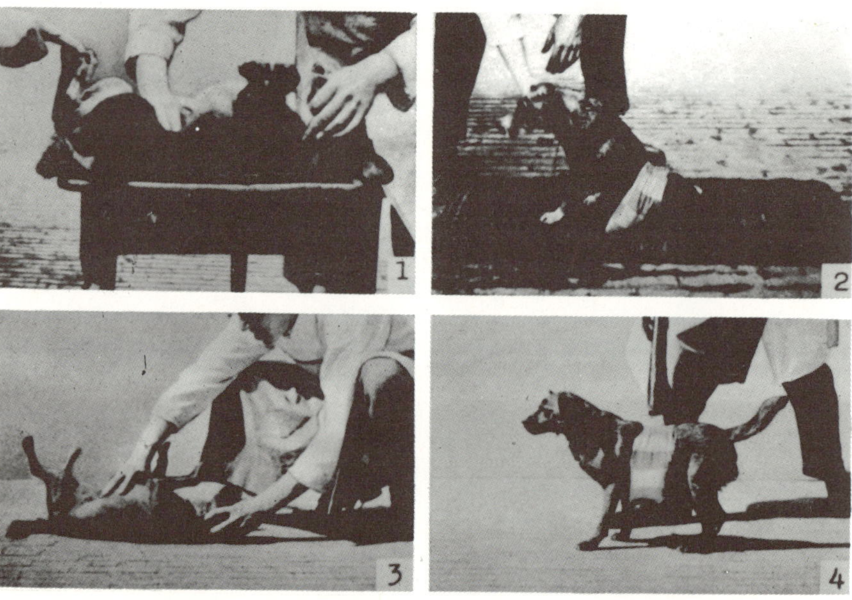

Fig. 259. Decorticate dog Miesel in a dorsal position, the muzzle pointing vertically upwards. The animal keeps the four legs totally flexed. Touching of the sole does not cause extension of the hindlegs. 2. The same dog. A sandbag of 1.7 kg was placed on the back of the standing animal. Because of the reduced force of supporting tonus, the hindlegs instantly gave way. Body weight of the animal, 6.7 kg. 3. Decorticate dog Fuchs in a dorsal position. 4. The same dog in a standing posture with a sandbag of 1.7 kg on the middle of its back. The hindlegs are just able to carry the load. (However, when the sandbag is laid on the caudal part of the back, they give way.) Body weight of the animal, 8.5 kg.

A third deviation becomes evident in an investigation of the proprioceptive supporting reactions of the hindlegs. When a decorticate dog is placed on its back, it keeps its four legs flexed in all postures of the head and neither touching nor static stress on the soles of the feet causes in these circumstances a distinct extension or fixed posture of the hindlegs. This, however, is also often the case in intact dogs under the same experimental conditions (Chapters V and XIV). On the other hand, there is a distinct difference when the animal is suspended in a ventral position in the air. Static stress on the hindlegs indeed causes extension with fixation also in decorticate dogs, but this fixation is considerably weaker than in intact dogs and in decerebellate dogs in the stage of permanent impairment. Immediately following extirpation of the cerebrum, the strength of supporting tonus and the proprioceptive supporting reactions are distinctly weakened in all four legs. While, however, the strength of supporting tonus is completely recovered in the forelegs, it remains almost always inferior in the hindlegs (with exception of the dog Bob; Fig. 259, No. 2). The abnormal weakness of the hindlegs of decorticate dogs in a standing posture already attracted the attention of Goltz (107). A detailed discussion of the impairments in extero- and proprioceptive supporting reactions and on the behavior of the force of supporting tonus after total and unilateral extirpation of the cerebrum can be found in Chapters VI and VII. The reduced strength of supporting tonus in the hindlegs supports shows that the proprioceptive muscular reactions do not occur in quite a normal manner.

The defects that occur in the leg-slackening reactions are probably also based on impairments of proprioceptive muscular reactions produced by stretch. The leg-slackening reactions, as we have seen in Chapter XIII, are retarded in decorticate animals and, moreover, are performed hypermetrically: the slack leg is raised a little too high and is displaced too far so that the hops are abnormally large (Fig. 210). The execution of reactions is much less abrupt than in decerebellate animals.

Usually, the most strongly impaired are the leg-slackening and hopping reactions inwards. The contraction of adductors as well as the raising and the inward transfer of the legs take place particularly late. The delayed contraction of the stretched adductors is probably the reason why the limbs of decorticate animals slide sideways so easily and frequently on a smooth floor. *The impairments of leg-slackening and hopping reactions are always more strongly marked in the hindlegs than in the forelegs* (see also Chapter XIII).

Probably based also on the impairment of stretch reflexes are the impairments appearing in other proprioceptive correcting movements (Fig. 213). These, too, are delayed and are carried out hypermetrically. It is true that this delay gradually decreases so that the animals never show the grossly abnormal paw positions seen early after extirpation of the cerebrum, but it never disappears completely, at least not in the hindlegs.

In the forelegs, a considerable time after extirpation, small passive alterations in position instantly produce correcting movements, especially when the animals are excited by pinching. In this respect, the animals even give the impression of overexcitability (also see Chapters IV and XIII).

Among the other reactions based mainly on stretch reflexes, leg-bracing and rocking reactions also show impairments. Leg-bracing reactions are not absent in decorticate animals; they are present in the forelegs as well as in the hindlegs (Fig. 182). However, while they show no considerable defects in the forelegs, they are usually distinctly impaired in the hindlegs, where they are retarded[1] and the force with which the hindlegs brace backwards or outwards is considerably reduced. The impairment of leg-bracing reactions in decorticate dogs is the reason that, in standing on a supporting surface which is slowly raised from one side, the adaptation of leg position to the position of the supporting surface takes place less promptly than in intact dogs (also see Chapter XIII).

In rocking reactions of the hindlegs, too, the extension of the leg carrying no weight on abduction of the supporting leg is executed with a much lesser force than in intact dogs (Chapter XI).

On the other hand, no distinct impairment could be observed in the pulling-up and propping-up reactions of the forelegs. Plantarflexion of the distal phalanges of the extremities produces typical negative supporting reactions, while an overstretching of the distal phalanges is accompanied by a withdrawal of the paws and, in the case of maintained overstretched posture, also by alternating flexion-extension movements in decorticate animals.

---

(1) It is extremely difficult to determine whether the reactions are delayed or not because they sometimes totally fail to appear in intact dogs under investigation due to "conditioned cerebral inhibitions." The appearance of the reactions sometimes occurs strikingly late in decorticate dogs in all four legs, but particularly in the hindlegs.

Thus, in summing up, one can say that in decorticate animals probably all stretch reflexes from first to sixth order are present and that, however, these reflexes do not proceed in a quite normal manner, as shown in the retarded occurrence of leg-slackening reactions, which often appear only on an abnormally strong stretch.

Concerning the influence of the position of the head, of tonic neck reflexes and of labyrinthine reflexes on the distribution of muscle tonus, the investigation showed a distinct influence of these reflexes only in standing posture of the animals. In this posture, raising and lowering of the head also produce in decorticate animals a distinct and typical increase and decrease of supporting tonus while, conversely, in the dorsal position the four legs remain in a more or less flexed position even under staticstress (see Chapters VII and VIII).

Alterations in supporting tonus due to labyrinthine rotating reactions produced by rotation round the longitudinal or bitemporal axis of the head are absent in the dorsal position of the animals, as in intact dogs; however, they appear in a ventral position and standing posture in a typical manner though less promptly. Their occurrence is usually distinctly retarded, especially in the hindlegs. Also, in standing on a supporting surface that is *rapidly* lifted or lowered from one of the four sides, the adaptation of limb position caused by these reflexes is not as prompt as in dogs with an intact cerebrum (Chapter IX). In summing up, in decorticate animals the following reactions among those participating in normal standing are impaired.

A. Reactions on stimulations from the body surface:

1. *Preparation for standing* (*placing*) and the correction of abnormal limb positions on stimulations from the body surface (as well as Munk's touching reflexes) are absent.

2. *The magnet reactions* are absent or at least are greatly weakened.

B. Reactions on proprioceptive stimulations produced by extension of muscles:

1. *The supporting reactions* only cause an inferior supporting tonus in the hindlegs; the force of supporting tonus is considerably reduced in the hindlegs (with the exception of dog Bob).

2. *The leg-slackening reactions* are retarded and hypermetric.

3. *The proprioceptive correcting movements*, at least those of the hindlegs, show the same impairments as the leg-slackening and hopping reactions.

534

4. *The leg-bracing reactions of the hindlegs due to* alterations in position of the supporting surface appear belatedly and are performed less forcefully.[1]

5. Certain reactions of the extremities, on labyrinthine stimulations produced by movements, occur less promptly, particularly in the hindlegs.

It is worth mentioning that investigations of reactions on stimulations from the body surface, beside the above-described impairments (A 1-2), resulted in the following observations. Just as in intact animals, in decorticate animals, too, the counterpressure of the supporting surface, in shifting the animals into a dorsal or lateral position, causes a reduction in supporting tonus and a drawing in of the legs into a flexed position (in lateral position also body righting reflexes on the body). A similar reduction in supporting tonus in all four legs also appears in decorticate animals on seizing a large fold of skin on the back (Fig. 219). Seizing a skinfold above the pelvis will only cause the hindlegs to give way and brings the animals into a squatting position, and seizing a skinforld above the shoulders is accompanied by a collapse of only the forelegs. Strange to say, a stroking of the skin of the back against the direction of the hair produces, on the contrary, almost always an especially obvious extension of all four legs, while stroking the skin above the pelvis only the hindlegs respond in that way. This extension reaction is particularly pronounced in decorticate cats and sometimes also in intact ones (Sherrington).

In the decorticate dog Fox, the following manifestations, produced by stimulations from the body surface, also became evident. In this animal, the left half of the cerebrum was removed first, after which it showed a tendency to turn the head towards the left and to run in anticlockwise circles to the left, as is usually the case in the first period after similar extirpations. Secondly, after removal of the right half of the cerebrum, it showed, as usual in the first period, a tendency to turn the head and make circular movements to the right. Sometime later, the animal ran again straight forwards very well, till one day it suddenly started again to run in circles to the left and continued to do so for several days. Examination revealed the cause of this turning of the head and the circular movement to the *left* : an eczema in the *right*

---

(1) The leg-bracing reactions due to pinching of the tail appear much more vividly and strongly.

external ear. As soon as the eczema was cured, the animal again ran straight forwards. When, however, a clamp was fastened to the right ear, it again showed turning of the head and a running in circles to the left (Fig. 260, No. 1). This was also the case when the clamp was fassened to the right side of the head or to the anterior part of the right tide of the neck; conversely, when the clamp was also fastened to the right side of the neck, but this time to the rear part of the neck or to a fold in the skin of the thorax, a turning of the head and a circular movement to the right appeared. In this case, the neutral point was situated approximately in the middle of the neck. When the clamp was fastened to the left side, opposite manifestations occurred (Fig. 260, No. 2 and 4).

Fig. 260. Decorticate dog Fuchs. 1. A spring-clip is attached to the *right* external ear; the animal shows a turning of the head and a circular motion to the *left*. 2. The clip is now fixed to the skin of the *right* flank; the animal now shows a turning of the head towards the right and, in running, deviates to the *right*. 3-4. On attaching the clip to the left side, opposite reactions appear, i.e., on attaching it to the left external ear, a turning of the head and a circular motion to the right (3), on attaching it to the left flank a turning of the head and circular motion to the left (4).

536

When a clamp was fastened to each of the two ears or both ears were seized with the hands, the head moved ventrally while the legs were braced forwards and the animal started to run backwards (Fig. 261, No. 1). The forwards bracing and backwards running also occurred when the muzzle was seized with one hand (Fig. 260, No. 2) and also when a clamp was fastened to the anterior part of the chest (Fig. 261, No. 3). However, the four legs instantly were braced backwards and the animal started running forwards when the clamp was fastened to the tail (Fig. 261, No. 4) or when the tail was grasped with the hand (Fig. 261, No. 5).

Fig. 261. Decorticate dog Fuchs. 1. Both ears of the animal are seized, each with one hand. The animal tries to liberate the head by a ventral flexion of the neck and, in addition, braces all four legs forwards. 2. Also when the muzzle is seized with one hand, the four legs brace forwards and the animal tries to liberate its head by pulling it backwards. 3. On attaching a clip to the median line of the anterior part of the chest, the four legs are also braced forwards and the animal starts to run backwards. 4. Conversely, attaching a clip to the tail causes a backwards bracing of the four legs, followed by a running forwards. 5. The same reaction is shown when the tail is seized with one hand.

We have already seen (Chapter XIII) that in decorticate animals traction backwards produces leg-bracing reactions backwards and a pulling forwards a bracing of the four legs forwards (Fig. 182). This however, is only the case when the traction acts on the trunk. When the pull is applied to the tail, leg-bracing reactions backwards always occur when the animal is pulled backwards as well as when it is pulled forwards by the tail. On the other hand, pulling forwards as well as a backwards pushing of the muzzle with the hand always produces a bracing forwards of the four legs and a running backwards.

When a spring-clip is fastened to the skin of one paw, this paw is drawn up and the animal runs on the other three legs; this is also the case when the sole of one paw is chafed from walking. On fastening the clip to the skin below the pubic bone, the entire posterior part of the body is sometimes lifted from the ground so that the animal stands only on the forelegs, while fastening the clip to the skin above the pelvis causes the hindlegs to give way and the posterior part of the body to sit down.

In addition to alterations in position, the fastening of a spring-clip to various parts of the skin produces other reactions. Thus, the animals sometimes react with utterances of pain as from pinching the skin. A clip fastened to the pinna of the ear often causes shaking of the head. A similar shaking of the head appears on sprinkling the skin of the head with water. When the animal is totally wet, it will perform shaking motions of the whole trunk like an intact dog, just as a limb is shaken in a typical way when it steps into water.

On fastening the clip to the skin of the thorax, the animal sometimes not only shows the turning of the head we have described, but also bites and, in so doing, the muzzle may indeed touch the clip, but it is not seized with the mouth and torn away as an intact dog would do. When the clip is fastened to certain parts of the skin of the chest, the animal sometimes raises the ipsilateral hindleg and performs scratching movements directed at the clamp which, however, are unsuccessful.

Among the reactions to other stimulations several are worth mentioning. Pinching the toes causes vivid ipsilateral flexion reflexes and crossed extension reflexes, sometimes accompanied by utterances of protest. On continuation of the pinching, the pinched toes carry out alternating flexion and extension movements as if attempting to liberate themselves.

When the muzzle of a decorticate dog is moistened, it will instantly lick off the moisture. When the animal touches a cold plate of glass with its muzzle, it will also sometimes begin to lick it. The decorticate dog Miesel instantly started to lick and to eat when the muzzle was placed passively into moist food. It was repeatedly observed that the animals licked the trunk and the paws clean.

When one restrains a running decorticate animal with a hand held against the upper side of the muzzle, the head is ventrally flexed and the animal tries to creep through below the hand. On holding it back with the hand on the lower side of the muzzle, the head is raised and on further restraint the animal raises up on its hindlegs. When the running animal reaches the corner of a room and can escape neither forwards nor to the right nor to the left, it will rise on the hindlegs and start scratching the wall alternately with the forelegs, just like an intact dog standing in front of a closed door, and will growl and snarl as if it were furious. As we have seen, grasping the tail causes a backwards bracing of the legs and a running forwards. When one continues to hold the tail, the animal will run faster and faster and finally break into a gallop. If one pulls the tail, the animal will also start to jump and to gallop and to growl and whine like an enraged dog. This is also the case but to a much stronger degree when at the same time the tail is pinched. As soon as the tail is released, the animal will then wag its tail and run away as if it were delighted.[1]

Sometimes, on releasing the tail, a backwards running occurred. Why the animals sometimes run forwards and sometimes backwards could not be explained.

When the animals in running made contact with the wall of the room with the right side of the head, they turned aside to the left and, on touching the wall with the front of the muzzle, they tried to keep on running and push themselves against the wall. On a forcible impact of the muzzle with the wall, the animals would run some distance backwards. Passive opening of the mouth often caused yawning. Sometimes, the animals also yawned spontaneously, especially when they were

---

(1) Several authors, among them Dusser de Barenne, assert that decorticate animals showed only negative expressions of emotion. But one of my decorticate cats instantly began to purr on a gentle stroking of the skin, while several decorticate dogs would wag their tails on stroking the skin.

aroused from sleep.[1] The animals showed alternating periods of wake-fulness and sleep. In a well-heated room they were more sleepy than in a cold one. Before eating they were restless and after eating they were sleepy. When an animal was not fed for a whole day, it grew more and more restless and finally started to howl like a dog standing at night in front of a closed door.

Food placed into the mouth was chewed and swallowed but hard and dry pieces of stale bread, however, were discarded even when mixed with chopped meat. Tickling of the mucous membrane of the nose caused sneezing. Bright, strong light produced narrowing of the pupils and often a blinking of the eyelids. They often reacted to certain noises by a pricking of the ears and by lifting up their heads. As shown by these observations, exteroceptive stimulations produce in decorticate animals varying behavior patterns, motor reactions and pseudo-affective reflexes, according to the kind (temperature, touching, pain and other stimulations), force, duration and point of application of these stimula-tions. These observations also showed that decorticate animals, by means of labyrinthine stimulations, stimulations from the muscles and interior organs or from the body surface, when these are brought to bear in an appropriate manner and on the appropriate spot, can carry out appropriate action, even very complicated ones and that by means of these stimulations any position (standing posture, standing on the hindlegs or on the forelegs, squatting or ventral position) can be pro-duced.

The decorticate animal can be distinguished from the intact one *firstly by the fact that in the former, a lesser number of stimulations is effective.* Since in the course of extirpation of the cerebrum the olfactory nerves were severed, all olfactory stimulations were ineffective. Among optical stimulations only bright light shone into the eyes caused reactions (narrowing of pupils, blinking). Movement of a hand in front of the eyes or the display of food no longer caused a reaction. Among acoustical stimulations only a few clearly perceptible ones produced reactions. It was remarkable that decorticate animals still reacted to the typically

---

(1) It was very remarkable that decorticate dogs rarely ever soiled the hay in their cages. When they were taken out of their cages and set down on the floor, they instantly started to urinate and defecate after running at first several times in circles in a typical way. Whether this was caused by the colder floor or some other reason could not be established.

sharp noises with which dogs are usually called, produced by sucking in air through tightly compressed lips, while Golts observed awakening in them to the "terrible noise" of a bicycle bell. Decorticate cats often reacted to the noise created when the soles of shoes rub on a stone floor. The decorticate animal, however, no longer reacts to calling or to other numerous acoustic stimulations.

A second difference is the fact that, in decorticate animals, the conditioned reflexes caused by stimulations from the body surface (like preparation for standing, Munk's avoiding reflexes to touch) *by visual, auditory and olfactory impulses* are absent. To certain stimulations, for instance grasping of the tail, the intact animal sometimes reacts like a decorticate one, i.e., with backwards bracing of all four legs and by running forwards, or it acts completely differently, for instance, by turning the head and by licking or biting the hand grasping the tail. *The subcortical reactions shown by decorticate animals are thus often inhibited or replaced by conditioned cerebral reflexes in intact animals.*

A third difference consists in the fact that certain unconditioned reactions show impairments in decorticate animals. Thus, there is:

1. a weakening of magnet reactions,

2. a reduction in the strength of supporting tonus in the hindlegs based on an impairment of the positive proprioceptive supporting reactions,

3. a retarded occurrence of leg-slackening and hopping reactions and of proprioceptive correcting movements, which take place only on abnormally strong alterations in position,

4. a less vigorous occurrence of leg-bracing reactions in the hindlegs,

5. a less prompt appearance of certain reactions of the extremities due to labyrinthine stimulations, especially in the hindlegs.

Thus, the muscles of the extremities, particularly those of the hindlegs, do not function quite normally in decorticate animals. One gets the impression that the muscles do not react as easily and do not contract as vigorously on exteroceptive (1) as well as on proprioceptive stimulations produced by muscular extension (2-4) or on labyrinthine stimulations. Whether these impairments are really caused by an absence of cerebral functions needs further clarification.

A decerebrate animal can be distinguished from a decorticate one as follows :

1. by the exaggerated extensor posture, the stiffness of the neck and of the back in the periods of manifest rigidity,

2. by the fact that in the decerebrate animal in a standing posture the stretch-reflex tonus gradually diminishes so that the legs give way after some time, usually after 10 minutes, while in the decorticate animal the legs under static stress can carry the trunk for hours,

3. by the incomplete supporting reactions,

4. by the absence of labyrinthine and body righting reflexes,

5. by the absence of hopping reactions, leg-slackening reactions, leg-bracing reactions, pulling-up and propping-up reactions,

6. by the absence of the inhibiting influence of the supine posture on the reactions of muscles of the extremities.

While in decorticate animals in a dorsal position almost no supporting tonus can be produced, the extremities of decerebrate animals in a dorsal position are particularly rigid due to the influence of tonic labyrinthine reflexes and the stretch reflexes (myotatic reflexes of Liddell and Sherrington) are especially vigorous.

XVIII. IMPAIRMENTS OF THE STANDING AND RIGHTING FUNCTIONS AFTER UNILATERAL EXTIRPATION OF THE CEREBELLUM

Observations in a number of animals in which the cerebellum was unilaterally removed[1] resulted in the following conclusions:

1. Immediately following extirpation, the manifestations were different in almost every animal.

2. In this stage, however, some of the deficiency symptoms could also be observed constantly in all animals and it was observed particularly that not one of the animals was able to hold itself upright and stand without being supported.

3. Conversely, in the stage of permanent impairment, there were almost no differences in the manifestations in the various animals. In this stage, they were again able to rise, stand and run freely.

A number of animals, after unilateral extirpation of the cerebellum, showed rolling movements, which were absent in the others. The direction of these rolling movements was always the same in the various animals; after extirpation of the right half of the cerebellum, they were always to the right (i.e, in the succession: ventral position — right-sided lateral position—dorsal position—left-sided lateral, etc.) and, after extirpation of the left half of the cerebellum, always to the left (i.e, ventral position — left-sided lateral position — dorsal position— right-sided lateral, etc.). Although in all animals the same half of the cerebellum was removed, the direction of the rolling movement was the same but the rest of the movements did not always correspond. After extirpation of the right half of the cerebellum, the rolling movements started in a number of animals with the transition from *the right-sided lateral position* via the dorsal to the left lateral position, while others started with the transition from *the left-sided lateral* position via the ventral to the right-sided lateral position. This difference was caused

---

(1) In addition to the four dogs Peter, Rollmops, Fox and Tip and the cat Josephine, which stayed alive for a considerable time after the unilateral extirpation of thet cerebellum (see p. 2), observations were made on a number of animals which were kep alive for only a short time.

by the difference in the position of the rest of the animals at the start of the rolling movment (Chapter XV).

In Chapter XV, we already discussed in detail the question whether perhaps an associated lesion or shock of the vestibular centers could be the essential cause of these movements. Similarly, the major role played by these movements in the various theories of cerebellar balance was discussed. In these theories, it was always assumed that these rolling movements are caused by the absence of balance reactions.

For this reason, we want to deal here more closely with the mechanism of these rolling movements and to discuss how well this assumption is founded.

The rolling movements appearing after a unilateral lesion of the vestibular system have been analyzed, among others, by Ewald (65), Winkler (335) and by Magnus and De Kleyn (143, 196). The latter two authors, based on cinematographic pictures of rolling movements after the unilateral labyrinthectomy in rabbits, came to the conclusion that the movements "can be interpreted as running and jumping movements of animals whose bodies are helically rotated." In the rotation from one lateral position to the other, one can, as emphasized by Magnus and De Kleyn, distinguish two phases: in the first phase, there occurs an increasing rotation of the head with a following helical rotation of the trunk and, in the second, a progressive movement (jumping movement). With the jumping movement, the helical rotation is totally or partly cancelled.

Magnus and De Kleyn go on to describe how, after a left-sided labyrinthectomy, the animal lies in a left-sided lateral position. At the start of the rolling movement, the head and then the anterior part of the body is rotated at first to the left, i.e., into a dorsal position, then the jumping movement takes place in which the entire animal rolls to a right-sided lateral position via its back. The head and the anterior part of the body then start to rotate again to the left and this rotation, too, is followed by a jumping movement in which the animal rolls to a left-sided lateral position via its belly. An entire rotation round the longitudinal axis thus is brought about in two jumps, one of them rotates the animal by 180° via its back and so does the other, via its belly. "Thus we are here dealing with very strong running movements of an animal whose body is helically rotated, which, however, do not advance the body forwards but rotates it through space." As is emphasized by Magnus and De Kleyn, every attempt to explain these rolling movments must start from the fact that unilaterally labyrinthectomized rabbits do not roll uninterruptedly after the operation but that these movements appear paroxysmally and, therefore, they regard the rolling as a combination of deficiency symptoms and paroxysmally occurring stimulation symptoms. The helical rotation of the body is a deficiency symptom caused by the unilateral absence of labyrinthine righting reflexes and the tonic labyrinthine reflexes acting on the neck muscles. The unilateral absence of these reflexes causes rotation of the head and, under the influence of the following neck righting reflexes, the helical rotation of the entire trunk. The running and jumping movements, in their opinion, however, are stimulation symptoms. They base their conception on the fact that running and jumping movements always, also in the intact animal, begin with some stimulation and on their observation that the rolling movements occur more forcefully and continue for a longer time, the more the labyrinth has been traumatized in the course of extirpation, and the more severe the bleeding that occurred. With increasing practice in the operative technique, rolling becomes less frequent and finally De Kleyn was able to perform a unilateral extirpation of the labyrinth so expertly that the animals made only a few rolling movements or none at all (sometimes in rabbits the rolling movements did not appear immediately but only a few days after the unilateral labyrinthectomy).

Magnus and De Kleyn's assertion that, in the production of rolling movements, the helical rotation of the trunk and the progressive movement produced by any kind of stimulation are participating, is of course correct and also the participation of stimulations emanating from the surgical wound is probable. Nevertheless, Magnus and De Kleyn's explanation does not seem to me to be sufficient since rolling movements appeared also on cocainization of one of the labyrinths. Furthermore, some time after the unilateral extirpation of the labyrinth, even those rabbits in which the postural effect had not diminished, and which, therefore, showed a particularly strong rotation of the head in a standing posture, were again able to run and jump without the occurrence of rolling movements. These movements failed to appear even when progression movements were produced by a strong pinching of the tail and were, therefore, performed especially forcefully and abruptly. These observations suggest that the rolling movements cannot be based only on progression movements of a helically rotated body but that other factors must also be participating.

When one places an *intact* rabbit in a left-sided lateral position and then tries to roll the animal by a rotation of the head to the left via the back into a lateral position to the right, one can easily succeed. For the production of half a rolling movement, thus, the rotation of the head is sufficient, although the body righting reflexes on the body will tend to rotate the body from a left-sided lateral position to a ventral one, i.e., to counteract this rolling movement. Therefore, a weakening or absence of these righting reflexes will facilitate the rotation of the trunk by way of the back. Investigation showed that, following a left-sided labyrinthectomy (rolling movements to the left), the body righting reflexes acting on the body from the left-sided lateral position were indeed weakened. When after a left-sided labyrinthectomy the animals were placed on the left side and the head was held fast in this position, they did not right the trunk. Magnus and De Kleyn also mentioned that, following a unilateral extirpation of the labyrinth, the body righting reflexes acting on the body were weakened when lying on the side of extirpation. We shall now examine the second half of the rolling movement, from the right-sided lateral position to a lateral position on the left *by way of the belly*. When an intact rabbit is placed into a right-sided lateral position and the head is again passively rotated to the left, the body indeed changes into a ventral position, but one will never succeed in rotating the trunk any further into a lateral position to the left. This will be prevented by leg-bracing and leg-slackening reactions of the left legs and by the body righting reflexes from a lateral position to the left. For this half of the rolling movement, the rotation of the head is not sufficient, even if the animal simultaneously makes kicking movements, but in addition to this, a weakening or absence of the just mentioned reactions would be necessary. We have already seen the weakening of body righting reflexes from the left-sided lateral position after extirpation of the left labyrinth. Specially performed experiments have shown that, after a left-sided labyrinthectomy, the leg-slackening and leg-bracing reactions of the left legs were also impaired to a high degree. Also, when in guinea pigs the functions of the left labyrinths were eliminated by injecting a 20% cocaine solution into the middle ear or by putting chloroform into the external auditory canal, the leg-slackening and leg-bracing reactions of the left legs showed severe defects, i.e., the leg-bracing reactions outwards disappeared while the leg-slackening reactions were either absent or so much delayed that they become totally ineffective. Moreover, the animals always kept the left legs in an adducted position (even when the head was passively placed symmetrically to the trunk) which also promoted falling into a lateral position to the left.   Magnus and De Kleyn had already noticed that, after a left-sided labyrinthectomy, the animals easily stumbled to the left. Left-sided labyrinthless rabbits in the stage of rolling movements sometimes further showed the following manifestation. When the animal was placed into a lateral position to the right and the head fixed in a normal posture (bitemporal axis horizontal), one almost always saw that the trunk was not only raised, as in intact animals, but would further rotate into a lateral position to the left (thus contrary to the influence of neck righting reflexes).   After a unilateral extirpation of the labyrinth, the trunk will sometimes perform half a rolling movement by way of the belly, even if the rotation of the head is prevented. Thus, this half of the rolling movement is not based on the combination of head rotation and progression movements, but the falling of the trunk from a ventral to a lateral position to the left is exclusively produced, in this case, by the impairment of those reactions which normally prevent falling, such as leg-bracing reactions.   In my opinion, the rolling movements after the unilateral extirpation of labyrinth can be interpreted as progression movements of an animal whose body is helically rotated and in which certain balance reactions are impaired unilaterally, on the side of the extirpated labyrinth. In other words, as progression movements of an animal in which the tonic labyrinthine reflexes and the labyrinthine righting reflexes are unilaterally eliminated, the leg-bracing and leg-slackening reactions and the body righting reflexes acting on the body are impaired.   The unilateral labyrinthectomy thus not only causes the unilateral elimination of labyrinthine functions but also temporarily an impairment of reflexes which have nothing to do with the vestibular system.[1]   These impairments are probably based on shock.   The observation by Magnus and De Kleyn that the rolling movements are reduced after careful operations in which the eighth nerve is left untouched as much as possible during removal of the vestibulae permits another explanation than that given by these investigators. A careful operation will surely cause a less severe shock to the brainstem centers and less severe impairments of the righting reflexes than an operation in which the eighth nerve is maltreated. Of course, with this assertion it still cannot be said that stimulations from injury do not participate in these rolling movements.

---

(1)   Although the body righting reflexes on the body are brought about by way of the same mesencephalic nuclei as the eliminated labyrinthine righting reflexes.

We have dealt so thoroughly with the different observations concerning the rolling movements of unilaterally labyrinthectomized animals because following the unilateral extirpation of cerebellum, in addition to a temporary head rotation and temporary and inconstant impairments of labyrinthine functions (Chapter XV) and leg-slackening and leg-bracing reactions as well as the body righting reflexes acting on the body, as we shall see presently, show impairments which participate in the inconstant occurring rolling movements.

Following a unilateral extirpation of the cerebellum (in the first 1-3 days), the animals *constantly* show the following impairments.

1. An abnormal distribution of tonus in the muscles of the extremities, neck and the spinal column on the side ipsilateral to the extirpation. The ipsilateral legs show an increased tendency to extension (Fig. 102) and the muscles of the neck and the spinal column on the side of the extirpation show a particularly strong tension and, when the animals are held in the air in a ventral position, cause a turning of the head towards the side of the operation and a concavity of the back towards this side.

    The increased tendency of the ipsilateral legs to extension can be observed when the animal is suspended in the air, whether head upwards or downwards, in dorsal, ventral or lateral position, and in a dorsal position on a supporting surface while lying on the side of the operation; it is, however, sometimes absent while lying on the nonoperated side.

2. Absence of preparation for standing and of magnet reactions in the ipsilateral legs.

3. Impairments in supporting reactions in the ipsilateral legs, similar to impairments after decerebration.

4. Absence of leg-slackening, leg-bracing and rocking reactions in the ipsilateral legs.

5. Absence of body righting reflexes acting on the body while lying on the side of extirpation.

Appearing inconsistently are:

1. A rotation of the head with the vertex of the cranium towards the side of extirpation.

After extirpation of the right half of the cerebellum, some animals, when suspended head downwards (with blindfolded eyes), showed a strong rotation to the right, and some others a slight one, while in others none or even a slight rotation to the left could be observed. It was also observed that, after waking up from narcosis, the head showed a turning to the left and an hour later a turning to the right. *After awakening, however, all animals constantly and continuously showed an increased tendency to turn the head towards the side of the operation when they were held in a ventral position in the air with closed eyes.*

2. Impairments of the labyrinthine righting reflexes.

In several cases, these reflexes were eliminated immediately following extirpation in both lateral positions as well as in the dorsal position; sometimes, this was the case only in one lateral position (after extirpation of the right half of the cerebellum from the right-sided lateral position). Sometimes, they were again present in both lateral positions soon after awakening. Impairments of the labyrinthine righting reflexes as well as of other inconstantly appearing labyrinthine manifestations, like deviation of the eyeballs, and nystagmus appeared only temporarily and soon disappeared.

3. Impairments of the body righting reflexes acting on the body while lying on the side of the intact half of the cerebellum.

From this lateral position, too, the body righting reflexes on the body sometimes failed to appear in the first days. Thus, from none of the lateral positions was the trunk raised into a ventral position if the head was held fixed. When the body righting reflexes on the body while lying on the side of the preserved half of the cerebellum were again present, the increased tendency to an extension posture of the ipsilateral legs could usually no longer be observed in this lateral position.

4. Impairments of magnet, positive supporting, leg-slackening, leg-bracing and rocking reactions in the limbs *opposite* to extirpation.

The inconstancy of these defects was reason for the great differences in symptoms appearing immediately after the unilateral extirpation of the cerebellum. Thus, the dog Peter, after extirpation of the right half of the cerebellum, was already able on the day of extirpation to right the head and to bring the anterior part of the body into a ventral position. The posterior part of the body, however, remained in a lateral position to the right and returned to that position instantly when placed passively in a ventral or left latral position. The animal showed no distinct rotation in basic posture and, in accordance with this, no rolling movements. It was not able to raise the posterior part of the body from the lateral position to the right, because of the absence of body righting reflexes on the body in this lateral position, and by the absence of magnet, supporting, leg-bracing and leg-slackening reactions in the right legs. Conversely, as we have seen, the posterior part of the body rolled immediately from the lateral position to the left into the same position to the right by way of the belly. This reaction resulted from the good functioning of the body righting reflexes from the left lateral position, while the following fall from ventral to a lateral position to the right was caused by the absence of the above-mentioned reactions in the right limbs. Similar manifestations were shown by the dog Fox following ablation of the right half of cerebellum. In this animal, too, the trunk could not tolerate the lateral position to the left and changed over instantly to a lateral position to the right, even when the head was fixed in the left lateral position. Here, too, this was brought about by the fact that the body righting reflexes as well as the balance reactions of the extremities were lively in the left legs while all these reactions were absent in the right legs (see Figs. 185 and 187).

Whereas the dogs Fox and Peter lay with the posterior part of the body or whole body in the lateral position to the right, the dog Tip in the earliest days after extirpation of the right half of the cerebellum was always found with the trunk in a lateral position to the left. In doing so, he always held the head flat with the lower jaw on the floor of the cage, i.e., rotated 90° to the right in relation to the trunk. This animal, contrary to the two former, showed rolling movements and a distinct basic rotation suspended head downwards, impairments of the labyrinthine righting reflexes, especially from lateral position to the right, an absence of body righting reflexes acting on the body from *both* lateral positions and, in addition to the absence of reactions of the extremities in the right legs, also a considerable impairment of these reactions in the left legs. Because of the basic rotation paired with the absence of labyrinthine righting reflexes and body righting reflexes from the lateral position to the right, the animal was not able to remain lying on its right side. When the animal was passively placed on this side, the head immediately rotated more and more to the right and, by so doing, went at first into a dorsal position (vertex of cranium down), followed by the anterior part of the body and finally, when the head continued to rotate to the right, the whole trunk rolled over the back to a lateral position to the left. Usually this was followed by the second half of the rolling movement since, due to the absence of leg-bracing reactions in

550

the right legs, the continued rotation of the head was followed by a
rotation of the trunk by way of the belly. When the animal had performed several of these complete rolling movements, it finally always
remained in a lateral position to the left and the last half of the
rolling movement, by way of the belly, failed to appear. The animal
succeeded in avoiding further movement by strongly pressing its lower
jaw to the floor of the cage which prevented a further rotation of the
head. Since in this animal, contrary to the dogs Peter and Rollmops,
the body righting reflexes were also absent from the left lateral position,
the trunk would remain lying on the left side.

Probably several other factors participated in this persistance of
the left lateral position after cessation of the rolling movement. Thus,
among others, the lateral position to the left is promoted by the turning
of the head to the right which is almost always shown by these animals
(Fig. 226). It is also possible to prevent a rolling over from the ventral
position by an extension of the limbs. Besides, the rotation of the
head become fatigued when the animals are exhausted by vigorous
rolling movements and, in this case, the unilateral body righting reflexes and reflexes of the extremities are also inhibited. It finally appeared that the head rotation was less strong in certain positions of
the trunk. Thus, the animals, when held suspended in the air head
downwards, sometimes showed a strong rotation of the head on the
trunk, together with peculiar, restless rotating movements of the head
and the anterior part of body, while in a ventral position in the air
the head was kept quite still and only showed a turning towards the
side of extirpation.

When, however, the animal was frightened, for instance by a loud
noise, a sudden lifting of the head associated with limb movements
appeared and it instantly rolled again over its belly to a lateral position to the right, followed by the other half of the rolling movement
over the back sometimes proceeding to several complete rolling movements.

The dog Rollmops, who also had the right half of the cerebellum
extirpated, showed other manifestations. This animal, as well as the
dog Tip, indeed showed rolling movements and these, too, were
directed towards the right, but these started in the first two hours
after extirpation with a rolling from a right-sided into a left-sided lateral
position by way of the back and the animal, after cessation of the rolling movement, remained in a lateral position to the *right*. When

brought into a lateral position to the left, it could not remain there and instantly rolled over its belly to the right side. This manifestation lasted for only two hours and, unfortunately, was not analyzed more closely. Later, the animal showed the same rolling movements as the dog Tip; the rolling movements now started on the left side and were also terminated in this position, while now the animal could not tolerate the lateral position to the right.

Thus, we see that, following a unilateral extirpation of the cerebellum, the symptoms may vary greatly, sometimes rolling movements appear and sometimes they are absent; they become evident only when, in addition to a distinct head rotation, the extirpation also caused impairments of body righting reflexes acting on the body and of reactions of the extremities, such as the leg-bracing reactions.

The symptoms changed considerably within the first week, and at the end of the week the various animals showed a much more comparable behavior. This was caused by the fact that the inconstantly appearing impairments were usually the first to recede. The animals no longer showed any rolling movements and were again able to right the head and to keep it freely raised in the air; the righting reflexes acting on the head were again present in both lateral positions. The right-sided decerebellate animals again showed body righting reflexes on the body from the left lateral position and magnet, supporting, leg-slackening and leg-bracing reactions could again be observed in the left legs; several of these reactions were then observed in a strikingly abrupt manner. At this stage, however, the animals were not yet able to stand freely. When set down on their legs and then released, they instantly fell into a lateral position on the right side. This was still the case even when, by placing the animals solely on their right legs, it could be shown that these limbs had supporting tonus sufficient to carry the trunk (see Chapter VI).

The right legs, when not under static stress, still showed an increased response to stretch and an increased resistance to passive flexion.

Falling to the right side is the result of the following defects:

1. That in setting down the animals on all fours, the left legs are extended and are strongly maintained in that position, but at the same time the right legs have decreased extension and supporting. Sometimes, while the left limbs are supporting, the right ones are drawn into a partially flexed posture (Fig. 187, No. 1 and 2).

2. That the left legs, in extending, at the same time brace strongly outwards forcing the trunk to the right, and the right legs go more and more into adduction. By this, the supporting tonus of the right legs further decreases (Fig. 187, No. 3).

3. That this passive adduction of the right legs produces neither a leg-bracing nor a hopping reaction outwards of these limbs and thus fail to prevent falling (Fig. 187).

4. That the hopping (rocking) reaction of the right legs is impaired so that a gradually increasing outwards bracing (abduction) of the left legs does not produce extension and increase in supporting tonus in the right legs.

It is noteworthy that in this stage, too, the posterior part of the body always turns over from the left-side to the right-side lateral position by way of the belly. When the animals are placed in a lateral position to the left, they raise their heads; then at first the anterior part of the body and then the whole trunk rotates into a ventral position. In doing so, as soon as the soles of the left feet touch the supporting surface, these legs are suddenly and sharply extended and braced outwards by which movement the animals are thrown into a lateral position to the right. From this position, the head and anterior part of the body indeed rise (due to labyrinthine and neck righting reflexes); the posterior part of the body, however, remains lying on the right side due to the absence of the body righting reflexes acting on the body from the lateral position to the right. Even when the head is fixed in the lateral position when the animal is lying on its left side, the trunk, on turning from the left into a ventral position (due to body righting reflexes), turns further to the right side by the abrupt extension and outwards bracing of its left hindleg.

The investigation of right-sided decerebellate animals shows that at this stage the following reactions are again present:

1. the labyrinthine righting reflexes from both lateral positions;

2. the body righting reflexes from the left lateral position;

3. the magnet, supporting, leg-bracing and leg-slackening reactions and the preparation for standing of the left legs.

Conversely, the following are absent or strongly impaired:

1. the body righting reflexes acting on the body from the lateral position to the right;

2. the leg-bracing, leg-slackening and rocking reactions and the readiness to stand on the right legs.

In addition, the magnet and supporting reactions of the right legs usually show distinct impairment, [1] while static stress on the left legs still causes abnormally strong decrease in stretch and supporting tonus in the right legs. At this stage, the animals start to crawl and, in doing so, however, the posterior part of the body remains in a lateral position to the right.

Gradually, usually in the second week, the body righting reflexes from the right lateral position return and also the magnet and supporting reactions in the right legs grow stronger. The animals are now able to bring the trunk into a ventral position from both lateral positions and also in crawling the posterior part of the body occupies this posture but still falls down each time on its right side due to the strong lateral bracing of the left legs. Even now, the animals are unable to stand without being supported, though the hopping, leg-bracing and leg-slackening reactions in the right legs have returned because the leg-slackening and hopping reactions are carried out too late to prevent falling.

Free standing and running is only possible when the retarded appearance of leg-slackening and hopping reactions has so much improved that these reactions occur just at the right moment to prevent falling.

After impairments have considerably receded and the animals are set down on a supporting surface from a ventral position in the air, the following manifestations are seen: the left legs brace outwards especially strongly and move the trunk towards the right. In so doing, lateral leg-slackening reactions of the right legs are produced, and these legs are raised and hop sideways so that the animal finally stands strongly, wide based, with all four legs in an abducted posture.

In the stage of permanent impairment, the animals are able to stand freely, to run and even to carry on their backs a sandbag equal to the weight of their own body (Fig. 105). The various reactions participating in the righting function and in standing are again present on both sides and when one sets down the animals alternately on each one of the four legs, it is seen that the force of supporting tonus is equally strong on both sides (see Chapter VI).

---

(1) On alternating setting down on the right and the left legs, the former still show an evidently weaker supporting tonus (PP. 157-158)

554

The various reactions, however, still show many defects. The leg-slackening reactions are carried out abnormally by the legs ipsilateral to the extirpation. *First of all, they are still retarded in appearance, i.e., only after an abnormally large alteration in position and, secondly, they are performed hypermetrically; the legs are raised abnormally high and transferred abnormally far from the intermediate position so that the hopping steps are almost twice as large as in intact dogs* (Fig. 203).

The leg-bracing reactions remain unequally strong on both sides, i.e., in the contralateral legs distinctly stronger than in the ipsilateral ones. This difference is always more evident in the hindlegs. *One gets the impression that the leg-bracing reaction in the contralateral legs is increased and in the ipsilateral ones reduced.*[1] These impairments of the leg-bracing reactions cause the following manifestations: First, that the animals, when passively pulled sideways, offer a less strong resistance to traction towards the side of extirpation than towards the other side (Fig. 195, No. 1 and 4). Second, when the animals are placed on a board that is raised or lowered on the right or the left side of the animal, a more prompt adaptation of the leg position occurs when the board is raised from the side of extirpation (or lowered from the intact side) than on raising the board from the side of the intact half of the cerebellum (or lowering on the side of extirpation; Fig. 195, No. 3 and 6). Third, these impairments cause in running movement a deviation of the trunk, particularly the posterior part of the body, towards the side of the operation (Fig. 188). When one observes unilaterally decerebellate animals from behind while they are running, one will clearly see how the contralateral hindleg strongly braces outwards each time it is set down and moves the posterior part of the body towards the side of extirpation. Due to this displacement, the raised ipsilateral hindleg is set down lateral to the starting point. Since, when the ipsilateral hindleg is put down, the posterior part of the body does not return to its original orientation, the posterior part of the body finally moves to an inclination almost at right angles to the direction of running (Fig. 188). Probably the asymmetrical contraction of the vertebral muscles also participates in this process.

---

(1) With the leg-bracing reactions outwards, firstly an abduction of the leg to the trunk and, secondly, an increase in fixation of the extended position occurs. Sometimes, especially after the unilateral extirpation of the *cerebrum*, one can see that the first component, abduction, appears especially prominent in the contralateral legs; the fixation in extended posture, however, is not as strong as in the leg-bracing reaction of the legs on the other side.

When the animals try to run away quickly in leaps and bounds, one can sometimes observe that the contralateral legs brace outwards so abruptly and forcibly that falling down can only be prevented by hopping steps of the ipsilateral legs. Sometimes, even in the stage of permanent impairment, the hopping occurs so late that the posterior part of the body falls to the side of extirpation because the strongly adducted hindleg on this side gives way and comes to lie below the belly. Moreover, the especially strong contraction of the abductors on setting down the contralateral hindleg often causes an obvious and abnormally strong rotation of the pelvis towards the side of the intact half of cerebellum, by which the ipsilateral half of the pelvis and the hindleg are raised abnormally high. The asymmetry of the leg-bracing reaction appears particularly evident when the animals are stood on the hindlegs alone and alternately moved to the right and to the left. As was discussed in detail in Chapter XIV, p. 445, in pulling the animal towards the side of extirpation the asymmetry causes a much stronger resistance and the legs are displaced in quite another rhythm than in a movement to the other side. *In moving towards the sides of extirpation the hindlegs make running movements sideways and on moving towards the intact side the hindleg contralateral to the extirpation will at first brace with remarkable strength and then make a hop.* By means of the remarkably strong leg-slackening and hopping reactions in the contralateral side, the hindlegs, on being pulled towards the side of the intact half of cerebellum, are displaced in the same way as in totally decerebellate dogs (and also by normal but very obstinate dogs),but this time by pulling towards both sides.

In running thus, the ipsilateral legs, when raised, are abnormally strongly flexed; the hindlegs, at the same time, moved abnormally outwards and the forelegs abnormally inwards.

At the stage of permanent impairment, the legs on both sides again show readiness to stand on stimulations from the body surface as well as on optical stimulation. The magnet and positive supporting reactions, too, are again very active on both sides. It is worth mentioning that, although in intact and in decorticate dogs in a dorsal position no distinct magnet and supporting reactions can be observed, *these appear particularly active in unilaterally decerebellate dogs even in a dorsal position, in the ipsilateral as well as in the contralateral legs.* In a dorsal position with the muzzle pointing upwards, both hindlegs show a lively extension on touching the soles of the feet as well as on static stress and considerable tonic fixation in extended posture. (The behavior of the strength of supporting tonus in the ipsi- and contralateral hindlegs in a dorsal position has been discussed extensively in Chapter VI.

As to the distribution of muscle tonus in the late stage of recovery, it appears that the tendency to extension of the ipsilateral legs has usually quite disappeared. Placed on their backs, the animals now hold the four limbs, as a rule, totally flexed. On passive movement of the legs without static stress, no distinct difference in resistance between right and left legs can be felt. Sometimes, however, in the to and fro shaking of the animals, a slightly increased extensor tendency in the ipsilateral legs can be detected. The patellar reflexes show no distinct difference.

Alterations in position of the head to the trunk or in space no longer caused distinct alterations in position of the legs without static stress; thus, the legs show no typical tonic neck or labyrinthine reflexes (while these, as we have seen, were clearly evident in the first stage after extirpation). *But a distinct impairment in distribution of muscle tonus appears when the blindfolded animals are held in a ventral position in the air, supported from the ventral side. In this case, the head always shows a turning towards the side of the operation while in the dorsal spine there is a concavity towards this side. Thus, the tension in the cervical and dorsal spine is in this position still stronger on the side of extirpation than on the other side.* When the animal is suspended head upwards or downwards, the asymmetrical contraction of the vertebral muscles can also be seen.

As discussed in Chapters VII and VIII, alterations in position of the head to trunk in the dorsal position of the animal still caused distinct alterations of tonus when the legs are under static stress. This is also the case in a standing posture. The decrease and increase in supporting tonus of the hindlegs on raising and lowering of the head are strongly pronounced and can also be sometimes observed in running.

Thus, in unilaterally decerebellate animals, several reactions which participate in standing and running show distinct impairment, as is evident from the preceding chapters. It is almost generally supposed that each half of the cerebellum exerts an influence only on the muscles of the ipsilateral half of the body. This supposition is not refuted by the observations we have mentioned. *Our observations do not agree, however, with the assertion that unilateral extirpation of the cerebellum causes impairment only in the side of the operation.* It is true that in my animals the impairments on the side of extirpation are much more significant and more easily observed. However, certain reactions of the contralateral legs, too, show deviations not only in the initial stage but also in the stage of permanent impairment. These must probably be ascribed to associated lesion of the preserved half of the cerebellum since a unilateral extirpation of the cerebellum is impossible to perform without any kind of colesion of the other half of the cerebellum. Besides, secondary haemorrhages, thrombus in the severed vessels, scarring processes, etc., can also have a detrimental effect on the preserved half of the cerebellum. Further, it could also be possible that the functional capacity of the preserved half of the cerebellum is injured by the absence of input from the other half.

But even with a normal functioning of the remaining half of the cerebellum, it is, a priori, not to be expected that the legs contralateral to the extirpation show completely normal reactions. As we have seen, several of the reactions of these legs are caused by stimulations emanating from the opposite side, and it is to be expected that the abnormally functioning ipsilateral limbs exert an abnormal influence on the contralateral ones.

Manifestations after the unilateral extirpation of the cerebrum prove that this supposition is totally justified. After removal of one half of the cerebrum, only the legs contralateral to the extirpation show distinct impairments (absence of preparation for standing, weakness of supporting tonus), which the ipsilateral legs do not. When animals are set down on the healthy ipsilateral hindleg and are then moved to and for, rocking reactions promptly appear and the affected, contralateral hindleg shows lively flexion-extension movements. Thus, the leg contralateral to the cerebral lesion reacts briskly to stimulations emanating from the unaffected leg. Conversely, rocking on the contralateral hindleg causes no, or only weak, reactions in the unaffected ipsilateral leg. The affected limb emits no, or only weak, stimulations; therefore, the healthy ipsilateral limb shows no, or only weak, reactions. In accordance with this fact, the occurrence of hopping reactions does not take place in the same manner on both sides after the unilateral extirpation of the cerebellum.

Finally, we want to point to the fact that the impairments shown by decerebellate animals are not all of the same nature. In some reactions, as, for instance, in lifting the ipsilateral legs while running, the *execution is exaggerated and hypermetric.* In others, as, for instance, leg-slackening reactions in the ipsilateral legs, a distinct hypermetria as well as *a distinctly retarded appearance* can be observed, while finally in others, as for instance the leg-bracing reactions of the contralateral legs, *a specially lively and uninhibited appearance* can be seen. The latter is also true for magnet and supporting reactions in the dorsal position of the animal. While magnet and supporting reactions in decorticate and usually also in intact animals in a dorsal position are totally inhibited, this is not the case in unilaterally decerebellate animals. Thus, we see the peculiar combination of retarded occurrence, uninhibited occurrence and exaggerated execution. Add to this an asymmetrical distribution of muscle tonus which is specially apparent when the animals are held in the air in a ventral position.

Thus, the most striking deviations in the stage of permanent impairment are:

1. the excessive execution of almost all movements of the extremities on the side of extirpation;

2. the retarded occurrence and the excessive execution of leg-slackening and hopping reactions of the ipsilateral extremities;

3. the uninhibited appearance of magnet and supporting reactions in both hindlegs in the dorsal position;

4. the particularly pronounced execution of leg-bracing reactions in the legs opposite to extirpation;

5. the weakening of leg-bracing reactions in the legs on the side of extirpation;

6. the stronger contraction of vertebral muscles on the side of extirpation in the various suspended postures in the air.

XIX. DEFECTS IN STANDING AND THE SO-CALLED "CERE-BELLAR ATAXIA" IN DECEREBELLATE ANIMALS

We have already described show, following the total extirpation of the cerebellum, some animals show rigidity while others show a pronounced hypotonia.

In *rigid* animals, the supporting reactions show the same impairments as in the decerebrate animal. Moreover, the behavior of the animals is similar to that after typical Sherrington's intercollicular decerebration. Rigid decerebellate animals show a similar distribution of muscle tonus, a pronounced extensor rigidity and a total absence of righting functions. The animals lie all the time in a lateral position and are not even able to raise their head. This is particularly striking since these animals, in opposition to decerebrate animals, can see and react on optical impulses with movements of the eyes, wagging of the tail, etc. They are, however, not able to right the head with the help of these impulses; the neck muscles do not react on optical stimulation with the contractions necessary for the righting of the head. Similarly, the muscles of the extremities fail to react to visual impulses with movement of the limbs, i.e., the optical readiness to stand is impaired (Fig. 14, No. 1). When the animals are propped up on their legs, these do not give way but are able, as in decerebrate animals, to carry the trunk. On altering the position of the head, the legs show typical tonic neck and labyrinthine reflexes. Sometimes, these reflexes can only be observed a few days after the extirpation when the rigidity has somewhat subsided ; immediately after extirpation, the extension rigidity of the legs is maximally strong in all positions of the head in these animals.

The *flaccid* animals show a totally different distribution of muscle tonus. They lie in their cages with flaccid, flexed legs. The flaccid neck muscles are usually not able to keep the head freely erect when the animal is suspended in the air. The head is, however, well rotated into a normal position as the animal lies, usually even on the day of the operation, so that it does not lie on its side as in rigid animals, but rests with the lower jaw flat on the ground. When the animals are held in the air in a lateral position, they also right the head; the labyrinthine righting reflexes are already usually distinctly evident on the day of the operation. Alterations in position of the head do not cause any perceptible tonic neck or labyrinthine reflexes. Static stress produces no distinct alterations in contraction of the muscles of the extremities. When the animal is set down in a standing posture, the legs instantly collapse under the weight of the trunk (Fig. 262).

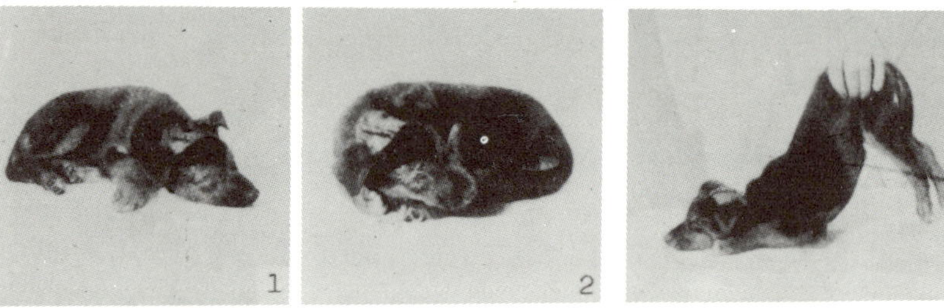

Fig. 262. Dog Pim on the second day after extirpation of the cerebellum. 1-2. Lying calmly, the head is rotated actively into a normal position, the flaccid legs lie flexed underneath the trunk. 3. On the attempt to set down the animal on its forelegs, these give way immediately.

Thus, greatly differing symptoms appear after the total extirpation of the cerebellum. However, rigid and flaccid animals correspond in that in both the body righting reflexes acting on the body, the magnet, leg-bracing and hopping reactions, the preparation for standing and also the rocking reactions are absent, while in both the neck righting reflexes are present.

Through the presence of cervical righting reflexes, the flaccid animals, already on the first or second day after extirpation, following the righting of the head, are also able to rotate the anterior part of the body into a ventral position (Fig. 262, No. 1), while the posterior part of the body remains in a lateral position due to the absence of body righting reflexes.

In flaccid animals, the readiness to stand, magnet and supporting reactions are the first to return; the strength of supporting tonus quickly increases so that first the forelegs and afterwards the hindlegs are able to carry the trunk (Figs. 65 and 59).

The return of supporting reactions occurs gradually and sometimes the following succession can be observed. At first the legs show a distinct alteration in contraction and increase in posture only when they are passively extended before applying static stress. In this case, the passively extended legs are able to carry the trunk and offer a distinct resistance to pressure on the soles of the feet. When, however, static stress is applied to the legs while they are still flexed, they do not go into extended posture, and they do not raise the trunk when the animals are set down on the supporting surface. In the following days, one sees that repeated passive flexion movements produce an extension of the flexed legs, and the same soon happens on simple static stress as well as on touching the soles (Fig. 65). When the soles of a flexed foot are now set down on a supporting surface, the legs extend immediately and lift the trunk. In rigid animals, the extension rigidity gradually decreases till finally the legs show a flexed posture in lateral and dorsal positions. In this case, alterations in position of the head no longer produce distinct tonic neck and labyrinthine reflexes. As the extensor rigidity disappears, the labyrinthine righting reflexes return and the animals are able to rotate at first the head and soon afterwards also the anterior part of the body into a normal posture. As soon as the limbs are less rigid, typical magnet and supporting reactions also appear.

Due to these alterations, the clinical pictures of rigid and flaccid animals resemble each other more and more. While, however, the flaccid animals already make this change within the first week, in the rigid ones this is usually the case only in the second to fourth week. At this stage, the animals begin to crawl; they also sometimes try to raise themselves on their legs, but fall down instantly since the leg-bracing and leg-slackening and hopping reactions as well as the body righting reflexes on the body are still absent. In crawling, the animals keep the legs totally flexed and, from this and by constantly falling, they give the impression of atonia. When, however, the animals are set down on their legs one can feel in pressing on their shoulders or pelvis that a considerable supporting tonus is already present and that the fact that the animals cannot yet stand and run freely cannot be ascribed to the condition of the supporting tonus. Due to the repeated falling down, in which they often hit the ground with an audible impact, they totally cease any attempt to rise. In this case, the extension of the legs is inhibited by cerebral influences, as shown by the following observations. Several animals which both in crawling and when held in the air gave an impression of atonia (Fig. 263, No. 1), appearing unable to stand, rose immediately on their legs when only a fold of skin was seized to guard them against falling down. In this case, they even showed an excessive extension and fixation while, on releasing the fold, the limbs again went into flexion (Fig. 263, No. 2). Others rose when they could lean on a wall on one side. They also dared to rise sooner when they were on soft ground, for instance on a lawn, than on a hard and smooth

562

supporting surface. At first, the posterior part of the body remained in a lateral position while crawling due to the absence of body righting reflexes. However, these reflexes also gradually returned and the posterior part of the body is then brought into a ventral position. Since the body righting reflexes were still weak and the leg-bracing reactions still absent, the neck righting reflexes were unopposed and every sudden turn or rotation of the head caused a tumble. A turning to the right causes the posterior part of the body to fall to the left and a rotation of the head to the right, however, a fall to a lateral position to the right. The rolling over of the posterior part of the body is further promoted by the abrupt execution of movements of the head and extremities.

Fig. 263. Bitch Suzanne, 2 weeks after extirpation of the cerebellum. The animal is not yet able to stand freely and in crawling she gives the impression that she lacks the ability to rise. 1-2. Even when held in the air in a ventral position, the animal makes an impression of atonia and asthenia. 3. When, however, the animal lying on the ground is seized by a fold of skin, she will rise and the legs go into a very exaggerated extensor posture while the back becomes curved and rigid.

When the animal is now brought into a standing posture and then released, it starts to wobble and rock, and every movement of the head causes a strong associated movement of the trunk which is not stabilized by leg-bracing reactions so that finally one or both legs of one side get into a strong adduction which does not yet cause a hopping reaction but only a decrease in supporting tonus. The legs collapse, the animal falls on its side. Sometimes, only the posterior part of the body falls down and, in doing so, the hindlegs collapse in an adduction posture and lie beneath the belly.

Gradually, in addition to the body righting reflexes, the leg-bracing and hopping reactions also return. The animals now attempt to rise more and more on their legs and run. Now, too, they fall continually on their side or tumble head forwards owing to the following factors:

1. the especially abrupt and hypermetric execution of all movements,

2. the exaggerated extension posture of the legs and the abnormal dorsal curvature of the back, causing a much more labile balance,

3. the abnormally strong postural fixation of legs and spinal column,

4. the abnormally large and abruptly performed alterations of supporting tonus and the exaggerated associated movement of the pelvis in following the active movements of the head,

5. through motor unrest, the "unrestrained movements" which appear in free standing,

6. the absence or impairment of leg-bracing reactions and

7. the impairments of hopping reactions.

The hopping reactions are now indeed present and are carried out appropriately but they still are too delayed to prevent falling.

In several animals, it was observed at this stage that, when the animal is set down on the hindlegs, the anterior part of the body is suddenly lifted from the ground and the animal may sometimes be thrown into a backward somersault. This is the case when the hindlegs are set down too far forwards or slide forwards. The cause of this phenomenon is that in decerebellate animals static stress produces abnormally strong contractions of the psoas (Fig. 31) which, in addition, become more abrupt when the hindlegs are moved forwards (Fig. 60). These exaggerated contractions by which the anterior part of the body is lifted from the ground and thrown backwards can be particularly observed when the animals gallop or attempt to jump over an object, for instance the rim of the cage.

Thus, at this stage, we see the animals in a continual irregular sequence of rising, falling, hitting the ground, stumbling, plunging, or tumbling forwards, backwards and sideways. This violent behavior is most evident in nervous and wild dogs which cannot desist from continually rising and running. Gradually, the leg-bracing reactions grow stronger and hopping reactions less retarded so that tumbling is less frequent. The animals indeed still stumble often, which causes them to change direction frequently when running and finally, after several hops, they either fall down again or recover their balance. They cannot yet take a straight course to a goal, but eventually reach it by maintaining an unstable equilibrium with frequent hops and jerks backwards, forwards and sideways.

The condition now gradually ameliorates; the leg-bracing reactions are so vigorous and the hopping reactions are so little delayed that the animals no longer fall down but usually react to each threatened fall with *one* appropriately performed hopping step.

The remaining defects no longer improve. The animals have reached the stage of permanent impairment (3-6 months after extirpation) and their characteristic symptoms are usually known under the name of "cerebellar ataxia."

The Manifestations in the Stage of Permanent Impairment

At this stage, the clinical picture of the symptoms is exactly the same in the various dogs and cats. Although the symptoms, on the whole, are the same in dogs and cats, there are yet several differences, as we shall see later, so that the clinical manifestations are even more constant in animals of the same species.

At this stage, the animals show a prompt (see Chapter IV) preparation for standing (placing) on optical impulses as well as on stimulations from the body surface. Also, in corrections of abnormal limb positions distinct deviations can no longer be observed.

Conversely, a number of other reactions show distinct and permanent impairment.

A. *Impairments of Magnet and Supporting Reactions*

A touching of the soles of the feet as well as static stress on the limb both cause *an abnormally strong extension of the limb* with a strong maintenance of extended position and also produce *an abnormal contraction of certain vertebral muscles* (Figs. 27, 31, 33, 34, 41, 60). These reactions are the reason that decerebellate animals stand with exaggeratedly extended legs and an abnormal dorsal convexity of the spinal column (Figs. 1 and 44).

Moreover, since the posture of the extremities, neck and back is constantly strongly maintained, decerebellate animals while standing and running give an impression of rigidity and hypertonia. Their posture resembles, superficially, that of decerebrate animals when they are set down on their legs (Fig. 264).

Fig. 264. 1. A decerebrate dog set down on its legs. 2. Dog Wolf, two years after extirpation of the cerebellum, in a standing posture.

In addition, the exaggerated extension posture of the legs and the dorsal convexity of the back cause the center of gravity of the animals to be situated abnormally high above the supporting surface. A slight shifting of the trunk is therefore sufficient to transfer the center of gravity beyond the supporting base. Thus, the decerebellate animal in a standing posture has a more labile balance and to preserve this balance requires an earlier transfer of the paws and an earlier occurrence of hopping reactions than in the intact animal. Finally, the abnormal contraction of the vertebral muscles probably also participates in the excessive associated movement of the pelvis with alterations in position of the head.

Thus, in decerebellate animals, contact with the soles of the feet as well as alterations in position of the distal phalanges (stretch reflexes first - fourth order) cause an abnormally intense contraction of extensors of the extremities and of the vertebral muscles. The abnormal contractions take place on exteroceptive as well as on proprioceptive stimulations.

As we have seen in Chapter V, in the magnet and supporting reactions of decerebellate animals not only the extensors but, as in intact animals, the other muscles of the extremities also take part in the postural contractions. Although in the freely unburdened standing decerebellate dog the continual strong fixation of the various joints is particularly striking and produces the impression of being intense, the maximum contraction does not give a remarkably high ability to withstand burdening the animal. It is true that the strength of supporting tonus is not inferior to that of intact animals (Chapter VI), which also are capable of carrying loads of more than their bodyweight;

but a considerable further increase in strength of supporting tonus in decerebellates was not established.

In addition to what has just been discussed, the investigation of magnet and supporting reactions showed other changes. While in a dorsal position with the muzzle pointing upwards in decorticate, and also in most intact dogs, neither a touching of the soles nor a static stress causes distinct reactions, *yet in decerebellate animals even in this position lively and extensive extension of the legs with considerable maintenance of extended posture occurs* (Figs. 28, 67) which continues as long as the touching or the static stress continues (45 minutes and longer). *Thus, a dorsal position produces in decerebellate animals less inhibition of magnet and supporting reactions than in decorticate and intact animals.*

An alteration in the latency or a retarded or premature appearance through other causes could not be established in the magnet and supporting reactions of decerebellate animals which, of course, does not mean that its appearance takes place as normally timed. Due to experimental conditions (cerebral inhibitions), the speed of appearance of these reactions in intact animals varies to such an extent that a normal latency period could not be measured.

The fact that negative supporting reactions in decerebellate animals cannot only cause a relaxation of the limb but sometimes a flexion and that flexion of the distal phalanges sometimes produces a lively raising and retraction of the limb to a flexed position has been discussed in detail in Chapter V.

B. *The Impairments of Leg-Bracing Reactions*

In decerebellate animals, the legs of both sides show strong inclination to brace outwards; the forelegs, in addition, forwards and the hindlegs, backwards.

When a decerebellate dog is set down on a supporting surface from a ventral position in the air, as soon as it is released it will transfer its legs in such a way that the forelegs are directed outwards and the hindlegs backwards and outwards (Fig. 44). If the attempt is made to shift the animal *instantly, especially strong leg-bracing reactions* appear and when it is pulled, for instance somewhat to the right, it does not make running movements directed laterally, but like an obstinate intact dog it will perform hopping steps sideways with the right legs.

Thus, the totally decerebellate dog when pulled sideways shows the same leg reactions as a unilaterally decerebellate dog when pulled towards the side of the intact half of the cerebellum.

The abnormally strong tendency to an outwards bracing of the legs can also be observed while running. The setting down of a fore- or hindleg is always followed by a strong outwards bracing of these legs by which the trunk is alternately pushed abnormally far to the right or to the left (Fig. 265) and the animal will run in a zigzag line.   When seen from behind, the gait produces an impression of ice-skating, owing to the alternating setting down and the following exaggerated outwards and backwards bracing of the left and the right hindleg (Fig. 265, A 8 B 6, B 9 — 11).

Since cerebellate animals, just like intact ones, usually set down the right hindleg and the left foreleg simultaneously, the posterior part of the body, in doing so, is displaced abnormally far to the left and simultaneously the anterior part of the body shifts abnormally to the right so that the trunk makes a strong rotating movement round the dorsoventral axis while running.

Thus, decerebellate animals show an increased tendency to an outwards bracing of the legs; moreover, this outwards bracing of the legs is carried out in an exaggerated manner.

C. *Impairments in Running*

On observing the running movements of intact animals, one will see that the foreleg when raised is, at the same time, adducted while the raising of the hindleg is accompanied by an abduction and, in addition, by a rotation of the pelvis to the other side.

Decerebellate dogs in running raise the legs too high. The raising of the foreleg is accompanied by too strong an adduction (Fig. 265, No. 16 and 20) and that of the hindleg by too strong an abduction and a too excessive rotation of the pelvis (Fig. 267). Moreover, the feet are set down too vigorously and with an audible impact and, after setting down, they are too strongly extended and abducted. Sometimes, particularly in decerebellate cats, the hindleg is put down first with a flexion and only then an excessive extension. In this case, the posterior part of the body, in addition to the right and left movements, makes marked up and down movements (Fig. 266). In the flexion of the hindleg, as it is put down, there is, in addition, an adduction posture produced by the outwards bracing of the opposite leg.

Fig. 265. Decerebellate dog Moor, one year after extirpation of the cerebellum, running with a sandbag of 5.3 kg on its back. 2. The animal sets down the left foreleg while it lifts the right one. 3-5. The animal braces the left foreleg more and more outwards, by which the anterior part of the body is more and more moved to the right and the animal turns to the right. Compare the running direction of 1 with that of 4 and 5. 4-5. The left hindleg is set down. 6-9. The left hindleg is more and more braced outwards and by doing so pushes the posterior part of the body to the right. 5-6. The right foreleg is set down, thus somewhat later than the left hindleg, and braced. 7-9. The right foreleg is also braced outwards so that the anterior part of the body goes to the left. By the movement of the posterior part of the body to the right (7-9) and of the anterior part to the left (7-9), the direction of running is again shifted more to the left (8-9) and nearly the same as in 1. 10-13. The left foreleg and the right hindleg are set down and then braced outwards so that the running direction again deviates to the right (13). 14-18. The right foreleg and the left hindleg are set down and then braced outwards so that the animal again runs towards the left (18-19). Thus this cinematographic picture shows in a typical manner the zigzag running of decerebellate animals caused by the alternating outward bracing of the left and the right legs. The other characteristic impairments of running, as, among others, the hypermetria of movements of the extremities, can also be distinctly observed. Thus, in 3, 4, 7, 13, 14 and 20, the abnormally high lifting of the forelegs and, in 16 and 20, the especially strong adduction while lifting can be seen. The picture further shows the strong compensatory movements of the tail and head. The tail is sometimes totally directed to the left (1, 2, 3, 6, 10, 11, 14, 21, 22)

Fig. 265 (cont.) and sometimes to the right (5, 7, 8, 15, 16, 17, 18, 19) while the head is also alternatively turned to the left (8-10, 17-19) and to the right (1-6, 12-15, 20-22). Finally, one can see that in this animal no trace of asthenia can be observed.

As for the succession of movements of the feet while running, observations showed that this is sometimes the same as in intact animals, but sometimes can also be quite irregular (Fig. 267), especially when the animals are restless and excited and try to run away quickly and also when they run over uneven ground.

How can these impairments in running movements be explained? Investigations carried out by Graham Brown made it probable that there are special centers for the running movements within the spinal cord.   It is possible that the muscles of the extremities do not react in a normal way but in an exaggerated and also perhaps in a retarded manner not only to other stimulations but also to impulses from these centers. However, the exaggerated extension on static stress, the strong contractions of certain flexors in negative supporting reactions and the excessive lateral bracing of the forelegs and the outwards and backwards bracing of the hindlegs surely contribute to the abnormal manner of running.

The hypermetria and the other impairments of running movements take place not only in running forwards, but also in running sideways and backwards, as can be observed especially when the animals are set down on a turntable with the fore- or hindlegs only in such a way that they have to displace their legs backwards and sideways.

D. *The Impairments of Hopping Rreactions*

The leg-slackening and hopping steps, too, *are carried out hypermetrically*. As we have seen in Chapter XIII, the legs are lifted abnormally high and set down abnormally far from the intermediate position. Since, moreover, the hop is retarded, i.e., only when the paw is set down abnormally far from the intermediate position, the hopping steps are usually twice as large as before extirpation of the cerebellum.

Fig. 266   Fig. 267

Fig. 266.   The cat Carolus, more than 1 year after extirpation of the cerebellum, running. A. Notice the strong up (A 2-3, 8) and down movements (A 1, 5, 10) of the posterior part of the body. B. The animal tries to run away quickly. Notice the exaggerated extensor position of the legs.

Fig. 267.   Dog Erik, $1\frac{1}{2}$ years after extirpation of the cerebellum, running. A 1-2. Just as in a normal dog, first the right hindleg and the left foreleg (1) are simultaneously lifted, and then the left hindleg and the right foreleg (2). A 6. Now also, the right hindleg and the left foreleg are lifted simultaneously, but in a strongly hypermetrical way. A 7. But the right hindleg is set down too late, so that it is still raised while the left foreleg is already set down on the ground and the right foreleg is gone high into the air. Thus, both right legs are now lifted at the same time. A 9. Again, in a normal way the simultaneous lifting of the left and right foreleg takes place. A 10. The right foreleg is now again set down and, at the same time, the left one is raised. The setting down of the left hindleg, however, is again retarded, so that now both left legs are raised from the ground. A 11. the animal now simultaneously lifts both forelegs. B 1. The animal, in a normal way, lifts at the same time the right hindleg and the left foreleg. B 2. While the left foreleg is again moved downwards, the right hindleg is lifted even higher. B 3. On setting down of the left foreleg, the right foreleg and left hindleg are raised simultaneously, and, since the right hindleg is still somewhat flexed, the animal supports itself only on the left foreleg, while the three others are held in the air. B 8. Both hindlegs are raised at the same time. On A 6, A 8, B 2, B 9, the exaggerated rotations of the pelvis can distinctly be observed. On A 8, B 6, B 9-11, it can be seen how closely the movements of the hindlegs resemble those of a person while ice-skating.

Although hopping appears late, it is carried out so appropriately that it almost always prevents falling. This is all the more remarkable as in decerebellate animals the center of gravity, due to its high position, falls outside the supporting base in the slightest alteration in posture. The fact that hopping reactions, nevertheless, are able to restore balance is due to the compensation of inappropriate retardation by the enlargement of the step. The effective occurrence of leg-slackening reactions followed by leg-bracing reactions makes it almost impossible to force

572

decerebellate animals to fall down, whether by rotating or tugging the head, pulling the tail, tugging a fold of skin of the trunk, or by pushing (Figs. 247, 248).

In running, it can sometimes be observed that the abrupt and excessive outwards bracing of the legs displaces the center of gravity beyond the supporting base and thus produces a hopping reaction in the opposite leg which nearly always restores the balance. The repeated occurrence of sideways directed hopping makes running seem even more abnormal and ataxic.

I consider the analysis of impairment of hopping reactions very important for cerebellar physiology. A number of reactions, like magnet reactions and reactions of the extremities produced by labyrinthine rotation, appear particularly lively, uninhibited and ample in decerebellate animals. Whether these reactions have abnormal timing could not be established.

The nature of the ratardation of hopping reactions is uncertain. A prolonged latency could not entirely explain the retarded appearance of leg-slackening reactions because, when a decerebellate animal is set down with one leg on a turntable and this is rotated faster and faster, the leg-slackening and hopping steps are also carried out faster and faster and follow each other in a faster succession so that they finally are carried out with a greater speed than in an intact animal standing on one leg on a disk rotation with a lesser speed. Under equal conditions with an equal speed of rotation, the decerebellate animal makes fewer steps than the intact one with each $360°$ rotation and, therefore, thesteps are performed more slowly. This is also the case with a high speed of rotation, although, in this case, the decerebellate animals, too, perform the steps more rapidly. This is caused by the fact that the leg-slackening phase of the hop in the decerebellate animal appears only when by the rotation of the disk the leg is drawn enormously far from the intermediate position.

For the occurrence of leg-slackening only after abnormally strong alterations in position, three explanations are possible. With increasing change in position, certain muscles are stretched more and more and from these muscles will emanate increasingly strong afferent stimulation. One can imagine that the elevators of the feet in decerebellate animals respond less easily and only on stronger stimulation (and even then with an excessive shortening) than in intact ones. The second possibility is that in decerebellate animals less vigorous stimulations are produced in the stretched muscles on a certain stretch and that these stimulations only on especially strong stretch reach the same intensity as in intact animals on a moderate extension and that the elevators react to these stimulations in an excessive manner.

The third possibility is that the abnormally strong contraction of the extensors of the supporting leg is the cause and that to overcome this abnormal contraction requires stronger stimulations and, therefore, occurs only on a greater alteration in position. The more vigorous stimulations at the same time explain the abnormally strong flexion once the extensor tension is overcome.

In any case, the impairments of hopping reactions (stretch reflexes fourth order) make it probable that after extirpation of the cerebellum, the reflexes produced by stretch of the muscles cannot appear in a normal way.

E. *The Impairments of Reactions Produced by Alterations in Position of the Head*

The reactions produced by alterations in position of the head also show impairments in decerebellate animals. Raising the head causes in them a particularly abrupt and strong decrease in supporting tonus in the hindlegs so that these give way, while lowering the head is followed by an abrupt, sometimes almost jumping extension of the hindlegs (Figs. 88 and 93).

Rotating and turning the head causes in decerebellate animals abnormal associated movements of the pelvis and the entire posterior part of the body. Sometimes, these movements are so strong that they produce hopping reactions sideways to prevent falling of the posterior part of the body. With very strong and abrupt turnings of the head, the hopping reactions are sometimes carried out too late and the rear part of the body falls. The tonic labyrinthine reflexes in the legs under static stress also appear especially uninhibited (Figs. 115-119) in decerebellate animals. In addition, labyrinthine stimulations produced by

574

rapid rotations of the head round the longitudinal or bitemporal axis also cause sudden alterations in supporting tonus and contractions in the vertebral muscles (Chapter IX, Figs. 122, 123, 128) and these reactions, too, are not abolished by the dorsal position in decerebellate animals.

It is, of course, evident that the rapid alterations in supporting tonus and the excessive movements of the pelvis on altering the position of the head contribute to make running clumsy and to complicate the symptom picture even more while running.

With a strong to and fro shaking of the head, for instance in shaking off some water sprinkled on the head, it can sometimes be observed that all four legs are abducted and flexed to prevent falling. Sometimes, flexion goes so far that the animal lies down in a ventral position. These reactions, which are very prominent in decerebellate animals, can also be observed in intact and decorticate animals. Thus, these are not "conditioned" reactions. How the simultaneous flexion is brought about, whether eventually by labyrinthine or by other stimulations, could not be established.

Summing up, the following factors participate in the abnormal running in decerebellate animals:

1. the exaggerated extension which appears every time a paw is set down (excessive magnet and positive proprioceptive supporting reactions);

2. the abnormal dorsal convexity and strong fixation of the back which are produced when the animal supports itself on one or several legs;

3. the excessive raising of the legs;

4. the excessive adduction of forelegs and excessive abduction of hindlegs during lifting of the legs;

5. the exaggerated outwards bracing of the legs as they are set down and the zigzag movement of the trunk caused by it;

6. the exaggerated backwards movements and backwards bracing of the hindlegs;

7. the too abrupt and uninhibited alterations in supporting tonus and the excessive associated movement of the pelvis on altering the position of the head;

8. the interruption of running by hopping reactions which are produced by the sudden outwards bracing of the opposite legs or by a too strong associated movement of the posterior part of the body;

9. the abnormal execution of hopping reactions produced in this manner.

In addition to these, other factors also play a role. In the discussion of rocking reactions (Chapter XI), we have seen that, in decerebellate animals, abduction and adduction of the supporting leg causes lively flexion and extension of the opposite leg taking no weight and that the flexion of the leg taking no weight is carried out excessively. Similar manifestations are shown by the leg taking no weight when the supporting leg is moved forwards or backwards in relation to the trunk (Fig. 163). Moreover, a passive movement forwards and backwards of the forelegs under static stress produces a particularly uninhibited increase and decrease in supporting tonus of the hindlegs (even in a dorsal position, Figs. 169, 170, 171).

The uninhibited reciprocal influence on the extremities, producing excessive flexion in the one with extension of the other, also makes running different. This difference appears particularly evident when one of the feet is injured. As in intact animals, they will try, in this case, to run on three legs and to keep the injured leg drawn up. It succeeds badly, however, as the flexed leg constantly makes associated movements, sometimes touching the paw on the ground.

Besides, decerebellate animals show another impairment which influences standing and running in abnormal way, i.e., the appearance of "unrestrained movements."

F. *The Unrestrained Movements*

   a) Tremor

   b) the true, unrestrained movements.

Decerebellate animals are not able to stand quietly when they stand freely but, in standing, constantly show a motor unrest. The head shows a continual shaking and wobbling, makes slight oscillating movements from left to right, up and down or performs slightly rotating trembling movements round the longitudinal or dorsoventral axis. The trunk, too, is sometimes moved up and down, forwards and backwards or sideways. The legs alternately show a slight extension and slight flexion alternating with slight abduction and adduction movements. They are often somewhat raised and again set down on the same spot and sometimes displaced and soon again replaced. In the interphalangeal and metacarpophalangeal joints, alternating flexion and extension movements can sometimes be observed, through which fingers and toes perform peculiar movements. When the animals stand on a hard stone floor, for instance, these movements are so strongly pronounced that the animals may seem to dance.

This phenomenon was already observed by Luciani, who called it "astasia." In his opinion, these movements are based on a slowing down of rhythm of single impulses. According to Luciani, the cerebellum normally accelerates "the rhythm of single impulses by which the different voluntary, automatic and reflex activities cooperate and bring about their normal fusion and regular continuity (static effect)."

Since these movements are neither produced voluntarily nor can apparently be stopped voluntarily, these were called by us "unrestrained movements."

In decerebellate dogs, two kinds of unrestrained movements can be observed, firstly the tremor, secondly cruder movements with larger divergences and a different rhythm, the true unrestrained movements which are sometimes termed by other authors, intentional wobbling or intentional tremor.

a) Tremor

The discussion of tremor in decerebellate animals may be preceded by some remarks on tremor in general. As is known, the appearance of tremor in man is not always a pathological manifestation. Tremor can also be caused under physiological conditions, among others by fright, by excitement, by a cold environment (shivering tremor). It also appears when one attempts to fix a part vigorously, the arm for instance, in a certain posture by simultaneous strong contraction of agonists and antagonists.

In animals, a similar physiological tremor can be observed. However, the tendency for tremor is different in the various species of animals. In dogs, it is present to a high degree. When intact dogs are permitted to run around for some time and one plays with them so that they become excited, on placing them on their back, a characteristic tremor of the legs will appear in a high percentage of them. In rabbits in a dorsal position, a tremor of the feet can also be established, though less frequently. Conversely, this is never, or only very rarely, the case in intact cats.[1]   Thus, cats are least suited for investigations on the central mechanism of tremor.

In several intact dogs which showed a distinct tremor in the dorsal position, the cerebrum was extirpated. In these animals even after a total transverse section of the midbrain just rostral to the point of exit of the nervi oculomotorii, in which the red nuclei were surely injured,[2] a continuation of the tremor was observed. In this animal, the caudal part of the midbrain was also further removed by means of a transverse section caudal to the red nuclei, which caused total disappearance of the tremor. After the first transverse section, the animal showed only a weak decerebrate rigidity and after the second a very pronounced one.

In other animals decerebrated caudal to the red nuclei, no tremor was found.

In the first days after extirpation of the cerebellum, the tremor was also absent, both in rigid as well as in flaccid animals. Conversely, the animals showed a vigorous trembling of the whole trunk and the legs due to cold or fever. This was also observed several times in rigid animals.

In the stage of permanent impairment, the liability to tremor is increased. At this stage, in all decerebellate dogs without exception, a distinct tremor can be observed. Besides, the tremor in these animals appears more generalized and can also often be established in the head and tail, in addition to the legs.

Just as in intact animals, the tremor sometimes is intermittent and sometimes more regular. In the first case, runs of heavier beats alternate with runs of finer ones, while in the second case the beats are more equal (Fig. 268, A and B). In decerebellate dogs, the tremor is not always present and not under all circumstances.

---

(1) Although this was regularly investigated, I have never yet succeeded in ascertaining a distinct tremor in cats. Kinnier Wilson states that he could observe a distinct tremor in one of the forelegs of a cat which had been unilaterally decerebrated by Bazett and Penfield.

(2) The role which Gordon Holmes, Kleist and others ascribe to the red nuclei in the production of tremors was discussed in detail in a previous publication (234).

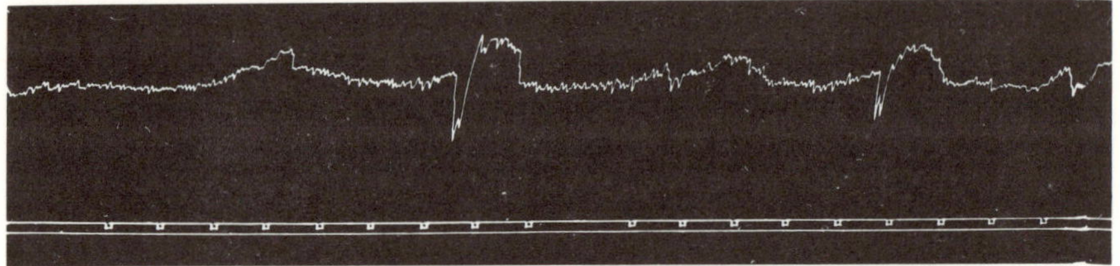

Fig. 268 A.

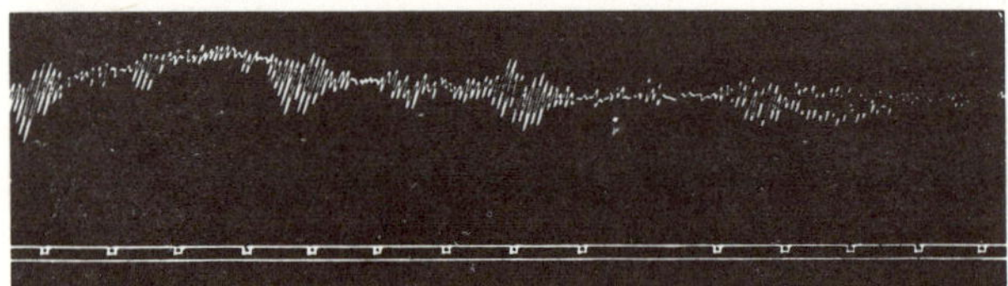

Fig. 268 B.

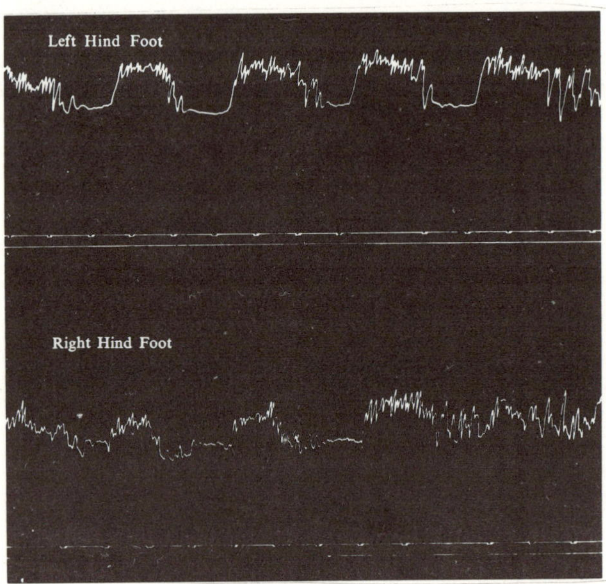

Fig. 268 C.

The tremor of the hindlegs is most distinctly present when the animals are brought into a dorsal position, the muzzle directed vertically upwards or ventrad. As already mentioned, the legs in this position show a flexed posture; this, however, is usually not maximal, the paws are not flaccid on the trunk but can be passively flexed further. This is particularly the case when the animals are more or less excited and, in this case also, the tremor is most lively. When, however, the animals behave calmly and lie there quite flaccid, sometimes no trace of a tremor can be found. When, the legs are led into an extended position, be it by touching or by static stress, the tremor always ceases. The

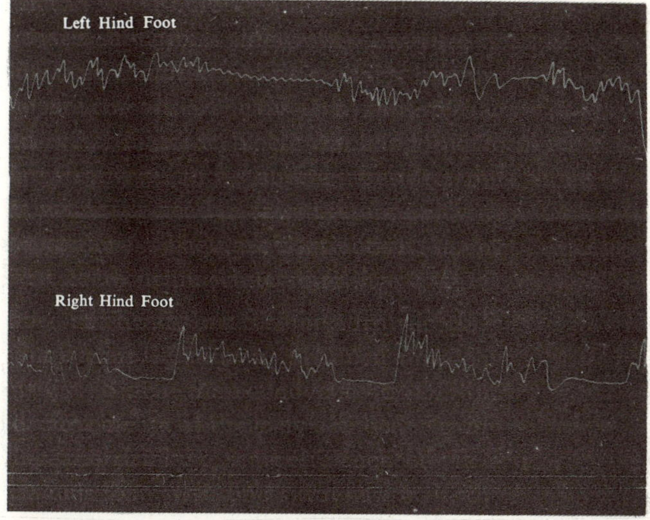

Fig. 268 D.

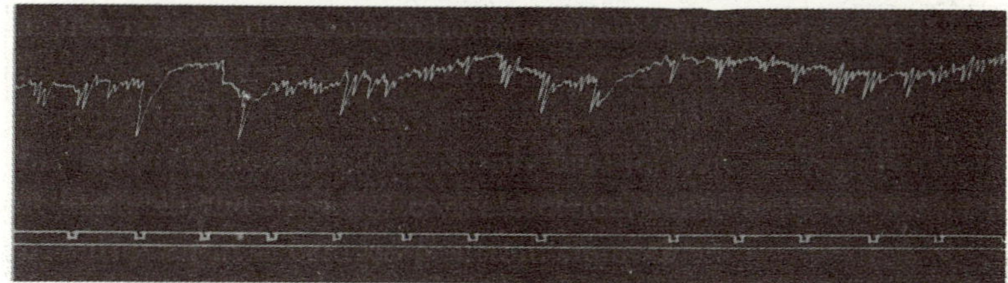

Fig. 268 E.

Fig. 268 A-E. Tremor in the intact, unilaterally decorticate, totally decorticate and decerebellate dog in a dorsal position, the muzzle pointing vertically upwards. A. Intact dog. Tremor of the hindlegs with uniform character. B. Intact dog. Tremor of the hindlegs with intermittent character. C. Dog Schwatzweisser, 3 months after extirpation of the left cerebral hemisphere. Tremor of the left and right hindleg. D. Dog Fuchs, 3 months after total extirpation of the cerebrum. Tremor of both hindlegs. E. Dog Piccolino, 2 years after extirpation of the cerebellum. Tremor of the hindlegs.

580

tremor of the hindlegs is also influenced by alterations in position of the head. When the muzzle is directed ventrally or vertically upwards (±90°), it is lively and shows large beats; on the dorsal movement of the head, it decreases so that it is either totally absent or insignificant, with very slight beats when the muzzle is finally directed downwards (Fig. 269). From these observations, it can be concluded that tremor appears only on certain distributions of tonus.

The tremor of the paws is absent when the animals lie quietly on the side with the head and trunk. When the animals are held in the air supported from the ventral side, it can also never be observed, neither with extended legs nor when the limbs are drawn into a flexed position. Whether it is present in a standing posture cannot be established due to the coarse dancing movements. In sleeping, it is always absent.

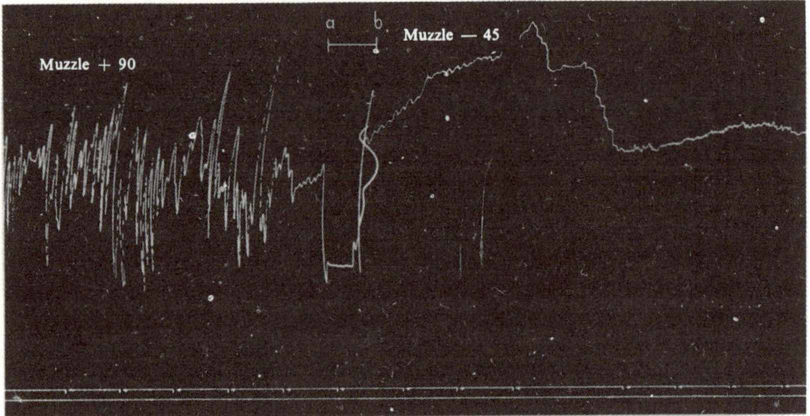

Fig. 269.   Influence of the head posture on the tremor of the hindleg. Decerebellate dog Erik in a dorsal position. At first, the muzzle is held vertically upwards (±90°). The hindleg shows a tremor with large beats. Then the head is moved backwards till the muzzle stands 45° below the horizontal (a-b). At this position of the head, the hindleg shows only a fine tremor.

The head tremor is most evident when the animals lie on their belly with raised head and stare at something.   In this case, the head sometimes shows a delicate pulsating vibration round the dorsoventral axis.

In normal hunting dogs, too, this trembling sometimes appears, for instance, when they intensely watch the movements of a rabbit.

Contrary to intact dogs, decerebellate dogs also sometimes show a tremor of the tail, but this tremor is not present in all of them. Just like the tremor of the paws, the tremor of the tail, too, can best be observed in a dorsal position of the animals, however, not when the tail lies flat on the supporting surface but when it is actively held in the air. Sometimes, when the animals are held in the air in a ventral position, a distinct, delicate pulsating tremor of the tail can also be seen.

In rabbits, after lesion of the cerebellum, a strikingly lively tremor of the hindlegs can often be established. On the other hand, in decerebellate cats as well as in intact ones, tremors of any kind are lacking. This gives the impression that in dogs and rabbits, extirpation of the cerebellum causes an increased liability to tremors.

b) The True Unrestrained Movements

These movements, too, do not appear immediately after extirpation of the cerebellum; in the first days they are absent in flaccid as well as in rigid animals. Even when the animals at this stage are passively brought into a standing posture, no typical "unrestrained" movements can be observed. These become apparent only when the animals start to carry their head freely in the air and when the supporting and magnet reactions have returned. They can be observed most strongly when the animals start to rise from the floor and try to erect themselves into a standing posture. In this case, the movements are particulary strong and uninhibited.

At this stage, a strange phenomenon sometimes appears which later ceases to occur, namely that as soon as the animal has risen into a standing posture the totally extended forelegs simultaneously begin a violent to and fro movement sideways. In this case, the movement of the legs is similarly directed, both forelegs together go alternately to the left and to the right and the claws audibly scratch the ground. These movements are probably produced by alternating contractions of the muscles of the right and the left side of the trunk. It is true that these contractions can still be observed later, but since at this stage the soles of the forelegs are held more flat on the ground and the animals no longer stand on the tips of their toes, tremor no longer causes a shifting of the legs, but a sideways movement of the trunk. When the animal is now lifted from the ground by the head (with one hand below the lower jaw) and sometimes by the tail, the to and fro movement of the legs can again be observed. In this case, the trunk sometimes makes peculiar rotating movements by which the legs go alternately to the left and to the right. When only the anterior part of the body is lifted from the ground with one hand beneath the lower jaw (as in Fig. 68) so that the animal stands only on the hindlegs, the anterior part of the body sometimes shows these rotary movements by which the forelegs are again moved sideways to and fro.

At the stage of permanent impairment, the various "unrestrained" movements can always be still clearly observed, though somewhat less strong, *in all decerebellate dogs and in decerebellate cats*. In standing, they can even occur *constantly*. Head and neck show a typical wobbling and are sometimes moved up and down, sometimes sideways, to and fro or perform rotary movements round the longitudinal axis. On palpation, one can distinctly feel that these movements are caused by alternating contraction of the right and left cervical muscles. These wobbling movements of the head and neck are strong in the standing posture, but they are sometimes also evident when the head is raised from a lying position to a ventral or lateral position. They are particularly strong when the animals attempt to eat or to drink while standing.

In this case, one will see how the head shakes to and fro violently and moves up and down so strongly that the muzzle is dipped into the food and withdrawn constantly. Since these movements of the head produce strong associated movements of the pelvis and a flexion and extension of the legs, eating and drinking scarcely succeed in a standing posture. When the animals lean with one side on a wall, the wobbling movements of the head are slighter, but even then eating and drinking are usually not possible and the animal often falls head forwards into the feeding bowl. For this reason, the animals become accustomed to lying down in a ventral position in front of the feeding bowl. In this position, the wobbling movements of the head and neck are less strong so that the animals are able to eat and drink. This is only true for the stage of permanent impairment. At the stages of convalescence, sometimes eating does not succeed even in a ventral position. Sometimes, the head shakes so violently from side to side in this posture also that an already seized piece of food is flung out of the mouth and the animal continually falls down. Furthermore, the muzzle dips so deeply into the food or knocks so violently against the rim of the feeding bowl that it gets chafed and sore.[1] In the immediate postoperative stage, eating

---

(1) Decerebellate cats also sometimes show at this stage a violent hammering with the head in the course of which the nose hits the ground when they go over from a lateral to a ventral position. These manifestations appeared particularly violent in the cat Peggy in which both labyrinths were removed first, and later the cerebellum.

was only possible when the animal was lying totally in a lateral position, the head also, and the food was pushed directly in front of the muzzle. Thus, while eating, decerebellate animals show especially violent wobbling movements of the head, while the intended head movements were strongly hypermetric.

In the vertebral column, too, sometimes unrestrained movements, rotating movements and alternating left- or right-sided concavity can be observed in the decerebellate animal and these movements are accompanied by an alternating contraction of the left and the right vertebral muscles. When the animals are lying in an active lateral position (with raised head), slight alternating flexion and extension movements can often be seen in the legs. In this case, the legs which are kept in a flexed posture are alternately a little extended and flexed. In a standing posture, too, the legs show similar movements, the totally extended legs continually flex a little, to go again instantly into a maximally extended posture.

Moreover, the legs are alternatively abducted and adducted in the shoulder and hip joints in relation to the trunk and moved forwards and backwards. They are also sometimes raised and again set down on the same spot or on a different one. However, whether these latter movements can also be interpreted as unrestrained movements is not yet certain. We have seen that decerebellate dogs stand with maximally extended, backwards directed hindlegs and that these legs continually bend a little and are extended again.

Due to those rapidly performed flexions and extensions, the trunk is alternately moved backwards and forwards and this can bring about secondary leg-slackening reactions, i.e., a lifting and setting down of one or both forelegs. Moreover, each movement of the head is accompanied by strong associated movements of the posterior part of the body and eventually by leg-slackening reactions of the hindlegs caused by them.

When one rubs with the hand alternately along the left and the right side of the neck of a standing decerebellate dog, the pelvis will make such strong lateral to and fro movements that the hindlegs are alternately displaced to the left and to the right and the posterior part of the body also dances to the left and to the right.

As we have seen, the unrestrained movements appear constantly and vigorously when the animals stand on all four legs. They are even stronger when the animals keep one hindleg raised while urinating. They are much increased when the toes are passively held to the supporting base; on restraining the toes of the forepaws the flexion-extension movements of the hindlegs are often carried out so abruptly that the posterior part of the body is lifted off the ground with the extension phase of movement. The unrestrained movements are neither abolished by blindfolding the eyes nor are they distinctly intensified. They also do not disappear by placing a burden on the back, just as they still occur obviously when the head or the pelvis is immovably kept still with both hands (when the head and pelvis are fixed simultaneously, they are indeed less strong). Finally, they are still distinctly present when the anterior or posterior part of the body is passively lifted from the ground so that the animal stands only on the fore- or hindlegs.

It is remarkable that unrestrained movements can also be brought about by visual stimulation. When a decerebellate dog is held in the air in a ventral position supported by the thorax, it will hang quite calmly and no unrestrained movements are to be observed. But when the animal is now slowly moved downwards, one will see that it becomes restless as soon as it nears the ground, starts to shake its head and to make rotary movements in the shoulder girdle, by which the forelegs are moved to and fro. This is not the case with blindfolded eyes in which case the unrestrained movements only appear when the paws touch the ground.

In the cat Peggy which had the labyrinths as well as the cerebellum removed, the unrestrained movements were much stronger and more violent than in exclusively decerebellate cats, giving the impression that theywere increased by extirpation of the labyrinths.

The unrestrained movements are reduced when the animals, while standing, lean against a wall. Also, while actively lying in a ventral and lateral position, they are lessened, though they can still be clearly observed. They distinctly decrease in the standing animal when a large fold of skin of the back is vigorously seized with both hands. Whether in this case the unrestrained movements are directly inhibited by stimulations from the skin or whether the disappearance is only an indirect result produced by the reduction in supporting tonus (Fig. 219) is not yet established. By pinching in other areas, for instance by pinching the tail or the ears, the unrestrained movements are not inhibited. Conversely, they are strikingly reduced when the head is passively

ifted and the trunk is simultaneously pressed caudally so that the sup-
lporting tonus of the hindlegs decreases and these give way.

The unrestrained movements are absent : first, immediately after
extirpation of the cerebellum, secondly, in sleep and, thirdly, when
the animal lies completely still and passive with head and trunk lying
flat on the supporting base.

While in dogs the resting position is either lying in the lateral or the
curled-up position with the head resting flat on the ground, cats often
rest in a ventral position. In doing so, they indeed sometimes keep their
head lifted, but with strongly retracted necks. In this position, too,
distinct unrestrained movements still appear. This probably is the
reason that decerebellate cats at rest almost always place themselves
against a wall or lie down in a ventral position with the posterior part
of the body against a corner of the room to fix the pelvis from both
sides.

In the fourth place, the unrestrained movements are absent when
the animals are held in the air, even when the animals are held in the
air by the posterior part of the body in a dorsal position (as in Fig. 270,
No. 2-4).

This is remarkable for the reason that under these conditions the
labyrinthine righting reflexes cause a contraction of the ventral flexors
of neck and anterior part of the body, so that these parts of the body
are actively raised. Nevertheless, the head is kept still and shows no
unrestrained wobbling movements. Also, when the animals are lifted
up by the hindlegs and held in the air in a ventral position (Fig. 270,
No. 1), no unrestrained movements are present, although, owing to the
contraction of the dorsal vertebral muscles, the head and the anterior
part of the body are now also kept raised.

In the fifth place, the unrestrained movements are absent when the
animals are placed in a dorsal position on a supporting surface. In
the flexed paws sometimes a termor can be observed, but not the typically coarse, unrestrained movements. It is remarkable, too, that these
movements also fail to appear when the legs are brought into an
extended position by touching or by static stress or when the conditions
of the standing posture are imitated as much as possible by placing a
loaded board over the four extended paws while the animal lies on its
back. Why the unrestrained movements fail to appear under these
conditions is not known. One could perhaps explain it by the fact that
the counterpressure exerted by the supporting surface on the skin of the

back produces stimulations which either directly inhibit the unrestrained movements or indirectly inhibit it by reducing the supporting tonus. As we have seen through the influence of counterpressure, the strength of supporting tonus of the legs in a dorsal position is never as strong as in a standing posture. Stimulations from the skin of the back are capable of reducing or eliminating the unrestrained movements as we have already noted on seizing a big fold of skin on the back of a standing animal. We should not forget that in a free standing posture only the soles of the feet are fixed on the supporting base and that in a dorsal position and with loaded paws there is a passive fixation of the head, back and feet.

Thus, decerebellate animals show unrestrained movements which are accompanied by alternating contractions of agonists and antagonists and which only appear when the animals are active, carry out movements or try to fix parts of the body in an active position.

How are these unrestrained movements brought about and what specifically produces the dance-like standing posture? On this, we have only conjectures. From the absence of movements in an active position in the air, we infer that it is not caused by an inability of the centers to inervate the muscles smoothly, as asserted by Luciani. Normal human beings and animals are not able to keep an extremity extended in the air absolutely motionless, or even to stand for long quite motionless. A normal person standing erect always shows slight fluctuations of the trunk in which the lower legs move alternatively forwards and backwards in the ankle joints. These movements are hindered by alternating contractions of the ankle joint extensors and flexors. The checking of the movement is probably brought about by stretch reflexes first order of the flexors and extensors of the ankle joints. When patients with cerebellar atrophy attempt to stand still, these alternating movements back and forth are sometimes exaggerated (see, among others, Thevenard, 298) and are accompanied by strongly alternating shortening contractions of the extensors and flexors of the ankle joint. In the backwards movement of the body, the passively extended ankle joint flexors contract so strongly in these patients that the tendons of the muscles protrude beneath the skin, the toes are moved dorsally and even the anterior part of the foot is lifted from the ground. By this contraction, the body is pulled forwards and the forwards movement is so abrupt and strong that the body moves forwards beyond the intermediate position. With this forwards movement, the passively stretched peroneal muscles contract so strongly that the bellies of the muscles visibly arch up beneath the skin of the calf and the heels are lifted from the ground. An excessive movement in the opposite direction appears again. Sometimes, this countermovement fails to appear and the leg is displaced to prevent falling.

We thus see that in normal, standing human beings, fluctuations occur which are prevented by contraction of the passively stretched muscles (stretch reflexes first order) and transformed into a movement into the opposite direction, i.e., towards the intermediate position. In patients with cerebellar atrophy, the movement in the opposite direction appears retarded only when the body has moved abnormally far from the intermediate position and is carried out in an exaggerated and particularly abrupt way. Thus, in this case, impairments appear similar to those in hopping reactions in the decerebellate dog; retarded appearance is followed by an excessive execution of movements.

One gets the impression, firstly, that the unrestrained movements in decerebellate animals appear only when the body parts, by simultaneous contraction of agonists and antagonists, are fixed in a certain position and this position is maintained by alternating increase and decrease in the tensions of agonists and antagonists; and, secondly, that the unrestrained movements are stronger the stronger the simultaneous contraction. In the standing decerebellate animal, all muscles of extremities, agonists and antagonists, are abnormally strained. Probably in the physiological to and fro movements of the trunk the passively extended muscles react with a late and exaggerated contraction. In the course of the to and fro movement, the extension of certain muscles is accompanied by a relaxation of antagonists. Perhaps this relaxation is delayed or incomplete in decerebellate animals standing with abnormally stressed muscles, resulting in greater delay in compensatory movement. The mechanism of unrestrained movements still needs numerous thorough investigation before this concept could serve as a working hypothesis.

Summing up, decerebellate animals show the following impairments:

1. *an increased liability to tremor* (in dogs and rabbits, but not in cats);

2. *unrestrained movements* (but only under certain conditions);

3. *an abnormal distribution of muscle tonus while standing;*

4. *an abnormal execution of almost all movements and actions, whether depending on cerebral cortex or directed subcortically.*

588

These impairments are caused, at least partly, by:

1. *a reduction of the inhibiting influence of various factors* as, for instance, the inhibiting influence of the dorsal posture on magnet and supporting reactions and of labyrinthine reactions of the extremities;

2. *the retarded initiation of certain reactions*, particularly the leg-slackening and hopping reactions;

3. *the excessive muscular contractions in various cortical as well as subcortical reactions*, as well as the abnormally energetic contractions of the vertebral muscles and extensors of the extremities on touching or static stress of the soles of the feet causing the dorsal curvature of the back and the exaggerated extensor posture of the legs while standing; also, the abnormally strong contraction of flexors of the extremities, by which the legs are lifted abnormally high in running, hopping and rocking reactions; the abnormally strong contraction of extensors of the digits in visually directed reactions, as was shown by the monkey Corrie in grasping an object;

4. *the strong increase in postural tension of various muscles on certain stimulations* as well as the singularly strong contraction of muscles of all extremities, agonists as well as antagonists, and of the vertebral muscles, in magnet reactions and static stress on the soles of the feet.

Factors 1-4 all contribute to the occurrence of unrestrained movements, tremors and abnormally strong associated movements.

G. *Coordination*

The defects in function observed after extirpation of the cerebellum led to a heated dispute between the various cerebellar investigators on the question of coordination.

The concept that the cerebellum is an organ for coordination was expressed for the first time by Flourens and defended by Bechterew, Bolk, Jelgersma, Pol, Barany and Dusser de Barenne; on the other hand, it is violently contested by Luciani, Rynberk and others. In surveying the literature on this concept, what attracted our attention in this controversy was not the question of the function of the cerebellum nor the impairments following extirpation of the cerebellum, but its relation to the three following problems :

(1) What is an organ for coordination ? (2) Which are the consequences of destruction of an organ for coordination? (3) Which

are the symptoms by which one can establish incoordination of a move-ment, or in other words, what is the definition of incoordinated movement?

Flourens (1822) observed that decerebellate animals were still capa-ble of executing muscular contractions and complicated movements, but not of standing without being supported, nor of running and jumping. He, therefore, arrived at the conclusion that "the ability to produce contractions in simple muscular groups was situated in the spinal cord and that the ability to coordinate the movements for walking, jumping, flying and standing emanated exclusively from the cerebellum." Flourens thus considered walking, jumping, etc., as coor-dinate movements and concluded from the absence of these movements that the cerebellum is an organ of coordination.

Thus, the word "coordination" was introduced into cerebellar phys-iology, never to disappear again, even after it was proved that decere-bellate animals can very well stand, run and jump.

For the other investigators, the concept that the cerebellum is an organ for coordination is not based on observations of a deficiency, but on the perception of the abnormal execution of various movements. According to Hulshoff Pol (228), the abnormal running movements that appear after cerebellar lesions, like the goose-step or the cock's step, are not incoordinate movements because these movements, though involuntary and abnormal, are regularly repeated and always carried out in the same manner. According to Hulshoff Pol, incoordination is only present when *the involuntary and abnormally performed movements always differ from one another.* Therefore, according to him, the rotary tremor of the head is not an expression of incoordination. Since, how-ever, Hulshoff Pol observed in dogs in which he destroyed a certain part (i.e., the sublobulus of lobulus med. post. Bolk) that on jumping they carried out abnormal movements with the hindfoot which differed each time and thus indeed assumed that the cerebellum is an organ for coordination.

The observation that a single movement under particular circum-stances (in jumping) is poorly coordinated is sufficient, according to Hulshoff Pol, for the assumption that the cerebellum is an organ for coordination, even when other movements are coordinated. According to Luciani and to Rynberk, the cerebellum is not an organ for coor-dination because the abnormal execution of the movements can be entirely explained by atonia, asthenia and astasia. Moreover, these authors consider the fact that decerebellate animals are well able to swim (according to Dusser de Barenne, however, not quite normally) as the crucial experiment to demonstrate that coordination is not impaired in decerebellate animals and that the cerebellum is not an organ for coordination.

Thus, according to these authors, an organ for coordination is an organ which regulates the normal progress of movements by other means than by tonic, sthenic and static effects. Furthermore, according to them, an organ for coordination exerts an influence on all actions and the observation that one action is carried out in a normal manner excludes the fact that coordination is abolished and the organ for coordination was removed.

Bechterew, too, considers the cerebellum as an organ for coordination. Based on observed impairments in movement, he feels impelled to assume that the harmony of movements with the position of the body is disturbed (See pp. 7 - 9). According to him, an organ for coordination has the function (in my opinion vaguely explained) of maintaining the harmony of movements with the position of the body. Jelgersma also considers the cerebellum as an organ for coordination, i.e., according to his definition, an organ which corrects certain actions. Jelgersma is the only person to state in detail how he imagines the mechanism for cerebellar coordination. According to his opinion, only *the higher coordinations acquired by training, brought about by way of the cerebrum*, such as speaking, running, standing, jumping, swimming, climbing are corrected by the cerebellum. These higher coordinations are acquired under the influence of conscious optical, acoustical and cutaneous stimulations, but are later produced by subconscious stimulations, from the muscles, joints and labyrinths. By means of these subconscious stimulations, the movements are normally constantly corrected and this correction, according to Jelgersma, is brought about by way of the cerebellum.

In the lower animals with less developed cerebellum, these "subconscious muscular and labyrinthine stimulations" go to the cerebellum and from there by way of centrifugal tracts back to the motor centers in the spinal cord. This mechanism is also preserved in higher mammals. In addition, these possess a cerebro-cerebellar system of coordination, in which case, according to Jelgersma, the subconscious stimulations go to the cerebellum at first and then by way of superior peduncle to the cerebrum, return to the cerebellum via the pons, and from there are finally led to the motor centers of the spinal cord.

There are not only several objections[1] to Jelgersma's assertions but information is totally lacking on the reasons for Jelgersma's proposition for impairment of coordination and why he considers running movements of patients with cerebellar disease and decerebellate animals as incoordinate movements.

Does Jelgersma, with his statement that running, jumping, etc., are higher coordinations, want to assert at the same time that each impairment of these functions, i.e., each impairment of these higher coordinations, is an impairment of coordination even if this impairment of function is caused by, for instance, hypotonia or any other abnormal muscular component?

Our review showed that every author who considers the cerebellum as an organ for coordination has his own concepts as to the function and mode of action of an organ for coordination and the question: what is understood by a coordinated movement; concepts that deviate considerably from the concepts of other authors. The dispute on the question of coordination is, therefore, mainly an argument about definitions. It is not my intention to express my view on this dispute, except to say that in my opinion cerebellar physiology is scarcely furthered by designation of the cerebellum as an organ of coordination.

Whether one designates the abnormally strong magnet reactions in a dorsal position, the exaggerated extensor posture of the legs in a standing one and the other defects as impairments of coordination or as a result of dystonia (Dusser de Barenne) or of anisosthenia (André Thomas) is only a question of opinion or nomenclature.

---

(1) These objections are :

1. that decorticate animals can still run very well and therefore running, at least in animals, is not a coordination brought about by the cerebrum;
2. that in decerebellate animals, as we have seen, numerous other reactions are impaired which are surely not produced by way of the cerebrum;
3. that the extirpation of the cerebellum also produces considerable impairments in decorticate animals;
4. that in decerebellate animals the various labyrinthine reflexes are still obviously present;
5. that in decerebellate animals the correction produced by proprioceptive "subconscious muscular stimulations" are not absent; these stimulations indeed produce excessive correction movements in these animals.

As long as there is so little agreement as to which symptoms the destruction of an organ of coordination *must* absolutely bring about and by what symptoms an incoordinate movement can be distinguished with certainty from an impaired but still coordinate one, the term "organ of coordination," in my opinion, makes little sense.

Moreover, animals not only show impairments in the execution of movements, but also while standing (incoordinate standing ?). Besides, it should be remembered that, as already pointed out by Munk, various movements of decerebellate animals, under some circumstances, show much fewer deviations and are executed much more appropriately than under other circumstances. In a free standing posture, they carry out a number of movements clumsily and inappropriately, for example, scratching movements are almost normal when the standing animal leans against a wall or lies on the ground. Corresponding observations on eating and drinking while standing or lying down have already been mentioned.

As stated by Nothnagel, v. Monakow, Strumpell, André Thomas and several other authors, cerebellar patients sometimes also perform movements rapidly and accurately when lying in a dorsal posture which they could not carry out at all, or only very imperfectly, while standing.

With reference to Jelgersma's assertion that the cerebellum plays a role especially in the coordination of speech, it should be mentioned that decerebellate dogs are able to bark, growl, whine and that decerebellate cats were heard to spit, to purr, to hiss, to cry and to mew in various keys. We do not attempt, of course, to compare these animal noises with human speech and even less to assert that the cerebellum does not normally influence speech; concerning this animal, experiments clearly cannot enlighten us.

By the analysis of manifestations discussed above, however, Luciani's concept should be considered as refuted.

The fact that decerebellate animals show neither atonia nor ashtenia was discussed in detail in the preceding chapters. The various impairments, therefore, cannot be related to atonia or asthenia as asserted by Luciani. The animals not only do not show any atonia in standing and running (Figs. 1, 44, 256, 265-267) but also in a ventral and dorsal position in the air (Fig. 270) as well as in a dorsal position on a supporting surface (Fig. 67).

Since animals, when held in the air by the hindlegs in a ventral or dorsal position, not only actively raise the head and the anterior part of the body, but keep them freely raised *quite calmly* without unrestrained movements, it is clear that the muscles which maintain this posture must be tonically (tetanically) contracted. Under these circumstances, there can be no question of a "slowing down of the rhythm of single impulses" and the ability to contract the muscles tonically is surely not abolished.

Asynergy

To what extent do observations made on decerebellate animals correspond to those of cerebellar patients?

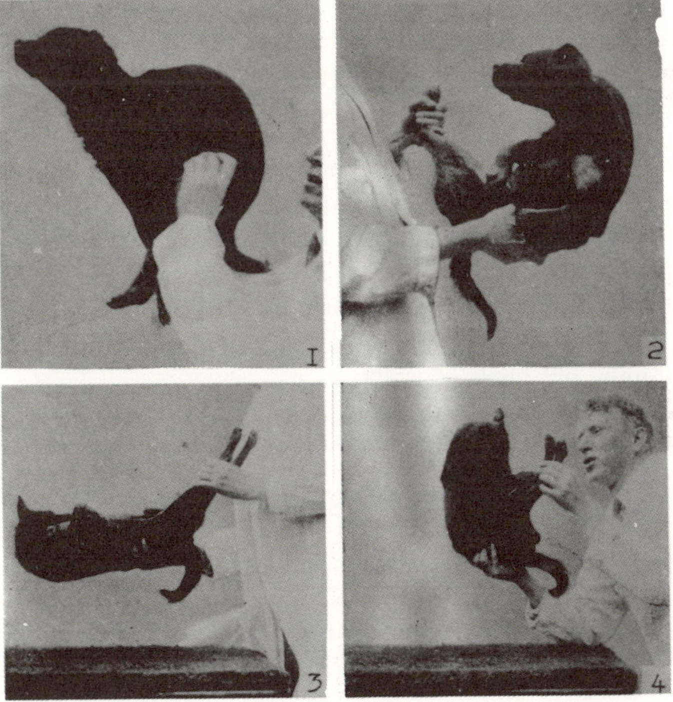

Fig. 270. Decerebellate dog Moor held in the air in a ventral position by the hindlegs. 2-4. Decerebellate dog Moor (2) and dog Erik (3 and 4) held in the air in a dorsal position. The animals keep the head lifted actively and show no trace of atonia.

According to Babinski, cerebellar patients show asynergy. We have seen that, in animals in the first stages after extirpation of the cerebellum, the associated movements occur in an abnormal manner. For instance, when in the first days after extirpation of the cerebellum, if in a ventral position the head of the animal is rotated towards the right, the whole trunk promptly goes into a lateral position to the right (Fig. 225), while, under similar circumstances in intact animals as well as in decerebellate animals in the stage of permanent imparment, a rotation of the pelvis to the left occurs. In the first days, the trunk entirely obeys the neck righting reflexes, due to the absence of body righting reflexes acting on the body and of the leg-bracing reactions. Thus the animals, to a certain extent, show an asynergy or dyssynergy, i.e., they show the absence of certain reactions and the unhindered occurrence of other physiological associated movements, namely the neck righting reflexes. Later on, however, when the leg-bracing reactions have returned, the defect is absent. In this case, the decerebellate animals show the same synergies shown by intact ones, i.e., a helical rotation of the trunk and the typical reactions of the legs which prevent falling (Fig. 101). But even now these synergies do not appear in a normal way but in an abnormally strong excessive manner.

In other reactions, too, as for instance the hopping reactions, an excessive execution of associated movements was observed. Thus, decerebellate animals in the stage of permanent impairment show, not asynergy but to some extent a hypersynergy. As discussed in detail on page 306 'it is not yet precisely shown that the symptoms which Babinski regards as expressions of asynergies are really based on asynergies and not on hypersynergies.

Adiadochokinesia

The ability to carry out actively alternating movements in quick succession is naturally very difficult to investigate in animals. When a piece of meat is moved to and fro in front of the muzzle of the animal, one gets the impression that the animal, particularly in a standing posture, is not able to move the head alternately to the left and to the right as rapidly as in intact animals.

Cerebellar catalepsy

Babinski describes this impairment as follows: when a cerebellar patient lying in a doral position lifts the legs in the air so that the hip and knee joints are flexed $\pm$ 90°, the legs show no fluctuations while, on the contrary, in normal patients, under similar conditions with the onset of fatigue, gradually increasing fluctuations can soon be observed. Other investigators (see among others André Thomas) never observed this cerebellar catalepsy in any cerebellar patient and Babinski recognized later on the rare occurrence of this kind of catalepsy. He, however, added that it is often perceived that the fluctuation of legs in cerebellar patients, in the above-described position, is no greater than in normal persons, although they often show strong fluctuations in standing and other actions. The difference in the behavior of a cerebellar patient and of a tabetic in these conditions is nevertheless remarkable, so that, in his opinion, a relative catalepsy can often be established. In Babinski's cerebellar patient who showed the catalepsy in full development, a post mortem examination showed that in addition to the cerebellar disease a tumor of the pons was also present which partly destroyed the pyramidal tracts.

Since it was not possible to train animals to lift their 'paws on command when in a dorsal position, Babinski's experiment could not be exactly imitated. However, investigations of totally and unilaterally decerebellate animals resulted in some observations which seem important in relation to cerebellar catalepsy. As we have seen, immediately following extirpation of cerebellum some of the animals are rigid. When these animals are placed on their back with the muzzle pointing upwards, they keep their legs extended into the air. In this case, the legs show no fluctuations and fatigue appears only after a long time. After unilateral extirpation of the cerebellum, only the legs ipsilateral to the extirpation show the same posture (Fig. 70). However, these results cannot be compared to Babinski's findings since these animals show no fluctuations when they are set up on their legs.

More important, therefore, are the observations in the stage of permanent impairment when strong fluctuations appear in standing. In this case, the animals in a dorsal position keep their limbs more or less flexed and these sometimes show a slight tremor, but no coarse fluctuations. When the legs are now brought into an extended posture by a slight touch of the soles of the feet, they remain extended as long as the soles are touched and, even when the touching lasts for 45 minutes or longer, the legs are kept completely still and show neither tremor nor fluctuations. On stressing the limbs with a loaded board, fatigue fails to appear for a long period and, no fluctuations can be established (Fig. 67), while in intact dogs passively extended legs, when stressed under the same conditions, almost instantly give way.

These manifestations correspond in some degree with those described by Babinski. This is even more the case with the manifestations shown

596

by the dog Vici, in which, in addition to the cerebellum, the right half of cerebrum was also extirpated (compare the post mortem findings of Babinski's patient!).   This animal, in a dorsal position and with a certain position of the head, kept the left hindleg rigidly extended and the posture continued as long as the position of the head was not altered. All fluctuations were absent here (Fig. 271).

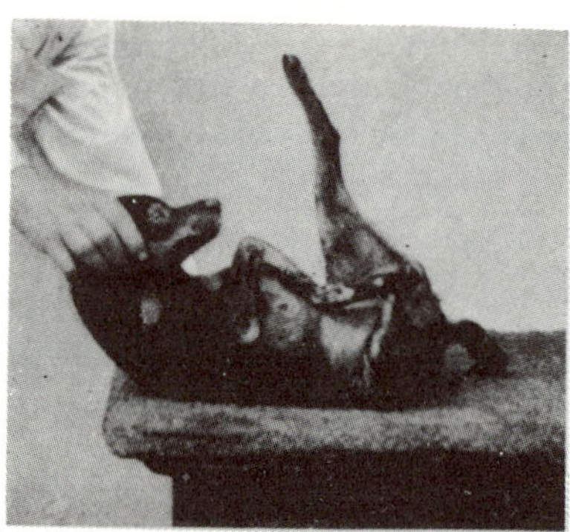

Fig. 271.   Decerebellate, right-sided decorticate dog Vici. In a dorsal position, with ventrally directed muzzle, the left hindleg is kept extended in the air. The leg shows no fluctuations and for a long time no manifestations of fatigue. Cerebellar catalepsy?

Hypermetria

Another impairment observed in cerebellar patients is hypermetria of movement.   This impairment was also observed in running movements, hopping reactions and rocking reactions, and in decerebellate animals and has been discussed in detail. We only wish to mention one more observation of hypermetria which corresponds exactly to observations made in man. André Thomas and Jumentié noted that cerebellar patients, on trying to grasp an object, for instance a drinking glass, open the hand too widely before grasping ("épreuve de la préhension," Fig. 272, No. 1.).   Exactly the same manifestation could sometimes be seen in the decerebellate monkey (Corrie, Fig.272, No.2). From these observations, apart from the conformity of manifestations, it can be seen that after extirpation of the cerebellum, the muscles react in an excessive manner in cerebral direction of movement.

Retarded execution of the various movements

It is stated that cerebellar patients initiate various movements, for example running movements, in a retarded manner. When a decerebellate animal is set down on a turntable on one foot, it will make fewer steps than an intact one with each rotation of the disk by 360°. We have

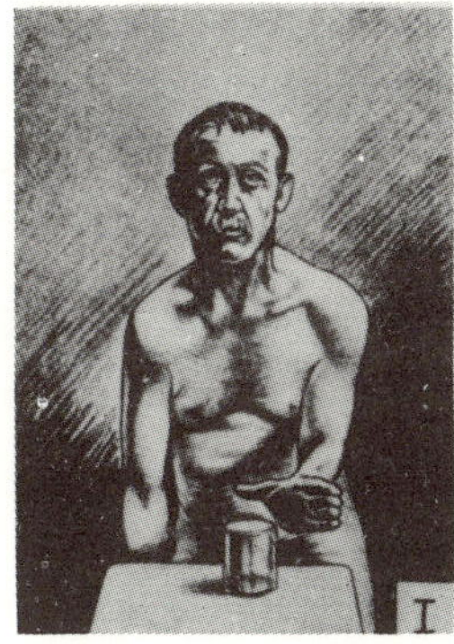

Fig. 272. 1. Epreuve de la préhension in a cerebellar patient. The hand is excessively opened. From : H. Claude and Levy-Valensi, Maladies du Cervelet, Fig. 27. 2. Decerebellate monkey Corrie. The animal tries to grasp the piece of cake offered. The animal, too, opens the hand far too much. 3. The animal eats up the cake held fast with both hands.

already described the delay in initiation of each step. With an equal rotation speed of the disk, it carries out the steps more slowly. When, however, the rotation speed is increased more and more, the steps of the decerebellate animal also grow more rapid so that they are finally carried out with great speed. This is also the case when the animals are set down on the rotating table with both fore- or hindlegs and one lets them run forwards, backwards or sideways. Thus, decerebellate animals can also carry out movements with rapidity. One obtains the impression, however, that the maximum speed which these aimals are able to reach is inferior to that of intact ones. Moreover, in running freely, the animals usually carry out the movements much more slowly than under the circumstances described above. This is especially the case when they run on a hard or smooth floor. When the animals run away to avoid being caught, however, they run very rapidly and one hardly succeeds in catching them. When they play with each other or pursue a rabbit, their movements are also very rapid. In retardation of movements, therefore, some "conditioned" cerebral influences must surely play a great role.

Stewart and Holmes' " rebound " symptom

When one orders a cerebellar patient to flex the elbow as vigorously as possible while one restrains the forearm and then suddenly releases it, the flexion will not stop instantly as in normal persons but hand and forearm fling upwards still further so that the patient sometimes hits his face with the hand. In the overfling which occurs after releasing, the elbow extensors are passively stretched. Thus, in cerebellar patients, these stretched muscles react too late with a modulating contraction. When decerebellate animals were seized by the tail and they tried to liberate themselves by pulling forwards, they were not able to stop the forwards movement as quickly as an intact animal when the tail was suddenly released.

Stewart and Holmes' symptom is probably based on abnormal delay in triceps contraction. As we have seen, a retarded occurrence of certain reactions can also be established with certainty in decerebellate animals.

Passivité

Passivité is the term used by André Thomas for the decrease in resistance against passive movement due to "hypostheny" of antagonists. This "passivité" appears, according to André Thomas, among others in the following impairments:

1. in the abnormally large flinging movements of the arms when the trunk, in a standing posture, is alternatively rotated passively to the left and to the right;

2. in the abnormally large flinging movements of the arms when these are at first lifted passively forwards and then suddenly thrown backwards;

3. in the abnormally strong flinging movements of the relaxed hands at the wrist joints when the forearm is passively moved up and down;

4. in the manifestation shown when a passively sideways lifted arm is suddenly released. In cerebellar patients, the arm, like an inert object, rebounds back several times from the thigh before lying quite still;

5. in the pendular knee-jerks;

6. in the reduced resistance to passive alternating movements of the head forwards and backwards, to the right and to the left;

7. in the too abundant flexion of the leg on tickling the sole of the foot; in my opinion, this manifestation is based on an excessive contraction of the knee and hip joint flexors and has nothing to do with hyposthenia of antagonists;

8. in Stewart and Holmes' symptom (as already discussed, in my opinion, this manifestation is probably based on a retarded contraction of elbow extensors, which are stretched on releasing the forearm);

9. in the manifestations which appear when a basket is held in the hand, with the elbow joint flexed at a right angle, and suddenly a weight of 1 kg is thrown into the basket. In a healthy person, the forearm will not move or only scarcely so, while in a cerebellar patient it is at first distinctly lowered and then returned to its initial position. In my opinion, this manifestation is probably caused by a retarded appearance of reflex contractions of the passively stretched elbow flexors (stretch reflexes first order).

Thus, we see a number of very different impairments which, according to André Thomas, are all due to only *one* cause. This conception seems to me at least highly contestable.

Only a part of these manifestations is present in decerebellate animals in the stage of permanent impairment and most of them are absent. Thus, among others, animals when suspended in the air head upwards keep their limbs firmly retracted against the trunk when the latter is alternatively rotated to the left and to the right, and show no abnormal flinging movements of the type mentioned above (1).

The feet do not fling in an abnormal manner when the forearm is passively moved to and fro (3). Neither does one succeed in bringing the feet into an oscillating movement and in flinging them vigorously backwards and forwards (2). Similarly, pendular movements on tapping the patellar tendon are absent (5) and passive movements of the head, particularly in a standing posture, encounter a very vigorous resistance (6).

Conversely, very obvious reactions on pinching and touching the sole of the paw are indeed present (7) and, as we have already seen, together with manifestations which correspond more or less to Stewart and Holmes' symptom (8).

Thus, very important differences become evident. One must, however, also consider that these manifestations can by no means be observed in all cerebellar patients. André Thomas also states that, in localized cerebellar lesions, the reduced resistance of certain muscles is often accompanied by an increased resistance of other muscles and of antagonists, which does not agree with several of the above-mentioned manifestations.

600

In any case, as long as discussion is limited to uncomplicated, shock-free cerebellar lesions, the above impairments do not prove that the results of experimental extirpation of the cerebellum cannot be applied to human pathophysiology.   In several of the defects specified by André Thomas it does not seem proven that they are indeed caused by "passivité," i.e., reduced muscular resistance, and not by a retarded appearance of reactions or by hypermetria.

Astasia

As we have seen, distinct unrestrained movements were also present in decerebellate animals, through which these animals were not able to stand still.

Nystagmus

The so-called "spontaneous" nystagmus is always absent in decerebellate animals in the stage of permanent impairment (see also Chapter XV; B).

Past Pointing

This and similar impairments are difficult to examine in animals. The decerebellate monkey Corrie, in the early stage, sometimes showed a defective grasp, which, however, could no longer be observed later on.

Impairments in running

The impairments in running are similar in many respects to those seen in cerebellar patients.

Atonia and asthenia

These manifestations were totally absent in decerebellate animals. As we have seen, hypotonia is also lacking in cerebellar patients and they even sometimes show a high degree of hypertonia and rigidity.

These manifestations which appear in totally or unilaterally decerebellate animals correspond to many of those observed in cerebellar patients but there are also considerable differences. Conformity appears, above all, in the hypermetria of movements, in the cerebellar ataxic running and the occurrence of unrestrained movements, which appear in man as well as in animals after cerebellar lesions.

As to the difference, it must be pointed out that, firstly, cases of uncomplicated cerebellar diseases or injuries are extremely rare and, secondly, that cases of tumors, inflammations, abscesses and haemorrhages cannot be compared with decerebellate animals in the stage of permanent impairment. In these patients, pressure and shock symptoms are almost always present by which the clinical picture is altered and in some cases resembles more the symptom picture which is seen in animals in the period immediately after extirpation of the cerebellum. In this period, as we have seen, the animals are sometimes very flaccid and show a high degree of hypotonia. Cases of cerebellar injuries accompanied by inflammation or intercurrent diseases should also not be compared w th it. In decerebellate animals, it could often be perceived that insignificant complications such as suture suppuration, purulent conjunctivitis or rhinitis can greatly delay the return of various functions. It could frequently be observed that by the occurrence of feverish disease, numerous symptoms disappeared, among them the excessive extension and fixation of legs and the abnormal dorsal curvature of the back on static stress of the limbs. Furthermore, it should not be forgotten that, in many cerebellar patients, the function of the cerebrum remained intact and, therefore, is able to inhibit numerous manifestations, just as in decerebellate animals. We have seen that decerebellate animals sometimes progress on a smooth floor not with over-extended legs as on a soft surface, but with strongly flexed limbs with the bely close to the ground.

Based on these different factors, it is conceivable that one finds in the literature descriptions of cases of cerebellar diseases in which the patient shows a pronounced hypertonia, even a rigidity and a symptom picture just like in Parkinson's disease. In other cases, however, a clear hypotonia could be established, while in still other cases, particularly those of cured cerebellar abscesses, which had destroyed a considerable part of the cerebellum, no, or almost no, symptoms could be observed.

XX. PHENOMENA AFTER COMBINED EXTIRPATION OF THE CEREBELLUM AND THE CEREBRUM

A. THE DECEREBELLATE AND UNILATERALLY DECORTICATE ANIMAL

In the dog Vici in which the cerebellum and the *right* half of cerebrum were simultaneously extirpated, there immediately appeared after this combined extirpation an extensor rigidity exactly like that after a typical decerebration. This condition gradually changed. First, the extensor rgidity in the right legs disappeared, while the left ones remained rigid much longer and even at the stage of permanent impairment still showed an increased tendency to an extended posture, in the various positions in the air (Fig. 273) as well as in the dorsal position on a supporting surface.

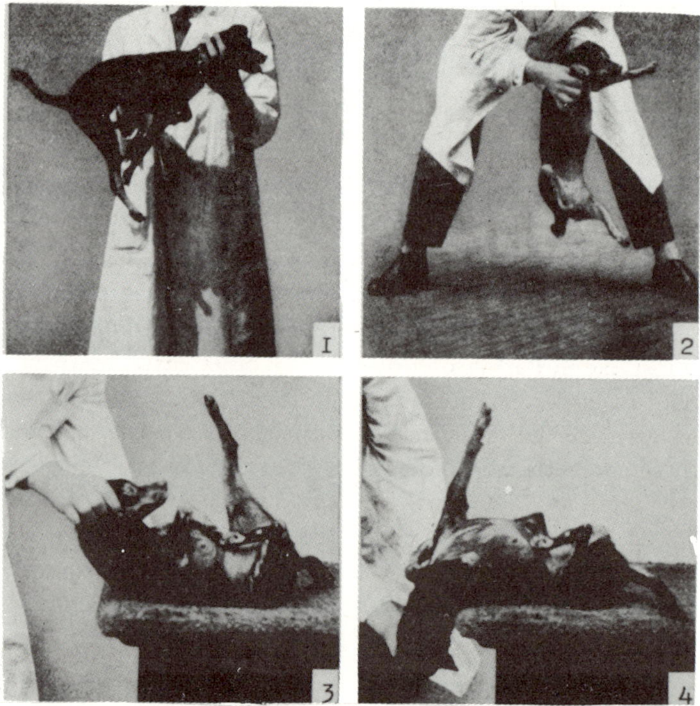

Fig. 273.   Dog Vici after extirpation of the cerebellum and the *right* half of the cerebrum. 1. In a ventral position in the air the animal keeps the *left* legs extended, the right ones flexed. 2. Similarly suspended head upwards. 3 and 4. In a dorsal position with ventrally directed muzzle the left hindleg, with dorsally directed muzzle the left foreleg is kept extended up in the air.

In the neck, too, the distribution of muscle tonus was asymmetrical. Suspended head downwards (with eyes closed), the asymmetrical contraction of the neck muscles usually caused a slight turning and rotation of the head towards the left, while in a dorsal position, due to a vigorous contraction of the left cervical muscles, a concavity of the neck to the left appeared.

At the stage of permanent impairment, a typical combination of manifestations which appear after an exclusive unilateral extirpation of the cerebrum and after the exclusive extirpation of the cerebellum could be observed. We have seen that after extirpation of the cerebrum as well as of cerebellum, the leg-slackening reactions appear retarded and only occur after a stronger change in position. In agreement with this, the leg-slackening reactions in the dog Vici remained impaired for a longer time in the left legs than in the right ones, though in the latter they were also distinctly retarded in onset. This delay in onset in the left legs was so severe that the leg-slackening reactions of these legs were even totally unsuitable for the maintenance and restoration of balance, so much so that the animal never learned to stand and to run freely. Sometimes, with a jumping movement, the animal rose on its four legs but, after quickly doing several steps, soon fell down again. In this case, the animal fell head forwards because the forelegs had not been displaced quickly enough and far enough forwards. At the same time, it nearly always fell to the left and only sometimes on its right side.

From the lateral position, the animal instantly lifted its head freely in the air, rotated the anterior part of the body into a ventral position, while the posterior part remained lying on its side. When the posterior part of the body was lying on its left side, after some time it almost always led into a lateral position to the right by turning the head to the left, via its belly. Thus, the usual posture of the animal was: posterior part of the body in a lateral position to the right, anterior part in a ventral position head freely lifted in the air, usually turned a little to the left (Fig. 274). The animal, however, could also keep the head straight forwards and move it to the left and to the right, forwards and backwards.

In this position, the animal continually crawled around the room, over obstacles, and once it had set itself a goal it was usually able to reach it, as for instance the feeding or drinking bowl.

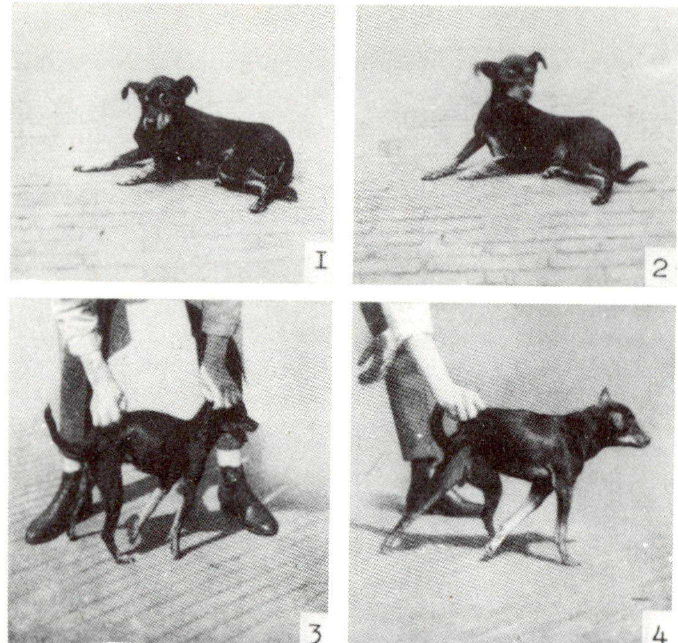

Fig. 274.   Decerebellate, right-sided decorticate dog Vici. 1, 2.   The animal
in its usual position; head raised freely in the air, anterior part of the
body in a ventral position, posterior part of the body in lateral position to the right.   3-4.   The animal held up in a standing posture.

Due to the extirpation of the right side of the cerebrum, the animal showed
the following manifestations.

1. Hemianopsia.

2. Absence of preparation for standing (placing) in the *left* legs to visual
   stimuli as well as on stimulations from the body surface. On contact
   of the lower jaw with the edge of a table, only the right foreleg was
   set down on the table, while the left remained hanging in the air
   directed backwards with extended elbow and wrist joints (Fig. 16).

   When the animal was placed on a lattice and the legs were
   pulled through, only the left legs were pulled back and set down
   with the feet on the lattice bars. When it was lowered from being
   suspended head upwards, until the tip of the tail or the posterior
   part of the body touched the ground, the animal only brought the
   right hindleg into a standing posture (Fig. 275).

3. Reduction of strength in supporting tonus of the *left* legs, although
   the *left* legs showed a higher tendency to an extended position, for
   instance suspended head upwards, a distinct stretch tonus. The *right*

legs, however, showed a distinct flexor tonus, yet the strength of supporting tonus in the left legs was inferior to that in the right ones. When the animal was set down on a supporting surface with only the right foreleg, this leg would not give way on adding a burden on the shoulder with a load of 5 kg, while the left leg, in the same position, would already give way under the load of 1 kg on the shoulder.

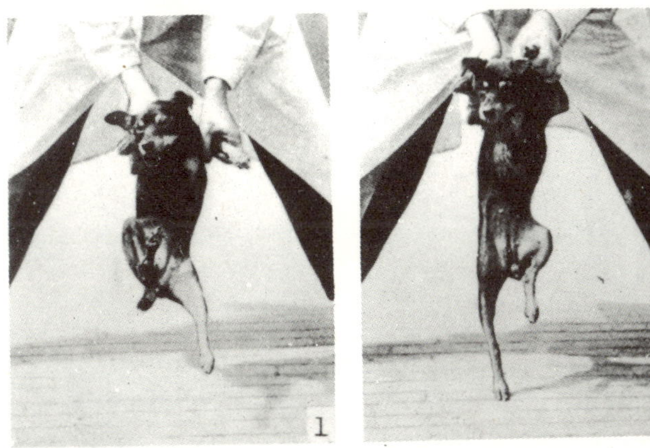

Fig. 275. Decerebellate, right-sided decorticate dog Vici. 1. Suspended head upwards. The animal holds the left hindleg extended, the right drawn into a flexed position. 2. On contact of the posterior part of the body with the ground, the animal sets down only the right hindleg in a standing posture, thus only the right hindleg shows readiness to stand. (Notice the alteration in position of the left hindleg.)

4. Much stronger impairment of the leg-slackening and hopping reactions and of the proprioceptive correcting movements in the left legs than in the right legs. In the left legs, these reactions only appeared after a much more extensive alteration in position than in the right ones. This difference must be interpreted as a result of a unilateral extirpation of the cerebrum. However, the fact that the impairments were so much stronger than usual after a unilateral extirpation of the cerebrum and that these reactions also appeared distinctly delayed in the right legs (Fig. 276) must be caused by the extirpation of the cerebellum.

5. It must also be interpreted as a result of right-sided extirpation of the cerebrum that in rocking to and fro, when the animal supports itself on the left legs, no or almost no rocking reactions could be observed in the right legs. Conversely, on standing on the right

606

legs, the rocking produces particularly violent flexing and extension movements (result of extirpation of the cerebellum!) of the left legs (Fig. 161).

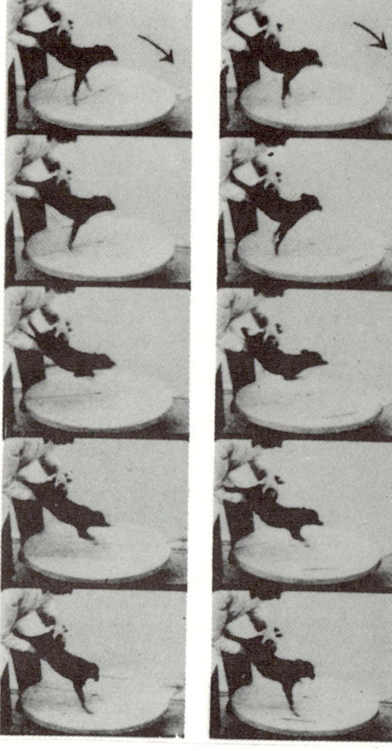

Fig. 276. Hopping reaction of the right foreleg. Notice the retarded appearance and the hypermetric performance of these reactions. (See also Fig. 201.)

As a result of extirpation of the cerebellum, the animal further showed:

1. unrestrained movements like wobbling of the head and trunk, especially when it was set down on its legs,

2. a pronounced hypermetria of all movements of the head as well as of the extremities (Fig. 276),

3. a particularly energetic occurrence of magnet and supporting reactions in all four legs, *also in the dorsal position.*

The lively occurrence of magnet reactions in the left legs is remarkable because this reaction is either absent or only weakly present in animals in which only the right half of the cerebrum was extirpated. In the dog Vici, however, touching of the soles of the left legs immediately caused a lively extension with distinct maintenance in extended posture. The strengthening influence of the extirpation of the cerebellum thus not only compensated the reducing influence of the extirpation of the cerebrum, but even caused an abnormally exaggerated, uninhibited appearance of this reaction (Fig. 277, No. 2).

The combined loss of cerebellar and cerebral functions probably caused the extensor posture shown by the left legs, even without being under static stress, in dorsal and in other positions. Similarly, it presumably caused the particularly strong and automatic influence on the muscle tonus of the left legs due to alterations in position of the head, the contralateral legs, or the pelvis. We have seen that in the dorsal position with the muzzle directed ventrad, the left hindleg was totally extended (Fig. 273) and offered a distinct resistance against passive, not static, flexion. This extended posture and resistance disappeared, among others, with the following manuevers:

a dorsal directed alteration in position of the head (Fig. 278, No. 1 and 2),

on pinching of the tail,

on a dorsal movement of the pelvis in the lumber joints,

on retraction of both forelegs under static stress and

on static stress or touching the sole of the opposite hindleg (Fig. 278, No. 2 and 3).

Fig. 277. Decerebellate, right-sided decorticate dog Vici. Magnet reaction of the left legs. 1. The animal suspended head upwards. On touching the sole of the right hindlegs, this leg is extended (magnet reaction). The other left hindleg is flexed at the same time. 2. On touching the sole of the left hindleg, this leg, too, goes over into an extended posture.

608

When in the dorsal posture of the animal, the right hindleg under static stress was passively moved at the hip joint, through which the pelvis was similarly moved due to rigidity, the left hindleg sometimes showed a flexed posi ion and sometimes an extended one. On abduction of the right hindleg, the left one was extended, while on the following adduction of the right hindleg the left one returned to its flexed posture, i.e., behaved as in rocking reactions. When the right hindleg was alternatively moved forwards and backwards, the following coul be seen:

on passing the intermediate position: flexion of the left hindleg (Fig. 279, No. 2).

on being strongly retacted: extention of the left hindleg (Fig. 279, No. 1).

on being strongly protracted: also extension of the left hindleg (Fig. 279, No. 3).

Fig. 278. Decerebellate, right-sided decorticate dog Vici in a dorsal position. 1. With a dorsally directed position of the head the left hindleg shows a flexed position (the left foreleg is extended). 2. On a ventral flexion of the neck, the left hindleg went over into an extended position. 3. On touching the sole of the right hindfoot, the left hindleg has returned into its flexed position.

When a strong pressure was exerted on the sole of the right hindfoot so that the leg was slightly bent, an extension of the left hindleg took place, while on pulling the right hindleg upwards the left one returned into its flexed posture (Fig. 280). The same passive movements of the left hindleg did not produce any alterations in position of the right one.

The muscle tonus of the left hindleg was influenced not only by alterations in position of the head, the pelvis, the forelegs and the opposite hindleg, but also by alterations in position of the leg itself.

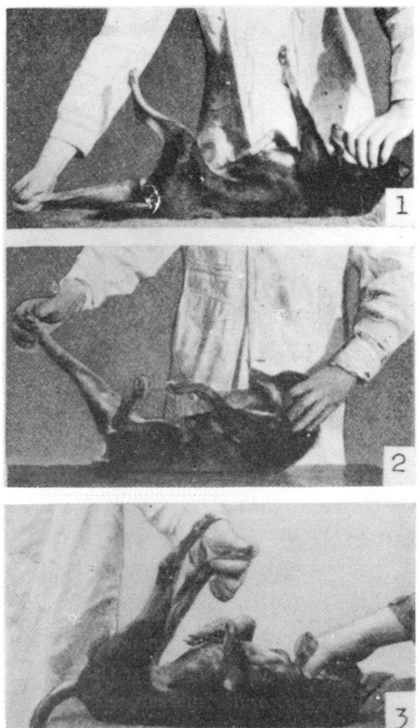

Fig. 279. Decerebellate, right-sided decorticate dog Vici in a dorsal position, the muzzle directed ventral. 1. On moving the extended, right hindleg backwards, the left hindleg is also extended. 2. On moving the right hindleg forwards, the left one goes into flexion as soon as the former nears the intermediate position (see also Fig. 277, No. 3) to return into its extended posture. 3. With a forwards directed position of the right hindleg in which the pelvis is also directed forwards. Notice also the reactions of the left foreleg.

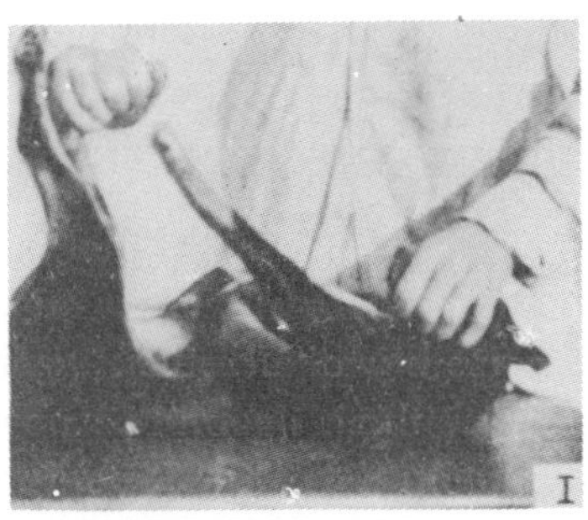
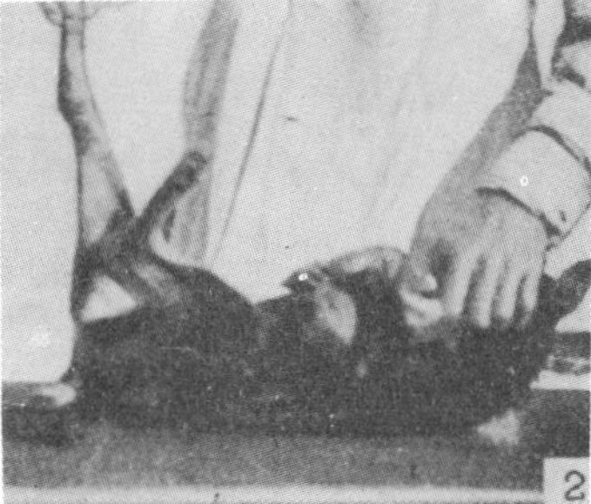

Fig. 280. Decerebellate, right-sided decorticate dog Vici in a dorsal position. 1. A strong pressure is exerted on the sole of the right hindfoot. The left hindleg is extended. 2. The right hindleg is pulled upwards. The left hindleg falls into a flexed position.

610

When the leg was adducted by a pressure on the outside of the thigh, it
went into flexion and,conversely, on abduction by pressure on the inside
of the thigh it was extended.

The left foreleg showed almost similar manifestations; it was also
extended when the opposite leg was moved backwards or abducted.
On altering the head position, however, it reacted contrary to the
hindleg, as it went into flexion on a ventral flexion of the head and on
a dorsal movement it was extended (Fig. 278, No. 1 and 2). Moreover,
it is worth mentioning that the left foreleg was also extended when a
strong pressure was applied on the sole of the right *hindfoot* (Fig. 280)
and also on a strong backwards movement of this right leg (Fig. 279,
No. 1).

Thus, we see that not only the forelegs exert an influence on
the hindlegs, but also the hindlegs on the forelegs.

In the dog Vici, we found a symptom picture that could hardly be
unravelled, but which became much clearer in comparison with
animals, on the one hand, which had only one half of the cerebrum
removed and, on the other hand, which only had the cerebellum extirpated. Particularly striking in this animal was the quite mechanical
influence of the position of the head, the pelvis and the right legs on
the muscle tonus and the position of the left leg. We have described
corresponding observations in decorticate and decerebellate animals.
These animals, however, showed these influences clearly only when
the legs were under static stress, but in the dog Vici the various reactions
were evident in the left legs when held freely in the air, even when the
animal was in a dorsal position. The cat Esperance and the dog
Däumling, in which one half of cerebellum were also removed, showed
symptoms which, on the whole, correspond with those described above.

B. THE DOG WITHOUT CEREBRUM AND CEREBELLUM

In the dog Robbie, the left half of the cerebrum was extirpated on
the 9th of December, 1926, the right half on the 29th of December,1926,
and the cerebellum on the 7th of February, 1927. The animal died of
distemper encephalitis on the 17th of March, 1927. The post-mortem
examination revealed a purulent bronchitis and pneumonia and an
empyema of the right frontal sinus.

After extirpation of the left half of the cerebrum, the animal showed the
usual manifestations such as impairment of preparation for standing

(placing) on the right side, hemianopsia, running circles to the left, etc. The right legs also showed the typical impairments of leg-slackening reactions and the strength of supporting tonus in the right legs, especially in the right hindleg, was distinctly reduced.

After extirpation of the second half of cerebrum, the right one, running in circles to the left, strange to say, continued, contrary to other animals which, in this case, would run in circles to the right for a time after extirpation. Running to the left is all the more remarkable as the animal, while standing still, always turned the head somewhat to the right and when suspended head downwards showed no distinct basic rotation but kept its head symmetrical to the trunk. However, as soon as the animal started to run, the head turned to the left and this position was maintained while running in circles to the left.

Later on, the animal would run straight forwards over a considerable distance and showed a deviation sometimes to the left and sometimes to the right, according to circumstances. A distinct asymmetry could no longer be observed at this stage, neither on the part of the various reactions of the extremities nor in the distribution of muscle tonus. This is particularly worth mentioning because in the course of extirpation of the cerebrum the left corpus striatum was also removed, but the right one was not (Figs. 281 and 282).

The animal now showed the typical manifestations of a decorticate dog. The preparation for standing was totally absent (Fig. 258) as well as the other conditioned reflexes on visual and acoustical impulses and on stimulations from the body surface. I would like to mention a few more manifestations as a comparison with the behavior following extirpation of the cerebellum.

The animal was well able to run and showed no abnormal extensor posture of the legs while standing or in running (Figs. 257, 283). Abnormal positions of the feet given passively were well corrected, although the "conditioned" correction on stimulations from the body surface was lacking.

However, the correction on proprioceptive stimulations distinctly occurred. Touching the soles of the feet produced no clear magnet reactions. But the proprioceptive supporting reactions were evidently present and, in a ventral position of the animal, static stress caused a distinct fixation in extended posture of the fore- as well as of the hindlegs. However, maintenance of posture in the hindlegs was less strong

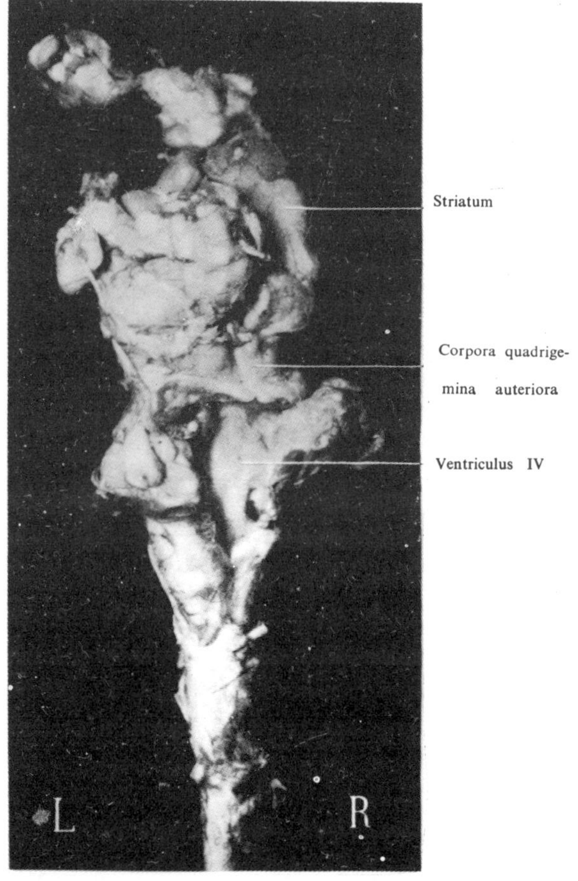

Fig. 281. Dog Robbie. The parts of the central nervous system left intact seen from the dorsal side. (For the microscopic investigation on serial sections, see Nos. 208 and 240 of the list of references.)

in this animal than before extirpation of the cerebrum: the strength of supporting tonus in the hindlegs was evidently reduced.

Strength of supporting tonus on the 5th of February, 1927, i.e., 38 days after extirpation of the cerebrum:

Bodyweight of animal   Strength of supporting tonus in the forelegs

6.8 kg   > 14 kg   < 16 kg

Strength of supporting tonus in the hindlegs

> 3 kg   < 4 kg

Due to the reduced strength of supporting tonus, the hindlegs instantly gave way when a sandbag of 1.7 kg was placed on the back of the animal. These impairments of muscle tonus were probably the reason for the fact that the hindlegs showed no distinct rocking reactions

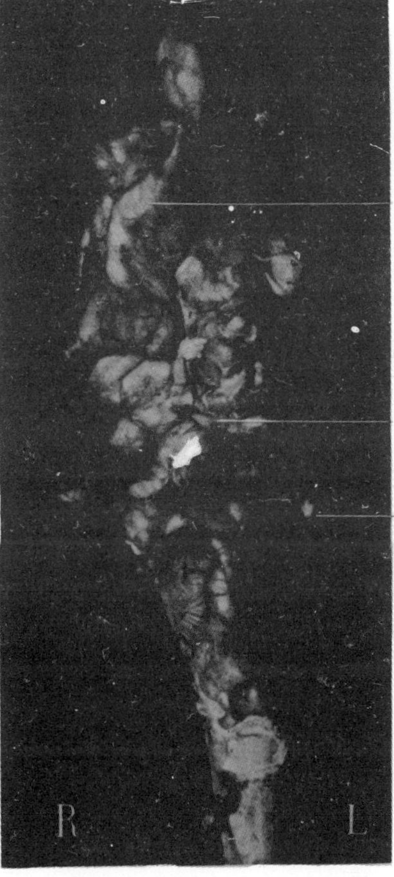

Fig. 282.   Dog Robbie. The parts of the central nervous system left intact seen
from the ventral side. (For the microscopic investigation on serial
sections, see Nos. 208 and 240 of the list of references.)

and that the adaptation of leg position to the position of the supporting surface did not occur promptly while standing on a board which was slowly moved tilted and downwards at the head- or tailend.

The strength of supporting tonus was evidently stronger when the anterior part of the body was passively raised and the animal supported itself only on the hindlegs which were strongly directed backwards at the hip joints. In this case, the hindlegs did not give way, although almost the entire weight of the trunk rested on them (6.8 kg). They also failed to give way when the animal, on its hindlegs, actively supported itself against a wall.

On flexion of the distal phalanges, the supporting tonus completely disappeared.

In the dorsal (supine) posture of the animal, the limbs were kept flexed in all positions of the head and, in this position, static stress applied to the soles of the hindfeet caused no extension and no, or only slight, fixation of the passively extended hindlegs even when the muzzle was directed vertically upwards or ventrad. Thus, the dorsal position, as is usually the case in intact dogs, exerted a distinctly inhibiting influence on the supporting reactions.

The animal showed prompt leg-slackening and hopping reactions of fore- and hindlegs forwards, backwards, outwards and inwards. When the animal was set down with one leg on a turntable, these reactions also appeared lively and showed only a slight delay. The various righting reflexes were also obviously present.

In running, it showed a normal coordination and the paws were always set on the ground in an appropriate manner.

Extirpation of the cerebellum, on the 7th of February, 1927, caused in this decorticate dog, as frequently in intact dogs, the appearance of the symptom picture of decerebrate rigidity. Immediately after extirpation of the cerebellum, the animal would lie on its side totally rigid with stiffly dorsiflexed neck, upwards curved back, dorsally extended tail and totally extended, rigid fore- and hindlegs. The animal made no attempt to rise, the labyrinth and body righting reflexes were absent.

The extremities offered a strong resistance to passive flexion and the tonus of the extremities showed a typical plasticity. With passive flexion, one could feel that the resistance relaxed suddenly (clasp-knife phenomenon and cog-wheel phenomenon). In holding the hand against the passively flexed paw, one could feel that the return of extensor tonus also occurred in jerks and sometimes with each respiratory movement. Passive flexion of the wrist joint encountered a strong, elastic resistance and caused a slight associated flexion of the elbow joint. However, the passive flexion of wrist joint and digits caused no relaxation of extensor tonus; with flexed wrist and phalangeal joints, a passive flexion of the elbow still encountered a considerable resistance. In the hindlegs, too, the extensor tonus did not disappear on flexion of the toes. Thus, the negative supporting reactions now showed distinct impairments.

In response to repeated pushes on the soles of the feet, the rigidity increased. Thus, static stress still had a distinct result. If the animal was passively set up on its legs, the limbs showed an exaggerated extensor posture, while the muzzle was kept pointing upwards in the air owing to the strong dorsiflexion of the neck. The legs did not give way under the weight of the trunk.

When the animal lay in the dorsal position, too, the legs were rigid and were kept extended upwards and the execution of pressure pushed on the soles of the feet also caused in this position a distinct increase in rigidity. Thus, the dorsal position no longer exerted an inhibiting influence either on the extensor tonus or on the reactions produced by static stress.

The rocking, leg-bracing and leg-slackening reactions as well as other proprioceptive correction movements were absent. Conversely, a number of other reflexes were clearly present. Thus, touching of the hair of the ear causes a lively reflex movement of the pinna and touching the interior of the nose, a sneezing reflex. The corneal reflex, the eyelid reflex on touching the eyelashes, the ipsilateral flexor reflex and the crossed extensor reflex on pinching of the toes also appeared distinctly.

The animal swallowed well meat put into its mouth (from the third day on) and urinated and defecated spontaneously. The urine showed no abnormalities, no sugar and no albumin, even on the day of the operation and on the first days after extirpation.

The animal had attacks of restlessness in which it made alternate running movements, in which the movements were carried out mainly in the shoulder and hip joints, and hardly any flexion of elbow and knee joints could be observed.

The rigidity of the animal gradually decreased, but, at the end of the first week, a distinctly increased tendency to extensor postures was still present (Fig. 283, No. 3-4).

Even now, the animal in the dorsal position still held its legs extended upwards (Fig. 283, No. 4) and no distinctly inhibiting influence could yet be observed in the dorsal position. There usually even appeared, under the influence of tonic labyrinthine reflexes, a distinct increase in rigidity of the neck and extremities on shifting from a ventral to a dorsal position. Also, when in a lateral position, in which it no longer showed a maximum rigidity, if the head was passively rotated into dorsal position, the tonic labyrinthine reflexes caused a distinct increase in extensor posture and rigidity in all four extremities. Moreover, in this case, a distinct influence of tonic neck reflexes could be

Fig. 283. Dog Robbie. 1—2. The animal, 38 days after extirpation of the cerebrum, and 2 days *before the extirpation of the cerebellum,* standing freely. The legs do not show an excessive extensor posture. 3-4. The animal, 6 days *after extirpation of the cerebellum.* It is distinctly rigid. In a standing posture (3) the legs, especially the forelegs, show an exaggerated extensor posture. Head and tail are kept actively raised. In a dorsal position (4), too, the legs are now held extended upwards.

observed (the increase was less in the cranial legs than in the mandibular legs) but the influence of the tonic labyrinthine reflexes predominated.

In addition to the tonic neck and labyrinthine reflexes, a distinct influence of the curvature of the back on the extensor tonus of the hindlegs could now be observed. As already mentioned, the animal would lie continually in a lateral position with extended hindlegs and the back showed a dorsal convexity. When in this position, if the legs were passively flexed and in so doing any alteration in position of the back was carefully avoided, the legs when released, soon returned to an extended posture. When, however, on passively flexing the hindlegs, the back was simultaneously made hollow, the legs, when released, remained in a flexed position. Although rigidity was reduced, the legs could still carry the trunk (Fig. 283, No. 3); the forelegs did not even give way when a sandbag of 5 kg was placed on the shoulders (Fig. 284, No. 1). The hindlegs, however, now instantly collapsed with a very slight pressure on the pelvis. When the anterior part of the body was passively lifted, the extensor tonus of the hindlegs also gave way; these legs could no longer carry the weight of the trunk alone, which they could when the animal was decorticate before extirpation of cerebellum. When the animal was set down with the hindlegs on a balance and the anterior part of the body was then lifted and a pressure exerted on the pelvis, it was seen that the hindlegs already give way at a counterpressure of 2 kg (Fig. 284, No. 3), while the forelegs did not give way with a counterpressure of 9 kg (Fig. 284, No. 2) but did so at a pressure of 11 kg.

Strength of supporting tonus:

of both forelegs in standing posture	+ 9 kg	failed at 11 kg
of both hindlegs in standing posture	+ 1 kg	failed at 2 kg
of both hindlegs with the animal in a lateral position	+ 1/8 kg	failed at 1 kg

Thus, the strength of supporting tonus now was less than before the extirpation of the cerebellum, although the animal was still distinctly rigid. As we have seen, after extirpation of the cerebellum a temporary decrease in the strength of supporting tonus could also be observed in intact animals.

When the forelegs of the dog Robbie were set down with the dorsum of the feet on the balance, the limbs could withstand a counterpressure of 7 kg, but not of 8 kg. Thus, with flexed wrist joints, the resistance of the limb to passive flexion was slightly less. A typical negative supporting reaction, however, did not appear, even then.

  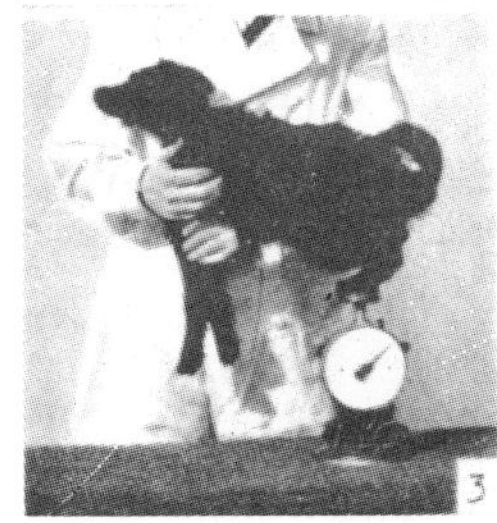

Fig. 284. Decerebellate and decorticate dog Robbie. 1. The forelegs do not give way on loading the shoulders with a 5 kg sandbag. 2. They give way just as little when a counterpressure of 9 kg. is exerted on the soles. 3. The hindlegs, however, already give way at a counterpressure of 2 kg.

In the second and third week after extirpation of the cerebellum, the rigidity gradually decreased and at the same time the labyrinthine reflexes returned. In the third week, the animal always rotated the head into a normal position, from a lateral position to the right as well as to the left, so that the head now always lay flat with the lower jaw on the supporting surface. Sometimes, it now also brought the whole anterior part of the body from a lateral position to the right into a ventral position and would then lie the head turned and curled up on the left side like an intact dog. In this position, the animal kept the four legs flexed, but usually the elbows did not yet show such a strongly flexed posture as seen in intact dogs in the same position. The animal often made attempts to raise itself up entirely from the lateral position on the right side, but the posterior part of the body, as soon as it was somewhat lifted up, immediately fell back into a lateral position, so that the attempt to raise itself up usually produced only clockwise movements to the left. When the animal became very restless, it sometimes made a sudden jumping movement by which the whole trunk was lifted from the ground, to fall back into a lateral position to the left.

From this lateral position, it also rotated the head into a normal posture, but this rotation of the head was not followed by a rotation of the anterior part of the body into a ventral position. In this lateral position, too, the legs were now kept somewhat flexed but, especially in the forelegs, an abnormal extensor tonus was still present and, in fact, more clearly pronounced than in the lateral position to the right.

The extensor rigidity still appeared in a typical manner when the animal was placed on its back or was lifted into the air in a ventral position. In this case, the head was instantly retracted, the legs were extended, the animal became very restless and usually started to make violent running movements. When it was lifted up, it sometimes barked like an angry, impatient dog. The animal, when lying on its side, was not yet able to lift its head calmly and freely in the air; it still had no control of its head movements. The raised head was thrown to the right and to the left, pulled backwards and again to the side by moving forwards. Sometimes, these movements followed each other in a regular manner and thus there appeared very peculiar complicated rotating movements. Sometimes, on this occasion, the head hit the floor or wall of the cage with an impact which produced an enraged howling. This head movement was also shown by the animal when it was held in the air. However, when one waited till it grew calm, it kept its head rotated 45° to the normal position (labyrinthine righting reflexes) and this in the left as well as in the right lateral position in the air.

In the fourth and fifth week after extirpation of the cerebellum, the increased extensor tendency had even further diminished and the animal now showed typical positive and negative supporting reactions. The paws were now totally flaccid with flexed distal phalanges, while the legs under static stress could only be flexed with an effort. The strength of supporting tonus had distinctly increased. This was apparent when the animal was set down on its hindlegs and the anterior part of the body was lifted from the ground. The hindlegs now no longer gave way and were again capable of carrying the trunk, but even now the strength of supporting tonus in the hindlegs was distinctly weaker than in intact dogs, and there was no longer any difference from the strength of supporting tonus before extirpation of the cerebellum. Now also the animal lacked magnet reactions. When the animal, standing on its hindlegs, was moved backwards, the legs alternately hopped backwards; with similar steps in moving the trunk to the left or to the right, even a sideways running of the hindlegs took place. When the animal also stood with only one leg on a supporting surface and was then moved passively in several directions, it displaced the hindleg forwards, backwards, outwards and inwards. Thus, the various hopping reactions had returned in the hindlegs, though their appearance was too delayed for the maintenance of balance. In the forelegs, these reactions could not yet be established and it was in general very difficult to set down the animal on its forelegs because the forelegs, on static strain, were abnormally extended and strongly retracted at the shoulders so that the animal fell head forwards and the feet slid backwards. When the animal was moved downwards from being suspended head upwards till it reached an inclined board, the hindlegs showed adaptation reactions. With the downwards movement, the hindlegs were passively moved forwards at the hip joints by the board, which produced running movements backwards, so that the posterior part of the body ran up and down the board, according to the position of the animal to its supporting surface (Fig. 285).

It was striking that the tonic labyrinthine reflexes appeared so particularly distinct even at this stage. While a rigidity could hardly be observed either in a ventral and lateral position or in suspended position head upwards or downwards, the shifting of the animal into a dorsal position always produced a lively extension of the extremities and a distinct contraction of dorsiflexors of the neck. An investigation of the presence of the various reactions at this stage brought the following results.

Preparation for standing (placing)	—
Munk's touching reflex	—
Correcting movements of stimulations from the body surface	—
Magnet reactions	— (?)
Positive and negative supporting reactions	+
Leg-slackening reactions	+ (only in the hindlegs)
Correcting movements on proprioceptive stimulations	+ (only in the hindlegs)
Leg-bracing reactions	—
Tonic neck reflexes	+
Body righting reflexes on the body	—
Labyrinthine righting reflexes	+ (from both lateral positions)
Tonic labyrinthine reflexes	+
Horizontal eye deviations and eye rotary nystagmus	+

The animal showed alternating periods of restlessness in which it made attempts to raise itself up and periods of complete repose in which it gave the impression of sleeping. In this case, it lay motionless with closed eyes and made quiet regular breathing movements. When it was shaken to and fro, it gave the impression of waking up; it opened its eyes, sometimes yawned and shook itself like an intact dog.

The animal chewed meat very well when it was put into its mouth but would spit out pieces of stale bread even when mixed with minced meat while it swallowed the meat. On tickling the mucous membrane of the nose, the animal sneezed and then licked its nose. When sprinkled with water, it would shake off the water and made typical shaking movements with the whole trunk like a wet intact animal.

The strong susceptibility to certain noises was very remarkable. To a sibilant inspiratory whistle[1] it reacted not only by pointing the external ears, but also by lifting the head from the supporting surface

Fig. 285. Decerebellate, decorticate dog Robbie. 1. The animal is lowered from a position suspended head upwards till the hindlegs touch a slanting board. 2. With a further movement downwards, the board has pressed the legs passively ventrally, causing running movements backwards and the posterior part of the body has *run downwards* a few steps on the board. 3. Again the animal is lowered from a position suspended head upwards, but now its back is directed towards the upper end of the board. 4. Now also backwards directed running movements were performed by the legs as soon as the hindlegs were displaced ventrally by the board and the posterior part of the body has *run upwards* on the board for a few steps.

(1) By "Zwitschern" (twittering, chirping in English) Rademaker says, "I mean the sound produced by sucking in air through a small cleft in the tightly closed mouth with pursed lips."

and movements of the legs. This observation is very remarkable for two reasons: firstly, because the animal reacts so strongly just to this sound with which intact dogs are usually called, while it did not react or reacted only slightly to other noises, even very loud ones, as for instance, the noise produced by vigorously hitting a table with the flat of the hand; secondly, because the reactions to this whistling noise were much more prominent than before extirpation of the cerebellum.

The animal did not show any tremor or any of the typical unrestrained movements of head and legs even when it was set down on all fours.

The regulation of temperature seemed normal and the metabolism, too, showed no noticeable impairment.

The animal was not yet able to run and stand freely. When set down on its legs, it instantly rolled to its side, when not laterally supported. Unfortunately, in the sixth week after extirpation of the cerebellum, it fell ill with distemper. It started to cough, developed a high fever and shivering attacks, purulent rhinitis and conjunctivitis. From the nose flowed a purulent secretion which the animal continually licked off. Palpation of the region of the frontal sinus caused a distressed howling. The post-mortem examination showed an empyema of the right frontal sinus. The animal died on the 17th of March, 1927, i.e., 38 days after extirpation of the cerebellum.

The lifespan of 38 days after extirpation of the cerebellum, as was shown by observations in animals in which the cerebellum was alone extirpated, is not sufficient to presume that the dog Robbie had already reached the stage of permanent impairment. The more so in that the animal fell ill already in the third week and developed a high fever which, as is known, inhibits the return of functions. Also, the signs of recovery seen almost daily before the animal fell ill so seriously, are proof that the stage of convalescence was not yet terminated. Thus, it surely did not show the maximum achievements that can be reached by an animal without a cerebrum and a cerebellum. From the difference in behavior of this animal and that of decorticate dogs with an intact cerebellum it should therefore not be concluded that this difference is due exclusively to the loss of cerebellar functions of the former. Although the experiments made with dog Robbie did not reach the desired conclusions, they nevertheless resulted in important facts:

1. That extirpation of the cerebellum also causes in decorticate dogs an extension rigidity with typical plasticity just like that after

decerebration and that this rigidity remains in full development for nearly a week.

2. That the rigidity caused by extirpation of the cerebellum also gradually decreases in the decorticate dog and almost totally disappears.

3. That, in the decorticate dog, too, the decrease in rigidity is accompanied by a return of supporting reactions and that because of this return the hindlimbs under static stress supported heavier loads than the stiffly extended hindlimbs in the stage of rigidity in standing posture of the animals.

4. That the strength of supporting tonus was indeed at first reduced after extirpation of the cerebellum, but reached again the same magnitude as before extirpation.

5. That also in the decerebellate dog some reflexes, such as labyrinthine righting reflexes and hopping reactions, were totally absent in the first days after extirpation but gradually returned and that the return of balance reactions after extirpation of the cerebellum *is not based, in any case, exclusively on compensating actions of the cerebrum* as is asserted by some authors.

6. That, however, the returned hopping reactions show a singularly retarded appearance.

7. That in the decorticate dog, too, extirpation of the cerebellum caused a reduction in the inhibiting influence of the muscle tonus produced by lying in the dorsal position. (Even shortly before its death, shifting the animal into a dorsal position produced an increase in extensor tonus caused by tonic labyrinthine reflexes).

8. That in the decorticate dog Robbie certain reactions appeared more clearly after extirpation of the cerebellum than before.

The manifestations shown by the decorticate dog Robbie after extirpation of the cerebellum thus strikingly correspond, on the whole, with those manifestations which could be observed after cerebellar extirpation in intact animals.

Various authors have asserted that the cerebellum mainly exerts its influence by way of the cerebrum, that it almost exclusively influences cerebral reactions, and they assume, based on this concept, that extirpation of the cerebellum would produce fewer alterations in a decorticate animal than in an intact one.

Other authors even expressed the supposition that the decerebellate thalamus animal would show fewer impairments than the decerebellate animal with intact cerebrum. According to the opinion of these authors, the various actions such as running, etc., in decerebellate animals are brought about voluntarily by cerebral stimulations since, according to them, extirpation of the cerebellum has destroyed the subcortical mechanism by which these actions are normally carried out. They point to the fact that habitual actions like riding a bicycle are carried out clumsily and with excessive movements as long as they are performed by way of the cerebrum, that, however, they are carried out adroitly as soon as they are performed without thinking, subconsciously and subcortically.

By the observations on the dog Robbie, these suppositions and expectations have proved not to be correct, at least with regard to the acute manifestations. The most important result of the experiments with the dog Robbie, in my opinion, is the fact that a dog without a cerebrum and a cerebellum can live more than 38 days if no intercurrent diseases occur. Repetition of these experiments should promote and further our knowledge relative to the functions of the cerebellum and the brainstem.

Based on the observations discussed in the last three chapters, one can assume for certain that the cerebellum exerts a regulating influence on the various motor reactions; on the "conditioned" reactions which, by optical and acoustical impulses and by stimulations from the body surface, are brought about by way of the cerebrum, as well as on subcortical reactions produced by extero- and proprioceptive stimulations.

After extirpation of the cerebellum, these reactions were constantly and permanently impaired, i.e., the various stimulations caused an abnormally extensive phasic and tonic shortening contraction of the muscles; moreover, several reactions also showed a distinctly delayed appearance, in which the retardation did not depend(at least not exclusively) on a prolonged latent period.

Moreover, after extirpation of the cerebellum, the inhibiting influence exerted by certain circumstances as, for instance, lying in the dorsal posture on the muscular reactions of the extremities seen in intact and decorticate animals, is constantly reduced.

APPENDIX

	Decerebellate dog	Decorticate dog	Decerebellate and decorticate dog	Decerebrate dog
Preparation for standing (placing)	+	—	—	—
Munk's touch reflexes	+	—	—	—
Correction movements on stimulations from the skin	+	—	—	—
Positive supporting reactions	+	+	+	+?
Negative supporting reactions	+	+	+	impaired
Correction movements on proprioceptive stimulations	+	+	?	— —
Rocking (hopping) reactions	+	+	?	— —
Leg-slackening and hopping reactions	+	+	+	— —
Leg-bracing reactions	+	+	?	— —
Pulling-up and propping-up reactions	+	+	?	— —

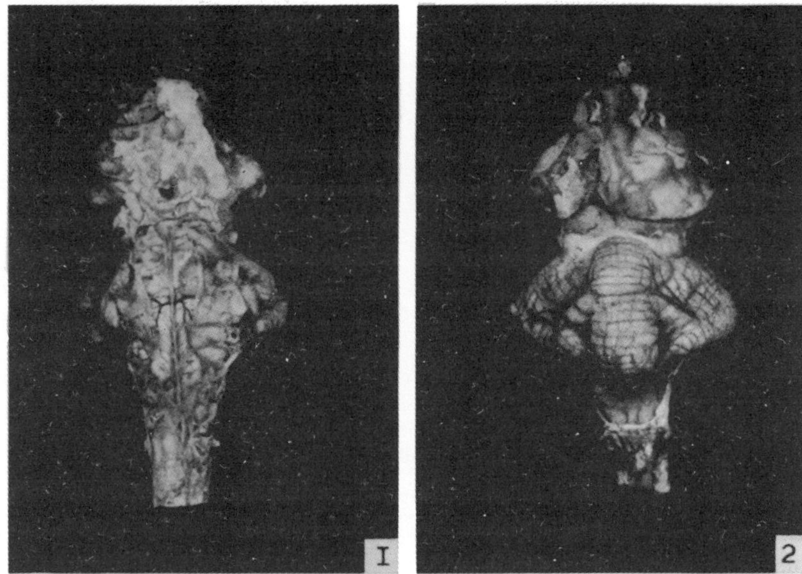

Fig. 286. Post-mortem findings in the decorticate dog Fuchs. Extirpation of the left half of the cerebrum on the 28th January, 1927; extirpation of the right half of the cerebrum on the 21st February, 1927. Death on the 8th November, 1927. 1. From the ventral side. 2. From the dorsal side.

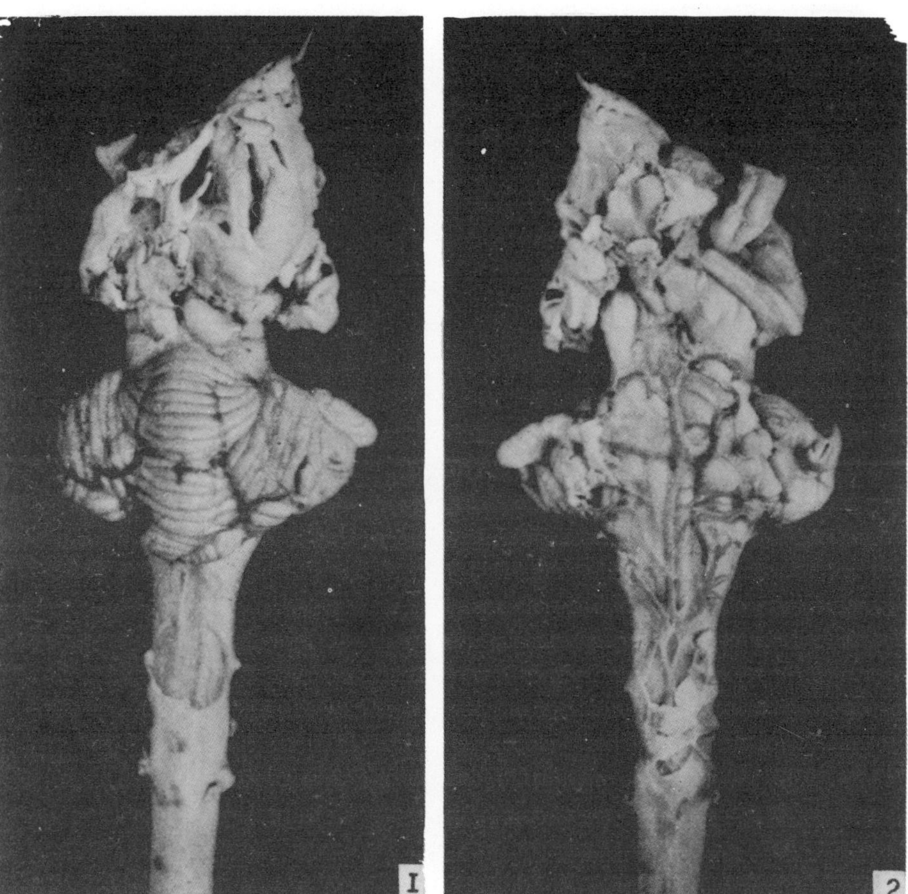

Fig. 287. Post-mortem findings in the decorticate dog Miesel. Extirpation of the left half of the cerebrum on the 18th December, 1926. Extirpation of the right half of the cerebrum on the 7th January, 1927. Death on the 8th November, 1927. 1. Seen from the dorsal side. 2. From the ventral side.

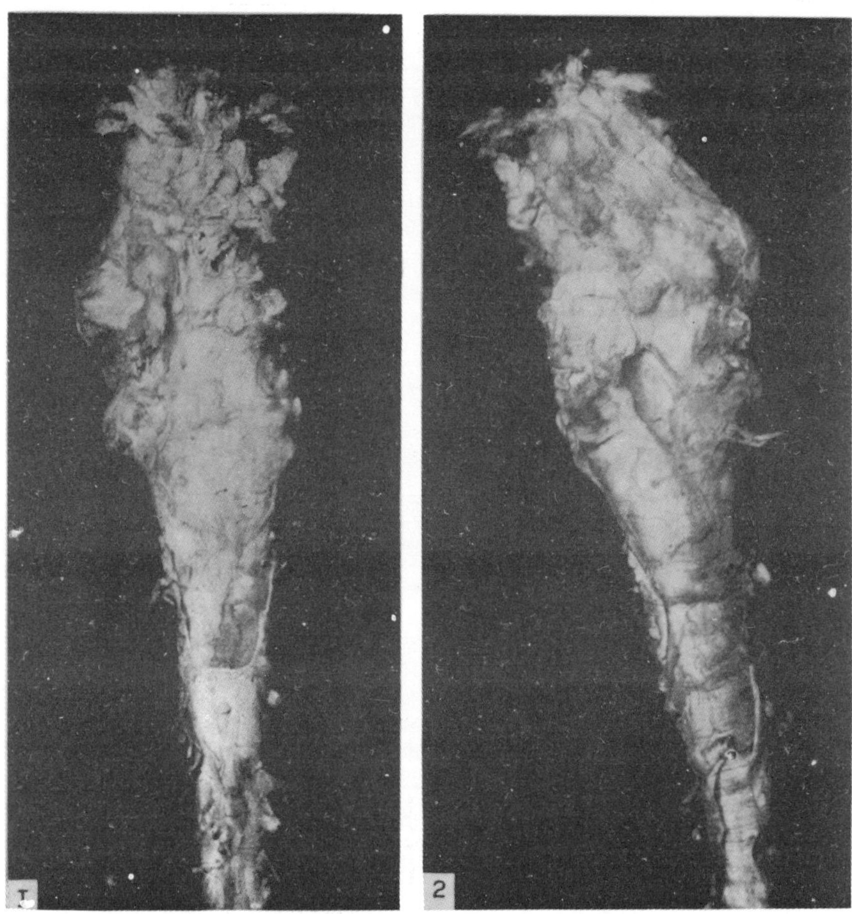

Fig. 288.   Post-mortem findings in the decerebellate and decorticate dog Daumling. Extirpation of the cerebellum on the 30th December, 1925; extirpation of the right half of the cerebrum on the 20th February, 1926, extirpation of the left half of the cerebrum on the 13th May, 1926. Death on the 10th June, 1926.   1. Seen from the ventral side. 2. From the dorsal side.   The cerebellum and the cerebral hemispheres are totally removed.   On the microscopic investigation of serial sections, see no. 208 of the list of references.

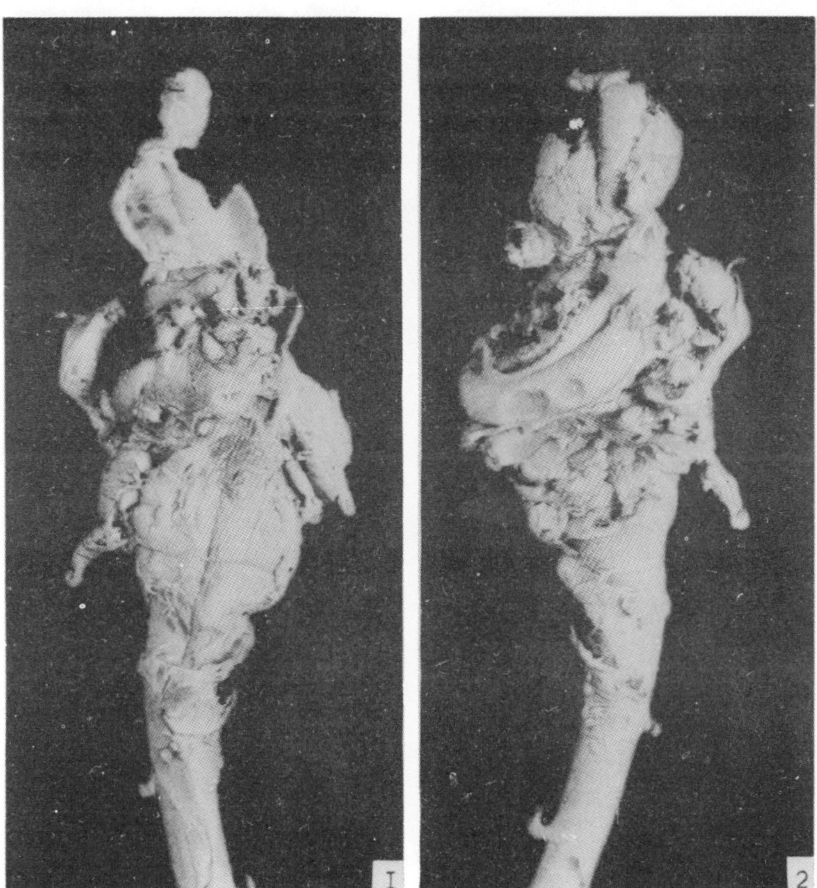

Fig. 289.　Post-mortem findings in the dog Vici. Extirpation of the cerebrum and of the right half of the cerebellum on the 7th April, 1925; extirpation of the left half of the cerebellum on the 19th October, 1926; death on the 19th October, 1926. 1. Seen from the ventral side. 2. From the dorsal side.

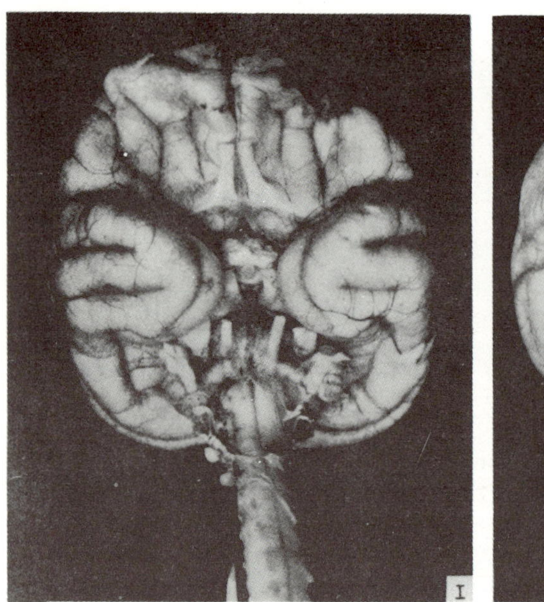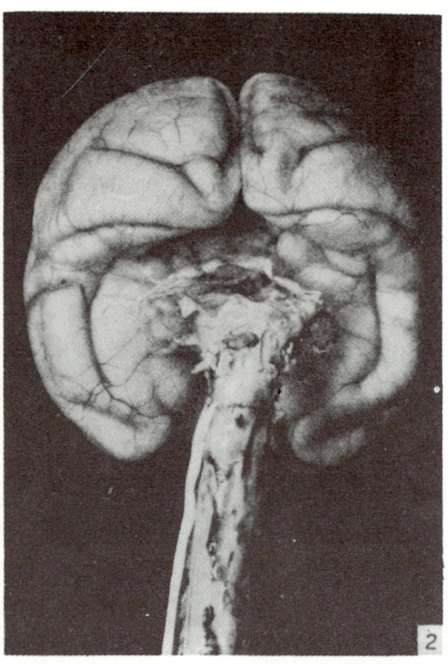

Fig. 290. Post-mortem findings in the decerebellate monkey Corrie. Extirpation of the cerebellum on the 13th January, 1926; died on the 29th April, 1927. 1. Ventral view. 2. Seen from the dorsal side. The cerebellum is almost totally removed; only on the right side a few small lamellae are preserved.

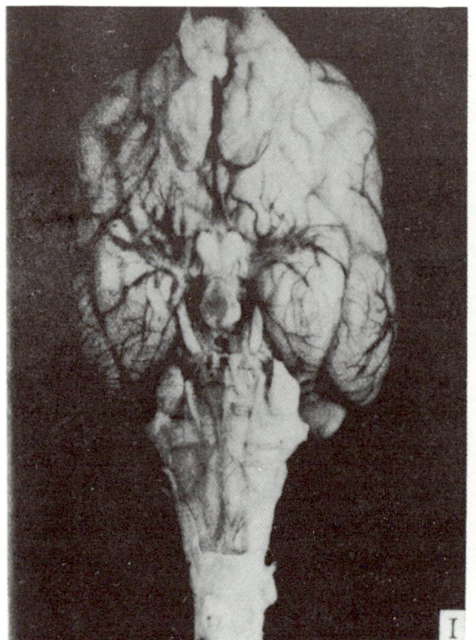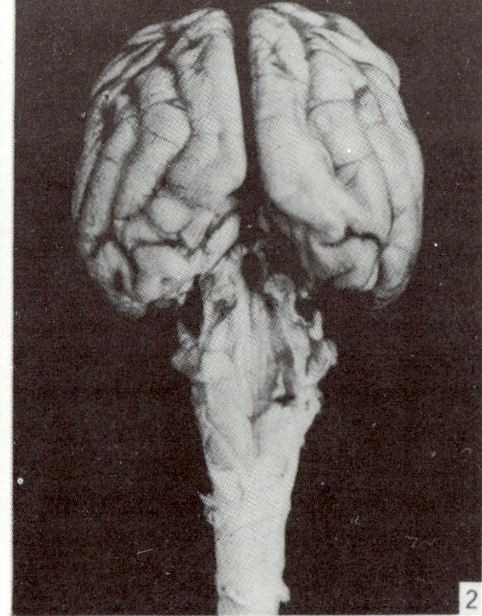

Fig. 291. Post-mortem findings in the decerebellate dog Piccolino. Extirpation of the cerebellum on the 24th December, 1924; death on the 7th January, 1929.

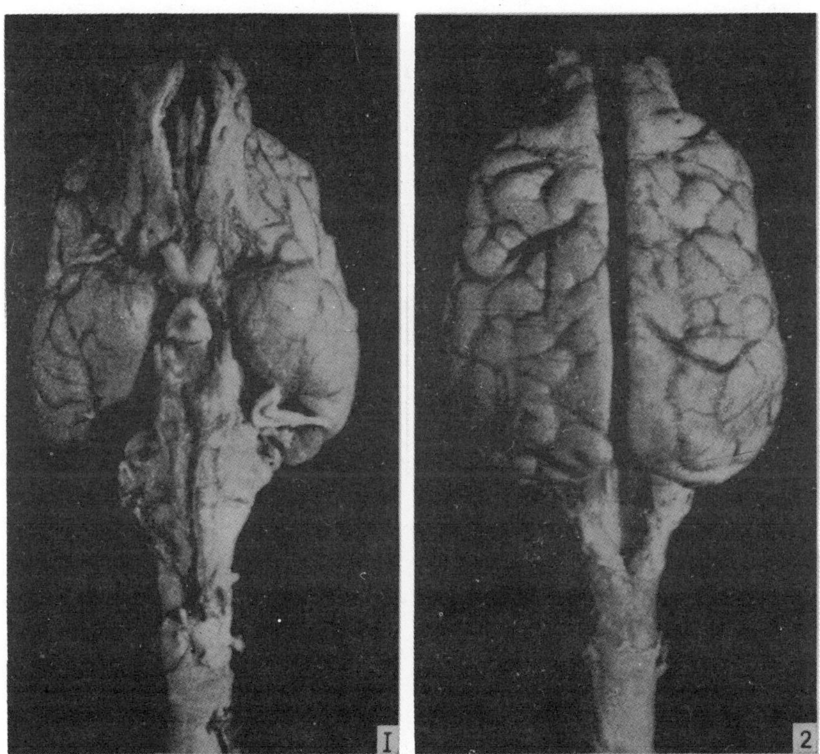

Fig. 292. Post-mortem findings in the decerebellate dog Cäsar.   Extirpation of the cerebellum on the 3rd February, 1925.   Death on the 15th March, 1927.

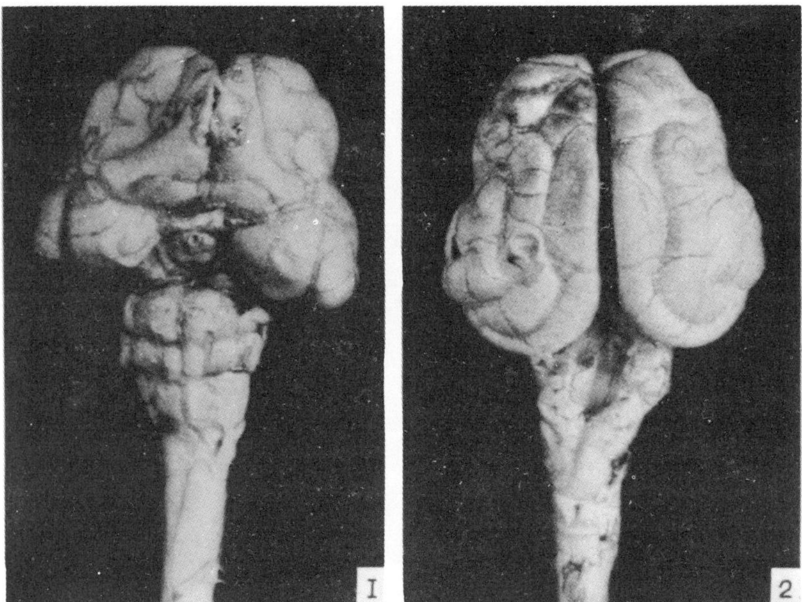

Fig. 293. Post-mortem findings in the decerebellate cat Pierrette. Extirpation
of the cerebellum on the 27th February, 1925; death on the 1st June,

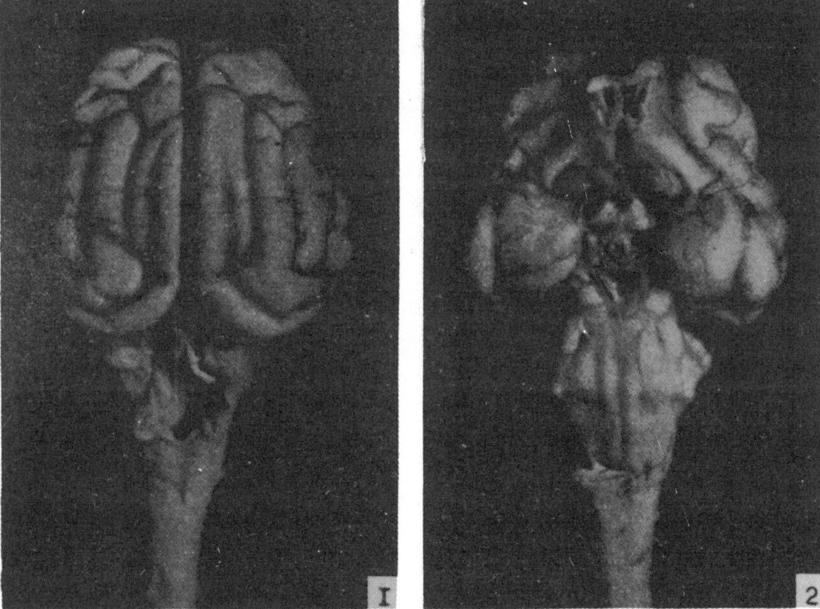

Fig. 294.   Post-mortem findings in the decerebellate and labyrinthectomized
cat Peggy. Extirpation of labyrinths on both sides on the 17th May,
1926; extirpation of the cerebellum on the 8th June, 1926. Death
on the 13th August, 1926.

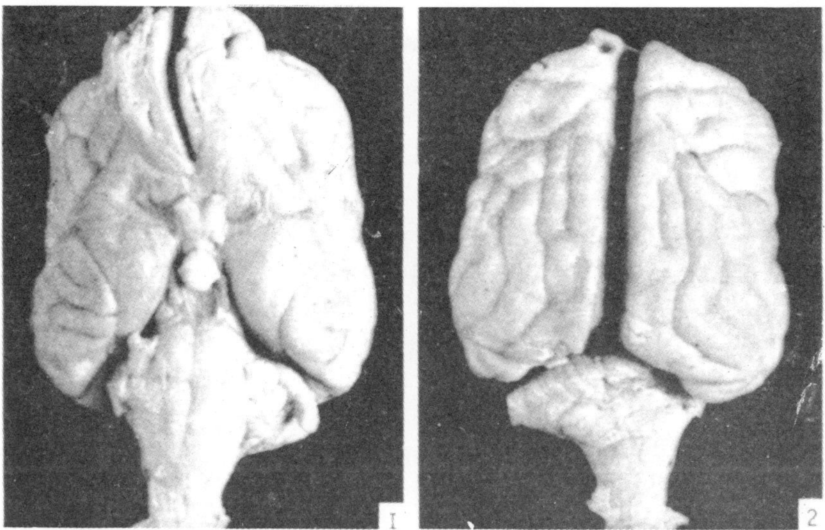

Fig. 295. Post-mortem findings in the unilaterally decerebellate dog Mops.
Unilateral extirpation of the cerebellum on the 11th February, 1926;
death on the 8th September, 1927.

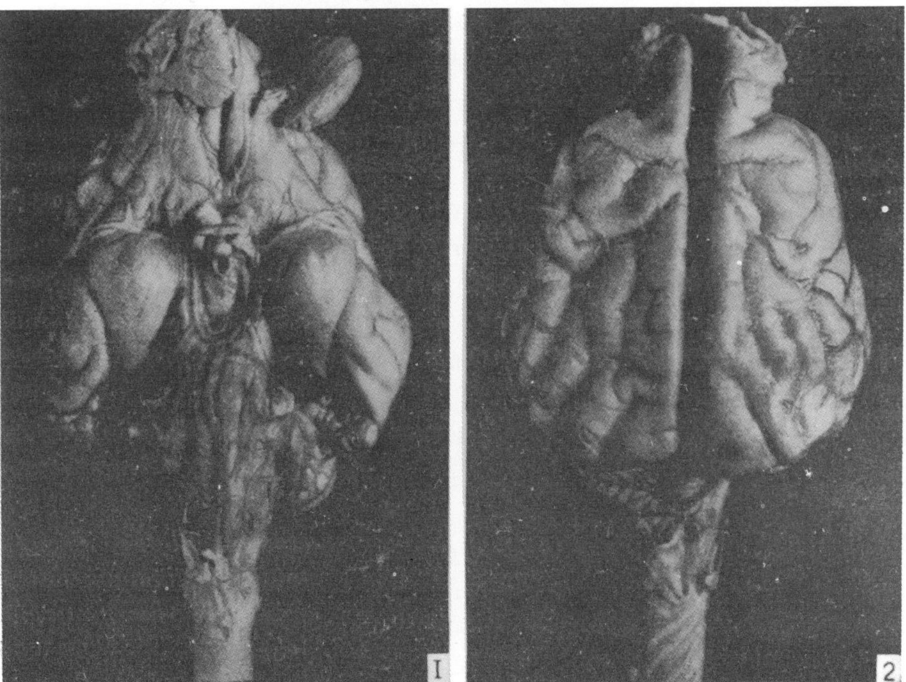

Fig. 296. Post-mortem findings in the unilaterally decerebellate dog Peter.
Unilateral extirpation of the cerebellum on the 4th February, 1926;
death on the 12th April, 1927.

LITERATURE

1. ABRAHAMS., E. J. : Hyptokinesis by een tumor in den Aquaeductus sylvii. Diss. Utrecht 1925.

2. ALAJOUANINE u. D. PETIT-DUTAILLIS: Volumineux abcès du cerveau gauche, douze ans après une blessure de guerre. Guérison sans séquelles après intervention. Considérations cliniques et thérapeutiques. Bull. Soc. méd. Hôp. Paris **1926**, Nr 36, 1630.

3. BABINSKI, J.: De l'asynergie cérébelleuse. Revue neur. **1899**, 806; **1901**, 260, 422; **1913**.

4.— Sur le rôle du cervelet dans les actes volitionnels nécessitant une succession rapide de mouvements; adiadococinésie. Ebenda **1902**.

5.— Équilibre volitionnel statique et cinétique. Ebenda **1902**, 470.

6.— Asynergie et inertie cérébelleuse. Ebenda **1906**.

7.— et JABKOWSKI: Étude de la raideur musculaire dans un cas de syndrome parkinsonien consécutif à une encéphalite, réaction des antagonistes. Ebenda **1920**, 564.

8. BARANY, R.: Beziehungen zwischen Bau und Funktion des Kleinhirns nach Untersuchungen am Menschen. Wien. klin. Wschr. **1912**.

9.— Kleinhirnzyste mit Ausfall der vestibulären Zeigereaktion bei den oberen Extremitäten nach abwärts. Ebenda **1912**.

10.— Lokalisation in der Rinde der Kleinhirnhemisphäre des Menschen. Ebenda **1912**, 2033.

11.— L. REICH u. J. ROTHFELD: Experimentelle Untersuchungen über die vestibulären Reaktionsbewegungen an Tieren. Neur. Zbl. **31**, 1139 (**1912**).

12.— Lokalisation in der Rinde der Kleinhirnhemisphären. Dtsch. med. Wschr. **1913**, 637.

13.— Direkte reizlose, temporäre Ausschaltung der Kleinhirnrinde nach der Methode von TRENDELENBURG, durch den Zeigeversuch nachweisbar. Lokalisation in der Kleinhirnrinde. Mschr. Ohrenheilk. **1916**.

13A. BARENNE, J. G. DUSSER DE : Siehe DUSSER.

13B. BAUM, H.: Siehe ELLENBERGER, W., u. H. BAUM.

14. BASTIAN, CHARLTON: Complete transverse softening involving the mid dorsal region of the spinal cord. Quain's Dictionary of the medicine, S. 1480 (1882).

15.— On the symptomatology of total transverse lesions of the spinal cord with special reference to the condition of the various reflexes. Medical chirurgical transactions 73, 151 (1890)(published by the Roy. Med. a. Chir. Soc. Lond).

16. BAZET, H. C., u. W. G. PENFIELD: A Study of the Sherrington decerebrate animal in the chronic as well as in the acute condition. Brain 45, 185 (1922).

17. BAUER, J., u. R. LEIDLER: Über den Einfluß der Ausschaltung verschiedener Hirnabschnitte auf die vestibulären Augenreflexe. Arb. neur. Inst. Wien 19, 155 (1911).

18. VON BECHTEREW, W.: Die Funktionen der Nervenzentra. H. II, 763. Kleinhirn und übrige Organe der statischen Koordination. Jena: Gustav Fischer 1909.

19. BENEDEK, L., u. E. DE THURZO: Du réflexe paradoxal des triceps et de sa localisation segmentaire. Revue neur. 2, 463 (1927).

20. BERITOFF, J. S., u. R. MAGNUS: Pflügers Arch. 159, 249 (1914).

21.— On the mode of origination of labyrinthine and cervical tonic reflexes and on their part in the reflex reactions of the decerebrate preparation. Quart. J. exper. Physiol. 9, 199 (1915).

22. BICKEL, A.: Untersuchungen über den Mechanismus der nervösen Bewegungsregulation. Stuttgart: Ferdinand Enke 1903.

23. BÖHME, A., u. W. WEILAND: Einige Beobachtungen über die Magnusschen Hals- und Labyrinthreflexe beim Menschen. Z. Neur. 44, 94 (1910).

24. DU BOIS-REYMOND, R.: Spezielle Bewegungslehre mit überblick über die Physiologie der Gelenke. Nagels Handbuch der Physiologie 4, 564. Braunschweig 1909.

25. BOLK, L. : Over de physiologische beteekenis van het cerebellum. Haarlem: De Erven Bohn 1903.

26.— Das Cerebellum der Säugetiere, eine vergl eichend-anatomische Untersuchung. Jena Gustav: Fischer 1907.

636

27.— Over functielocalisatie in het cerebellum. Nederl. Tijdschr.
Geneesk. **1908**, 1.

28. BREMER, F.: Contribution à l'étude de la physiologie du
cervelet. La fonction inhibitrice du paléo-cérébellum. C. r.
Soc. Biol. Paris **86**, Nr. 16, 29. April 1922.

29.— La fonction inhibitrice du paléo - cérébellum. Arch. internat.
Physiol. **19**, H. 2, 189 (1922).

30.— Recherches sur la physiologie du cervelet chez le pigeon.
C. r. Soc. Biol. Paris **90**, 381 (1924).

31.— et R. LEY: Recherches sur la physiologie du cervelet chez le
pigeon. Bull. Acad. Méd. Belg., Séance du 29. Januar, S. 60.

32. BRONDGEEST. P. Q.: Onderzoekingen over den tonus der
willekeurige spieren. Diss. Utrecht 1860.

33. BROUWER, B., u. L. Coenen: Untersuchungen über das
Kleinhirn. Psychiatr. Bl. (holl.) **1921**, 201.

34.— Über Querläsion des Rückenmarks beim Menschen und das
Bastinesche Gesetz. Ebenda **19**, 377 (1915).

35. DE BRUIN, J.: Enkele neurologische gevallen uit de kinder-
praktyk. II: Een gecompliceerd geval van Idiotica amauro-
tica progressiva familiaris infantilis (Tay-Sachs). Nederl.
Mschr. Verlosk. Vrouwenziekten en Kindergeneesk. **22**, 167
(1919).

36. BRUNS. L.: Über einen Fall traumatischer Zerstörung des
Rückenmarkes an der Grenze zwischen Hals- und Dorsalmark.
Arch. Psychiatr. u. Neur. **25**, 759 (1893).

36A. BROWN: Siehe DENNY-BROWN.

37. BROWN, GRAHAM: Die Groβhirnhemisphären. Handbuch
der normalen und pathologischen Physiologie **10**, 418.
Berlin: Julius Springer 1927.

38. CAJAL., RAMON Y.: Los ganglios centrales del cerebelo
de las aves. Trab. Labor. Invest. biol. Univ. Madrid **6**, 143
(1908).

39. CATE, J.: Zur Frage über die funktionelle Lokalisation im
Kleinhirn. Abstracts of Communications to the XII. Physiological Congress held at Stockholm, August 3-6, 1926.

40. CHRISTIANI, A.: Zur Physiologie des Géhirnes. S. 14.
Berlin: Otto Enslin **1885**.

41. CLAUDE, H., et J. LHERMITTE: Étude anatomo-pathologique d'un cas de section totale de la moëlle. Recherches sur la réflectivité. Buil. Soc. Méd. Hôp. Paris, 15. Februar 1916.— Ann. Méd. **1916**, H. 4.

42.— — Le tétanos médullaire par effraction. Paris méd., 2. November 1918.

   Siehe auch LHERMITTE, J.: La section totale de la moëlle dorsale, S. 75. Paris 1919.

43.— — Les paraplégies cérébello-spasmodiques consécutives aux lésions bilatérales des lobules paracentraux par projectiles de guerre. Bull. Soc. méd. Hôp. Paris, 26. Mai 1916.

44.— et LÉVY-VALENSI: Maladies du cervelet et de l'isthme de l'encéphale. Nouveau Traité de médicine et de thérapeutique, fasc. 32. Paris: J. B. Baillière et fils **1922**.

45. COBB, S., A. A. BAILEY, u. P. R. HOLTZ: On the genesis and inhibition of extensor rigidity. Amer. J. Physiol. **44**, 239 (1917).

46. COBB, STANLEY: Review on the tonus of skeletal muscle. Physiologie. Rev. **4**, 518 (1925).

47. DALTON, C.: On the cerebellum, as the centre of coordination of the voluntary movements. Amer. J. med. Sci. **1861**, 83.

48. DÉJÉRINE, J., et A. THOMAS: L'atrophie olivo-pontocérébelleuse. Nouvelle Iconographie de la Salpétrière, S. 350 (1900).

49.— et A. et J. MOUZON:Sur l'état des réflexes dans les sections complètes de la moëlle épinière. Revue neur. **1915**.

50. DELMAS-MARSALET, P.: Les réflexes de posture élémentaires. Paris: Masson et Cie., Éditeurs, **1927**.

51. DENNY-BROWN, D. E., a. E. G. T. LIDDELL: Observations on the motor twitch and on reflex inhibition of the tendon-jerk of M. supraspinatus. J. of Physiol. **63**, 70 (1927).

52.— — The stretch reflex as a spinal process. Ebenda **63**, 144(1927).

53.— — Extensor reflexes in the fore-limb. Ebenda **65**, 305 (1928).

54.— On the nature of postural reflexes. Proc. roy. Soc. Lond. (B), **104**, 252 (1929).

55. DOLLINGER, A.: Zur Klinik der infantilen Form der familiären amaurotischen Idiotie (Tay-Sachs). Z. Kinderheilk. **22**, 167 (1919).

56. DRESEL: Die Funktionen eines groβhirn- und striatumlosen Hundes. Klin. Wschr. **3**, 2231 (1924).

57. DREYFUSS, R.: Experimenteller Beitrag zur Lehre von den nichtakustischen Funktionen des Ohrlabyrinths. Pfugers Arch. **81**, 604 (1900).

57A. DUCCESCHI e SERGI: I sensomusculare nelle lesioni del cerveletto. Arch. di Fisiol. **2**, 233 (1904).

58. DUCHENNE (de Boulogne), G. B.: Physiologie des mouvements, Paris: J. -B. Baillière et fils **1867**.

59. DUCHENNE, G. B.: Physiologie der Bewegungen. Cassel u. Berlin: Theodor Fischer **1885**.

60. DUSSER DE BARENNE, J. G.: Über eine neue Form von vestibulären Reflexen beim Frosch. Psychiatr. Bl. (holl.), WINKLER-Festschrift, S. 258. 20. September 1918.

60A.—Recherches expérimentales sur les fonctions du système nerveux central, faites en particulier sur deux chats, dont le néopallium a été enlevé. Arch. néerl. Physiol. **4**, 31 (1920).

60B.— Proefondewindelijke physiologie van het zenuwstelstel. Leerboek der Zenuwziekten BOUMAN-BROUWER 1 A, 282 (1922). Haarlem: De Eryen F. Bohn.

61.—Die Funktionen des Kleinhirns. Handbuch der Neurologie des Ohres. Herausgegeben von G. ALEXANDER u. O. MARBURG. Berlin u. Wien: Urban u. Schwarzenberg. 1924, S. 590.

62. EDINGER, L.: Über das Kleinhirn und den Statotonus. Zbl. Physiol. **26**, 618 (1912).

63.—u. B. FISCHER: Ein Mensch ohne Groβhirn. Pflügers Arch. **152**, 1 (1913).

64. ELLENBERGER, W., u. H. BAUM: Anatomie des Hundes. Berlin: Paul Parey **1891**.

65. EWALD, R.: Physiologische Untersuchungen über das Endorgan des Nervus octavus. Wiesbaden **1892**.

66. FERRIER, D.: The Functions of the Brain (1876). Deut-

sche Augabe: Braunschweig **1879**. Französische Ausgabe: Paris **1878**.

67.— a. W. A. TURNER: A record of experiments illustrative of the symptomatology and degenerations following lesions of the cerebellum and its peduncles and related structures in monkeys. Philosophic. Trans. roy. Soc. Lond. **85**, 722 (1894).

68. FISCHER, E., u. W. STEINHAUSEN : Allgemeine Physiologie der Wirkung der Muskeln im Körper. Handbuch der normalen und pathologischen Physiologie **8**, 619. Berlin: Julius Springer 1925.

69. FISCHER, M. H., u. R. H. KAHN : Ein bisher unbekanntes Vestibularisphänomen. Pflugers Arch. **216**, 555 (1927).

70. FOERSTER, O.: Physiologie und Pathologie der Koordinaten (1902).

71.— Schlaffe und spastische Lähmung. Handbuch der normalen und pathologischen Physiologie **10**, 893. Berlin: Julius Springer 1927.

72. FOIX, CH.: Réflexes toniques de posture. Revue neur. **1921**, 840.

73.— Contracture plastique. Ebenda **1921**, 1130.

74.— L'automatisme médullaire. Questions neurologiques d'actualité. Paris: Masson et Cie. **1922**.

75.— et A. THÉVENARD: Les réflexes tendineux dans la maladie de Parkinson. Inexcitabilité temporaire post-réflexe. Contracture posturéo-réflexe. Tonus de posture et Tonus d'action. Leurs rapports avec les contractures pyramidales et extra-pyramidales. Revue neur. **1922**, 948.

76.— — Les réflexes de posture. Ebenda **1923**, 449.

77.— et LAGRANGE: Tonus de posture locale. Tonus de posture général ou mieux tonus d'attitude. Tonus d'action, leur dissociation chez un tabétique hémiplégique. Ebenda **1**, 260 (1924).

78. FOIX, CH.: Sur le tonus et les contractures. Ebenda **2**, 1 (1924).

79.— et A. THÉVENARD: Le tonus, les contractures, la contraction musculaire volontaire et réflexe, étudiées par l'électromyo-

640

graphie et la phonomyographie (Courants d'actions, son musculaire). J. Physiol. et Path. gén. **23**, 309, 332 (1925).

80.— — Réflexes de posture et réflexes d'attitude. Posture générale, le phénomène de la poussée, et la posture locale. Presse méd., 30. December 1925.

81. FREEMAN, WALTER, u. PAUL MORRIN: L'influence des réflexes toniques du cou sur les syncinésies. Revue neur. **1923**, 452.

82.— — Réflexes d'automatisme mésencéphaliques. Les syncinésies, les réflexes cervicaux et les réflexes vestibulaires. L'Athétose. Ebenda **1924**, 158.

83. v. FREY, M.: Die Gliederung des Tastsinns. Verhandlungen der Gesellschaft deutscher Nervenärzte, 17. Jahresversammlung gehalten zu Wien vom 15. bis 17. September 1927, S. 71. Leipzig: F. C. W. Vogel **1928**.

84. FREUND, C. S.: Störung der Schwereempfindung. Jahresversammlung der Gesellschaft deutscher Nervenärzte. Breslau **1913**.

85. FREUSBERG, A.: Siehe GOLTZ. FR. u. A. FREUSBERG.

86. FULTON, J. F. a. E. G. T. LIDDELL: Electrical responses of extensor muscles during postural (myotatic) contraction. Proc. roy. Soc. Lond. (B) **98**, 577 (1925).

87.— Muscular contraction and the reflex control of movement. Baltimore: Williams and Wilkins Company **1926**.

88.— The mechanism of the postural contraction (tonus) of the skeletal muscle. Proc. Soc. exper. Biol. a. Med. **22**, 700 (1926).
88A.— Siche J. PI-SUNER u. J. F. FULTON.

89.— a. J. PI - SUNER: A note concerning the probable function of various afferent endorgans in skeletal muscle. Amer. J. Physiol. **83**, 554 (1928).

90. GAMPER, ED.: Klinische Beobachtungen an einem Fall von Arhinenzephalie und Mitteilung des anatomischen Befundes. Verh. Ges. dtsch. Nervenärzte, 14. Jahresvers., gehalten zu Innsbruch 1924, S. 224. Leipzig: F. C. W. Vogel **1925**.

91.— Bau und Leistungen eines menschlichen Mittelhirnwesens (Arhinchzephalie mit Encephalocele). Zugleich ein Beitrag zur Teratologie und Fasersystematik. II. Klinischer Teil. Z. Neur. **104**, 117 (1926).

92. VAN GEHUCHTEN, A.: Les centres nerveux cérébro-spinaux, S. 215 (1908).

93.— Le mécanisme des mouvements réflexes. Compte Rendû des Travaux du I Congrès International de Psychiatrie, Neurologie etc. Amsterdam 1908.

94. GILDENMEISTER : Z. Biol. **63**, 173 (1913).

95. GIRNDT, O.: Physiologische Beobachtungen an Thalamuskatzen. Mitt. II.: Die phasischen Extremitätenreflexe der Thalamuskatze im akuten Versuch. Pflügers Arch. **213**, 427 (1926).

96. GOLDFLAM, S.: Paradoxe Kontraktion. Z. Neur. **76**, 516 (1922).

97.— Dehnungskontraktion der Antagonisten. Ebenda **76**, 521 (1922).

98. GOLDSTEIN, K.: Über Störungen der Schwereempfindung bei gleichsegtier Kleinhirnaffektion. Neur. Zbl. **1913**, 405.

99.— u. F. REICHMANN: Beiträge zur Kasuistik und Symptomatologie der Kleinhirnerkrankungen, im besonderen zu den Störungen der Bewegungen, der Gewichts- , Raum- und Zeitschätzung. Arch. f. Psychol. **56** (1916).

100.— Über Halsreflexe beim Menschen. Zbl. Neur. **30**, 413 (1922).

101.— u. W. RIESE : Über induzierte Tonusveränderungen beim Menschen. Klin. Wschr. **1923**, 1201.

102.— Über induzierte Tonusveränderungen beim Menschen (sogenannte Halsreflexe, Labyrinthreflexe usw.). II. Mitt. Z. Neur. **89**, 303 (1924).

103.— Neuere Erfahrungen zum Problem der sogenannten induzierten Tonusveränderungen usw. Zbl. Neur. **41**, 715 (1925).

104.— Über die Funktionen des Kleinhirnes. Klin. Wschr. **1924**, 1255.

105.— Das Kleinhirn. Handbuch der normalen und pathologischen Physiologie **10**, 222. Berlin: Julius Springer 1927.

106. GOLTZ, FR., u. A. FREUSBERG: Über die Funktionen des Lendenmarkes des Hundes. Pflügers Arch. **8**, 460 (1874).

107.— Der Hund ohne Groβhirn. Ebenda **51**, 570 (1892).

108. GRAHE, K.: Bogengangsreflexe auf die Extremitäten beim Kaninchen. Ebenda **204, 421** (1924).

109.— Über den Einfluβ von Becken- und Extremitätenbewegungen auf die Stellung der ubrigen Körperteile. **Z. Hals-, Ohren-,** Nasenhlk. **15**, 465 (1926).

110. GROEBBELS, F.: **Die Lage- und Bewegungsreflexe der** Vögel. Mitt. IX: Die Wirkung von Kleinhirnläsionen und ihre anatomisch-physiologische Analyse. Pflügers Arch. **221**, 15 (1928).

111.— Die Lage- und Bewegungsreflexe der Vögel. Mitt. X: Die Analyse der Beziehungen zwischen Labyrinth und Kleinhirn. Ebenda **221**, 41 (1928).

112.— Die Lage- und Bewegungsreflexe der Vögel. Mitt. XI: Die Analyse der Stützreaktion. Ebenda **221**, 50 (1928).

113. GUILLAIN, GEORGES, P. MATHIEU, u. I. BERTRAND: Étude anatomo-clinique sur deux cas d'atrophie olivo-ponto-cérébelleuse avec rigidité. Ann. Méd. **20, H. 5, 417** (1926).

114. DE HAAN, P.: Otolithenreflexen by den Mensch. Nederl. Tijdschr. Geneesk. **1927**, 2237.

115. HANSEN, K., u. W. RECH: Beziehungen des Kleinhirns zu den Eigenreflexen. Dtsch. Z. Nervenheilk. **87**, 207 (1925).

116. HEAD, HENRY, a. G. RIDDOCH: The automatic **bladder,** excessive sweating and some other reflex conditions in gross injuries of the spinal cord. Brain **40** (1918).

117. HOFF, H., u. P. SCHILDER: Die Lagereflexe des Menschen. Wien: Julius Springer **1927**.

117 A. HOFFMANN, P.: Eigenreflexe. Berlin: Julius Springer **1922**.

118. HÖGYES, A.: Über den Nervenmechanismus der assoziierten Augenbewegungen. Mschr. Ohrenbeilk. **46**, 809 (1912).

119. HOLMES, GORDON: On certain tremors in organic lesions. Brain **27**, 327 (1904).

120.— The symptoms of acute cerebellar injuries due to gunshot injuries. Ebenda **40**, 461 (1917).

121.— Croonian Lectures on Cerebellum. Lancet **202**, 1177, 1231; **203**, 59, 111 (1922).

122. HOWARD, C. P., a. C. E. ROYCE: Progressive lenticular degeneration associated with cirrhosis of the liver (Wilson's Disease). Arch. int. Med. **24**, 497 (1919).

123. HUNT, J. RAMSAY: Théorie statosynergique de la fonction cérébelleuse. Revue neur. **1927**, 11, 445.

124. INGVAR, SVEN: Zur Phylo- und Ontogenese des Kleinhirns nebst einem Versuche zu einheitlicher Erklärung der zerebellaren Funktion und Lokalisation. Haarlem: De Erven Bohn **1918**.

125. JACOB, A.: Die extrapyramidalen Erkrankungen. Berlin: Julius Springer **1923**.

126. JARKOWSKI, J.: La réaction des antagonistes dans le syndrome parkinsonien. Revue neur. **1921**, 613.

127.— "Kinésie Paradoxale" des Parkinsoniens. Paris: Masson et Cie. **1925**.

128. JELGERSMA, G.: De physiologische beteekenis van het Cerebellum. Amsterdam: Scheltema en Holkema's Boekhandel **1904**.

129.— De functie der kleine hersenen. Psychiatr. Bl. (holl.) **19**, 214 (1915).

130. DE JONG, H.: Experimente zur Tremorfrage. Verh. Ges. dtsch. Nervenärzte, 16. Jahresvers., gehalten zu Düsseldorf 1926, S. 148.

130.A— Over rhythmische verschijnselen van het normale en het zieke zenuwstelsel. Nederl Tijdschr. Geneesk. **1**, 2345 (1927).

131.— Weitere Beiträge über rhythmische Erscheinungen im Nervensystem. Dtsch. Z. Nervenheilk. **103**, 1 (1928).

132.— Étude sur les phénomènes oscillatoires dans les perturbations de la fonction cérébelleuse chez l'homme et chez le chien. Proceed. koninkl. Akad. v. Wetensch. Amsterdam, **32**, 1 (1929).

133. JONKHOFF, D. J.: Een geval van halsreflexen van Magnus en De Kleyn. bij een mensch en haar belangrijkheid voor de prognose. Nederl. Tijdschr. Geneesk. **1920**, 307.

134. JONKHOFF, D. J.: De Invloed van eenige geneesmiddelen op de labyrinthreflexen van konijnen, caviae en katten. Diss. Utrecht 1921.

134A. Siehe auch: MAGNUS, R.: Körperstellung, S. 67, 677.

135. KAPPERS, C., u. U. ARIENS: Die vergleichende Anatomie des Nervensystems der Wirbeltiere und des Menschen 2, 682. Haarlem : De Erven F. Bohn 1921.

136. KARPLUS, J. P., u. A. KREIDL: Über Totalexstirpation einer oder beider Großhirnhemisphären an Affen (*Macacus rhesus*). Arch. f. Anat. **1914**, 155.

137. KAUSCH: Über das Verhalten der Sehnenreflexe bei totaler Querschnittsunterbrechung des Rückenmarks. Mitt. Grenzgeb. Med. u. Chir. **7** (1901).

138. KLEIST, K.: Gegenhalten (motorischer Negativismus), Zwangsreifen und Thalamus opticus. Mschr. Psychiatr. **65**, 317 (1927).

139. DE KLEYN, A., u. R. MAGNUS: Die Abhängigkeit des Tonus der Extremitätenmuskeln von der Kopfstellung. Pflügers Arch. **145**, 455 (1912).

140. — — Die Abhängigkeit des Tonus der Nackenmuskeln von der Kopfstellung. Ebenda **147**, 403 (1913).

141. — — Die Abhängigkeit des Körperstellung vom Kopfstande beim normalen Kaninchen. Ebenda **154**, 163 (1913).

142. — — Ein weiterer Fall von tonischen Halsreflexen beim Menschen. Münch. med. Wschr. **1913**, Nr 46.

143. — — Analyse der Folgezustände einseitiger Labyrinthexstirpation mit besonderer Berücksichtigung der Rolle der tonischen Halsreflexe. Pflügers Arch. **154**, 178 (1913).

144. — Zur Analyse der Folgezustände einseitiger Labyrinthexstirpation beim Frosch. Ebenda **159**, 218 (1914).

145. — u. R. MAGNUS: Weitere Beobachtungen über Hals- und Labyrinthreflexe auf die Gliedernmuskeln des Menschen. Ebenda **160**, 429 (1915).

146. — — Über die Unabhängigkeit der Labyrinthreflexe vom Kleinhirn und über die Lage der Zentren für die Labyrinthreflexe im Hirnstamm. Ebenda **178**, 14 (1920).

147.— — Beiträge zum Problem der Körperstellung. Mitt.   IV: Optische Stellreflexe bei Hund und Katze. Ebenda **180**, 291 (1920).

148.— — Labyrinthreflexe auf Progressivbewegungen. Ebenda **186**, 39 (1921).

149. LABORDE: Les fonctions du cervelet. Études de critique expérimentale. C. r. Soc. Biol. Séance du 25 Janvier, 1890.

150.— Traité élémentaire de physiologie du système nerveux. Kapitel **17** (1892).

151. LANDAU, A.: Über einen tonischen Lagereflex beim alteren Säugling. Klin. Wschr. **2**, 1253 (1923).

152.— Über motorische Besonderheiten des zweiter Lebenshalbjahres· Mschr. Kinderheilk. **29**, 555 (1925).

153.— Zur Motorik des alteren Säuglings. Zbl. Neur. **40**, 372 (1925).

154. LANGE, BOGUMIL: Inwieweit sind die Symptome, welehe nach Zerstörung des Kleinhirns beobachtet werden. auf Verletzungen des Acusticus zurückzuführen? Pflügers Arch. **50**, 615 (1891).

155. LANGELAAN, J. W.: Over den bouw en de verriehtingen der Kleine hersenon. Nederl. Tijdschr. Geneesk. **1907**, 1.

156.— On Congenital Ataxia in a cat. Verh. Kon. Akad. Wetensch. Amsterd., Sectie II, **13**, Nr 3, 1 (1907).

157. LEIRI, F.: Beitrag zur Pathophysiologie des Kleinhirns auf Grund eines Falles von Erweichung in der einen Kleinhirnhemisphäre nebst Status lacunaris cerebri. Acta Soc. Medic. fenn. Duodecim **5**, H. 1, 1 (1924).

158.— Über posturale Reflexe. Acta med. scand. (Stockh.) **63**, 184 (1925).

159.— Le cervelet, un organe servant à l'innervation des antagonistes dans l'activité musculaire. Acta oto-laryng. (Stockh.) **6**, 516 (1924).

160.— Einige Beobachtungen über die Beeinflussung der Enthirnungs starre durch nicht proprioceptive Erregungen. Pflügers Arch. **212**, H. 3/4, 455 (1926).

161. LEJONNE, P., et G. LHERMITTE: Atrophie olivo-rubro-cérébelleuse. Essai de classification des atrophies du cervelet. L'Encéphale **1909**.

646

162. LÉRI, ANDRÉ: Un phénomène réflexe du membre supérieur: Le "signe de l'avant-bras." Revue neur. **25**, I, 277 (1913).

163.— Hémiplégie. Sémiologie Nerveuse, herausgegeben von Ch. Achard, A. Baudoin, Laignel-Lavastine, A. Leri und L. Lévi. Paris: J. B. Baillière et fils **1925**.

164. LEVEN et OLLIVER: Recherches sur la physiologie et la pathologie du cervelet. Arch. gén. Méd. **1862/1863**.

165. LEY, R. A.: Forme atypique d'atrophie cérébelleuse ayant évolué en syndrome rigide. Arch. internat. Méd. expér. **1**, 277 (1924).

166.— Contribution à l'étude des localisations cérébelleuses et du Syndrome cérébelleux. Ebenda **2**, 5 (1925).

167. LEWANDOWSKY, M.: Über die Verrichtungen des Kleinhirns. Arch. f. Physiol. **1903**, 129.

168.— Die Funktionen des Nervensystems. Jena: Gustav Fischer **1907**.

169.— Handbuch der Neurologie. Teil I, S. 358 (1910).

170. LEWY, F. H.: Die Lehre vom Tonus und Bewegung. Berlin: Julius Springer **1923**.

170A. LHERMITTE, J.: Siche CLAUDE, H., u. J. LHERMITTE.

171.— La section totale de la moëlle dorsale. Bourges, Tardy-Pigelet et fils **1919**.

172. LIDDELL, E. G. T., a. C. S. SHERRINGTON: A comparison between certain features of the spinal flexor reflex and of the decerebrate extensor reflex respectively. Proc. roy. Soc. Lond. (B) **95**, 299 (1923).

173.— a. Sir C. SHERRINGTON: Reflexes in response to stretch (myotatic reflexes.) Ebenda **96**, 212 (1924).

174.— — Further observations on myotatic reflexes. Ebenda **97**, 267 (1925).

174 A.— Siehe D. E. DENNY-BROWN u. E. G. T. LIDDELL und J. F. FULTON u. E. G. T. LIDDEL.

175. LILJESTRAND, G., u. R. MAGNUS: Über die Wirkung des Novocains auf den normalen und den tetanusstarren

Skelettmuskel und uber die Entstehung der lokalen Muskelstarre beim Wundstarrkramph. Pflügers Arch. **176**, 168 (1919).

176. LONGET, F. A.: Traité de Physiologie 3. Paris 1873.

177. LÖWENBERG: Über die nach Durchschneidung der Bogengänge des Ohrlabyrinthes auftretenden Bewegungsstörungen. Arch. Ohrenkeilk. **3**, 1 (1873).

178. LOTMAR, F.: Ein Beitrag zur Pathologie des Kleinhirns. Mschr. Psychiatr. **24**, 217 (1908).

179.— Die Stammganglien und die extrapyramidalt - motorischen Syndrome. Berlin: Julius Springer **1926**.

180.— LÖWENTHAL, M., a. V. HORSLEY: On the relations between the cerebellum and other centers (namely cerebral and spinal) with special reference to the action of antagonistic muscles. Proc. roy. Soc. Lond. **46**, 20 (1897).

181. LUCIANI, L.: Das Kleinhirn. Leipzig: Eduard Besold **1893**.

182.— Physiologie des Menschen 3, Kap. VII: Das Hinterhirn, S. 437. Jena: Gustav Fischer **1907**.

183. LUSSANA, F.: Leçons sur les fonctions du cervelet. J. Physiol. de l'Homme **5**, 18 (1862).

184.— Physiopathologie du cervelet. Arch. ital Biol. **7**, 145 (1886).

185. LUYS, I.: Études sur l'anatomie, la physiologie et la pathologie du cervelet. Arch. gén. Méd. **22** (1864).

186. MAAS, O.: Störung der Schwereempfindung bei Kleinhirnerkrankung. Neur Zbl. **1913**, 405.

187. MAGENDIE, F.: Précis élémentaire de physiologie. II. Ausgabe, 1, 335. Paris 1825.

188. MAGNUS, R.: Experimentelles und Klinisches über tonische Reflexe. Handelingen v. h. XIIIe Nedrel. Natuur- en Genees kundig Congres, S, 317 (1911).

188A.— Siehe DE KLEYN, A., und. R. MAGNUS, Nr. 139.

189.— u. C. G. L. WOLF: Weitere Mitteilungen über den Einfluß der Kopfstellung auf den Gliedertonus. Pflügers Arch. **149**, 447 (1913).

189A.— Siehe DE KLEYN, A., u. R. MAGNUS, Nr. 140—143.

190.— u. W. STORM VAN LEEUWEN: Die akuten und die dauernden Folgen des Ausfalles der tonischen Hals- und Labyrinthreflexe. Pflügers Arch. **159**, 157 (1914).

191. MAGNUS, R.: Welche Teile des Zentralnervensystems müssen für das Zustandekommen der tonischen Hals- und Labyrinthreflexe auf die Körpermuskulatur vorhanden sein? Ebenda **159**, 224 (1914).

191A.— Beiträge zum Problem der Körperstellung. Mitt. I. Ebenda **163**, 405 (1916).

191B.— Siehe BERITOFF, J. S., u. R. MAGNUS, Nr. 20.

191C.— Siehe KLEYN, A. DE, u. R. MAGNUS, Nr. 144 u. 145.

192.— Beiträge zum Problem der Körperstellung. Mitt. II. Ebenda **174**, 134 (1919).

193.— u. J. C. DUSSER DE BARENNE: Beiträge zum Problem der Körperstellung. Mitt. III. Ebenda **180**, 75 (1920).

194.— u. A. DE KLEYN: Beiträge zum Problem der Körperstellung. Mitt. IV. Ebenda **180**, 291 (1920).

195.— Körperstellung und Labyrinthreflexe beim Affen. Ebenda **193**, 396 (1922).

196.— Körperstellung. Berlin: Julius Springer **1924**.

197.— Cameron prize lectures on some results of studies in the physiology of posture. Lancet **1926**, 531, 585.

198. MARIE, PIERRE, et CH. FOIX: Deux cas d'hémiplégie cérébelleuse syphilitique avec autopsie. Revue neur. 1912.

199.— — Les syncinésies des hémiplégiques. Ebenda **1**, 3 (1916); **2**, 145 (1916).

200.— FOIX et ALAJOUANINE: De l'atrophie cérébelleuse tardive à prédominance corticale. Ebenda **1922**, 849, 1082.

201. MARINESCO, G., et A RADOVICI: Idiotie amaurotique et rigiditée décérebrée. L'Encéphale **18**, 145 (1914).

202.— — Contribution à l'étude des réflexes profonds du cou et des réflexes labyrinthiques. Revue neur. **1924**.

203. MAYER, C., u. O.REISCH: Über die Widerstandsbereitschaft des Bewegungsapparates (Gegenhalten Kleists) und über krankhafte Greifphänomene. Verh. Ges. dtsch. Nervenärzte, 17. Jahresvers., gehalten zu Wien, vom 15. bis 17. September 1927, 258. Leipzig : F. C. W. Vogel **1928**.

204. MEYER, G. H.: Die Statik und Mechanik des menschlichen Knochengerustes (1873).

205. MINKOWSKY, M.: Sur les mouvements, les réflexeset les réactions du foetus humain de 2 à 5 mois et leur relations avec le système nerveux foetal. Revue neur. **1922**, 1105.

206.— L'état actuel de l'étude des réflexes. Paris: Masson et Cie. **1927**.

207. V. MONAKOW, C.: Gehirnpathologie. Nothnagels spezielle Pathologie und Therapie. Wien **1897**.

208. MORRISSON, L. R.: Anatomical studies of the central nervous system of dogs without forebrain or cerebellum. Haarlem: De Erven F. Bohn **1929**.

209. MOURGUE, R.: L'activité statique du muscle. Encéphale **1921**, 297.

210.— Un cas typique de spasme de torsion consécutif à l'encéphalite léthargique. Gaz. Sci. méd. Bordeaux **1922**, 254.

211.— Le syndrome clinique de la rigidité de Wilson étudié dans un cas de spasme de torsion cٖensécutif à l'encéphalite épidémique. Arch. Suisses, Neurol. et Psychol. **2**, 163 (1922).

212. MUNK, H.: Über Groβhirnexstirpation beim Kaninchen. Pflügers Arch. **34**, 470 (1894).

213.— Über die Funktionen des Kleinhirns. Sitzgsber. preuβ. Akad. Wiss., 26. April 1896; 17. Januar 1907; 12 März. 1908.

214.— Über die Funktionen von Hirn und Rückenmark. Gesammelte Mitteilungen. Neue Folge, Kap. 14: Über die Funktionen des Kleinhirns, 286. Berlin: August Hirschwald **1909**.

215.—McNALLY, W. J., a. TAIT, JOHN: Ablation experiments on the labyrinth of the frog. Amer. J. Physiol. **75**, 155 (1925).

215A.— Siche TAIT, JOHN, u. W. J. McNALLY.

216. OPPENHEIM, H.: Die Geschwülste des Gehirns. Nothnagels Spezielle Pathologie und Therapie 9, Teil 2. Wien **1896**.

217. PAVLOV, I. P.: Die höchste Nerventätigkeit (das Verhalten) on Tieren. München : J. F. Bergmann **1926**.

218. PAVLOV, I. P.: Conditioned Reflexes. An investigation of the physiological activity of the cerebral cortex. Oxford University Press: Humphrey Milford **1927**.

219.— Les réflexes conditionnels. Étude objective de l'activité nerveuse supérieurs des animaux. Paris: Librairie Felix Alcan **1927**.

220.— Leçons sur l'activité du cortex cérébral. Paris: Amédée Legrand, Editeur, **1929**.

221. PEIPER, A., u. H. ISBERT: Über die Körperstellung des Säuglings. Jb. Kinderheilk. **115**, 142 (1927).

222. PETTE, H.: Klinische und anatomische Studien zum Kapitel der tonischen Hals- und Labyrinthreflexe beim Menschen. Dtsch. Z. Nervenheilk **86**, 193 (1925).

223.— Die Stützreaktion beim Menschen. Zbl. Neur. **48** (1927).

224. PHILIPPSON, M.: L'autonomie et la centralisation dans le système nerveux des animaux. Bruxelles **1905**.

225. PIÉRON, H.: Les formes et le mécanisme nerveux du tonus. Revue neur. **1920**, 986.

226. PI-SUNER, J., a. J. F. FULTON: The influence of the proprioceptive nerves of the hind limbs upon the posture of the fore limbs in decerebrate cats. Amer. J. Physiol. **83**, Nr 2, 548 (1928).

227. POINCARÉ: Leçons sur la physiologie normale et pathologique du système nerveux 2. Paris **1864**.

228. POL D. J. HULSHOFF: Cerebellaire Ataxie Psychiatr. Bl. (holl.) **13**, 273 (1909).

229.— Cerebellaire functies in verband met hun localisatie. Ebenda **19**, 181 (1915).

230. POURFOUR DU PETIT: Nouveau système du cerveau. Rev. obstétr., anat. et chir. Paris **1766**.

231. PRITCHARD, E. A. BLAKE: Die Stützreaktion. Pflügers Arch. **214**, 148 (1926).

232. PROBST, M.: Über Anatomie und Physiologie des Kleinhirns. Arch. f. Psychiatr. **35** (1902).

233. RADEMAKER, G. G. J.: La signification des noyaux rouges et du reste du mésencéphale pour le tonus musculaire, les attitudes normales et les reflexes labyrinthiques. Rev. d'Oto-Neuro-Ocul. **3**, 1 (1925).

234.— Die Bedeutung der roten Kerne ünd des ubrigen Mittelhirns für Muskeltonus, Körperstellung und Labyrinthreflexe. Berlin: Julius Springer **1926**. (Holländische Ausgabe : Leiden **1924**.)

235.— Démonstration de 2 chats décérebellés, de 2 chiens décérebellés et d'un chien ayant subi l'ablation, outre du cervelet, de la moitié droite du cerveau. Onzième réunion annuelle de physiologistes néerlandais. 12 December 1925.

236.— I. Körperstellung und Statik kleinhirnloser Tiere sechs oder mehr Monate nach der Operation. II. Motilitätsstörungen Kleinhirnlosser Tiere. Abstracts of communications to the XIIth internätional physiological Congress held at Stockholm, 3. bis 6. August 1926. Skand. Arch. **49** (1926).

237.— Statik und Motilitätsstörungen kleinhirnloser Tiere. Verh. Ges. dtsch. Nervenärzte, 16. Jahresvers., gehalten zu Düsseldorf vom 24. bis 26. September 1926, S. 144.

238.— Iets over de physiologie en pathologie van het staan. Nederlandsche Algemeene Zicktekundige Vereeniging, 14de algemeene vergadering op 19. Juni 1926. Nederl. Tijdschr. Geneesk. **1927**, H. 13, 1641.

239.— On the physiology of reflex-standing. Proc. Kon. Akad. Wetensch. Amsterd. **30**, H. 7 (1927).

240.— ·a. C. WINKLER: Annotations on the physiology and the anatomy of a dog, living 38 days without both hemispheres of the cerebrum and without cerebellum. Ebenda **31**, 332 (1928).

241. RAYMOND et RAYMOND CESTAN: Sur un cas d'endothéliome épithéloide du noyau rouge. Revue neur. **10**, 463 (1902).

242.— — Sur un cas de papillome épithéloide du noyau rouge. Ebenda **14**, 81 (1902).

243. RICHTER, C. P., a. L. H. BARTEMEIER: Decerebrate Rigidity of the Sloth. Brain **49**, 207 (1926).

243A. RIDDOCH, G.: Siehe HEAD, H., und G. RIDDOCH.

244.— The reflex functions of the completely divided cord in man, compared with those associated with less severe lesions. Brain **40**, 264 (1918).

245. RIESSER, O.: Der Muskeltonus. Handbuch der normalen

652

und pathologischen Physiologie **8**, 1, 192. Berlin: Julius Springer 1925.

246. RIJNBERK, G. VAN: Tentativi di localizzazioni funzionali nel cerveletto. Ia nota preventia. Il lobulus simplex. Arch. di Fisiol. **1**, 569 (1904).

247.— Tentativi di localizzazioni funzionali nel cerveletto. IIa nota preventia. Il centro per gli anteriori (crura prima lobuli ansiformis Bolk). Ebenda **2**, 18 (1904).

248.— Over functioneele localisatie in het cerebellum. Nieuwe verhandelingen v. h, Bataafsch Genootschap v. Proefondervindelijke Wijsbegeerte te Rotterdam, Ile Reeks **6**, Teil 2, S. **1** (1906).

249.— Die neueren Beiträge zur Anatomie und Physiologie des Säuger. Fol. neurobiol. **1** (1908).

250.— Das Lokalisationsproblem im Kleinhirn. Ergebnisse der Physiologie von Asher u. Spiro. S. 653 (1908).

251.— Weitere Beiträge zum Lokalisationsproblem im Rleinhirn. Fol. neurobiol., Erg.- H. S. 143 (1912).

252.— De jongste onderzoekingen over den bouw der kleine hersenen, vooral in verband met het localisatievraagstuek. Nederl. Tijdschr. Geneesk. **1924**, 516.

253. ROSSI, G.: Sulle localizzazioni cerebellari corticali e sul loro significato in rapporto alla funzione del cervelletto. Arch. di Fisiol. **19**, H. 5 (1921).

254. ROTHMANN, M.: Der Hund ohne Groβhirn. Mendels neurol. Zbl. **28**, 1045 (1909).

255.— Demonstrationen zur Lokalisation im Kleinhirn. Neur. Zbl. **28**, 1289 (1909); **29**, 389 (1910); **30**, 168 (1911).

256.— Über die elektrische Erregbarkeit des Kleinhirns und ihre Leitung zum Rückenmark. Ebenda **29**, 1084 (1910).

257.— Zur Funktion des Kleinhirns. Ebenda **29**, 1205 (1910).

258.— Zur Kleinhirnlokalisation. Berl. klin. Wschr. **1913**, 336.

259.— Die Funktion des Mittellappens des Kleinhirns. Mschr. Psychiatr. **35** (1913).

260.— Kleinhirnlokalisation. Neur. Zbl. **1913**, 702.

261.— Demonstration zur Rindenexstirpation des Kleinhirns. Ebenda **33**, 1010 (1914).

262. ROTHMANN, H.: Zusammenfassender Bericht über den Rothmann schen großhirnlosen Hund nach klinischer und anatomischer Untersuchung. Z. Neur. **87**, 247 (1923).

263. ROUSSET et GIRAUD: Destruction du cervelet sans symptômes cérébelleux. Revue neur. **1909**.

264. RUSSELL, J. S. RISIEN: Experimental researches into the functions of the cerebellum. Philosophic. Trans. roy. Soc. Lond. **185**, 722 (1894).

264A. SAMOJLOFF, A., u. M. KISSELEFF: Zur Charakteristik der zentralen Hemmungsprozesse. Pflügers Arch. **215**, 699 (1927).

264B. — — Die Verkürzungs- und Verlängerungsreaktion des knieextensors der decerebrierten Katze. Ebenda **218**, 268 (1928).

265. SCHALTENBRAND, G.: Normale Bewegungs- und Lagereaktionen bei Kindern. Dtsch. Z. Nervenheilk. **87**, 24 (1925).

266.— Über die Entwicklung des menschlichen Aufstehens und dessen Störungen bei verschiedenen Nervenkrankheiten. Ebenda **89**, 82 (1925).

267.— Enthirnungsstarre. Ebenda **100**, 165 (1927).

268. SCHIFF, J. M.: Lehrbuch der Physiologie des Menschen **1**, 331 (Lahr 1858/1859).

269. SCHOEN, R.: Die Stützreaktion. Mitt. I. Pflügers Arch. **214**, 21 (1926). — Mitt. II. Ebenda **214**, 48 (1926).

270. SCHWAB, OTTO: Über Stützreaktionen (Magnus) beim Menschen. Z. Neur. **108**, 585 (1927).

271. SERRES, E. R. A.: Anatomie comparée du cerveau, dans les quatres classes des animaux vertébrés, appliquée à la physiologie du système nerveux 2. Paris **1826**.

272. SHERRINGTON, C. S.: Cataleptoid reflexes in the monkey. Proc. roy. Soc. Lond. (B) **60**, 411 (1897).

273.— Decerebrate rigidity and reflex coordination of movements. J. of Physiol. **22**, 319 (1898).

274.— The parts of the brain below cerebral cortex, viz., medulla oblongata, pons, cerebellum, corpora quadrigemina, and

654

region of thalamus. Sharpey-Schäfer's Textbook of Physiology, **2**, 884 (1900).

275.— On the proprioceptive system, especially in its reflex aspect. Brain **29**, 467 (1906).

276.— The integrative action of the nervous system. I. Ausgabe London: Humphrey Milford 1906; II. Ausgabe London **1920**.

277.— Strychnine and reflex inhibition of skeletal muscle. J. of Physiol. **36**, 185 (1907).

278.— On plastic tonus and proprioceptive reflexes. Quart. J. Exper. Physiol. **2**, 109 (1909).

279.— Flexion-reflex of the limb, crossed extensionreflex, and reflex stepping and standing. Ebenda **40**, 28 (1910).

280.— Postural activity of muscle and nerve. Brain **38**, 191 (1915).
280A.—Siehe LIDDELL u. SHERRINGTON.

281.— Problems of muscular receptivity. Nature (Lond.) 21 und 28 June 1924.

282. SIMONELLI, G.: Sulla funzioni dei lobi medi del cerveletto; il lobo posteriore (pyramis, uvula, nodulus) secondo Ingvar. Arch. di Fisiol. **19**, H. 5 (1921).

283.— Vérification anatomique des cervelets opérés de destruction. du lobus posterior (pyramis, uvula, nodulus) et considérations sur la doctrine cérébelleuse de Ingvar. Revue neur. **1**, 432 (1924).

284. SIMONS, A.: Kopfhaltung und Muskeltonus Sitzgsber. Berl. Ges. Psych. u. Nervenkrankh. 3. Dezember 1919 und 12. Januar 1920. Ref.: Zbl. Neur. **39**, 132, 256 (1920).

285. — Kopfhaltung und Muskeltonus. Z. Neur. **80**, 499 (1923).

286. SPAMER, C.: Experimenteller und kritischer Beitrag zur Physiologie der halbkreisförmigen Kanäle. Pflügers Arch. **21**, 479 (1880).

287. SPATZ, H.: Physiologie und Pathologie der Stammganglien. Handbuch der normalen und pathologischen Physiologie **10**, 318. Berlin: Julius Springer 1927.

288. SPIEGEL, E. A.: Der Tonus der Skelettmuskulatur. Berlin: Julius Springer **1927**.

288A. — u. W. J. BERNIS: Arb. neur. Inst. Wien **27**, 197 (1925).

289. STARLING, E. H.: Principles of human physiology. 4. Ausgabe. London: J. u. A. Churchill **1926**.

290. STRASSER, H.: Lehrbuch der Muskel- und Gelenkmechanik. Berlin **1908**.

291. V. STEIN, ST.: Appareil servant à dèterminer les déviations des fonctions statiques du labyrinthe de l'oreille et sa démonstration. Moscou **1893**.

292. STENVERS, H. W.: Klinische studie over de functie van het cerebellum en de diagnostick der cerebellum- en bruggehoektumoren. Diss. Utrecht 1920.

293.— Un "Stellreflex" du bassin chez l'homme. Arch. néerl. Physiol. **2**, 669 (1918).

293A. STEINHAUSEN, W.: Siehe FISCHER, E., u. W. STEINHAUSEN.

294. STERNBERG: Die Schnenreflexe und ihr Bedeutung für die Pathologie des Nervensystems. Leipzig u. Wien **1893**.

295. STEWART, G., u. GORDON HOLMES: Symptomatology of cerebellar tumours: a study of forty cases. Brain **1904**.

296. TAIT, JOHN, a. W. J. McNALLY: Rotation and acceleration experiments, mainly on frogs. Amer. J. Physiol. **75**, 140.

296A.— Siehe McNALLY, W. J., a. JOHN TAIT.

297.— Ablation experiments on the labyrinth of frogs. The Laryngoscope, S. I. Oktober 1926.

298. THÉVENARD, A.: Les dystonics d'attitude. Paris: Gaston Doin et Cie., Éditeurs, 1926.

299. THIELE, F. H.: On the efferent relationship of the optic thalamus and Deiters' nucleus to the spinal cord, with special reference to the cerebellar influx of Dr. Hughlings Jackson and the genesis of the decerebrate rigidity of Ord and Sherrington. J. of Physiol. **32**, 358 (1905).

300. THOMAS, ANDRÉ: Le cervelet, étude anatomique, clinique et physiologique. Paris: Steinheil, Éditeur, **1897**.

301.— et JUMENTIÉ: Sur la nature des troubles de la motilité dans les affections du cervelet. Revue neur. **1901**.

302.— — Atrophie lamellaire des cellules de Purkinje. Ebenda **1905**.

303.— La fonction cérébelleuse. Paris: Octave Doin et fils **1911**.

304.— et A. DURUPT: Les localisations cérébelleuses (vérification anatomique). Fonction des centres du lobe latéral. Revue neur. **1913**.

305.— — Recherches expérimentales sur les fonctions cérébelleuses, dysmétrie et localisations. Encéphale **1913**, H. 7.

306.— — Localisations cérébelleuses. Paris: Vigot fréres, Éditéurs, **1914**.

307.— Étude sur les blessures du cervelet. Paris: Vigot frères, Éditeurs, **1918**.

308.— Extensibilité et réflexe antagoniste. Paris méd. **1922**, 323.

310.— Pathologie du cervelet. Nouveau Traité de Méd., G. H. Roger, F. Widal, et P. J. Teissier. vol. 39, p. 755. Paris : Masson et Cie. **1925**.

311. TURNER, VIOLET: A case of prolonged hyperpyrexia in a child with a midbrain tumour. Brit. J. Childr. Dis. **13**, 261 (1916). Zitiert nach S. A. K. Wilson, Nr. 332.

312. VON UEXKÜLL, J.: Studien über den Tonus. VI. Die Pilgermuschel. Z. Biol. **58**, 305 (1912).

313. URECHIA, C. I., et S. MIHALESCO: Un cas de rigidité congénitale avec autopsie. Bull. et Mém. Soc. Méd. Hôp. Paris **1926**, Nr 40, 1778.

314. VINCENT, CLOVIS, E. BERNARD, et J. DARQUIER: Tumeur cérébelleuse avec rigidité parkinsonienne et lenteur de l'idéation. Revue neur. **2**, 31 (1923).

315. VULPIAN, A.: Leçons sur la physiologie générale et comparée du systéme nerveux, S. 532. Paris 1866.

316. WACHHOLDER, K.: Erregungsverteilung zwischen Streckern und Beugern in der Enthirnungsstarre. Pflügers Arch. **221**, 66 (1928).

317. WALSHE. F. M. R.: A case of complete decerebrate rigidity in man. Lancet **205**, Nr 13, 644 (1923).

318.— On certain tonic or postural reflexes in hemiplegia with special reference to the so-called "associated movements". Brain **46**, 1 (1923).

319.— On variations in the form of reflex movements, notably the Babinski plantar response under different degrees of spasticity and under the influence of Magnus and De Kleyn's tonic neck reflexes. Brain **46**, 281 (1923).

320.— La rigidité décérebrée de Sherrington et ses relations avec la rigidité musculaire d'origine pyramidale et extra-pyramidale chez l'homme. Encéphale **1925**, H. 2, 73.

321.— The significance of the voluntary element in the genesis of cerebellar ataxy. Brain **50** (1927).

322. WEBER, ED. u. W.: Mechanik der menschlichen Gehwerkzeuge. Göttingen **1873**.

323. WEED, L. H.: Observations upon decerebrate rigidity. J. of Physiol. **48**, 205 (1914).

324. WEILAND, W.: Münch. med. Wschr. **1912**, 2539.

325. WEIR-MITCHELL, S.: Researches on the physiology of the cerebellum. Amer. J. med. Sci. **1869**, 320.

326. WEISENBURG, T. H., a. B. J. ALPERS: Decerebrate rigidity following encephalitis. Arch. of Neur. **1927**.

327. WERTHEIM - SALOMONSON, J. K. A.: On a shortening reflex. Proc. kon. Akad. Wetenschappen Amsterd., 22. Juli 1913.

328.— Tonus and the reflexes. Brain **43**, 369 (1920).

329.— Methoden van onderzoek. Tonus, Reflexen, Evenwicht. Leerboek der Zenuwziekten, Algemeene gedeelte B., S. 128, 171, 303. Haarlem: De Erven F. Bohn **1923**.

330. WESTPHAL: Arch. f. Psychiatr. **9**, 788 (1878).

331.— Ebenda **14**, 87 (1883).

332. WILSON, S. A. KINNIER: On decerebrate rigidity in man and the occurrence of tonic fits. Brain **45**, 220 (1922).

333.— Croonian Lectures on some disorders of motility and of muscle tone, with special reference to the corpus striatum. Lancet, 4., 11. u. 25. Juli; 1. u. 8. August 1925.

334. WILSON, J. G., a. F. H. PIKE: The effects of stimulation and exstirpation of the labyrinth of the ear and their relation to the motor system. Philosophic. Trans. roy. Soc. B **203**, 127 (1913).

335. WINKLER, C.: The central course of the nervous octavu[s] and its influence on motility. Proc. Kon. Akad. Wetensch. Amsterd., **11**, 14, Nr 1 (1907).

335A.— Le cervelet. Anatomie du système nerveux 3. Haarlem: De Erven F. Bohn **1927** .

335B.— Siehe G. G. J. RADEMAKER u. C. WINKLER (240) u. C. R. MORRISSON (208).

336. ZELIONY, G. P.: Observations sur des chiens auxquels on a enlevé les hémisphères cérébraux. C. r. Soc. Biol. Paris Jg. 65, **1**, 707 (1913).

337. ZINGERLE, H.: Über Stellreflexe und automatische Lageänderungen des Körpers beim Menschen. Klin. Wscrh. **3**, 41 (1924).

338.— Klinische Studie über Haltungs- und Stellreflexe, sowie andere automatische Körperbewegungen beim Menschen. J. F. Psychiatr. **31**, 330 (1926).

INDEX OF NAMES

INDEX OF SUBJECTS

Leg-bracing reactions

Rolling movements

Shortening reaction
Signe d'avant-bras
Spontaneous reflexes
Standing reflexes
Static reactions
Stewart-Holmes sign
Strabismus
Stretch reflexes (Dehnangs)

Supporting reactions, negative

Supporting tonus

— caloric 276

— compensatory 446,449

Dyssynergy 437,594
Dystonia 591

Extensor reflex, crossed 15,17,24,27,60,99,143,378,465
Extensor thrust 70,115,146
Eye after nystagmus 279,446,448,457

— after-reactions 270,275,279,446,448
— nystagmus 270,279,448
— reactions 279,448

Grasping movement (grasping reflex) 75,468

Head nystagmus, rotary 447
Hopping reactions (see also leg-slackening reactions) 138,310,369,381,
   384,392,398,404,508,552
Hypermetria 372,381,382,441,459,459,485,522,557,596,600,606
Hypersthenia 429,496
Hypersynergy 306,437,594
Hypertonia 17,133,168,178,179,492,495,512,515,600
Hyposthenia 429,496,598
Hypotonia 16,19,133,134,492,495,512,517,521,591,600

Incoordination 589,591,592

Labyrinthine righting reflexes 63,141,210,236,299,392,433,447,620

— behavior of — after total extirpation of the cerebellum 433
— behavior of — after unilateral extirpation of the cerebellum 433
Labyrinthine reflexes, behavior of — in decerebellate animals-tonic
                                   11,23, 142,236,265,286, 310,
                                   423,440

— due to rectilinear movement 533
— caused by rotation round the bitemporal axis 265,392,445
— caused by rotation round the longitudinal axis 269,392,445
— caused by rotation round the dorsoventral axis 274,275,276,280,445

Modern Egyptian Press